Vanadium

Vanadium

Extraction, Manufacturing and Applications

Edited by

Baoxiang Yang
Jinyong He
Guifang Zhang
Jike Guo

ELSEVIER

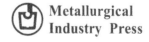

Metallurgical Industry Press

Based on an original Chinese edition:
《钒基材料制造》
Copyright © 2021 Metallurgical Industry Press. All rights reserved.
(冶金工业出版社负责中国大陆地区销售，Elsevier负责中国大陆地区以外销售)

图书在版编目(CIP)数据

钒基材料制造＝Vanadium：Extraction, Manufacturing and Applications：英文/杨保祥等编著. —北京：冶金工业出版社，2021.1

ISBN 978-7-5024-8236-7

Ⅰ.①钒… Ⅱ.①杨… Ⅲ.①钒基合金—金属材料—制造—英文 Ⅳ.①TG146.4

中国版本图书馆CIP数据核字(2021)第017336号

出 版 人　苏长永
地　　址　北京市东城区嵩祝院北巷39号　邮编　100009　电话　(010)64027926
网　　址　www.cnmip.com.cn　电子信箱　yjcbs@cnmip.com.cn
责任编辑　于昕蕾　美术编辑　彭子赫　版式设计　孙跃红
责任校对　王永欣　责任印制　李玉山

ISBN 978-7-5024-8236-7

冶金工业出版社出版发行；各地新华书店经销；北京虎彩文化传播有限公司印刷
2021年1月第1版，2021年1月第1次印刷
191mm×235mm；35.25印张；858千字；548页
299.00元

冶金工业出版社　投稿电话　(010)64027932　投稿信箱　tougao@cnmip.com.cn
冶金工业出版社营销中心　电话　(010)64044283　传真　(010)64027893
冶金工业出版社天猫旗舰店　yjgycbs. tmall. com
(本书如有印装质量问题，本社营销中心负责退换)

Contents

Preface ix
Abstract xiii

Chapter 1: Introduction 1

Chapter 2: Vanadium and its compounds 9
    2.1 Vanadium 10
    2.2 Physical properties of vanadium 12
    2.3 Chemical properties of vanadium 12
    2.4 Thermodynamic properties of vanadium 13
    2.5 Vanadium compounds 14
    2.6 Compound salts of vanadium 24
    2.7 Physiological effects of vanadium 27
    2.8 Diseases from vanadium pollution 30
    2.9 Biological effects of vanadium in humans and animals 30
    2.10 Biological effect of vanadium on plants 32

Chapter 3: Vanadium mineral resources 33
    3.1 Metallogenic characteristics of vanadium ore 34
    3.2 Minerals of vanadium 46
    3.3 Main vanadium minerals for vanadium extraction 51
    3.4 Distribution of vanadium resources 55

Chapter 4: Extraction of vanadium from titanoferrous magnetite: mineral processing and enrichment methods 59
    4.1 V-bearing titanoferrous magnetite 60
    4.2 Main mineral characteristics of V-bearing titanoferrous magnetite 62
    4.3 Typical mineral processing methods 65

vi Contents

    4.4 V-containing titanoferrous magnetite outside of China 77
    4.5 Main equipment 82

Chapter 5: Extraction of vanadium from V-bearing titanoferrous magnetite by smelting to produce V-containing iron 89
    5.1 Blast furnace smelting 90
    5.2 Ironmaking by direct reduction 108
    5.3 Main production equipment 116

Chapter 6: Extraction of vanadium from V-containing titanoferrous magnetite: enrichment of vanadium slag 123
    6.1 Enrichment of vanadium slag 124
    6.2 Production of V-containing steel slag from molten iron by steelmaking 136
    6.3 Pretreatment of V-containing molten iron for slag blowing 137
    6.4 Enriched vanadium slag from V-containing molten iron by sodium treatment and oxidation 140
    6.5 Comparison of various vanadium extraction methods 141
    6.6 Main equipment 145

Chapter 7: Extraction of vanadium from V-containing titanoferrous magnetite: extraction of V_2O_5 from vanadium slag 149
    7.1 Sodium salt roasting 151
    7.2 Extracting vanadium pentoxide from vanadium slag by calcium salt roasting ... 175
    7.3 Extracting vanadium pentoxide from Na-treated vanadium slag 181
    7.4 Extracting vanadium from V-containing steel slag by acid leaching 182
    7.5 Extracting V_2O_5 from vanadium slag by sulfide addition 184
    7.6 Main raw materials 184
    7.7 Main vanadium extraction equipment 193

Chapter 8: Extracting vanadium from shale 201
    8.1 V-bearing carbon shale 202
    8.2 Extracting vanadium pentoxide from stone coal 207
    8.3 Comparison of vanadium extraction processes from stone coal 224

Chapter 9: Extracting vanadium pentoxide from minerals and composite materials containing vanadium 227
    9.1 Extracting vanadium pentoxide from V-bearing titanoferrous magnetite 228
    9.2 Extracting vanadium pentoxide from V-containing oil residue 232

9.3 Extracting vanadium pentoxide from carnotite 233
9.4 Extracting vanadium pentoxide from alumina-rich slag 234
9.5 Extracting vanadium pentoxide from vanadinite 234
9.6 Extracting vanadium pentoxide from uranium molybdenum vanadium ore 235
9.7 Extracting vanadium pentoxide from kakoxene 235
9.8 Extracting vanadium pentoxide from V-bearing limonite 236
9.9 Extracting vanadium pentoxide from oil refining slag 236
9.10 Extracting vanadium pentoxide from bauxite 236
9.11 Vanadium extraction from spent catalysts 237
9.12 Vanadium extraction from fly ash 239

Chapter 10: Ferrovanadium 243
10.1 Electro-silicon thermal process 244
10.2 Aluminum thermal method 255
10.3 Direct alloying of vanadium slag to produce ferrovanadium 262
10.4 Ferrosilicon–vanadium alloy 266

Chapter 11: Applications of vanadium in the steel industry 267
11.1 Classification of steel 268
11.2 Main elements in steel and their effects 270
11.3 Role of vanadium in steel 275
11.4 Applications of vanadium in steel 280
11.5 Vanadium in cast iron 319

Chapter 12: Vanadium metal 333
12.1 Metallic vanadium 333
12.2 Vanadium refining 350
12.3 Application of vanadium metal 357

Chapter 13: Characteristics and technical quality standards of main vanadium products 359
13.1 Vanadium slag 360
13.2 Vanadium trioxide 363
13.3 Vanadium pentoxide 365
13.4 Vanadium pentoxide standard 367
13.5 Ammonium vanadate 373
13.6 Ferrovanadium 375
13.7 Vanadium nitride 389

13.8 Vanadium metal..........391
13.9 Vanadium aluminum alloy..........391

Chapter 14: Vanadium series products and functional materials..........395

14.1 Vanadium halides..........396
14.2 Vanadates..........399
14.3 Vanadium oxides..........404
14.4 Vanadium pigments..........407
14.5 Functional materials..........408

Chapter 15: Vanadium catalysts..........415

15.1 Catalyst types..........416
15.2 Manufacturing methods..........417
15.3 Sulfuric acid production and flue gas desulfurization catalysts..........420
15.4 Flue gas denitrification catalysts..........429
15.5 Vanadium catalysts for organic synthesis..........439

Chapter 16: Vanadium batteries..........445

16.1 Technical background of vanadium cell development..........446
16.2 Vanadium battery systems..........449
16.3 Applications of vanadium batteries..........456
16.4 Key materials for vanadium batteries..........462
16.5 Vanadium battery assembly..........468

Chapter 17: Vanadium carbonitride and vanadium–aluminum alloy..........471

17.1 Vanadium carbide..........472
17.2 Vanadium nitride..........479
17.3 Nitrided ferrovanadium..........482
17.4 Aluminum–vanadium alloy..........483

Chapter 18: Development of the vanadium industry..........487

18.1 Development of vanadium industry technology..........489
18.2 South African vanadium industry..........494
18.3 Russian vanadium industry..........501
18.4 New Zealand vanadium industry..........505
18.5 China's vanadium industry..........508

Important schedule..........521
References..........529
Index..........533

Preface

The global vanadium industry has developed greatly in four stages: recognition of the industrial value of vanadium, development and optimization of industrial processes, streamlining of the vanadium industrial chain, and expansion of the market and applications. Competition and cooperation has occurred in the industry, which has embraced world resources, technology and equipment, products, environment, and the market. The development of the vanadium industry began when the value of vanadium for various applications was recognized. In the initial stages, little attention was paid to vanadium. The value of vanadium was demonstrated for various applications, including catalysis and steel alloying. In addition, increasing awareness of the economic and social benefits of vanadium resulted in the growth and development of the vanadium industry. In the case of shortages of the resource, new technology and equipment for exploration and enrichment of vanadium resources were developed. Innovation in the field of vanadium extraction technology has accelerated the development of regional competition and resources cooperation.

South Africa, Russia, and China (Panzhihua) have exploited V-bearing titanoferrous magnetite resources and successfully overcome shortages in vanadium resources. The range of vanadium resources has been further diversified to include petroleum waste, stone-like coal, and recycled materials, which has facilitated the rapid expansion of vanadium development and application. Hence, the production capacity and market have subsequently increased in size. New and highly advanced technologies were gradually introduced into the vanadium industry, affecting the development and application of this material. Panzhihua developed high-strength V-based steel rail, which enabled the development of China's high-speed railways. Alloys of vanadium and nonferrous metals are special materials for aerospace applications. Vanadium is also a functional catalyst used in the production of chemicals and for environmental remediation. The radiation shielding effect of vanadium has contributed to its use in the nuclear industry. In addition, an emerging application is vanadium-based batteries, which are expected to facilitate clean energy storage. After the vanadium resource shortage was addressed, a circular economy was implemented in the vanadium industry, with a focus on safe production. To ensure clean production targets and extend the industrial chain, while facilitating stable development of high-efficiency industries, new highly advanced technologies were implemented to update industrial equipment and methods.

I have been studying vanadium extraction and its applications since 1985, focusing on innovation and the development of the vanadium industry. Vanadium has been my major area of research at Panzhihua, giving me the opportunity to witness the development and growth of Chinese vanadium extraction and application technologies. Since then, vanadium and titanium have been increasing sought after on a global scale. The Chinese government heavily invested in and developed the Panzhihua region via key metallurgy projects in the 1960s. The construction sites of these projects covered the banks of the Jinsha river and the Lan volcanoes. V-bearing titanoferrous magnetite from the Panzhihua-Xichang region has been mined with explosive methods. Like an ivory sculpture, a steel city has emerged in Nongnongping of Panzhihua. In parallel, iron has been produced, which has promoted large-scale development of the vanadium industry in China. In addition, the characteristic atomized vanadium extraction process has made a generation of Panzhihua people intoxicated with their invention. The scenery connected with the construction of the big three lines has made people travel back and forth. After more than 40 years of gathering the achievements of the world, a 10-mile city focused on steel, vanadium, and titanium production has been created. Currently, the construction of the "vanadium and titanium capital of China" in Panzhihua is accelerating. In addition, the national innovation platform for the comprehensive use of vanadium and titanium resources in Panxi is increasing. There is a commitment to developing functional vanadium and titanium, along with refined steel. This reflects the trend of vanadium being used as an engineering structural material and functional material, which is affecting the global development of new highly advanced technology.

There is a significant body of literature regarding vanadium, which discusses vanadium from different angles. We have read most of these sources and benefited greatly from their insight. I joined Panzhihua Iron and Steel Company in 1985 and have collaborated with many people in the study of vanadium and titanium over the last 30 years. I have accumulated a lot of data over this time and developed my knowledge in this field. Therefore I was motivated to write a book based my data and experiences, to disseminate information regarding vanadium and titanium to general academics, and how these can better serve for producing high-end products, and provide insight into the hi-tech-oriented vanadium and titanium industries.

Jinyong He is the chairman of Panzhihua Yinjiang Jinyong Industry and Trade Co., Ltd. and our partner. He has been committed to vanadium battery research and demonstration. His team has collaborated with the Chinese Academy of Engineering Physics, and they have achieved three large-scale demonstrations of vanadium battery technology with improved docking cohesion and promoted the transformation of the vanadium battery and energy storage industries. This group also studies preparation of high-purity vanadium metal by thermal reduction and achieved 98.7% pure metal.

Dr. Guifang Zhang is professor at Kunming University of Science and Technology, as well as our research partner, and is committed to teaching basic theory and researching magnetite−iron−titanium−vanadium ore. She provided a wide range of model data for reference for this book and has contributed to increasing the relevant knowledge regarding manufacturing of V-based materials.

"Vanadium extraction, manufacturing and application" consists of five parts. The first introduces the basic properties of vanadium, the second part addresses vanadium extraction, the third part covers vanadium applications, the fourth part focuses on development of the world vanadium industry, and the fifth part presents vanadium manufacturing data. The discussion of basic vanadium characteristics includes the properties of vanadium and vanadium compounds, vanadium resources, and vanadium minerals. The vanadium extraction part discusses vanadium pentoxide extraction from different minerals using different extraction and separation processes to produce various products. The vanadium applications are discussed considering the vanadium pentoxide product grades, focusing on chemical applications such as practical catalysts. In addition, we focus on ferrovanadium alloys and their application in vanadium bearing steel and nonferrous metal alloys. Vanadium compounds are discussed to highlight the properties of functional materials. Vanadium battery application is discussed to highlight the liquid phase transformation of the vanadium system. The discussion of the development of the world vanadium industry highlights the focused development of an industry within a region. We discuss the classification and use of characteristic materials, and the relationship between vanadium technology research and industry. We present an overview of the development of the world vanadium industry, which is the best example of innovative development and growth of China's vanadium industry. The data related to vanadium manufacture mainly focuses on its relevance and providing recognition for it.

This book is expected to build an industry framework and provide a comprehensive reference for professionals. According to the key-development issues in the vanadium industry, a layout is given which is expected to contribute to developing knowledge in the research, industry, and academic fields. This book was written by referring to many sources, including scientific reports, journals and books, some internal data, and information from our network. There was also some contribution from teaching colleagues. There are several review articles related to the study of vanadium, and all of these have been cited and discussed here. We have performed a multilevel analysis to form our conclusions. We would like to say thank you and ask forgiveness for the omission of references. Due to time constraints and the limited academic level, some information has been accumulated for a long time, and studies in other languages provided multiple translated versions which are difficult to cite and confirm. There are varying opinions regarding the vanadium industry due to the different fundamental ideas and levels of understanding of those in the field. Due to early scarcity of vanadium resources, few people have a detailed understanding of the industry. If our book contains some omissions or errors, please feel free to contact us. We invite experts, professors, and colleagues with differing views to discuss them further with us.

"Vanadium extraction, manufacturing and application" embodies the accumulation of knowledge and the consensus achieved by a group of people during writing the book. We consider it a sublimation of a process and an era. We do not evaluate the limitations of the past opinions of scholars, nor their impact on the development of the industry, since only now has it been possible to gain a true perspective of vanadium extraction, manufacturing and application. Hongfei Hu, Shengyou Zhu, Huijun Miao, Huai-bin Wang, Yi Peng, Maojiang Niu, Yangjun Yang, Xiaozhe Cheng, Guoju Song, Yuchang Zhou, LiXia Gong, Zibi Fu and Ping Pan from the Panzhihua Iron and Steel Research Institute provided their professional perspectives and confidential analyses. Ming Du, Kaihua Li, Qingdong Miao, Bing Song, Yafei Chen, and Cuifen He from the Iron and Steel Research Institute provided professional chart support. Professor Jun Zhu from Xian University of Architecture and Technology, senior engineer Haidu Jing from Panzhihua Iron and Steel Group, and professors Shaoli Yang and Jinhua Luo from Panzhihua University, Yunfang Yang, Feng Wei from Xian International Study University, Peicheng He from The Chinese University of Hongkong also made contributions to the book. We thank the contributors and our friends in the industry for providing us with information, arguments, correction, illustrations, editing, and support!

From a professional point of view, "Vanadium extraction, manufacturing and application" is suited for the following audience: (1) experts and scholars in the fields of vanadium resources, technology and equipment, products, environment, and marketing; (2) technical and management personnel working in commercial vanadium plants; (3) sales personnel promoting vanadium products; (4) decision-making consultation agency personnel in the vanadium industry; (5) government economic planners and technical managers in areas related to the vanadium industry; (6) teaching, research, and design personnel in colleges and research institutes, along with postgraduates and students participating in research; and (7) others who have a personal interest in the topic.

Baoxiang Yang

Abstract

In this book the process of vanadium production is comprehensively and systematically introduced, along with the key factors influencing the development of the vanadium industry. Various topics are reviewed and analyzed, including the status of vanadium resources, technologies for extracting vanadium from the various resource types, development of vanadium extraction during periods of shortage, specific regional development, structure of the vanadium industry, and the endogenous evolution of vanadium products with respect to application requirements. The main contents of this book include a background discussion regarding vanadium, its extraction processes, applications of the material, development of the industry, along with data related to vanadium manufacturing applications.

The subjects of vanadium production and application were written with the following professional audience in mind: (1) experts and scholars in the fields of vanadium resources, technology and equipment, products, environment, and marketing; (2) technical and management personnel working in commercial vanadium plants; (3) sales personnel promoting vanadium products; (4) decision-making consultation agency personnel in the vanadium industry; (5) government economic planners and technical managers in areas related to the vanadium industry; (6) teaching, research, and design personnel in colleges and research institutes, along with postgraduates and students participating in research; and (7) others who have a personal interest in the topic.

CHAPTER 1

Introduction

Vanadium is an important component of nonferrous metals, sometimes referred to as ferrous metals. Metallic vanadium and its compounds and alloys have unique and excellent properties, which have contributed to a great era of change in the global industrial civilization. Vanadium has also facilitated the rapid development of modern industries, in particular the steel, chemical, petroleum, energy, and nuclear industries, along with the fields of nonferrous metals, architecture, and environmental protection. Therefore vanadium will be remembered for its role in human history.

Vanadium was discovered in 1801 when the Spanish mineralogist A.M. del Rio was studying lead ores, and it was identified incorrectly as a compound of chromium. In 1830, the Swedish chemist N.G. Sefström separated an element in the process of smelting iron and named it vanadis. The German chemist F. Wöhler proved that the new element found by Sefström was the same as that discovered by del Rio [1−4,7−9,14,16,17]. In 1867 Henry Enfield Roscoe first obtained pure vanadium in the form of hydrochloric acid vanadium(III) by hydrogen reduction.

In 1882 basic slag was treated with 0.1% hydrochloric acid at low temperature and used to successfully prepare vanadium products. In 1894 a method for reduction of vanadium oxides was invented by Jose Maweng. Later, in 1897, Gordon Schmitt invented the alumina-thermal reduction process for vanadium oxide. In 1906 patronite was discovered in Peru, and the United States mined and extracted this mineral for 50 years, which accounted for a quarter of the global production at the time. In the United States, this process was mainly used to extract vanadium from carnotite. In 1911 American ford first successfully used vanadium in a steel alloy. In 1912 Bleecker published a patent regarding the use of sodium salt roasting combined with a water leaching process to recover vanadium. Before 1936 the Soviet Union obtained vanadium slag from poor vanadium containing iron ore and used the slag to prepare vanadium products. In the 1970s they successfully applied the Ca-salt treatment process to extract vanadium. In 1955 Finland began to extract vanadium from magnetite. In 1957 South Africa began to produce vanadium slag and vanadium oxide from titanoferrous magnetite. In the 1960s and 1980s, the United States, Japan, Canada, and Peru began extracting and recovering vanadium from oil ash and waste catalyst materials, and the monopoly of regional vanadium resources was gradually broken.

The production of vanadium in China was started by the Jinzhou Iron Institute in the period of Japanese control, where vanadium concentrate was used as the raw material for vanadium extraction. After the founding of the People's Republic of China, the administration of the iron and steel industry within the heavy industry ministry issued scientific research tasks in 1954 including "extracting vanadium and smelting ferrovanadium from titanoferrous magnetite containing vanadium of Chengde great temple by pyrometallurgy." Projects completed in 1955 provided the basis for the construction of China's ferrovanadium (FeV) industry. In 1958 the Jinzhou ferroalloy plant resumed production of FeV from concentrates of Chengde V-bearing titanoferrous magnetite. Since these early times, vanadium and titanium have been increasingly valued globally. The Chinese government heavily invested in and developed the Panzhihua region in key metallurgy projects in the 1960s. The construction sites of these projects covered the banks of the Jinsha River and the Lan volcanoes. V-bearing titanoferrous magnetite from the Panzhihua-Xichang region has been mined with explosive methods. Like ivory sculpture, steel city has emerged in Nongnongping of Panzhihua. Iron has been turned out. Vanadium and titanium have been developed for special applications, steel refined. In 1978 the Panzhihua Iron & Steel Company (Hereinafter referred to as "Pan Steel") introduced an original atomization method for producing vanadium slag. This resulted in China becoming one of the key global players in vanadium production. Pan Steels original atomization technology enabled the Chinese history of vanadium imports, a lot of precious foreign exchange saved for the country. Panzhihua changed the distribution of the vanadium extraction technology and vanadium resources, along with the economic geography of China and the world. Panzhihua changed the vanadium industry, and vanadium changed Panzhihua, which attracted attention globally. The Jinzhou ferroalloy plant, Nanjing ferroalloy plant, and Panzhihua Steel and Emei Ferroalloy factory were constructed in succession. All of these plants used vanadium slag as the raw material to produce vanadium pentoxide. Panzhihua steel adopted gas reduction of ammonium vanadate to produce vanadium trioxide for smelting ferrovanadium and preparing vanadium nitride and vanadium carbide. They also developed specialty steels by direct alloying of steel slag and vanadium slag. Such materials were a major development in construction steels, which enabled the production of high-strength V-containing steels, which reduce the weight of steel structures. Pan steel developed a series of steel rails based on microvanadium alloying, which increased the strength of the steel and enhanced the anti-wear and shock absorption properties, facilitating development of heavy load and high-speed railways. In the 1970s and 1980s, several small vanadium plants were built in southern China to process rich stone−like coal resources, which partially resolved the shortage of vanadium resources at the time and promoted widespread development of the vanadium industry in China [1−3,10,11,20,38,39,45,67,80,84].

Vanadium oxide has good catalytic performance and vanadium-based catalysts show unique behaviors. The emergence of vanadium pentoxide greatly increased the catalytic production of sulfuric acid and facilitated scale-up of this industry by replacing noble metal catalysts.

Vanadium-based catalysts proved effective in the petrochemical industry, accelerating chemical reaction processes, and resulting in an increase in the reliability and stability of organic synthesis reactions. In 1880 vanadium catalysts were identified and experimental research in this field began in 1901. In 1913 vanadium catalysts were introduced for the first time in processes in the Baden aniline soda company, Germany, and by 1930, they were commonly used in the factory. The popularity of vanadium catalysts increased throughout the 1930s as a replacement for platinum catalysts used in sulfuric acid production. Vanadium catalysts mainly consist of vanadium compounds, such as oxides, chlorides, and complexes, as well as a variety of forms, such as salts of heteropoly acids. The most common active component contains one or more of the following V_2O_5 additives. V_2O_5 is the main active component of most catalysts for oxidation reactions. Vanadium compounds are important industrial catalysis for oxidation and are widely used in the sulfuric acid industry, and for the synthesis of organic chemicals, such as phthalic anhydride and maleic anhydride, and for catalytic polymerization, alkylation reactions, and oxidative dehydrogenation. Vanadium catalysts have been used to perform double catalysis for desulfurization and denitrification, as well as for environmental purification for removing inorganic and organic chemicals. Recently, vanadium batteries have promoted the development of clean energy applications. Vanadium has been an "industrial MSG," boosting the industrial economy [5,12,13,25,35].

In 1889 Prof. Aroud at the University of Sheffield began studying the special role of vanadium in steel. In the early 20th century, Henry Ford discovered a new way of applying vanadium steel, and the performance of key components of Ford motors was improved by vanadium alloying. Vanadium pentoxide can be used during the smelting of FeV in the steel industry. The addition of vanadium imparts complex functions to the steel products; grain refinement of the steel matrix can be achieved, which comprehensively improves the performance of the steel product, such as strength, toughness, ductility, and heat resistance. Vanadium-containing steels are suitable for railways, bridges, construction, and other applications as they can satisfy the strength requirements. The use of higher strength steels allows the use of less material, achieving a weight reduction that is advantageous for constructing high buildings and allows safe passage of heavy vehicles and high-speed locomotives over metal structures. Vanadium nitride has been successfully used in the promotion and application of steel alloying, characterized by dissolved and fixed nitrogen in the steel. Some of the vanadium content in the steel can be substituted by nitrogen, allowing the V content to be reduced [4,6,14,16,17,25]. In addition, V−Al alloys are used in the manufacture of Ti-based alloys (e.g., Ti6Al4V). Vanadium microtreatment has been used to improve the strength of Al-based alloys, cast Cu and Ti, and enhance the microstructure of Cu-based alloys. Since Roscoe first obtained pure vanadium by hydrogen reduction of vanadium chloride, the application of vanadium metal has expanded rapidly, where its special properties are also applied in the areas of radiation shielding and superconductivity.

Vanadium does not occur in simple minerals and is mainly found in the form of associated minerals with other metals. The most common extraction methods for vanadium include the (1) acid method, where the ore is treated with sulfuric acid or hydrochloric acid to produce $(VO_2)_2SO_4$ or VO_2Cl, respectively; (2) alkali method that produces $NaVO_3$ or Na_3VO_4 after treatment with sodium hydroxide or sodium carbonate; and (3) chloride roasting that involves roasting and calcining the ore with salt to obtain $NaVO_3$. Extraction of vanadium from minerals is generally aimed at producing standard vanadium pentoxide products. Technological design covers the entire extraction process of extracting standard vanadium pentoxide products from the various V-containing raw materials, setting technical process parameters, and considering the equipment processing capacity. Standard vanadium pentoxide products are used as intermediate materials for producing various vanadium products.

Vanadium extraction from raw materials can be divided into primary and secondary processes. Primary processes are focused on effective vanadium extraction to enrich the vanadium content of the ore as much as possible. Secondary processes involve vanadium extraction from byproducts of other enrichment processes. The extraction process also depends on the selection of the initial chemical additives, generally divided into alkali and acid treatments, where the alkali methods are further divided into sodium and calcium salt methods. The additive can be a multicomponent system depending on the specific composition of the vanadium ore or slag. The roasting process requires no additive. The extraction of vanadium from the slag can be achieved using alkali leaching, acid leaching, or leaching in hot water. The leaching liquid is purified to form a precipitate containing unstable vanadate or vanadium hydrate. Standard vanadium pentoxide materials are obtained by calcining the vanadate or vanadium hydrate. This process can also be used to produce stable vanadate salts, such as ferric vanadate.

The recovery of vanadium from V-containing iron ore began in the early 19th century. During the First World War, Germany and France recovered vanadium from Lorin iron ore, which contained 0.06%–0.10% V, followed by processing in a blast furnace and use of a Thomas converter. The V-containing iron ore was used as a raw material to produce cast iron in a blast furnace, where the vanadium oxide was reduced, producing a hot metal-containing 0.10%–0.15% V. When pig iron was used to make steel, vanadium in pig iron was oxidized to form Thomas slag with a V content up to 0.5%. The Thomas slag had no direct value because of its low-V content. German steel mills mixed low-V slag and V-containing byproducts from steelworks, such as roof coatings of workshops containing 0.8%–0.8% V, which was smelted in another blast furnace to obtain cast iron containing vanadium and a high phosphorus content. This material was treated in a Thomas converter to obtain high-V slag, which can be directly used as a raw material for chemical plants in the production of vanadium. However, the high phosphorus content in the metal after V extraction can be problematic for steel making. In 1882, low-temperature sulfuric acid treatment was used to extract vanadium oxides from Bessemer slag with V1.1% in the Le Creuset steel plant, which was used as a raw material in dye factories. In 1924, R.V. Seth reported a Swedish process of extracting vanadium from iron ore by producing pig iron containing vanadium, and then vanadium slag. The V_2O_5 was then extracted from the slag, followed by smelting of the FeV.

The vanadium extraction process is complicated by the different vanadium contents of mineral resources (e.g., regional differences) and the target material in different stages of the process. For example, the vanadium oxide in raw material can be of a low grade and have many different characteristics and behaviors. The vanadium industry includes three types of vanadium extraction from low-grade raw materials, high-grade raw materials, and extraction of both precious metals and vanadium. Various process configurations have been demonstrated, including continuous enrichment, balance, transformation, and recycling processes. Vanadium extraction includes characteristics of both chemical metallurgy and metallurgical chemical.

During vanadium extraction, it is necessary to choose suitable process parameters. The vanadium component needs to be converted into compounds that are soluble in salt, alkali, acid, and/or water. Considering the solubility of sodium vanadate, vanadium extraction requires a roasting and calcination process to convert low-valence vanadium oxides into high-valence ones. The reaction between Na and vanadium oxide occurs when these compounds are mixed and treated at high temperature. Conversion and structural transformations occur, and soluble vanadates as stable intermediate compounds are formed. It is possible to separate soluble vanadium and other mineral components into a liquid phase, while the insoluble materials are retained in the slag. This process is suitable for extracting vanadium as the main target of vanadium-containing raw materials.

The vanadium content of minerals and comprehensive materials is generally low, while some vanadium containing raw materials are rich in precious metals or other valuable materials. Hence, vanadium extraction needs to be comprehensively planned to include recovery of both vanadium and other valuable metals. Vanadium generally occurs in mixed compositions, where some vanadium is in the form of the primary mineral, while the rest occurs as secondary V-containing raw materials. Hence, it is difficult to balance different process functions of recycling and extraction. Acidolysis and acid leaching processes can produce a unified liquid phase, which can be separated by precipitation of different metal salts for inorganic and organic extraction processes, depending on the properties of the liquid. All of these processes can be successfully used for extraction and enrichment of vanadium products, and recovery of valuable metals. Vanadium extraction is a chemical metallurgy process that can achieve high recovery in a short processing time, although it requires a high grade of raw materials. It played an important role in the vanadium industry before the widespread development of South Africa's vanadium industry in the 1960s.

The application areas of vanadium are continuously expanding as an understanding of its unique properties under various conditions is improving. For all applications and storage of vanadium, the metal must be solidified and stabilized (with respect to its properties and shape). All unstable vanadium products should be recalled and recycled, and closed-loop production systems should be implemented for this purpose. All energy and materials are balanced, and intermediate vanadium products and tailings must be protected during storage. Wastewater from vanadium extraction plants must be discharged according to environmental standards.

Extraction of vanadium pentoxide from vanadium slag is a selection process and considered the most important step in extracting vanadium. The characteristics of this process need to be systematically and theoretically analyzed and applied to production experience. This is expected to result in a new era for the industry and achieve the advanced development of vanadium products. The process of extracting vanadium pentoxide from vanadium slag should reflect the core goal by selecting appropriate technology and parameters for the entire process to optimize the production of the valuable industrial products and resources. The development of the vanadium industry should be considered comprehensively, from the supply of raw materials to the equipment design (considering both the advanced technology required, and the cost and controllability), and, finally, to satisfy market demand for high-end products. These goals all need to be achieved while ensuring environmental protection and safety in the vanadium industry.

Vanadium products can be divided into primary, secondary, and tertiary levels. Primary products are mainly raw materials for vanadium extraction, including vanadium minerals, vanadium concentrates, oil refining catalysts, V-containing slag, scrap, and other residues (e.g., vanadium steel slag), which are considered high quality as they containing up to 25% V_2O_5. The secondary products (or intermediates) are vanadium oxide and vanadium trioxide, which are used as catalysts for the production of sulfuric acid and petrochemicals, or the development of other vanadium products. The tertiary products (or consumption materials) are FeV, V–Al alloys, and vanadium compounds. These materials are used in clean energy development and are being investigated for application in hydrogen storage alloys and vanadium flow batteries. The addition of 6% Al and 4% V to a Ti-based alloy (Ti6Al4V) can significantly improve the stability, weldability, and fatigue strength; hence, this alloy is widely used for casting and forging. Ti6Al4V has been widely used for fabricating the outer shell and wings of aircraft engines. V-based alloys are also used as coatings for nuclear materials. Furthermore, vanadium is used as cathode, gate, X-ray target, and vacuum tube heating filament materials in the electronics industry. Vanadium silicate and gallium vanadium are good superconducting materials for intermetallic compounds.

There is a significant body of literature, including other books, which discusses vanadium from different angles. We have read most of them and benefited greatly from their insight. I joined Pan steel in 1985 and have collaborated with many people in the study of vanadium and titanium over the last 30 years. I have accumulated a lot of data over this time and developed my knowledge in this field. Therefore I was motivated to write a book based on my data and experiences, to disseminate information regarding vanadium and titanium to general academics, better served for high-end products, and the hi-tech-oriented vanadium and titanium industries. I have focused on innovation and development in the vanadium industry, which has been my most important field of research experience. I have witnessed the development of the vanadium industry in Panzhihua, China and the growth of Chinese vanadium extraction technology.

Jinyong He is the chairman of Panzhihua Yinjiang Jinyong Industry and Trade Co., Ltd., and our partner. His field of research is vanadium batteries. His team has had collaboration with the Chinese academy of engineering physics, and they have achieved three large-scale demonstrations of vanadium battery technology with improved docking cohesion, and promoted the transformation of the vanadium battery and energy storage industries. This group also studies the preparation of high-purity vanadium metal by thermal reduction and has achieved 98.7% pure metal. Dr. Guifang Zhang is a professor at Kunming University of Science and Technology, as well as our research partner, and is committed to teaching basic theory and researching titano ferrous magnetite containing vanadium. She provided a wide range of model data and references for this book and has contributed to increasing the relevant knowledge regarding the manufacturing of V-based materials.

This book regarding Vanadium extraction, manufacturing and application consists of five parts. The first introduces the basic properties, compounds, resources, and products of vanadium. The second part addresses vanadium extraction, focusing on V_2O_5 extraction, and discusses the different minerals, processes, and products made from V_2O_5. The third part covers vanadium applications, specifically vanadium pentoxide products, and focuses on chemical applications such as catalysts, with emphasis on vanadium alloy applications in the vanadium steel and nonferrous metal alloy fields. The properties of the materials are highlighted by vanadium compounds. A representative application of liquid-phase conversion of vanadium in vanadium batteries is presented. The fourth part discusses the development of the vanadium industry in the world, highlighting production in one area. Finally, the fifth part gives relevant vanadium manufacturing data.

This book is expected to build an industry framework and provide a comprehensive reference for professionals. According to the key development elements of the vanadium industry, a layout is given which is expected to contribute to developing knowledge in the field in research, industry, and academic. The vanadium extraction, manufacturing and application process reference book was written by referring to many sources, including scientific reports, journals, and books, some internal data, and information from the network. There was also some contribution from teaching colleagues. There are several review articles related to the study of vanadium, and all of these have been cited and discussed here. We have performed a multilevel analysis to form our conclusions. We would like to say "thank you" and ask for forgiveness for the omission of references. Because of time constraints and the limited academic level, some information has accumulated for a long time, and studies in other languages have multiple translated versions that are difficult to cite and confirm. There are varying opinions regarding the vanadium industry because of the different fundamental ideas and levels of understanding of those in the field. Because of early scarcity of vanadium resources, some people have a detailed understanding of the industry. If our book contains some omissions or errors, please feel free to contact us. We invite experts, professors, and colleagues with differing views to discuss them further with us.

Many experts, professors, and colleagues have collaborated and made great contributions to the vanadium industry. The fields of vanadium production and application have benefited from much innovation, development of knowledge, and consensus, and developed processes in an era of growth and redirection. In this book, we do not evaluate the limitations of previous opinions in the field, nor discuss the impact of this on the development of the industry. Instead, we focus mainly on the current and future perspectives of manufacturing V-based materials. We are grateful for the contribution of all colleagues and supportive friends in the industry who provided us with information, argumentation, supporting evidence, calibrations, illustrations, and assistance with editing.

From a professional point of view, "vanadium extraction, manufacturing and application" is suited for the following audience: (1) experts and scholars in the fields of vanadium resources, technology and equipment, products, environment, and marketing; (2) technical and management personnel working in commercial vanadium plants; (3) sales personnel promoting vanadium products; (4) decision-making consultation agency personnel in the vanadium industry; (5) government economic planners and technical managers in areas related to the vanadium industry; (6) teaching, research, and design personnel in colleges and research institutes, along with postgraduates and students participating in research; (7) others who have a personal interest in the topic.

CHAPTER 2

Vanadium and its compounds

Chapter outline
2.1 Vanadium 10
2.2 Physical properties of vanadium 12
2.3 Chemical properties of vanadium 12
2.4 Thermodynamic properties of vanadium 13
2.5 Vanadium compounds 14
    2.5.1 Vanadium pentoxide 15
    2.5.2 Vanadates 19
    2.5.3 Compounds of vanadium and nonmetallic elements 21
    2.5.4 Compounds of vanadium and metallic elements 22
    2.5.5 Intermediate compounds 22
    2.5.6 Ternary vanadium compounds 22
    2.5.7 Ions of vanadium in solution 23
2.6 Compound salts of vanadium 24
    2.6.1 Pentavalent salt 24
    2.6.2 Tetravalent salts 25
    2.6.3 Ammonium salts of vanadium 25
    2.6.4 Sodium salts 26
    2.6.5 Vanadium oxygenates 26
    2.6.6 Vanadium sulfates 26
    2.6.7 Vanadium bronze 27
    2.6.8 Vanadium peroxide 27
2.7 Physiological effects of vanadium 27
    2.7.1 Vanadium deficiency 28
    2.7.2 Food sources 28
    2.7.3 Metabolic absorption 28
    2.7.4 Physiological functions 28
    2.7.5 Physiological requirements 29
    2.7.6 Overdose 29
    2.7.7 Physiological processes 29
    2.7.8 Vanadium use by human organs 29
2.8 Diseases from vanadium pollution 30
2.9 Biological effects of vanadium in humans and animals 30
2.10 Biological effect of vanadium on plants 32

Vanadium is widely distributed as a trace element, accounting for about 0.02% of the matter in the Earth's crust. Vanadium is a rare, weak, and viscous transition metal, with a total of 31 isotopes, including one stable one (^{51}V). The term "vanadium" is derived from the name of the beautiful goddess Venus (or Vanadis). In 1801 Spanish mineralogist A.M. del Rio discovered vanadium in lead ores and misnamed it as a compound of chromium. It was not until 1830 that it was separated during the process of smelting iron by Swedish chemist N.G. Sefström and named vanadis. The German chemist F. Wöhler proved that the new element found by Sefström was the same as that discovered by del Rio. In 1867 British scientist Henry Enfield Roscoe first obtained pure vanadium by hydrogen reduction of vanadium(III) chloride [1,2,14–16,18,22,26,39,40].

2.1 Vanadium

English name: vanadium
Elemental symbol: V
Atomic number: 23
Group number: 5
Atomic weight: 50.9415 g/mol
Element type: metal
Density: 6.11 g/cm^3
Atomic volume: 8.78 cm^3/mol
Concentration in the Sun: 0.4 ppm
Concentration in the Earth's crust: 160 ppm
Number of protons: 23
Number of neutrons: 27
Period: 3
Group category: transition metal
Group number: VB
Electronic layer distribution: 2−8−11−2
Crystal structure: body-centered cubic cell with two metal atoms
Oxidation state: predominantly V^{3+}, V^{4+}, and V^{5+}. Less common: V^{3-}, V^{1-}, V^0, V^{1+}, and V^{2+}
Cell parameters: $a = 303$ pm, $b = 303$ pm, $c = 303$ pm, $\alpha = 90°$, $\beta = 90°$, $\gamma = 90°$
Ionization energy (kJ/mol)
$M-M^+$ 650
M^+-M^{2+} 1414
$M^{2+}-M^{3+}$ 2828
$M^{3+}-M^{4+}$ 4507
$M^{4+}-M^{5+}$ 6294

$M^{5+} - M^{6+}$ 12,362
$M^{6+} - M^{7+}$ 14,489
$M^{7+} - M^{8+}$ 16,760
$M^{8+} - M^{9+}$ 19,860
$M^{9+} - M^{10+}$ 22,240
Propagation rate of sound in V: 4560 m/s
State at room temperature: solid
Melting point: 2183K (1919°C ± 2°C)
Boiling point: 3673K (3000°C–3400°C)
Triple point: unknown
Critical point: unknown
Heat of fusion: 21.5 kJ/mol
Heat of vaporization: 459 kJ/mol
First ionization energy: 650.3 kJ/mol.
Second ionization energy: 1413.5 kJ/mol
Third ionization energy: 2828 kJ/mol
Electronegativity: 1.63
Electron affinity: 50.7 kJ/mol
Specific heat: 0.49 J/g K
Heat of atomization: 514 kJ/mol atoms
Valence electron configuration: $3d^3 4s^2$
Minimum oxidation number: −1
Maximum oxidation number: 5.
Minimum common oxidation number: 0
Maximum common oxidation number: 5
Crystal structure: cubic
Mohs hardness: 7 (soft)
Color: bright white
Toxic: yes
Reaction with air and air response: mild, w/ht ≥ , V_2O_5, VN
Reaction with 6 M HCl: none
Reaction with 15 M HNO_3: mild
Reaction with 6 M NaOH: none
Number of isotopes: 2
Hydrides: VH, VH_2
Oxides: VO, V_2O_3, VO_2, V_2O_5
Chlorides: VCl_2, VCl_3, VCl_4
Atomic radius: 134 pm
Ionic radius (1^+, 1^- ion): unknown

Ionic radius (1− ion): unknown
Ionic radius (2+ ion): 93 pm
Ionic radius (2− ion): unknown
Ionic radius (3+ ion): 78 pm
Thermal conductivity: 30.7 J/(ms °C)
Electrical conductivity: 39.371 mΩ cm.
Polarizability: 12.4 Å^3

2.2 Physical properties of vanadium

Vanadium is a ductile, hard, and nonmagnetic metal. Its main physical properties are shown in Table 2.1.

Table 2.1: Main physical properties of vanadium.

Density (20°C)	6.1	g/cm^3
Melting point	1929	°C
Boiling point	3410	°C
Specific heat (0°C–100°C)	498	J/(kg K)
Evaporation heat	457.2	kJ/mol
Thermal conductivity (0°C–100°C)	31.6	W/(m K)
Resistivity (20°C)	19.6	μΩ cm

The mechanical properties of industrial vanadium metal are shown in Table 2.2.

Table 2.2: Mechanical properties of industrial metal vanadium.

Properties	Industrially pure products		High-purity products
Tensile strength σb (MPa)	245–450	210–250	180
Ductility (%)	10–15	40–60	40
Vickers hardness HV (MPa)	85–150	60	60–70
Elastic modulus (GPa)	137–147	120–130	
Poisson's ratio	0.35	0.36	
Yield strength (MPa)	125–180		

2.3 Chemical properties of vanadium

The valence electron structure of the vanadium atom is 3d34s2, where all five valence electrons can participate in bonding to produce compounds with oxidation states of +2, +3, +4, and +5, including those stable with the compound of V_2O_5. Vanadium compounds with V^{5+} states have oxidizing properties. Compounds with low valence have reducing behavior, where stronger reduction performance is observed for lower valence states. Metallic vanadium at room temperature is stable, and no reactions with air, water, or alkali materials occur. Vanadium is resistant to hydrochloric acid, dilute sulfuric acid, alkali solutions, and seawater corrosion, but it can be corroded by nitric acid, hydrofluoric acid, or concentrated sulfuric acid. The corrosion resistance of vanadium is shown in Table 2.3.

Table 2.3: Corrosion resistance of vanadium.

Medium	Corrosion rate [mg/(cm² h)]	Corrosion rate (nm/h)	Material
10% H_2SO_4 (boiling)	0.055	20.5 (70°C)	Vanadium plate
30% H_2SO_4 (boiling)	0.251		
10% HCl (boiling)	0.318	25.4 (70°C)	Vanadium plate
17% HCl (boiling)	1.974		
	Corrosion rate (35°C) (μm/a)	Corrosion rate (60°C) (μm/a)	Material
4.8% H_2SO_4	15.2	53.3	
3.6% HCl	15.2	48.3	
20.2% HCl	132	899	
3.1% HNO_3	25.4	1100	
11.8% HNO_3	68.6	88,390	
10% H_3PO_4	10.2	45.7	
85% H_3PO_4	25.4	160	
	Corrosion rate [mg/(cm² month)]		
Liquid Na (500°C)	0.2		

At high temperatures, vanadium metal easily reacts with oxygen and nitrogen. When the metal is heated in air, vanadium is oxidized to brown and black trioxides, blue V_3O_4, and hyacinth-colored vanadium pentoxide. When vanadium is heated to 900°C–1300°C in nitrogen, vanadium nitride is produced. Vanadium can react with carbon at high temperature to produce vanadium carbide. When vanadium is heated with silicon, boron, phosphorus, or arsenic in a vacuum or inert atmosphere, the corresponding silicide, boride, phosphite, and arsenide can be formed. At room temperature, pycno vanadium is stable under oxygen, nitrogen, and hydrogen atmospheres. When vanadium is heated in air, it can be oxidized into the dark brown vanadium trioxide (V_2O_3), a blue-black vanadium tetroxide (V_2O_4), or orange vanadium pentoxide (V_2O_5). At lower temperatures, vanadium can react with chlorine to produce vanadium tetrachloride (VCl_4). At higher temperatures, vanadium carbide (VC) and vanadium nitride (VN) can be produced by reaction with carbon and nitrogen, respectively.

2.4 Thermodynamic properties of vanadium

The heat capacity (C_p), enthalpy change ($H_T^\ominus - H_{298}^\ominus$), and entropy $S_T^\ominus$ of metallic vanadium are shown in Table 2.4 [3,4,27].

Table 2.4: Heat capacity (C_p), enthalpy change ($H_T^\ominus - H_{298}^\ominus$), and entropy $S_T^\ominus$ of metallic vanadium.

T (K)	C_p (J/mol K)	$H_T^\ominus - H_{298}^\ominus$ (J/mol)	$S_T^\ominus$
298	24.74	0	29.35
300	24.79	46	29.52
400	26.08	2596	36.76
500	26.92	6380	42.66
600	27.55	8000	47.65
700	28.05	10,760	51.96
800	28.68	13,610	55.73
900	29.52	16,500	59.16
1000	30.56	19,510	62.34
1100	31.61	22,610	65.27
1200	32.78	25,830	68.08
1300	34.00	29,180	70.76
1400	35.29	32,660	73.35
1500	36.59	36,260	75.82
1600	37.89	40,030	78.21
1700	39.18	43,920	80.55
1800	40.53	47,940	82.81
1900	41.78	52,080	85.03
2000	42.96	56,310	87.21
2100	44.00	60,710	89.35
2200	39.77	82,480	99.44
2300	39.77	86,460	101.19
2400	39.77	90,430	102.87
2500	39.77	94,410	104.50
2600	39.77	98,390	106.05
2700	39.77	102,370	107.56
2800	39.77	106,340	109.02
2900	39.77	110,320	110.41
3000	39.77	114,300	111.75

The heat of evaporation $\Delta H_T^\ominus$, the Gibbs free energy change for evaporation, and the vapor pressure of vanadium are shown in Table 2.5 [3,4,27].

Table 2.6 shows the heat of evaporation $\Delta H_T^\ominus$ and Gibbs free energy change $\Delta G_T^\ominus$ of vanadium oxides [3,4,27].

Table 2.7 shows the heat of evaporation $\Delta H_T^\ominus$ and Gibbs free energy change $\Delta G_T^\ominus$ of VC [3,4,27].

2.5 Vanadium compounds

The valence electron configuration of vanadium is 3d34s2, which can form compounds with oxidation numbers, including +5, +3, and +2, where the most important compounds for

Table 2.5: Heat of evaporation $\Delta H_T^\ominus$, Gibbs free energy change for evaporation, and vapor pressure of vanadium.

T (K)	$\Delta H_T^\ominus$ (J/mol)	$\Delta G_T^\ominus$ (J/mol)	P (Pa)
298.15	514,770	469,000	5.49×10^{-78}
400	514,770	453,220	5.39×10^{-55}
500	514,560	438,060	1.53×10^{-41}
600	514,350	422,810	1.42×10^{-38}
700	513,930	407,550	3.85×10^{-26}
800	513,510	392,290	2.24×10^{-21}
900	513,100	377,240	1.20×10^{-17}
1000	512,680	262,190	1.15×10^{-14}
1100	512,050	347,160	3.16×10^{-12}
1200	511,420	332,100	3.32×10^{-10}
1300	510,590	317,260	1.72×10^{-8}
1400	509,750	302,420	5.02×10^{-7}
1500	508,700	287,580	9.29×10^{-6}
1600	507,450	272,740	1.20×10^{-4}
1700	505,990	258,110	1.12×10^{-3}
1800	504,320	243,690	8.15×10^{-3}
1900	502,650	229,270	4.81×10^{-1}
2000	500,970	214,850	2.35×10^{-1}
2100	499,090	200,640	9.9×10^{-1}
2190	497,420	186,850	3.37
2190	467,310	186,850	3.37
2200	467,100	185,590	3.39
2400	472,970	159,470	3.29×10
2600	470,040	133,340	2.04×10^2
2800	467,110	107,630	9.55×10^2
3000	464,400	82,140	3.64×10^3
3200	461,680	56,640	1.17×10^4
3400	459,170	31,560	3.22×10^4
3600	456,870	6480	7.91×10^4
3652	456,250	0.000	9.8×10^4
3800	454,780	−18.600	1.76×10^5

applications have the oxidation number +5. Certain compounds of vanadium have catalytic and physiological functions. The standard free energy of formation of vanadium oxidation is shown in Table 2.8.

2.5.1 Vanadium pentoxide

Vanadium pentoxide (V_2O_5) is an orange-to-brick-red solid, tasteless, toxic (vanadium compounds are generally toxic), and slightly soluble in water. The aqueous solution of V_2O_5 is generally yellow and acidic. Currently, V_2O_5 is extracted from vanadium slag (including $FeO \cdot V_2O_3$), stone like coal minerals, and coproduction of vanadium during the process of steelmaking with V-containing iron ore.

Table 2.6: Heat of evaporation $\Delta H_T^\ominus$ and Gibbs free energy change $\Delta G_T^\ominus$ of vanadium oxides.

T (K)	VO $\Delta H_T^\ominus$ (kJ/mol)	VO $\Delta G_T^\ominus$ (kJ/mol)	V_2O_3 $\Delta H_T^\ominus$ (kJ/mol)	V_2O_3 $\Delta G_T^\ominus$ (kJ/mol)	V_2O_4 $\Delta H_T^\ominus$ (kJ/mol)	V_2O_4 $\Delta G_T^\ominus$ (kJ/mol)
289.15	−418.0 (±20.9)	−390.8 (±20.9)	−1239.0 (±25)	−1160.0 (±27)	−1432 (±16.5)	1323.0 (±19)
345	−	−	−	−	−1432.0	−1306.5
345	−	−	−	−	−1423.5	−1306.5
400	−418.0	−381.0	−1237.0	−1132.5	−1421.5	−1287.5
500	−416.6	−372.6	−1235.0	−1105.5	−1419.5	−1254.0
600	−418.6	−464.3	−1233.0	−1080.5	−1417.5	−1220.5
700	−414.5	−355.9	−1231.0	−1055.5	−1413.0	−1189.0
800	−414.5	−347.5	−1227.0	−1028.5	−1411.0	−1155.5
900	−412.4	−339.1	−1224.5	−1005.0	−1409.0	−1124.5
943	−	−	−	−	−	−
943	−	−	−	−	−	−
1000	−412.4	−330.8	−1222.5	−982.0	−1404.5	−1093.0
1100	−410.3	−322.4	−1220.5	−956.5	−1400.5	−1063.5
1200	−408.2	−314.0	−1218.5	−933.6	−1398.5	−1032.5
1300	−408.2	−305.6	−1216.5	−910.5	−1394.5	−1003.0
1400	−406.1	−299.4	−1214.0	−885.5	−1392.0	−971.5
1500	−404.0	−291.0	−1200.0	−862.5	−1390.0	−942.0
1600	−404.0	−282.6	−1206.0	−839.5	−1388.0	−912.5
1700	−401.9	−276.3	−1203.0	−816.5	−1385.5	−883.5
1800	−401.9	−268.0	−1201.5	−793.5	−1383.5	−852.0
1818	−	−	−	−	−1383.5	−848.0
1818	−	−	−	−	−1270.5	−848.0
1900	−399.8	−261.7	−1199.5	−772.5	−1266.5	−829.0
2000	−399.8	−253.3	−1195.5	−751.5	−1260.0	−806.0
2100	−397.7	−244.9	−1193.5	−726.5	−1256.0	−783.0
2185	−397.7	−237.8	−1170.5	−708.0	−1252.0	−764.0
2185	−415.3	−237.8	−1208.0	−708.0	−1287.5	−764.0
2200	−414.9	−236.6	−1208.0	−703.5	−1287.5	−760.0
2300	−412.4	−228.6	−1206.0	−681.0	−1281.0	−737.0
2400	−412.4	−220.2	−1204.0	−658.0	−1277.0	−712.0
2500	−410.3	−212.3	−1205.5	−637.0	−1270.5	−688.5
2600	−410.3	−203.9	−1197.5	−613.5	−1264.5	−663.5
2700	−408.2	−195.9	−1195.5	−590.5	**−1260**.0	−640.5
2800	−408.2	−187.6	−1193.5	−507.5	−1254.0	−615.5
2900	−406.1	−179.6	−1189.0	−544.5	−1247.5	−592.0
3000	−406.1	−171.7	−1187.0	−521.5	−567.5	−567.5

T (K)	V_6O_{13} $\Delta H_T^\ominus$ (kJ/mol)	V_6O_{13} $\Delta G_T^\ominus$ (kJ/mol)	V_2O_5 $\Delta H_T^\ominus$ (kJ/mol)	V_2O_5 $\Delta G_T^\ominus$ (kJ/mol)
289.15	−4455.0 (±63)	−4105.0 (±73)	−1561.5 (±21)	−1430.0 (±23)
345	−	−	−	−
345	−	−	−	−

(Continued)

Table 2.6: (Continued)

T (K)	V₆O₁₃ $\Delta H_T^\ominus$ (kJ/mol)	V₆O₁₃ $\Delta G_T^\ominus$ (kJ/mol)	V₂O₅ $\Delta H_T^\ominus$ (kJ/mol)	V₂O₅ $\Delta G_T^\ominus$ (kJ/mol)
400	−4438.0	−3988.0	−1559.5	−1386.0
500	−4420.5	−3877.0	−1557.5	−1342.0
600	−4402.5	−3770.5	−1553.5	−1300.5
700	−4385.5	−3668.0	−1550.0	−1258.5
800	−4366.5	−3566.0	−1547.0	−1217.0
900	−4350.0	−3466.5	−1545.0	−1175.0
943	−	−	−1543.0	−1157.5
943	−	−	−1478.0	−1157.5
1000	−4333.5	−3368.5	−1476.0	−1139.0
1100	−4733.0	−3268.5	−1471.5	−1105.5
1200	−4716.5	−3167.5	−1467.5	−1072.0
1300	−4699.5	−3067.5	−1463.5	−1038.5
1400	−4681.0	−2966.5	−1461.5	−1007.5
1500	−4664.0	−2866.5	−1457.0	−975.5
1600	−4647.5	−2765.8	−1455.0	−942.0
1700	−4628.5	−2665.3	−1453.0	−910.5
1800	−4612.0	−2564.8	−1471.0	−879.5
1818	−	−	−	−
1818	−	−	−	−
1900	−4595.0	−2464.3	−1451.0	−846.0
2000	−4576.0	−2364.0	−1449.0	−814.5
2100	−4559.5	−2263.7	−1449.0	−783.0
2185	−4545.0	−2178.0	−1446.5	−753.6
2185	−4649.5	−2178.0	−1482.0	−753.6
2200	−4647.5	−2160.8	−1481.5	−748.5
2300	−4630.5	−2047.8	−1479.2	−715.0
2400	−4615.5	−1943.0	−1478.0	−682.0
2500	−4603.5	−1821.7	−1476.4	−648.5
2600	−4586.5	−1708.6	−1474.7	−615.5
2700	−4572.0	−1595.6	−1473.0	−582.0
2800	−4557.5	−1482.5	−1471.2	−548.6
2900	−4540.5	−1377.5	−1469.5	−515.0
3000	−4528.0	−1256.0	−1465.4	−482.0

Under heating conditions the reaction of V_2O_5 with a pure alkali compound is:

$$4FeO \cdot V_2O_3 + 4Na_2CO_3 + 5O_2 \rightarrow 8NaVO_3 + 2Fe_2O_3 + 4CO_2 \uparrow \quad (2.1)$$

$NaVO_3$ is leached from the calcined material and dissolved in water. Ammonium sulfate is added to achieve a pH of 5−6. With acid neutralization, the liquor containing $NaVO_3$ is adjusted to pH = 2−3, and ammonium hexavanadate is precipitated, and then roasted and converted to V_2O_5.

Table 2.7: Heat of evaporation $\Delta H_T^\ominus$ and Gibbs free energy change $\Delta G_T^\ominus$ of VC.

T (K)	$\Delta H_T^\ominus$ (J/mol)	$\Delta G_T^\ominus$ (J/mol)	T (K)	$\Delta H_T^\ominus$ (J/mol)	$\Delta G_T^\ominus$ (J/mol)
298.15	−175,600 (±400,000)	−149,500 (±400,000)	1800	−168,800	−24,600
400	−176,000	−141,000	1900	−168,800	−16,300
500	−175,600	−132,800	2000	−168,800	−7900
600	−175,000	−124,500	2100	−168,800	0
700	−174,600	−116,500	2185	−168,880	+7100
800	−174,600	−108,000	2185	−186,000	+7100
900	−174,000	−99,500	2200	−186,000	+8350
1000	−173,000	−92,000	2300	−186,000	+17,500
1100	−172,200	−83,000	2400	−186,000	+26,400
1200	−171,500	−74,500	2500	−186,000	+35,100
1300	−171,000	−70,500	2600	−186,000	+44,000
1400	−170,200	−58,200	2700	−186,000	+52,600
1500	−169,500	−49,800	2800	−186,000	+61,500
1600	−168,800	−41,500	2900	−186,000	+70,100
1700	−168,800	−33,000	3000	−186,000	+79,300

Table 2.8: Standard free energy of formation of vanadium oxidation.

Reaction	A (kJ/mol)	B [kJ/(mol K)]	Temperature range T (K)
$V(s) + 1/2O_2(g) = VO(s)$	−412.8	0.0817	298–2000
$2V(s) + 3/2O_2(g) = V_2O_3$	−1220	0.2364	600–2000
$V(s) + 2O_2(g) = V_2O_4(\beta)$	−1402	0.3066	600–1818
$6V(s) + 13/2O_2(g) = V_6O_{13}(s)$	−4368.4	1.0042	600–1000
$2V(s) + 5/2O_2(g) = V_2O_5(s)$	−1554.6	0.4224	298–943

V_2O_5 is an amphoteric oxide (mainly acidic) and soluble in a strong base (such as NaOH):

$$V_2O_5 + 6OH^- \rightarrow (cool)2VO_4^{3-} + 3H_2O \text{ (vanadate radical, colorless)} \quad (2.2)$$

$$V_2O_5 + 2OH^- \rightarrow ()2VO_3^- + H_2O \text{ (metavanadate, yellow)} \quad (2.3)$$

V_2O_5 is also soluble in strong acids (such as H_2SO_4), where only the VO^{2+} ions (not the V^{5+} ones) are observed in the yellowish solution:

$$V_2O_5 + 2H^+ \rightarrow 2VO_2^+ + H_2O \text{(yellowish)} \quad (2.4)$$

V_2O_5 is a strong oxidant; in the reaction with HCl, V^{5+} can be reduced into V^{4+} and release chlorine gas:

$$V_2O_5 + 6H^+ + 2Cl^- \rightarrow 2VO_2^- + Cl_2\uparrow + 3H_2O\text{(blue)} \quad (2.5)$$

V_2O_5 is a catalyst in the sulfuric acid industry and is used as a corrosion inhibitor for equipment in the petrochemical industry.

The main properties of the most common vanadium oxides are shown in Table 2.9.

Table 2.9: Properties of the most common vanadium oxides.

Property	VO	V_2O_3	VO_2	V_2O_4	V_2O_5
Crystallographic system	Face-centered cubic	Rhombic	Monoclinic	α	Trapezoidal
Color	Light gray	Dark	Blue		Orange yellow
Density (kg/m³)	5550–5760	4870–4990	4330–4339		3252–3360
Melting point (°C)	1790	1970–2070	1545–1967		650–690
Decomposition temperature (°C)					1690–1750
Sublimation temperature (°C)	2063	2343	1818		943
Heat of formation $\Delta H^\ominus_{298}$ (kJ/mol)	−432	−1219.6	−718	−1428	−1551
Absolute entropy $s^\ominus_{298}$ [J/(mol K)]	38.91	98.8	62.62	102.6	131
Free energy $\Delta G^\ominus_{298}$ (kJ/mol)	−404.4	−1140.0	−659.4	−1319	−1420
Water solubility	None	None	Slightly soluble		Slightly soluble
Acid soluble	Soluble	HF and HNO_3	Soluble		Soluble
Alkali solubility	None	None	Soluble		Soluble
Oxidation–reduction	Reduction	Reduction	Amphiprotic		Oxidation
Acid–base properties	Base	Base	Base		Amphoteric

2.5.2 Vanadates

Vanadate occurs in many forms, where those containing V^{5+} ions include metavanadate, vanadate, pyrovanadate, and polyvanadate compounds. Among these, the most stable form is metavanadate, followed by pyrovanadate. Vanadate is generally not stable and hydrolysis reactions quickly occur, resulting in the formation of pyrovanadate. The physical and chemical properties of vanadates are shown in Table 2.10.

Under certain conditions, acid is added to the vanadate solution to gradually decrease the pH, and the vanadate will gradually dehydrate:

$$VO_4^{3-} \to (\text{pH } 12-10) V_2O_7^{4-} \to (\text{pH } 9) V_3O_9^{3-} \to (\text{pH } 2.2) H_2V_{10}O_{28}^{4-}$$
$$\to (\text{pH} < 1) VO_2^+ \text{ (vanadateradical)[polyvanadateradical]} \tag{2.6}$$

Table 2.10: Physical and chemical properties of vanadates.

Compound	Molecular formula	State	Appearance	Melting point (°C)	Solubility	ΔH_{298} (J/mol)	ΔS_{298} (J/mol)	ΔG_{298} (J/mol)
Metavanadic acid	HVO_3	s	Yellow		Soluble in acid/base			
Ammonium metavanadate	NH_4VO_3	s	Canary yellow	Decomposition at 200	Soluble lightly in water	−1051	140.7	−886
Sodium metavanadate	$NaVO_3$	s	Colorless monoclinic system	630	Soluble in water	−1145	113.8	−1064
		Aqueous solution				−1129	108.9	−1064
Potassium metavanadate	KVO_3	s	Colorless		Soluble in hot water			
Sodium vanadate	Na_3VO_4	s	Colorless hexagonal crystal system	850–856	Soluble in water	−1756	190.1	1637
	Na_3VO_4	Aqueous solution				−1685		
	NaH_2VO_4	Aqueous solution				−1407	180.0	−1284
Sodium pyrovanadate	$Na_4V_2O_7$	s	Colorless hexagonal crystal system	632–654	soluble in water	2917	318.6	−2730
Calcium metavanadate	CaV_2O_6	s		778		−2330	179.2	−2170
Calcium pyrovanadate	$Ca_2V_2O_7$	s		1015		−3083	220.6	−2893
Calcium vanadate	$Ca_3V_2O_8$	s		1380		−3778	275.1	−3561
Ferrum metavanadate	FeV_2O_6					−1899		−1750
Lead pyrovanadate	$Pb_2V_2O_7$	S		772		−2133		−1946
Lead vanadate	$Pb_3V_2O_8$	s		960		−2375		−2161
Magnesium metavanadate	MgV_2O_6	s				−2201	160.8	−2039
Magnesium vanadate	$Mg_2V_2O_7$	s		710		−2836	200.5	−2645
Manganese metavanadate	MnV_2O_6					−2000		−1849

VO_2^+ can be reduced by Fe^{2+} and oxalic acid to form VO^{2+}:

$$VO_2^+ + Fe^{2+} + 2H^+ \rightarrow VO^{2+} + Fe^{3+} + H_2O \qquad (2.7)$$

(Vanadyl ion)   (vanadium acyl ions)

$$2VO_2^+ + H_2C_2O_4 + 2H^+ \rightarrow (heating)2VO^{2+} + 2CO_2 + 2H_2O \qquad (2.8)$$

These reactions can be used to determine the vanadium content by oxidation–reduction reactions.

VO^{2+}, ZrO^{2+}, and HfO^{2+} and the previously encountered SbO^+, BiO^+, and TiO^{2+} ions are intermediate products of the corresponding high-valence cationic hydrolysis reaction, namely, the acyl ions. Vanadate exists as VO^{2+} in strong acids and is oxidizing. The standard electrode potential of vanadium in acid solution is as follows:

$$\frac{E_A^\circ}{V} \quad VO_2^+ \quad 1.000 \quad VO^{2+} \quad 0.337 \quad V^{3+} \quad -0.255 \quad V^{2+} \quad -1.13$$

Ion color (yellow) (blue) (green) (purple)

2.5.3 Compounds of vanadium and nonmetallic elements

Vanadium and hydrogen form compounds such as VH, VH_2, V_2H, and V_3H_2. Vanadium and oxygen form compounds such as (VO), (VO_2), dark brown (V_2O_3), blue-black tetroxide (V_2O_4), and orange vanadium pentoxide (V_2O_5). In the temperature range of 0°C–678°C, intermediates include V_2O_5, V_3O_7, and V_6O_{13}. At higher temperatures of 678°C–1542°C, intermediates include VO_2, V_4O_7, V_5O_9, V_6O_{11}, V_7O_{13}, and V_8O_{15}, while at 1542°C–1957°C, intermediates include V_2O_3 and V_3O_5. Vanadium and carbon form compounds, including vanadium carbides (VC, V_2C) and VCx polymers, while vanadium and nitrogen form vanadium nitride (VN, V_2N), VN: $VN_{0.71}$–VN_1, V_2N: $VN_{0.37}$–$VN_{0.5}$.

- compounds of vanadium and chloride: VCl_2, VCl_3, and VCl_4;
- compounds of vanadium, chlorine, and oxygen: VOCl, $VOCl_2$, $VOCl_3$, and VO_2Cl;
- compounds of vanadium and fluorine: VF_2, VF_3, VF_4, and VF_5;
- compounds of vanadium, fluorine, and oxygen: VOF, VOF_2, VOF_3, and VO_2F;
- compounds of vanadium and bromine: VBr_2, VBr_3, and VBr_4;
- compounds of vanadium, bromine, and oxygen: $VOBr_2$ and $VOBr_3$;
- compounds of vanadium and iodine: VI_2, VI_3, and VI_4;
- compound of vanadium and iodine and oxygen: VOI_2;
- compounds of vanadium and sulfur: V_2S_5, VS_2, V_2S_3, VS, VS_5, V_3S, V_5S_4, V_xS, V_3S_4, V_5S_8, V_3S_5, VS_4, and others;
- compounds of vanadium and silicon: V_3Si, V_5Si_4, V_5Si_3 和VSi_2, and polymers of VSi_x;

compounds of vanadium and phosphorus: VP, V_2P, V_3P, $V_{12}P_7$, V_4P_3, $V_{12}P_7$, $V_{2.4}P_9$, and VP_2;
compounds of vanadium and arsenic: V_3As, VAs_2, V_4As_3, V_5As_3, and V_2As;
compounds of vanadium and boron: V_3B_2, VB, V_3B_4 and VB_2; and
hydroxides formed by vanadium under alkaline conditions: $V(OH)_2$ and $V(OH)_3$.

2.5.4 Compounds of vanadium and metallic elements

Vanadium and cobalt compounds: VCo_3, VCo, and V_3Co;
Vanadium and nickel compounds: VNi_3, VNi_2, and V_3Ni;
Vanadium and aluminum compounds: VAl_3, VAl_{11}, VAl_6, VAl_7, and V_5Al_8;
Vanadium and lead compounds: V_3Pb;
Vanadium and antimony compounds: V_3Sb;
Vanadium and zinc compounds: V_4Zn_5 and VZn_3;
Vanadium and tin compounds: V_3Sn; and
Vanadium and gallium compounds: V_3Ga.

2.5.5 Intermediate compounds

$VOSO_4$ is formed in sulfuric acid medium.
$VOCl_3$ is formed in chlorine medium.
NH_4VO_3 is formed in ammonia medium.
$FeVO_4$ and $Fe_2V_4O_{13}$ are formed in the V_2O_5–Fe_2O_3 system.
NiV_2O_6, $Ni_2V_2O_7$, and $Ni_3V_2O_8$ are formed in the V_2O_5–NiO system.
$CrVO_4$ is formed in the V_2O_5–Cr_2O_3 system.
$AlVO_4$ is formed in the V_2O_5–Al_2O_3 system.
CaV_2O_6, $Ca_2V_2O_7$, and $Ca_3V_2O_8$ are formed in the V_2O_5–CaO system.
$K_2O \cdot 4V_2O_5$, $K_2O \cdot V_2O_5$, $16K_2O \cdot 9V_2O_5$, $\alpha 2K_2O \cdot V_2O_5$, $\beta 2K_2O \cdot V_2O_5$, and $3K_2O \cdot V_2O_5$ are formed in the V_2O_5–K_2O system.
NaV_6O_{15}, $Na_8V_{24}O_{23}$, NaV_6O_{15}, $NaVO_3$, $Na_4V_2O_7$, $Na_2V_5O_{13.3}$, and Na_3VO_4 are formed in the V_2O_5–Na_2O system.
$7MgO \cdot 3V_2O_5$, $2MgO \cdot V_2O_5$, $3MgO \cdot V_2O_5$, $2MgO \cdot 3V_2O_5$, $3MgO \cdot 2V_2O_5$, and $MgO \cdot V_2O_5$ are formed in the V_2O_5–MgO system.
$BaO \cdot V_2O_5$, $2BaO \cdot V_2O_5$, and $3BaO \cdot V_2O_5$ are formed in the V_2O_5–BaO system.
$V_2O_5 \cdot 9Nd_2O_5$ is formed in the V_2O_5–Nd_2O system.

2.5.6 Ternary vanadium compounds

V_3PC_{1-x}, V_2PC, $V_{5+x}P_3C_{1-x}$, V_4P_2C, and $V_6P_3C_{0.6}$ are formed in the V–P–C system.
CrV_2S_4 and VCr_2S_4 are formed in the Cr–V–S system.

Ca(VO$_3$)$_2$, Ca$_2$V$_2$O$_7$, Ca$_3$(VO$_4$)$_2$, Ca$_7$V$_4$O$_{17}$, Ca$_4$V$_2$O$_9$, Ca$_5$V$_2$O$_{10}$, FeVO$_3$, Fe$_2$V$_4$O$_{12}$, Ca$_2$Fe$_2$O$_5$, CaFe$_2$O$_4$, Ca$_4$Fe$_{14}$O$_{25}$ and CaFe$_4$O$_7$ are formed in the CaO–Fe$_2$O$_3$–V$_2$O$_5$ system.

2.5.7 Ions of vanadium in solution

2.5.7.1 Vanadic acid

Vanadic acid in aqueous solution forms vanadic acid anions or vanadate ions, which can exist in a variety of agglomerates and combine with other ions to form various vanadium- and oxygen-containing compounds. The existence of vanadic acid depends on the pH and concentration of vanadium in the solution. Under highly alkaline conditions, vanadic acid is present as VO$_4^{3-}$. Increasing the acidity of the solution results in hydrolysis of the vanadic acid. When the vanadium concentration is very low, the vanadate radicals are in the form of a single nucleus. Vanadic acid has very strong polymerization behavior, and polymerization reactions occur at high vanadium acid concentrations.

Vanadic acid can be combined with other acid radicals (such as W, P, As, and Si) to form compound salts. The vanadate precipitated in weak or strong alkaline solutions is orthovanadate or pyrovanadate, respectively. Under neutral conditions, V$_3$O$_9^{3-}$ and V$_4$O$_{12}^{4-}$ are precipitated. However, under acid conditions, polyvanadate is precipitated. At 40°C and pH of 2–8 the main form of vanadate in solution is V$_3$O$_9^{3-}$, V$_4$O$_{12}^{4-}$, HV$_6$O$_{17}^{3-}$, and HV$_{10}$O$_{28}^{5-}$. For pH < 2, decavanadate is converted to lauronitrile vanadate, as shown in the following equation:

$$6H_6V_{10}O_{28} = 5H_2V_{12}O_{31} + 13H_2O \tag{2.9}$$

Poly(H$_2$V$_{12}$O$_{31}$) is hydrated with vanadium oxide, where the protons can be replaced by other positive ions. The sequence of combination and substitution is as follows: K$^+$ > NH$_4^+$ > Na$^+$ > H$^+$ > Li$^+$. With increasing acidity the poly(H$_2$V$_{12}$O$_{31}$) reacts with protons to produce VO^{2+}, as shown in the following equation:

$$H_2V_{12}O_{31} + 12H^+ = 12VO_2^+ + 7H_2O \tag{2.10}$$

2.5.7.2 Hydrated ions

The bivalent and trivalent vanadium oxidation states form a simple hydrated ligand, either [V(H$_2$O)$_6$]$^{2+}$ or [V(H$_2$O)$_6$]$^{3+}$. The hydrated ligands are unstable and are oxidized to tetravalent and pentavalent vanadium.

2.5.7.3 Vanadium and oxygen ions

The simple hydrated ligands formed by the tetravalent and pentavalent vanadium oxidation states are generally complex. Vanadium and oxygen atoms are combined directly to form V=O bonds, resulting in tetravalent VO^{2+} and pentavalent VO^{2+}. Vanadium in the high oxidation states of a transition metal can coordinate with oxygen to form oxyacid salt anions. Vanadium is coordinated with oxygen to form focal vanadate, pyrovanadate, and vanadate with a single core. Multicore oxygen anionic condensation or polymerization tend to occur, producing polyacid radical anions and heteropoly anions. In aqueous solutions, vanadate is unstable and can be converted into pyrovanadate, which can be condensed into polymetavanadate (VO^{3-}) when boiled.

For vanadium oxyacid anions the vanadium and oxygen atoms form covalent bonds, where the main form is VO_4^{4-}. Free vanadium ions or the hydrated pentavalent form do not exist under these conditions. The polymerization of pentavalent vanadium in aqueous solutions depends on the concentration of vanadium, pH, and temperature. The reaction conditions for the various vanadate radicals are shown in Table 2.11.

Table 2.11: Reaction conditions for vanadate radicals.

Reaction	pH	C(V) (mol/L)	lg K	Temperature (°C)
$VO_4^{-3} + H^+ = HVO_4^{-2}$	14–11	0.00036	13.15	
$2VO_4^{-3} + H^+ = V_2O_7^{-3} + H_2O$	11–9	0.0001	25.05	20
$2V_2O_7^{4-} + 4H^+ = V_4O_{12}^{4-} + 2H_2O$	9–7	>0.0033	18.46	20
$4H_2VO_4^- = V_4O_{12}^{4-} + 4H_2O$		0.018–0.1000	10.10	40
$10H_2VO_4^- + 4H^+ = V_{10}O_{28}^{6-} + 12H_2O$	5.9	0.000386	~29	15
	4.0	0.000386	~23	90
	~6.2	0.0672	~26	15
	~4.5	0.0672	~21	90
$HV_2O_7^{3-} + 5H^+ = 2VO_2^+ + 3H_2O$	1.2	0.2000	6.1	
$HVO_3 + H^+ = VO_2^+ + H_2O$	3–2	0.025–0.020	3.30	25
$V_{10}O_{18}^{6-} + H^+ = HV_{10}O_{28}^{5-}$		0.0125–0.1	6.12	
$HV_{10}O_{28}^{5-} + H^+ = H_2V_{10}O_{28}^{4-}$		0.025–0.1	4.699	
$HVO_4^{2-} + 3H^+ = VO_2^+ + 2H_2O$		0.0001	16.17	20
$VO_2^+ + 2H^+ = VO^{3+} + H_2O$	2–1.4	0.003–0.005	2.79	18–20

2.6 Compound salts of vanadium

Vanadium salts occur in a range of colors, including green, red, black, and yellow. For example, bivalent vanadium salt is often purple, trivalent vanadium salt is green, and the tetravalent vanadium salt is light blue. The alkaline derivatives of tetravalent vanadium are often brown or black, while vanadium pent oxide is red. These colorful vanadium compounds are used to prepare bright pigments that are added to glass and used to make various types of ink.

2.6.1 Pentavalent salt

The pentavalent salts of vanadium include ammonium metavanadate, sodium metavanadate, potassium metavanadate, sodium orthovanadate, and sodium pyrovanadate.

2.6.2 Tetravalent salts

Tetravalent salts of vanadium include vanadyl sulfate, vanadyl oxalate, vanadium tetrachloride/halide, vanadium oxytrichloride, and halogen vanadium oxide. Vanadium oxides mainly include V_2O_5, VO_2, V_2O_3, and VO. The chemical stability of vanadium halide decreases with increasing valence of vanadium. The chemical stability of vanadium of the same valence decreases gradually with increasing atomic number (fluoride to iodide). The main properties of V^{3+} and V^{5+} halides are listed in Tables 2.12 and 2.13.

Table 2.12: Main properties of V^{3+} halides.

Property	VF_3 (S)	VCl_3 (S)	VBr_3 (S)	VI_3 (S)	VOCl (S)	VOBr (S)
Color	Green	Red	Gray, brown	Brown, black	Brown	Purple
Density (g/cm³)	3392 (292K)	2820 (293K)	2659 (293K)	5140 (293K)	3340 (298K)	4000 (291K)
Sublimation temperature T (K)	1679	698 (disproportionation)	673 (disproportionation)	553 (vacuum decomposition)	593 (vacuum decomposition)	753 (decomposition)
Enthalpy change $\Delta H^\ominus_{298}$ (kJ/mol)	−1334.7 ± 83.68	−426.656	493.712	−280.328	602.496	–

Table 2.13: Main properties of V^{5+} halides.

Property	VF_5 (S)	VF_5 (L)	VOF_3 (S)	$VOCl_3$ (L)	$VOBr_3$ (L)	VO_2F (S)	VO_2Cl (S)
Color	White	/	Yellow	Yellow	Dark red	Brown	Orange
Density (g/cm³)	>2500 (293K)	2508 (293K)	2659 (293K)	1830 (293K)	2993 (288K)	/	2290 (293K)
Melting point (°C)	292.5	/	283 (sublimation)	194.1	214	573	/
Boiling point (K)	321.3	/	/	400.2	406, 453 (decomposition)	/	453
Enthalpy change $\Delta H^\ominus_{298}$ (kJ/mol)	−1442.77	/	/	740.17	/	/	765.17

2.6.3 Ammonium salts of vanadium

The most common ammonium salts of vanadium are ammonium metavanadate (NH_4VO_3), ammonium vanadate ($NH_4)_3VO_4$, and ammonium polyvanadate $2(NH_4)_2O \cdot 3V_2O_5 \cdot n(H_2O)$, where NH_4VO_3 is considered the most important. Ammonium metavanadate is a white or light-yellow crystal with a density of 2304 kg/m³, which decomposes when heated above 473K. Ammonium metavanadate decomposes into NH_3 and V_2O_5 at 523K in oxygen-rich air. In air, at 523K-613K, it decomposes into $(NH_4)_2O \cdot 3V_2O_5$ and $V_2O_4 \cdot 5V_2O_5$, and into NH_3 and V_2O_5 at 693K–713K. V_2O_5 can be obtained by oxidation in pure oxygen at 583K–598K. Ammonium metavanadate heated in a hydrogen atmosphere generates $(NH_4)_2O \cdot 3V_2O_5$ at 473K, $(NH_4)_2O \cdot V_2O_4 \cdot 5V_2O_5$ at 593K, V_2O_{13} and V_2O_3 at 673K, and V_2O_4 at 1273K.

Ammonium metavanadate heated at 623K in a mixed atmosphere of CO_2, N_2, and Ar produces $(NH_4) \cdot V_2O_4 \cdot 5V_2O_5$ V_6O_{13} obtained at 673K–773K; when water vapor is introduced at 498K, $(NH_4)_2O \cdot 3V_2O_5$ is produced. Several polymers of ammonium vanadate occur, which are expressed as $\alpha[(NH_4)_2O] \cdot \beta[V_2O_5] \cdot \gamma[H_2O]$, where $\alpha = 1$, 2, or 3; $\beta = 0.3-5$; and $\gamma = 0-6$.

2.6.4 Sodium salts

The compounds in the V_2O_5–Na_2O binary system are NaV_6O_{15}, $Na_8V_{24}O_{63}$, $NaVO_3$, $Na_4V_2O_7$, and Na_3VO_4, where the most common ones are sodium metavanadate ($NaVO_3$), sodium pyrovanadate ($Na_4V_2O_7$), sodium orthovanadate (Na_3VO_4), and sodium polyvanadate ($NaVO_3$). $NaVO_3$ has a molecular weight of 121.93 and occurs as a white or light-yellow crystal with a relative density of 2.79, and melting point of 630°C. It has a solubility in water of 21.1 g/100 mL at 25°C and 38.8 g/100 mL at 75°C and is only slightly soluble in alcohol. $NaVO_3$ is used as a chemical reagent, catalyst, desiccant, dye fixative, and raw material for producing ammonium vanadate and potassium metavanadate. It is also used for medical photography, plant inoculation, and as a corrosion inhibitor.

2.6.5 Vanadium oxygenates

2.6.5.1 Vanadium oxygenate

Vanadates containing alkali metals and alkaline earth metals are soluble in water, but other salts are less soluble. Salts with sodium, ammonium, calcium, and iron are commonly formed. Considering the aggregation state of vanadium, vanadium oxygenate can be classified into orthovanadate (VO_4^{3-}), pyrovanadate ($V_2O_7^{4-}$), metavanadates (VO^{3-}, $V_4O_{12}^{4-}$), and polyvanadates ($V_4O_{12}^{4-}$, $V_6O_{16}^{2-}$, $V_{10}O_{18}^{2-}$, and $V_{12}O_{31}^{2-}$).

2.6.5.2 Thiovanadate

V_2S_5 can react with alkali metal sulfide solutions, such as $(NH_4)_2S$, which can form the thiovanadate or $(NH_4)VO_3$ and $(NH_4)_2S$. The reaction is as follows:

$$(NH_4)VO_3 + 4(NH_4)_2S + 3H_2O = (NH_4)_3VS_4 + 6NH_4OH \qquad (2.11)$$

Here, VS_4^{3-} is an orthothiovanadate, and pyro-orthovanadate can also be formed.

2.6.6 Vanadium sulfates

2.6.6.1 Vanadium sulfate

Bivalent vanadium and trivalent vanadium can easily form $V_2(SO_4)_3$ and VSO_4. $V_2(SO_4)_3$ that is insoluble in water and relatively stable. $VSO_4 \cdot 7H_2O$ is soluble in water, easily oxidized, and combines with $FeSO_4$ to form compound salts, which tend to be stable.

2.6.6.2 Vanadyl sulfate

Pentavalent vanadium and tetravalent vanadium easily form $VOSO_4$ and $(VO)_2(SO_4)_3$ compounds, which tend to be inactive and insoluble in water.

2.6.7 Vanadium bronze

Vanadium oxide can react with other metal oxides at high temperature to produce vanadates. In the $Na_2O-V_2O_5$ system at 600°C–700°C, $x(V_2O_5)/x(Na_2O) = (100-35)/(0-65)$ forms and four salts can be precipitated: NaV_6O_{15}, $Na_8V_{24}O_{53}$, $NAVO_3$, and $Na_4V_2O_7$. At 700°C–1300°C, $x(V_2O_5)/x(Na_2O) = (35-20)/(65-80)$ forms, where Na_3VO_4 is precipitated. NaV_6O_{15} belongs to the $Na_2O \cdot xV_2O_4 \cdot (6-x)V_2O_5$ ($x = 0.85-1.06$) complex salt system, while $Na_8V_{24}O_{53}$ belongs to the $5Na_2O \cdot yV_2O_4 \cdot (12-y) V_2O_5$ ($y = 0-2$) system. Vanadium bronze, which is insoluble in water, can be oxidized in air to form soluble vanadates. When the soluble vanadates are cooled and deoxidized, they can be converted into vanadium bronze.

2.6.8 Vanadium peroxide

Nonacidic aqueous solutions of metavanadate can be added to hydrogen peroxide to form vanadate peroxides. For example, meta-ammonium vanadate and hydrogen peroxide react to form ammonium vanadate peroxide $[(NH_4)_6H_4V_2O_{10}]$, which is a nonaqueous dissociant. In an acidic solution the vanadium ions react with hydrogen peroxide to form a brick red compound.

2.7 Physiological effects of vanadium

Vanadium is a necessary trace element in the human body, with a typical concentration of about 25 mg. Under the conditions of body fluids with a pH of 4–8, vanadium mainly occurs as metavanadate (VO^{3-}) ions, although orthovanadate (VO_4^{3-}) also occurs. because of the similar biological effects, all vanadates are the two of +5 oxidation ions. VO^{3-} ions migrate in the body via ion transport systems and freely enter cells; they are reduced to the vanadyl (VO^{2+}) ion with a +4 oxidation state by reduced glutathione in the cells. Phosphate and magnesium ions (Mg^{2+}) widely exist in cells. As VO_3 and phosphate have a similar structure, and VO^{2+} and Mg^{2+} ions have a similar size (ionic radius of 0.61 and 0.65 Å, respectively), they compete with phosphate and Mg^{2+} ions in the cells, resulting in ligand interference in biochemical reaction processes. For example, vanadium has extensive biological effects in the cells, including inhibiting ATP phosphoric acid hydrolase, RNA ribozyme fructose phosphate kinase, glyceraldehyde phosphate kinase, 6-phosphate glucose enzyme, and phosphoric acid tyrosine protein kinase reactions.

2.7.1 Vanadium deficiency

Vanadium is thought to be necessary for normal growth, where the tetra- and pentavalent states are the most biologically significant. Tetravalent vanadium occurs as vanadyl cations, which easily combine with proteins to form complexes, which prevent oxidation. Pentavalent vanadium is also an vanadyl cation that easily combines with biological materials to form complexes. In many biochemical processes, vanadate radicals can compete with or replace phosphate radicals. Vanadyl is reduced by vitamin C, glutathione, or NADH to VO^{2+} (+4 valence state), or vanadyl. The role of vanadium in human health is still not clear in the nutrition and medical communities and is a topic of further research. However, it is agreed that vanadium plays an important role and is thought to help prevent cholesterol buildup, lower blood sugar levels, prevent tooth decay, and help produce red blood cells. Some vanadium is lost through urine every day.

The most recognized vanadium deficiency was reported in a 1987 study of goats and rats, in which vanadium-deficient goats showed increased rates of miscarriage and decreased milk production. In rats, vanadium deficiency caused growth inhibition, increased thyroid-weight-to-body-weight ratio, and changed plasma thyroid hormone concentrations. Vanadium deficiency in humans has not yet been clarified; some studies suggested that it may lead to cardiovascular and kidney diseases, impaired wound regeneration, and neonatal death.

2.7.2 Food sources

Food sources of vanadium include cereal products, meat, chicken, duck, fish, cucumbers, shellfish, mushrooms, parsley, dill seed, and black pepper.

2.7.3 Metabolic absorption

Only a small fraction of vanadium consumed by humans is absorbed, estimated to be less than 5%, where most of the remainder is excreted in the feces. Vanadium is absorbed in the small intestine, forms complexes with low-molecular-weight materials, and then binds to plasma ferritin in the blood. Vanadium in the blood is then quickly transported to the various tissues. Generally, most tissues contain less than 10 ng/g of vanadium. In the human body, 80%–90% of the absorbed vanadium is excreted in the urine and can also be excreted by bile (with a typical vanadium content of 0.55–1.85 ng/g).

2.7.4 Physiological functions

Experimental results showed that vanadium regulates Na^+/K^+-ATPase, modulates phosphatase, adenylate cyclase, and protein kinase reactions, which are involved in hormone, protein, and lipid metabolism in vivo. It also inhibits the synthesis of cholesterol in the liver of young rats. It may also have the following functions: (1) preventing fatigue and heat stroke because of overheating, (2) promoting bone and tooth growth, (3) assisting in the normalization of fat metabolism, (4) preventing heart attacks, and (5) assisting in normal operation of nerves and muscles.

2.7.5 Physiological requirements

There is no specific data regarding the physiological requirements. In the diet, people can provide less than 30 μg of vanadium per day, which is more than 15 μg. Therefore it is necessary to consume a daily intake of 10 μg of vanadium from the diet. There is usually no need for special supplementation. It is important to note that ingesting synthetic vanadium can be toxic. In addition, smoking reduces the absorption of vanadium.

2.7.6 Overdose

Vanadium does not easily accumulate in the body, so it is very rare to be poisoned by intake from food. However, it can be toxic when ingesting 10 mg or more a day (or 10−20 μg/g of food). Overdose with vanadium can result in stunted growth, diarrhea, reduced intakes, and death.

2.7.7 Physiological processes

Vanadium is an essential component of some enzymes in living organisms. Some nitrogen-fixing microbes use V-containing enzymes to fix nitrogen in the air. Mice and chickens also need small amounts of vanadium to regulate growth and reproduction. In addition, V-containing hemoglobin exists in sea squirts. Some V-containing substances have insulin-like effects and may be used to treat diabetes.

2.7.8 Vanadium use by human organs

Vanadium occurs in very low concentration in the human body (less than 1 mg). It is mainly distributed in viscera, especially in the liver, kidney, thyroid, and other organs, as well as in bone tissue. The normal vanadium requirement of a human body is around 100 μg/day. The absorption of vanadium in the gastrointestinal tract is only 5%, which occurs mainly in the upper digestive tract. Vanadium in the environment can be absorbed by the body through the skin and lungs. About 95% of the vanadium in the blood is transported in the ionic state (VO^{2+}) and via the blood protein transferrin, allowing vanadium and iron to interact with each other in the body. Vanadium is important for normal development and calcification of bones and teeth and can strengthen the resistance of teeth to caries. Vanadium can also promote sugar metabolism, stimulate vanadate dependence of the NADPH oxidation reaction, enhance lipoprotein lipase activity, accelerate adenylate cyclase activation and amino acid conversion, and promote the growth of red blood cells. Therefore the development of teeth, bones, and cartilage can be impeded when vanadium is deficient. If liver phosphatide concentrations are low, benign edema and abnormal thyroid metabolism can occur.

2.8 Diseases from vanadium pollution

Air and water contaminated by vanadium pollution is harmful to the respiratory system and skin of humans, which can result in symptoms of contact and allergic dermatitis, characterized by swelling, redness, and skin necrosis.

2.9 Biological effects of vanadium in humans and animals

Study of the biology and toxicology of vanadium began in 1876 and this field was developed rapidly in the 1970s and 1980s. Studies found that the chemical properties of vanadium were the basis for determining its biological effects. The toxicity of vanadium compounds and the size of life effect are related to the total amount of vanadium, which depend on the specific vanadium compounds present. For example, VO^{2+} is biologically inactive, while VO^{3+} is easily absorbed. The toxicity of vanadium is very low, but its compounds have moderate toxicity to plants and animals, where the toxicity increases with increasing vanadium valence state. In the ecosystem, vanadium can exist in $(-1)-(+5)$ valence states and usually forms a wide range of polymers. Vanadium is usually present in the form of vanadate (VO_3^-, V^{5+}) and V−O cations (VO_2^+, V^{4+}) in the cells and bodily fluids. After entering the cells, vanadate is reduced by glutathione ($C_{10}H_{17}O_6N_3S$) and other materials, forming V−O cations, which combine to form stable protein, phosphate, citric acid, and lactic acid ligands.

Both the vanadate and V−O cations are required in optimal amounts to enhance the physiological function of the body, maintain growth of the organism and normal functioning of the cardiovascular system, inhibit synthesis of cholesterol, and promote hematopoietic function. Vanadium affects the insulin in tissues; promotes absorption, oxidation, and synthesis of glucose; and has insulin-like effects. In addition, it can promote protein tyrosine phosphorylation and absorption of potassium, while reducing the hydrolysis of triglyceride and protein degradation.

Due to the increase in vanadium environmental pollution in recent years, research regarding the biological effects of vanadium usually focuses on its toxicology. The accumulation of vanadium has obvious effects on the lung, liver, and other organs of mammals. For example, in the United States, the Alaskan whale (Cetaceans) has shown increased concentrations of vanadium in the liver from 0.1 to 1 μg/g. This is a clear indication of biological accumulation, where cumulative concentration was positively correlated with the animal's age and size. These results were also verified by analyzing vanadium contents in rat renin. The accumulation of vanadium in animals has medium-to-high toxicity effects, which results in changes to the respiratory system, nervous system, gastrointestinal system, hematopoietic system, and metabolism. It can also reduce the intake of food and cause diarrhea and weight loss. Changes in the metabolism and biochemical functions also occur, including inhibited reproduction and growth, and a reduction in the animal's ability to resist external stresses, toxins, and carcinogens, even resulting in death. Toxicity studies with rats showed that the toxic concentration of vanadium was 0.25 mg/L and the lethal concentration was 6 mg/L.

In the early 1970s vanadium was identified as an indispensable trace element for animals such as chickens and mice, which triggered the question whether vanadium was also critical for human beings. Nielsen believes vanadium is an essential element in the bodies of humans and higher animals, but at high contents of $1-2$ μg/g, the effects are not well known. Therefore it is still unclear whether vanadium is essential for the human body.

Studies have shown that normal adults have about 25 mg of vanadium in their bodies, which remains low in the blood and corresponds to about 0.78 mmol/L. The two main ways that vanadium enters the body are via the daily diet or the environment. As for many other trace elements, the diet is the most common intake method. The amount of vanadium from the diet is $10-20$ μg/day, where the maximum amount of vanadium needed for the human and animal body is about 20 μg/day. However, there are reports that in the United States, $10-60$ μg/day of vanadium is provided by the diet. Vanadium in the environment can enter the body via absorption by the skin and lungs, which is relatively rare for most other essential elements. The vanadium in the body accumulates in the stomach, kidneys, liver, and lungs and can also be stored in fat and plasma lipids. Although there is a good current understanding of the biochemical effect and function of vanadium, the metabolism of vanadium is still poorly understood and data are lacking. Therefore Sabbioni used the RNAA method to test the vanadium concentration in human blood, serum, and urine, which gave values of ~ 1 nmol/L in the blood and serum, and ~ 10 nmol/L in the urine. The measured values were independent of gender and vanadium content. However, it is also believed that due to the lack of appropriate reference values, the abovementioned data are only normal values under test conditions. The study also showed that when vanadium accumulated to a certain concentration in the human body, it showed toxicity. Vanadium can irritate the eyes, nose, throat, and respiratory tract, causing a cough. As vanadium competes with calcium, excess vanadium can result in Ca being dissociated and excreted from the body. Vanadium is a toxin that can be absorbed by the whole body and can affect the gastrointestinal tract, nervous system, and heart. It can cause severe vasospasm, gastrointestinal peristalsis, and other symptoms in the kidney, spleen, and intestines.

2.10 Biological effect of vanadium on plants

Although there is no clear evidence that vanadium is as essential element for plant growth, a study found that vanadium has important functions in the growth and development of plants, especially leguminous plants. A moderate amount of vanadium can promote the crop growth, nitrogen fixation, and greening of plants. However, vanadium compounds are also toxic to plants in high quantities. When 150 ppm of ammonium vanadate was applied to rice seedlings, it promoted growth, while 500 ppm showed effects of poisoning, and the seedlings died when 1000 ppm was applied. Soybean production was reduced by the application of 25 mg/kg of vanadium, and the soybean yield decreased significantly when the vanadium concentration reached 50 mg/kg. Excessive vanadium can also strongly inhibit the various ATP enzyme processes in the plant root cell membrane, causing plant dwarfing, a decrease in production, and a reduction in the absorption of nutrients such as calcium and phosphate. In the presence of excess vanadium, the absorption of calcium by the root tip of sorghum was reduced, extraction of phosphate by the maize root system was reduced, and the growth and development of soybean seedlings was inhibited.

CHAPTER 3

Vanadium mineral resources

Chapter outline
3.1 Metallogenic characteristics of vanadium ore 34
    3.1.1 Main genetic types of global vanadium deposits 35
    3.1.2 Vanadium deposits in China 35
    3.1.3 Types of vanadium deposits in China 36
    3.1.4 Chinese regions mining V-bearing minerals 38
    3.1.5 Vanadium deposits outside of China 40
3.2 Minerals of vanadium 46
    3.2.1 Vanadinite 47
    3.2.2 Roscoelite in sandstone 47
    3.2.3 Carnotite 48
    3.2.4 Patronite 48
    3.2.5 Tyuyamunite 49
    3.2.6 Barnesite 49
    3.2.7 Vanadium magnetite 49
    3.2.8 Vanadine 50
    3.2.9 Uvanite 51
    3.2.10 Vichlovite 51
    3.2.11 Montroseite 51
3.3 Main vanadium minerals for vanadium extraction 51
    3.3.1 V-bearing titanoferrous magnetite 51
    3.3.2 Stone-like coal 52
    3.3.3 Vanadinite 52
    3.3.4 Carnotite 55
    3.3.5 Roscoelite 55
    3.3.6 Patronite 55
    3.3.7 Organovanadium ore 55
3.4 Distribution of vanadium resources 55
    3.4.1 Type and distribution 55
    3.4.2 V-bearing titanoferrous magnetite 56
    3.4.3 Stone-like coal resources 56
    3.4.4 Global distribution of vanadium resources 56
    3.4.5 Global vanadium reserves 58

Vanadium is widely distributed in the Earth's crust with an average content of 0.02%. On average, there is one vanadium atom in every 10,000 atoms, more than copper, nickel, zinc, tin, cobalt, lead, and other metals. Natural vanadium exists in the form of compounds, mainly trivalent or pentavalent vanadium. Vanadium has a variety of valence states, some with metallic properties and others with metalloid properties, allowing it to produce a large variety of compounds and complexes. The ionic radius of trivalent vanadium is close to that of trivalent iron; therefore trivalent vanadium rarely appears alone but exists in minerals containing iron and aluminum minerals via isomorphism. This is also the main reason why vanadium is highly dispersed in nature. Pentavalent vanadium usually forms independent minerals, such as carnotite, and vanadium-bearing mica, asphalt, petroleum, and coal. As vanadium is widely dispersed globally, it is rare to find V-rich ores. Almost all locations have vanadium but at very low concentrations. Traces of vanadium have been found in marine organisms (such as sea urchins), magnetite, various kinds of bituminous minerals, coal ash, and meteorites. The sun's spectral lines also indicate the presence of vanadium [2–4,14–16,19–23,26].

3.1 Metallogenic characteristics of vanadium ore

The crystal structure of an ore depends on the crystallization conditions of the magma it was formed from. Ore with a the big grain disseminated structure is easily selected, while dense ores with well-connected mineral grains are hard to choose. Most V-containing minerals have exogenetic geological origins and contain mostly vanadate-containing VO_4 groups. During endogenetic geological processes, the few minerals that are formed enter the crystalline lattice of various petrogenetic compounds and rare compounds in the form of isomeric impurities. In the endogenous strata, vanadium is present in oxidation states of V^{3+} (igneous rock), and V^{4+} and V^{5+} (hydrothermal strata), which is different from the composition of exogenetic minerals. V^{5+} is not found in the minerals, and there are no V^{2+} compounds or metallic vanadium in nature. Fig. 3.1 shows the global abundance of vanadium.

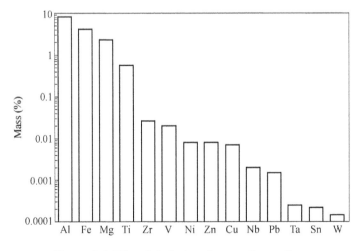

Figure 3.1: The global abundance of vanadium

3.1.1 Main genetic types of global vanadium deposits

Vanadium easily migrates during reversible chemical cycles occurring during geological events. In the process of magmatism, sedimentation, and metamorphism, a variety of useful mineral and nonmineral deposits are formed, and all contain cyclic elements in the form of oxides, hydroxides, silicates, carbonates, sulfates, chlorides, and other mineral compounds.

3.1.1.1 Magmatic deposits

There are two types of magmatic deposits, titanoferrous magnetite containing vanadium and titanium free magnetite containing vanadium. The magmatic deposits contain two kinds of V-bearing titanoferrous magnetite: one native to gabbro, gabbro—diabase, and ultrabasic rocks; and the other from late magmatic differentiation between plagioclase and gabbro. All vanadium is found in homogeneous V-bearing titanoferrous magnetite or V-bearing magnetite. Such deposits are found in South Africa, Russia, China, the United States, and Canada. Ti-free V-bearing magnetite has a low V_2O_5 content of 0.1%—0.3% and is found in, for example, New Jersey, the United States, and Kiruna, Sweden.

3.1.1.2 Sedimentary deposits

Sedimentary deposits include V-bearing iron ore, shale, and hydrocarbon minerals. Shale contains a high content of carbon and oil, mostly from organic matter. V-containing hydrocarbons are present in bitumen ores. Such deposits include carnotite, the black rock series, phosphorite, and petroleum bituminite deposits in the variegated rock series. Among these, vanadium in the black rock series and phosphorus block rocks are a very important vanadium resource in the United States.

3.1.1.3 Intrusive volcanic rock deposits

Intrusive volcanic rock deposits mainly contain alkaline-rich volcanic intrusive rocks.

3.1.1.4 Other deposits

Other types of V-containing deposits include sedimentary metamorphic deposits, magmatic metamorphic deposits, beach placers, lateritic clays, residual alluvial deposits, and vanadium deposits in modern volcanic cinders.

3.1.2 Vanadium deposits in China

The vanadium deposits in China are magmatic, sedimentary, and volcanic (porphyrite) types. Magmatic deposits are located in the geotectonic platform areas, edges of platforms, and geosynclinal fold zones controlled by regional deep faults. The formation mechanism and location of the deposit are always closely related to the base—ultrabasic rocks of different ages, such as in the Tianshan−Yinshan metallogenic belt, South Shaanxi−West Hubei metallogenic belt, and Sichuan−Yunnan metallogenic belt. There are magmatic deposits containing vanadium in Panxi (western Sichuan); Chengde, Xiangyang, and Yunyang (Hubei); and Hanzhong (Shaanxi). This type of deposit can be classified into late magmatic isomerism and late magmatic penetration types.

Sedimentary vanadium ore is characterized by certain strata and lithology. Vanadium deposits mainly exist within the black rock series. In China, they are mostly distributed on the Sino−Korean platform, Yangtze platform, and the sides of the Qinling Mountains/Qilian fold. They are also distributed in the Tarim platform and Southern China fold system. The lithology is characterized by Sinian−Early Cambrian, Silurian, and Permian silica−carbonaceous rocks, aluminum (clay) soil rocks, and phosphate rocks. Vanadium deposits with economic value are mostly found in the so-called "lower Cambrian black rock series" in the Yangzi platform and the Qinling Mountains/Qilian fold system. With regards to the spatial distribution, the north Chinese provinces of Shaanxi and Henan have a small fraction of black rock vanadium deposits, while most are found in southern China, for example, Sichuan, Guizhou, Guangxi, Hunan, Hubei, Zhejiang, Anhui, Jiangxi, and Guangdong. In particular, the black rock strata is distributed in extending over 1600 km through the Zhejiang, southern Anhui, southwest Jiangxi, Hunan, and northwest Guangxi regions. Vanadium deposits in intrusive volcanic rocks are mainly found in the Nanjing−Wuhu region in the middle and lower reaches of the Yangtze River. These porphyritic deposits were formed in the Jurassic−Cretaceous period.

3.1.3 Types of vanadium deposits in China

3.1.3.1 Magmatic deposits

The magmatic deposits include late magmatic differentiation and late magmatic injection types, both of which are related to basic−ultrabasic rocks. Such deposits are distributed in the Panxi and Chengde regions. V-bearing titanoferrous magnetite from the Panxi area is found in stratiform rock containing gabbro/olivine, gabbro/oliveolite, and gabbro/olivine/olivine structures. The V-bearing titanoferrous magnetite from Chengde occurs in the contact zone and fracture zone of plagioclase and gabbro.

Late magmatic differentiation V-bearing titanoferrous magnetite can be classified into gabbro (from Panzhihua−Xichang), gabbro/troctolite/olive gabbro (from Baima and Badong), gabbro/gabbro/pictolite (from Hongge), gabbro/norite (from Bijigou and Wangjiang of Shaanxi), and diabase (from Tiefoshi-Taoyuan of Shaanxi) types. The plagioclase−gabbro type from Heishan and Damiao of Hebei is a representative late magmatic injection deposit. Vanadium from these two types of deposits is found within titanoferrous magnetite.

3.1.3.2 Sedimentary deposits

The sedimentary deposits mainly occur in lower Cambrian black shale that is composed of black carbon shale, black carbon siliceous rock, and black carbon siliceous rock. Such deposits have high organic carbon contents of 8%–12%. According to the composition, black shale can be classified into stone-like coal seams, phosphorus block rock, vanadium deposit, and Ni–Mo multielement-enriched layers. It is called "stone-like coal." The content of vanadium oxide in each layer varies greatly, generally over the range of 0.13%–1%, where 60% of the deposit contains a content <0.50%. The lithologic sequence of vanadium ore layers from the bottom to top is black carbonaceous silica, black carbonaceous silty shale, vanadium deposit, and black carbonaceous shale. According to the combination of different elements, these layers can be divided as follows: V, V–Mo, V–Ni, V–Ga, V–U–Ni–Mo, V–U–Mo, Ni–Mo–V, and V–Cd. The silicate phase of vanadium (barium) is a result of absorption of vanadium from clay within an open shallow sea basin. The source of the elements in the ore layer includes continental debris from weathering of the crust, as well as from hot brine or submarine volcanic eruptions. The mineralization of vanadium begins at the sedimentary stage and is generally completed by the diagenetic stage. Mo, Co, Ni, Cd, Ag, Pt-group metals, and other elements occur in black shales. Therefore these stone-like coals are considered low-calorific and inferior coal but are low-grade ores for precious metals. Vanadium is adsorbed by carbon or clay minerals in black shale and exists as a single mineral (vanadium mica).

3.1.3.3 Volcanic deposits

Volcanic rock (porphyrite) is found in medium-base porphyrite, where the ore is characterized by breccia and vein types. Vanadium oxide is contained within magnetite and hematite minerals.

3.1.3.4 Other deposits

Vanadium in volcanic rock is affected by humid environments, weathering, and etching and can infiltrate clay minerals, resulting in residual vanadium in bauxite and red soil iron ore. The vanadium in clay can be etched and migrate to other deposits. V-bearing igneous rock or sediments under dry oxidative conditions and high temperature can form vanadium that is water soluble under certain pH conditions in the presence of a reducing agent. When combined with other cations, the precipitation of low concentrations of vanadates can occur. Coprecipitation with organic matter results in the formation of vanadium sulfide and hydrocarbons. Subsequent moisture evaporation and decomposition of the hydrocarbon structure result in a tar shale.

3.1.4 Chinese regions mining V-bearing minerals

3.1.4.1 Panzhihua deposit

The Panzhihua ore of V-bearing titanoferrous magnetite is found in the early basic—ultrabasic rock mass and belongs to the late magmatic differentiation deposit. The lithology of the ore-bearing rock, from top to bottom, is divided into three ore belts (gabbro, pyroxenite, and olivine), with five ore-bearing layers in the middle gabbro, lower gabbro, upper—middle—lower pyroxenite, and olivine—pyroxenite. The ore body mainly exists in two ore-bearing layers of pyroxenite, where the subsequent one is an ore-bearing layer of olivine. The occurrence of ore deposits is consistent with that of the rock mass.

The minerals present in the ore include metal oxides, sulfoarsenide, and gangue minerals. The metal oxides are mainly V-bearing titaniferrous magnetite, ilmenite, and ulvite, with some magnesium aluminate spinel, vanadium hematite, perovskite, anatase, rutile, chromite spinel, and leucoxene. The sulfoarsenides mainly consist of pyrrhotite, nicopyrite, chalcopyrite, and pyrite, with some linnaeite, cobaltite, violinite, sperrylite, and toxic sand. The gangue minerals are mainly pyroxene, olivine, and plagioclase, with some hornblende, black mica, and apatite. The altered minerals include chlorite, serpentine, and iddingsite.

The ore is mainly composed of sponge meteorite and inclusion structures, characterized by the disseminated structure, block structure, and mottled structure. From the point of view of iron ore, the ore is considered poor quality; however, the iron ore reserves are immense. In addition to iron, titanium, and vanadium, there are valuable elements, including cobalt, nickel, copper, chromium, manganese, gallium, selenium, tellurium, niobium, tantalum, phosphorus, sulfur, and platinum metals, although the ore has a low grade.

3.1.4.2 Chengde deposit

The Damiao deposit of V-bearing titanoferrous magnetite is located in the Xuanhua—Chengde—Beipiao deep-fault zone at the eastern end of the Inner Mongolia Earth axis, where the basic—ultrabasic rock intruded into the pre-Sinian stratum. Residual slurry separated from late ore-bearing slurry was formed by penetrating structural fractures. More than 50 V-bearing titanoferrous magnetite ore bodies were produced in lenticular, vein, or cystic forms from anorthosite or anorthosite contact parts of the fracture zone, and with a clear surrounding rock boundaries. The ore body in gabbro is usually disseminated or veined and has a gradual relationship with the surrounding rocks. The ore body is generally 10—360 m long, with a depth of 10—300 m. The ore has two types: dense block and disseminated. The main ore minerals include magnetite, ilmenite, hematite, and rutile. Magnetite and ilmenite are separated as solid solutions. Vanadium occurs as an isomorphism in V-bearing titanoferrous magnetite. The average grade of V_2O_5 ore is 0.16%—0.39%. In iron concentrate the V_2O_5 content is 0.77%.

3.1.4.3 Yaoshan deposit in Anhui

The mineralogical rock mass of the Yaoshan deposit in Anhui province contains V-bearing titanoferrous magnetite formed in the Jurassic−Cretaceous period. It is a meso-basaltic volcanic intrusive (subintrusive) rock mass. The ore-bearing rock mass is located in the Maanshan area of Anhui. The V-bearing diorite rock mass is 700−900 m long and 500 m wide, with a vertical depth of about 400 m extending to the northeast at an angle of 40−65 degrees. The ore is characterized by brecciated and vein structures with a V_2O_5 grade of 0.22%. Alteration of the surrounding rock is common, where the metallic minerals include magnetite, hematite, and limonite. Gangue minerals include actinolite, garnet, apatite, quartz, and carbonate minerals.

3.1.4.4 Xialan deposit in Guangdong

The Xialan deposit in the Guangdong province contains V-bearing titanoferrous magnetite in the lower middle part and a weathering crust in the Xialan basic complex. It can be divided into the weathered ore body and primary ore. The weathered ore bodies are exposed on the surface in a layered distribution with uniform mineralization. Due to long-term weathering, hydrolysis, and oxidation, most silicate minerals have been damaged, and the grade of the layered native ore bodies is low. Most of the V-bearing titanoferrous magnetite and ilmenite are weathered eluvial deposits and are well preserved, though some are weathered into hematite and limonite. The V-bearing titanoferrous magnetite and ilmenite from this mine are mostly coarse grained. The main ore body is close to Xingning city in the Guangdong province, where the ore from the original mine contains 0.17% V_2O_5.

3.1.4.5 Vanadium-silver deposit in Baiguoyuan, Hubei province

The vanadium-silver deposit in Baiguoyuan is located in Xingshan in the Hubei province. The auriferous strata are on top of the Sinian Doushantuo formation, where the ore deposit is on top of the Sinian Doushantuo formation. The lithology is black leaf dolomite mudstone with silty dolomite. The shape of the ore body is simple, and the formation of layered and surrounding rock is integrated. There are also rich vanadium and silver layers in the ore layer, sometimes in the vanadium deposit, with lamellar or lenticular silver ore bodies. The ore body is found on the ground surface and has a slight slope. The ore has a clay structure and an mm-grade rhyme structure. Vanadium is mainly contained within mica, and no simple vanadium minerals are present. Silver is mainly included in pyrite, with some also occurring in aguilarite, acanthite, a selenium variant of naumannite, and argyrodite. The selenium content of the ore is 0.005%−0.007%.

3.1.4.6 Vanadium phosphate uranium deposit in Fangshankou, Gansu

This vanadium phosphate uranium deposit belongs to a large sedimentary vanadium deposit located in the western end of the north mountain fold belt of the complex tectonic belt of the Yinshan–Tianshan mountains and the Dunhuang area of Gansu province. The geological structure of the area is short axis and anticline, where the core of the anticlinal nucleus is a part of the Jixian system and the two wings are Cambrian. The ore body is a lenticular and layered and was produced in the Cambrian strata. It has a uniform distribution, length of 100–300 m and thickness of 1–2 m. The vanadium occurs within both carbon argillite and carbon phyllite minerals. The average V_2O_5 content in this deposit is 0.826%. The distribution of vanadium in the various minerals is 74.9% vanadium mica, 11.5% vanadium tourmaline, 18.2% vanadium kaolinite, pyrite, 4.1% hematite, and 1.3% in other minerals. This deposit has a large vanadium mine, medium-sized phosphate rock mine, and small uranium mine.

3.1.4.7 Vanadium deposit in Yangjiabu

The Yangjiabu vanadium deposit is a sedimentary deposit formed by a lower Cambrian shallow sea bay. The structure is located in the Yangzi platform in the Danjiangkou area of Hubei, which is the north wing reversal of Yangjiabu–Tangjiashan. The monoclinal structure is composed of Cambrian strata, inclined to the south, with a slope of 15–30 degrees. The Cambrian stratum contains silica–silty sand, which has metamorphosed into killas. The ore-bearing bed is divided into three ledges from the bottom to the top: (1) ledge with a thickness of 0.9–10 m, containing siliceous rocks and silty killas ore, with a V_2O_5 content of 0.7%–1.3%; (2) deposits of the mother lode with a thickness of 1.4–22.4 m, containing stone-like coal, where siliceous rocks contain the ore with a V_2O_5 content of 0.8%–1.3%; and (3) ledge with a thickness of 1–8 m containing carbonaceous killas and a V_2O_5 content of 0.7%–1.3%. The main ore minerals include vanadium hydrate mica, barium–vanadium mica, calcium–vanadium garnet, and arsenite. Other valuable components include copper, nickel, molybdenum, cobalt, gold, silver, platinum, palladium, yttrium, and uranium. The distribution of the vanadium in the ore is 50% with illite, 20% in vanadium mica and chrome–vanadium garnet stone, and the remaining 30% evenly combined with organic matter as soluble salt containing vanadium, where the uranium content is generally lower than 0.003%.

3.1.5 Vanadium deposits outside of China

Typical vanadium deposits outside China occur with magmatic iron ore, especially V-bearing titanoferrous magnetite, in which some iron is replaced by vanadium. Such vanadium deposits are generally of higher grade than Chinese deposits.

3.1.5.1 V-bearing titanoferrous magnetite in the former Soviet Union

Rich V-bearing titanoferrous magnetite occurs in the former Soviet Union region that has a variety of mineral compositions, chemical compositions, and physical properties; however, all deposits belong to the endogenous vanadium-bearing type. The main ore deposits are distributed in the Ural, Kola Peninsula, Siberia, and the far east. The Catchikara deposit is giant pyroxenite with a disseminated ore body, few thin vein ore layer, and an ore logistics layer found in the ore body. The interstices between diophanes, common amphibole, and olivine (iron meteorite structure) are filled with its shape particles. Catchikara Mining and Dressing Co. LTD mine the Guchevo (Гусевгорское) deposit, Catchikara deposit (Качканарское) owned by the Central Ural Recovery Company and the first Ural open-pit mine (ПервоуАЛьское). There are many V-bearing titanoferrous magnetite deposits in the former Soviet Union that contain high TiO_2 contents ($TFe/TiO_2 < 10$), such as the Metaviev deposit (Медведевское), Volkov ore (Волковское), Copan ore (Копанское), pudoone mine (Пудожгорское), and the Zinnisk mine (Ченеиское). In addition, there is a considerable amount of V-bearing titanoferrous magnetite deposits with high TiO_2 content in Kusinsk, Ilsinsk, and Samchikan.

3.1.5.1.1 Guchevo (Гусевгорское) deposit

Considering the chemical composition (TFe 16.6%, V_2O_5 0.13%, TiO_2 1.23%), the Guchevo (Гусевгорское) deposit belongs to the group of impregnation ore bodies of titanoferrous magnetite containing low vanadium content. The content of titanoferrous magnetite is the highest in diallagite and the lowest in gabbro (TFe < 14%). The metal-containing minerals are magnetite and ilmenite, along with a small amount of hematite and sulfide. In ore deposits, coarse and medium disseminated ore are difficult to grade, where the lowest industrial grade ore contains a minimum of 16% TFe.

3.1.5.1.2 Catchikara (Качканарское) deposit

The Catchikara deposit is located in the north and northeast slopes of the Catchikara mountains. The main ore has TFe 16%–20%, while the remainder has TFe 14%–16%. In the peridotite the V_2O_5 content is 0.13%–0.14%, and TiO_2 content is 21.24%–1.28%.

3.1.5.1.3 The first Urals (ПервоуАЛьское) deposit

The first Urals (ПервоуАЛьское) deposit is located in the western slope of the central Ural mountains. The main metallic mineral is magnetite in the form of fine aggregate in particles of hornblende, which contains 3%–5% ilmenite. Nonmetallic minerals in this ore include amphibole, feldspar, chlorite, and epidote. The inner mine has a TFe content of 14%–35%, which is divided into poor impregnated ore (TFe 14%–25%), rich impregnated ore (TFe 25%–5%), and titanoferrous magnetite ores (TFe 35%). The most widely distributed poor impregnated ores contain 14%–16% TFe, ~0.19% V_2O_5, and 2.3% TiO_2.

The Caius deposit in Keliri-Cora is a type of immersed rock, which is mined by digging and land cutting. The main rocks are gabbro, anorthosite, alkaline rock, and ultramafic rocks, with an ore grade of TFe 29%–45%, TiO_2 5%–10%, and V_2O_5 0.15%–0.75%. The Chjing deposit in Keliri-Cora is mined by digging and land cutting to produce blocks.

The main rock is alkaline rock, magnesite, and ultramafic rocks, with a grade of TFe 36%, TiO_2 7%, and V_2O_5 0.26%. The Afrikaner deposit in Keliri-Cora is disseminated and mined by layered lens mass output. The major ore types are gabbro, plagioclase, and ferric rocks, and the raw ore has a grade of TFe 11%–18% and TiO_2 8%–18%. The deposit in Lake Jerut of Keliri-Cora is mined by digging and land cutting, where the main rocks are anorthosite, ferroite, and super ferric rocks with an ore grade of TFe 13%–37%, TiO_2 8%–26%, and V_2O_5 0.13%. The Pudozhans deposit in Keliri-Cora is mined by digging and land cutting, where the main rocks are gabbro, plagioclase, and alkaline and ultramafic rocks. The ore grade is TFe 13%–37%, TiO_2 8%–26%, and V_2O_5 0.13%.

3.1.5.2 South African vanadium titanomagnetite

There are igneous complex ores in Bushveld, Rooiwater, Mambula, and Usushwane in South Africa, in which V-bearing titanoferrous magnetite is disseminated and produced in blocks. The main rock types are gabbro, anorthosite, and alkaline rock. The Bushveld ore is the major V-bearing titanoferrous magnetite deposit in South Africa, where the Mapochs mine is the largest and lies north of the city of Rosenke in the east of Transvaal. The chemical composition of V-bearing titanoferrous magnetite (%) from this mine is shown in Table 3.1.

Table 3.1: Chemical composition of V-bearing titanoferrous magnetite from the Mapochs mine (%).

TFe	TiO_2	V_2O_5	Cr_2O_3	SiO_2	Al_2O_3
53–57	14–15	1.4–1.7	0.5–0.60	1.5–2.0	3–4

The Transvaal Alloy Company has two mines: one located in the Wabberskov and the other is Yolog, which is located 20 km north of Mapochs. The Kennedy Valley mine is located in the eastern part of Bushveld, where the ore contains about 2.5% V_2O_5. The Vanmeterko mine is located in Botswana, with an average V_2O_5 content after mineral processing of 2.0%. The Rhovan company is studying the development of promising mines in Botswana, similar to the Vanmeterko mine.

3.1.5.3 V-bearing titanoferrous magnetite in Canada

The Steel Mountain mine in Newfoundland, Canada has disseminated ore, and layered lenticular blocks are produced. The main rocks are gabbro, alkaline, mafic, and ultramafic rocks, where the ore grade is TFe 50%–55%, TiO_2 10%, and V_2O_5 0.4%. The Indian Hurd mine in Newfoundland is mined by digging and land cutting to produce blocks, where the main rock types are gabbro, alkaline, mafic, and ultramafic rocks with an ore grade of TFe 64%, TiO_2 2%–6%, V_2O_5 0.2%–0.7%. The Matava ore in Ontario is disseminated, mined by digging and land cutting to produce blocks. The main rock types are alkaline, mafic, and ultramafic iron rocks, where the raw ore grade is TFe 38%, TiO_2 8%, and V_2O_5 0.76%. The deposit at Cross Lake, Manitoba is a disseminated ore mined by digging and land cutting to produce blocks. The main rocks are anorthosite, alkaline rocks, mafic rocks, and ultramafic rocks, with an ore grade of TFe 28%−60%, TiO_2 3%−10%, and V_2O_5 0.02%−0.5%. The Banks Island mine in British Columbia is mined by digging and land cutting. The main rock types are anorthosite, alkaline rocks, mafic rocks, and ultramafic rocks, where the ore grade is TFe 20%−50%, TiO_2 1%−3%, and V_2O_5 0.07%−0.55%. The ore from Porcher Island in British Columbia is mined by digging and land cutting. The main rock is gabbro, plagioclase, alkaline rock, and ultramafic iron rock, where the grade of the raw ore is 25%, TiO_2 2%, and V_2O_5 0.2%−0.035%.

The ore type of the Seven in Quebec is layered lenticular type, mined by digging and land cutting, where the main rocks are alkaline, mafic, and ultramafic iron rocks, and the grade is TFe 11%−42% and TiO_2 3%−16%. The ore from the Magpies mountains in Quebec is layered lenticular, mined by digging and land cutting, where the main rocks are gabbro, mafic, and ultramafic iron rock, and the grade of the raw ore is TFe 43%, TiO_2 10%, and V_2O_5 0.2%−0.35%. Quebec's Saint Urbain mine produces blocks, where the main rock is gabbro, alkaline, mafic, and ultramafic iron rocks, and the grade is TFe 35%−40%, TiO_2 38%−45%, and V_2O_5 0.17%−0.34%. The Maureen mine in Quebec is mined by digging and land cutting. The main rocks are gabbro, alkaline, mafic, and ultramafic iron rocks, and the grade if TFe 25%−43%, TiO_2 19%, and V_2O_5 0.05%−0.34%. The ore in the Dolores Lake region of Quebec is an impregnated type, where the main rocks are gabbro, plagioclase, alkaline, and mafic rocks, and the grade is TFe 28%−53%, TiO_2 5%−8%, and V_2O_5 0.3%−1.0%.

The Lac tio mine in the Allard region of Canada contains hematite, ilmenite, and in which hematite enveloped by ilmenite, with lumpish ilmenite (TiO_2 .FeO) to hematite (Fe_2O_3) ratio of roughly 2:1. The gangue is mainly plagioclase, along with small amounts of pyroxite, black mica, pyrite, and magnetite. The lamellar lenticular structure is mined by digging and land cutting to produce blocks. The original mine contained 0.27% V_2O_5. The Miche camo Lake mine in Labrador province is mined by digging and land cutting to produce blocks, where the main rock is gabbro, alkaline rock, mafic rock, and ultramafic iron rock.

3.1.5.4 V-bearing titanoferrous magnetite in the United States

The V-bearing titanoferrous magnetite in the United States is very rich, where the regions yet to be exploited are mainly in Alaska, New York, Wyoming, and Minnesota. There are also deposits in Sanford Lake, New York, with a 1600-m long ore body, which was mineralized by a process similar to that which formed the Adirondack composition containing gabbro and plagioclase during the Precambrian period. The lower rocks are dense coarse-grained plagioclase, and the upper rocks are disseminated or dense, fine- to medium-grained gabbro. The upper and lower layers are parallel to each other with an inclination angle of 45 degrees. The average iron content of the ore is 34%, along with 18%–20% TiO_2 and 0.45% V_2O_5. The Deanna comprehensive deposit in New York is a layered lenticular type that is mined by digging and land cutting to produce blocks. The main rocks are gabbro, anorthosite, alkaline, and mafic rocks, where the ore grade is TFe 20%, TiO_2 7%, and V_2O_5 0.05%. The Van Sickle deposit, New Jersey is a disseminated and layered lenticular type, mined by digging and land cutting to produce blocks. The main rocks are gabbro, anorthosite, alkaline, mafic, and ultramafic rocks, where the ore grade is TFe 60%, TiO_2 6%, and V_2O_5 0.4%.

The Piedmont mine, North Carolina contains impregnated ore that is produced in block form. The main rock is gabbro, anorthosite, alkaline, and ultramafic iron rock, and the raw ore grade is TFe 40%–65%, TiO_2 12% ± , and V_2O_5 0.13%–0.38%. The mines in North Carolina and Tennessee Appalachian regions contain disseminated ore that is produced in blocks. The main rocks are gabbro, anorthosite, alkaline, and ultramafic rocks, with an ore grade of TFe 40%–60%, TiO_2 5%–7%. Wyoming's Iron Mountain mine contains a lamellar lenticular ore that is mined by digging and land cutting. The main rock is gabbro, alkaline, and ultramafic iron rock, and the grade is TFe 17%–45%, TiO_2 10%–20%, V_2O_5 0.17%–0.64%. The Ovens Lake mine in Wyoming is mined in block, where the main rock is gabbro, alkaline, and ultramafic iron rock, and the raw ore grade is TFe 29%, TiO_2 5%, and V_2O_5 0.2%. The Iron Mountain mine in Colorado is mined in block, and the main rock is plagioclase, alkaline, mafic, and ultramafic iron rock. The grade of the raw ore is Fe 40%–50%, TiO_2 14%, and V_2O_5 0.41%–0.45%. The San Gabriel Mountain mine in California is lamellar lenticular type, mined in blocks. The main rock is alkaline, mafic, and ultramafic iron rock, with a raw ore grade of TFe 46%, TiO_2 20%, and V_2O_5 0.53%. The Snathham mine in Alaska is layered lenticular type, mined by digging and land cutting, where the main rocks are gabbro, anorthosite, alkaline, and ultramafic rocks with an ore grade of TFe 19%, TiO_2 2.6%, and V_2O_5 0.09%. The Kliuke Van mine in Alaska is mined by digging and land cutting, where the main rocks are gabbro, plagioclase, alkaline, and ultramafic iron rock, and the raw ore grade is TFe 15%–20%, TiO_2 2.0%, and V_2O_5 0.05%. The Lake Illiamra deposit in Alaska is layered lenticular type, mined by digging and land cutting, where the main rocks are gabbro, anorthosite, alkaline, and ultramafic rocks, and the ore grade is TFe 12%–19%, TiO_2 1.3%, and V_2O_5 0.02%.

3.1.5.5 V-bearing titanoferrous magnetite in Northern Europe

There are deposits of V-bearing titanoferrous magnetite in Finland, Norway, and Sweden in Northern Europe; for example, the Otanmaki and Mustavaara mines in Finland. The Otanmaki mine is located in the north of Finland, which is impregnated ore mined by digging and land cutting. The main rock is gabbro, plagioclase, alkaline, and ultramafic iron rock. The Mustavara ore is also disseminated, with similar properties to those of Otanmaki. The grade of the Otanmaki ore is TFe 34%–45%, TiO_2 13%, and V_2O_5 0.45%. Table 3.2 shows the complete chemical composition of Otanmaki ores.

Table 3.2: Chemical composition of Otanmaki ore (%).

Chemical composition	TFe	FeO	Fe_2O_3	V_2O_5	TiO_2	SiO_2	Al_2O_3	CaO	MgO	Na_2O
Otanmaki ores	17.0	11.5	12.0	0.36	3.1	41.0	15.0	9.2	4.7	2.3

The Ternis deposit in Norway is the largest titanium mine in Europe. The ore is impregnated and lenticular, where the main rocks are gabbro, plagioclase, alkaline, mafic, and ultramafic iron rock. The total ore reserves are about 300 million tons. The original ore contains TFe 20% and TiO_2 17%–18%. There are deposits of V-bearing titanoferrous magnetite in Rhodesad, in which the TiO_2 content is low, disseminated and mined by digging and land cutting. The main rocks are gabbro, anorthosite, alkaline, and ultramafic rocks, with a grade of TFe 30%, TiO_2 4%, and V_2O_5 0.30%. The Selvag mine in Laverton is disseminated ore, mined by digging and land cutting, where the main rocks are anorthosite, alkaline, mafic, and ultramafic rocks, and the ore grade is TFe 35%, TiO_2 4.0%, and V_2O_5 0.4%. After iron concentration the grade is TFe 60%, TiO_2 5%, and V_2O_5 0.7%. The deposits of Rhodesad in Morey are disseminated and mined by digging and land cutting. The main rocks are gabbro, anorthosite, alkaline, and ultramafic rocks, with an ore grade of TFe 35%, TiO_2 6.0%, and V_2O_5 0.5%, which increases to TFe 62%, TiO_2 2%, and V_2O_5 0.9% after iron concentration. The Sorge mine in Morey is mined by digging and land cutting, where the main rock is alkaline, mafic, and ultramafic iron rock, and the grade is TFe 10%–30%, TiO_2 5.0%–50%, and V_2O_5 0%–1.0%. The Oslo deposit in Morey is disseminated and mined by digging and land cutting, and the main rocks are anorthosite, alkaline, mafic, and ultramafic iron rock. The raw ore grade is TFe 10%–30%, TiO_2 5.0%–50%, and V_2O_5 0.1%–1.0%. The Stolgongan deposit in Egersund is disseminated and layered lenticular type and is mined as blocks. The main rocks are gabbro, alkaline, mafic, and ultramafic rocks, with an ore grade of TFe 5%, TiO_2 17.0%, and V_2O_5 0.14%, or TFe 65%, TiO_2 5%, V_2O_5 0.73% after iron concentration.

There are deposits of V-bearing titanoferrous magnetite in Taberg and Kiruna in Sweden, which are disseminated types and mined by digging and land cutting. The main rock is gabbro, anorthosite, alkaline, and ultramafic rocks. The vanadium content in the Taberg ore can reach 0.7% V_2O_5.

3.1.5.6 V-bearing titanoferrous magnetite in Asia-Pacific

In addition to China, there are deposits of V-bearing titanoferrous magnetite in Australia, New Zealand, India, Sri Lanka, and other countries.

3.1.5.6.1 V-bearing titanoferrous magnetite in Australia

The deposits in Australia are mainly concentrated in Western Australia, such as the Coates, Barra, and Barambi mines. The Coates mine is in the coastal area of Wundowie, Australia. The ore body contains magnetite gabbro, and there is better composition of ore. The Balla mines in Australia are high in vanadium and titanium, similar to the Bushveld mine in South Africa, with a grade of TiO_2 15%, V_2O_5 0.7%, and TFe 26.0% in the Barambi and Gabapaninsa mines.

3.1.5.6.2 V-bearing titanoferrous magnetite in India

The deposits in India are mainly concentrated around the banks of Trivandrum in the southern state of Kerala, Mayurbhanj of Orissa, Singhbhum of Hal, and Udras of Tamil Nadu. Most of the mines are coastal with TiO_2 contents of 15%–30%. Other deposits in Dabra contain TiO_2 10.2%–28.7% and V_2O_5 1.45%–8.8%.

3.1.5.6.3 V-bearing titanoferrous magnetite in Sri Lanka

In addition to the northeastern coast of Pulmoddai, there is ilmenite in Sri Lanka from the northwest coast of Kurndilapmalai to the west coast of Kirinda in the south, which contains 53.61% TiO_2 and 31% Fe. In addition to the 70%–80% ilmenite (TiO_2 and FeO), there is also 10% TiO_2 (rutile) and 8%–10% of zircon ($ZrO \cdot SiO_2$).

3.1.5.6.4 V-bearing titanoferrous magnetite in New Zealand

The west coasts of both the north and south islands of New Zealand contain large deposits of V-bearing titanoferrous magnetite, which are one of the main areas of V-bearing titanoferrous magnetite extraction, with an average TFe content of 18.0%–20%, which fluctuates in the range of 4%–60%. Mineral sand has TFe 22.1%, TiO_2 4.33%, and V_2O_5 0.14%.

3.1.5.7 V-bearing titanoferrous magnetite in South America

The V-bearing titanoferrous magnetite in Maracas of Brazil is a typical high-vanadium deposit with 1.27% V_2O_5. The ore body is mainly coarse-grained gabbro/pyroxenite, which contains Ti and Pt-group elements, although the reserves are moderate. The Campo Alegre de Lourdes deposit in Brazil is ferromagnesite immersed, with 50% Fe, 21% TiO_2, and 0.75% V_2O_5, with a moderate reserve size.

3.2 Minerals of vanadium

Ninety-eight percent of the known global vanadium reserves are V-bearing titanoferrous magnetite. In addition to these deposits, vanadium resources also occur in phosphorus block rock ore, uranium-bearing sandstone, siltstone, bauxite, carbon-bearing crude oil, coal, oil shale, and bituminous sand. The few deposits with high-vanadium contents and/or rich accumulation are found and vanadium is associated with V-bearing titanoferrous magnetite, bauxite, coal, oil, and other mineral deposits. More than 65 vanadium-containing minerals are known. Vanadium metal is difficult to find in nature as many vanadium compounds are soluble in dilute aqueous solutions of acids and alkalis and are hence, often widely distributed, interspersed within rocks, and combined with several elements. Vanadium minerals can be produced from paragenic or complex ores.

Vanadium mainly occurs in V-bearing titanoferrous magnetite, where the V_2O_5 content can reach 1.8%. The great majority of vanadium products are derived from vanadium resources. Among the more important vanadium minerals are patronite ($V_2S + nS$), vanadinite [$PbCl_2 \cdot 3Pb_3(VO_4)_2$], sulvanite ($3Cu_2S \cdot V_2S_5$), titanium cerium uranium ore (Fe, U, V, Cr, and rare-earth titanate), carnotite ($K_2O \cdot 2UO_3 \cdot V_2O_5 \cdot 3H_2O$), and vanadium mica [$2K_2O \cdot 2Al_2O_3 \cdot (Mg, Fe)O \cdot 3V_2O_5 \cdot 10SiO_2 \cdot 4H_2O$]. There are abundant vanadium reserves in oil deposits and stone-like coal.

3.2.1 Vanadinite

Vanadinite is also known as "brown lead ore" and appears in chlorovanadate, an oxidized composite lead-containing ore. Vanadinite is a vanadium mineral belonging to the phosphochlorine lead ore series in the apatite mineral family. Vanadinite is a raw material for both vanadium and lead production; it is a major source of lead ore. Vanadinite is usually red, yellow, or brown, with a turpentine luster, and some sources are brighter than others. The minerals are mostly cylindrical, and many of them are hollow. The components of vanadinite include $Pb_5MO_{43}Cl$, 19.4% V_2O_5, and 16.2% minerals containing V_2O_5. The mineral structure can be a hexagonal system, with hexagonal prism-like crystals or needles, or fiber-like structures. The aggregate is a spherical crystalline cluster with bright red, orange red, light brown red, yellow, or light brown luster, luster with resin or adamantine luster. The mineral usually occurs in dense yellow to yellowish brown clumps with streaks of white to pale yellow. Its hardness is 2.5–3, with a specific gravity of 6.66–7.10. Vanadinite can be produced in the oxidation zone of lead ore deposits, and large amounts can be extracted from lead ore and used to produce vanadium.

3.2.2 Roscoelite in sandstone

The composition of roscoelite in sandstone is $KV_2[AlSi_3O_{10}](OH \cdot F)_2$. The homogeneity of the homoplasmy is, for example, Mg, Cr, Fe^{2+}, and Fe^{3+}. In Hubei province (China), roscoelite in sandstone contains SiO_2 48.05%, TiO_2 0.38%, Al_2PO_3 15.00%, V_2O_3 14.62%, Cr_2O_3 1.56%, Fe_2O_3 0.56%, MgO 4.32%, CaO 0.34%, BaO 1.28%, K_2O 6.19%, Na_2O 0.13%, P_2O_5 0.13%, F 0.05%, and H_2O 5.44%. The total of these components is 98.33%.

Roscoelite in sandstone belongs to a monoclonal crystal system, where the pure mineral contains 22.8% V_2O_5. They are complex aluminum silicates similar to white mica.

The physicochemical properties of roscoelite are given here. Most crystals of roscoelite in sandstone are bright green and in the form of fine fibers. Some are flaky, fine fiber one with a silky luster and soft texture, such as asbestos. The cleavage parallel {001} is extremely complete. The material has a hardness of 2.5 and relative density of 2.88. The green color of roscoelite is due to V^{3+} ions. With increasing V_2O_3 content the color changes from light to dark green, while Cr-containing minerals give a blue color. The roscoelite in sandstone is distinguished from other mica minerals as it is bright green and can be accurately identified by chemical analysis and differential thermal curve analysis.

Roscoelite in sandstone is green and polychromatic, and that from Hubei province (China) is deposited in a carbonaceous slate with high content of organic matter. There is a large amount of coal in the rock, which occurs with Cr–V hydromite, Cr–V dolomite, and other symbiotic minerals. Most crystals of roscoelite in sandstone are bright green, fine, and fibrous, with some flakes.

The physical properties of roscoelite in sandstone include a hardness of 2.5, specific gravity of 2.88 g/cm^3, clean cleavage along the {001} plane, with no breaking. The color of the material and its streaks changes from light to dark green with increasing V_2O_3 content and is blue if the composition contains Cr. Roscoelite is opaque, with a silky luster and is nonfluorescent.

3.2.3 Carnotite

Pure carnotite has the composition $K_2[VO_2]_2[VO_4]_2 \cdot 3H_2O$ and contains 42%–46% uranium and 11.3% vanadium. It has a monodiagonal structure that appears as tiny crystals, slices, or plates and is usually a powder block that is yellow or yellowish green with a glassy luster. Carnotite has a hardness of 2–2.5, specific gravity of 4.46 and is highly radioactive. It is soluble in dilute acid and is mainly distributed in uranium shale, sandstone, and limestone. It often exists in the organic matter of the weathering zone of sedimentary rocks (mainly sandstone) or in the oxidation zone of sedimentary uranium deposits. Carnotite is used as a raw material for extracting vanadium, uranium, and radium. It is an important source of uranium, in the form of hydrated potassium uranyl vanadate. Pure carnotite deposits can contain 53% uranium and 12% vanadium. It is always yellow and soft, some are small, and some are soil.

3.2.4 Patronite

Patronite mainly contains VS_4 with a V content of 28.4%. It has a monodiagonal crystal structure and vanadium sulfite occurs as an impurity. It can be dense or powdery, black, or dark gray with streaks of black, with a metallic luster. It has a hardness of 2, and density of 2.98 g/cm^3. The mineral is a bituminous shale containing acid vein rock and is used as a raw material for vanadium extraction.

3.2.5 Tyuyamunite

Tyuyamunite is also known as metatyuyamunite and is a uranium−calcium compound containing V_2O_5 with a molecular structure of $Ca(UO_2)_2(VO_4)_2 \cdot H_2O(3-5)$. The crystals can be shaped like plates or leaves and is flattened along the [001] plane and lengthened along the axes. The aggregate is powdery and has a golden yellow to pale green yellow color. The crystal has an adamantine luster, while the aggregate has a waxy luster.

3.2.6 Barnesite

Barnesite has a composition of $Na_2V_6O_{16} \cdot 3H_2O$, where the pure mineral has a V content of 46.2%.

3.2.7 Vanadium magnetite

The common name of vanadium magnetite, FeV_2O_4, is coulsonite, and it is one of the main V-containing minerals. A small amount of V^{3+} is substituted for Fe^{3+} in the composition, with a vanadium pentoxide content of 68.41%−72.04%. It is classified as a magnetite subspecies containing vanadium. It is cubic system with hexahedral crystals that are semi-self-granular or laminar. It appears blue or black, with a metallic luster, hardness of 4.5−5, and density of 5.15 g/cm^3. It is distinguished from magnetite by its inactivity to nitrohydrochloric acid and hydrochloric acid. It is produced in late magmatic deposits, magmatic differentiation deposits, and hypothermal deposit and is used for extracting vanadium and other vanadium compounds, as well as vanadium refining. The crystal system is isometric with the Fm3m space groups. The morphology is hexahedral crystals that are semi-self-granular or laminar.

The ore rock types include (1) mafic−ultramafic ferric layered intrusions composed of dunite, harzburgite, pyroxenite, norite, gabbro, and anorthosite; (2) layered intrusions composed of troctolite, norite, plagioclasite, and ferrodiorite; (3) layered intrusions composed of peridotite, pyroxenite, and gabbro; and (4) plagioclase−gabbro complex. Such rocks are often amorphous. The paragenic ore includes chromite, copper−nickel sulfide, and platinum-group element deposits.

Most of the important ore bodies are stratified parallel to the sedimentary strata of the layered rock and diagenetic facies that are mainly igneous accumulations composed of gabbro and plagioclase that are found in the bottom of each facies cycle. The surrounding rock is mainly gabbro or plagioclase–gabbro. The ore layer and the lower surrounding rock are mostly mutated contact, where the upper surrounding rock is mostly in gradual contact. In addition, vein- and tubular-like ore bodies are not integrated and injected into the rock facies belts but are in abrupt contact with the surrounding rocks. The ore minerals are mainly vanadium magnetite, V-bearing titaniferrous magnetite, sulfide, and ilmenite. The main gangue mineral are plagioclase, pyroxene, and olivine. The ores show heap crystal (product) structures, including the interstitial structure, poikilitic structure, sideronitic texture, and exsolution texture. The tectonic structures are disseminated, stripped, and block structures.

This type of deposit is a magmatic segregation deposit. The deep magma chamber in the mantle hot spot provides superior conditions for the full differentiation and mineralization of mantle mafic magma. The crystallization of magnetite and ilmenite is generally late but differs by rock mass. The magnetite crystals in this type of deposit in China were formed much later than pyroxene and plagioclase and have the structural features of late magmatic sedimentary deposit, where the vein penetrates the ore body. In Bushveld complexes, magnetite is crystalized later than the olivine and orthopyroxene, which corresponds to plagioclase. In this case the layered ore body is mainly distributed in rock zones (anorthosite is given priority) with top and bottom, where the host rock is early magma and determines the structural characteristics of the deposit. However, it occurs at each rock phase in the bottom and middle of the main ore zone and may come from the bottom of the host rock slurry injection of the ore body.

(1) It is large layered mafic plutonic complex rock body with deep fracture control in the continental plate; (2) the ore-bearing rock masses in China are mostly weakly alkaline sodium series, strongly alkaline series, TiO_2, V_2O_5, P_2O_5, low m/f, and SiO_2; (3) a geophysical indication of this mineral is that the body is heavy and shows magnetic anomalies and superposition.

3.2.8 Vanadine

Pure vanadine (V_2O_3) has a vanadium content of 68%. Aluminum bauxite, bauxite, also known as vanadine, alumina, or bauxite, is mainly composed of alumina that is simply a hydrated alumina containing impurities and is an earthy mineral. It is white or grayish white or can also be brownish yellow or light red due to iron impurities. It has a density of 3.9–4 g/cm^3, hardness of 1–3 and is opaque, brittle, and extremely difficult to melt. Vanadine is insoluble in water but soluble in sulfuric acid and sodium hydroxide solutions.

Aluminite is the scientific name of bauxite. Its composition is extremely complex, and this general name is given to a variety of minerals from different geological sources, such as boehmite, diaspore, and gibbsite ($Al_2O_3 \cdot 3H_2O$), while some deposits are composed of diaspore and kaolinite ($2SiO_2 \cdot Al_2O_3 \cdot 2H_2O$), and others are mainly kaolinite. With increasing kaolinite content the deposit it considered to be composed of general bauxite or kaolinite clay. Bauxite is generally formed by chemical weathering or exogenous processes, and there are few pure minerals; most minerals contain some impurities, and clay minerals, iron ore, titanium minerals, and clastic heavy minerals also occur.

3.2.9 Uvanite

Pure uvanite, $(UO_2)_2V_6O_{17} \cdot 15H_2O$, has a V content of 68%. Tyuyamunite has a chemical symbol of $Ca\{(UO_2)2[V_2O_8]\}$. It is an orthorhombic system and occurs as an aggregate of crystal slabs and plates. The color is lemon yellow, orange or brownish green, and the color changes to yellowish green in the sun. It has an oily sheen and glassy luster, Mohs hardness of 1–2, and specific gravity of 3.41–3.67. Complete cleavage occurs, although the fracture is uneven. This mineral can be soluble in acids. It is found in uranium mines and is a typical oxidation-zone mineral. The chemical composition is $Ca\{(UO_2)_2[V_2O_8]\} \cdot 8H_2O$, where calcium can be replaced by potassium.

3.2.10 Vichlovite

Pure vichlovite, $Pb_2 (Mn, Fe)[VO_4]2 \cdot H_2O$, contains 14.2% vanadium.

3.2.11 Montroseite

Pure montroseite, $(Fe, V)O(OH)$, contains 60.7% vanadium.

3.3 Main vanadium minerals for vanadium extraction

The minerals that can be used for vanadium extraction include vanadium ore, composite polymetallic vanadium ore, and organic vanadium deposits. Different countries and regions have various V-containing ores, for example, patronite in Peru; vanadite in South Africa; V-bearing titanoferrous magnetite in China, Russia, and South Africa; uvanite in the United States, Western Australia, and Uzbek; and putty of fuel power generation in the Middle East and Venezuela.

3.3.1 V-bearing titanoferrous magnetite

V-bearing titanoferrous magnetite is divided into two types: basite (gabbro) and basic and ultrabasic rocks (gabbro, pyroxenite–pyroxenite). Basite deposits are found in Panzhihua, Baima, and Taihe (China), while basic and ultrabasic rocks occur in Xinjie, Damiao, and Hongge (China). In general, the two types have similar geological characteristics, where pure basite is equivalent to that found in basic and ultrabasic rocks. Basite minerals mainly contain iron, titanium, and vanadium. The basic and ultrabasic rocks contain iron, titanium, and vanadium, accompanied by high contents of chromium, cobalt, nickel, and platinum-group elements. The V-bearing titanoferrous magnetite is not only an important source of iron but also contains high-value materials, including vanadium, titanium, chromium, cobalt, nickel, platinum, and scandium, giving it higher overall value.

3.3.2 Stone-like coal

Stone-like coal is formed in the ancient stratum, where the biological remains of bacteria and algae are transformed into sag and coal under the conditions provided by shallow seas, lagoons, and bays. Stone-like coal is not easily distinguished from limestone or carbon shale with the naked eye; it has high ash content (generally greater than 60%) and is a combustible organic mineral with deep deterioration. Stone-like coal is a new type of metallogenic vanadium deposit; such vanadium deposits are known as black shale types and are formed in marginal sea slope areas. The main vanadium mineral in stone-like coal is vanadium illite.

High-quality stone-like coal with high carbon content is black, with a semibright luster and few impurities, and a relative density of 1.7–2.2. Stone-like coal with lower carbon content is dark gray and dull; is mixed with pyrite, quartz vein, and phosphorus calcium nodules; and has a relative density of 2.2–2.8. The calorific value of stone-like coal is low (3.5–10.5 MJ/kg) for use as a fuel. Stone-like coal with high calorific value can be used as a fuel for power generation using improved combustion technology. It can also be used for making cement and chemical fertilizer, where the ash residue is used to make carbonized bricks. Vanadium pentoxide can be extracted from V-bearing stone-like coal.

There are currently more than 60 elements found in China's stone-like coal resources. Among them, vanadium has industrial value, although the V_2O_5 content in stone-like coal is low, generally ~1.0%. In stone-like coal, vanadium usually occurs as V(III), with some V(IV), and rarely V(V). The ionic radii of V(III) and Fe(II) are both 74 pm, while that of Fe(III) is similar (64 pm). Hence, V(III) forms few minerals on its own but forms isomorphs in ferrallite with the siloxane structure, such as roscoelite and kaolin, followed by silver, uranium, molybdenum, and phosphorus. Table 3.3 shows the occurrence of vanadium in stone-like coal in the various regions of China.

3.3.3 Vanadinite

Vanadinite is also known as "brown lead ore." Vanadinite is a V-bearing mineral belonging to the pyromorphite series of the apatite minerals group. Vanadinite is the main source of lead ore, in addition to being a source of vanadium.

Table 3.3: Occurrence of vanadium in stone-like coal in different regions of China.

Region	Vanadium mineral		Mineral content of ore (%)	V_2O_5 content (%)	V_2O_5 rate of distribution (%)
Hubei	Organic asphaltite			17.45	15
Yangjiabu	Silica–aluminate salts	Illite		7.00	50
		Roscoelite		1.98	1
	Silicate minerals	Titanium-vanadium garnet		16.00	16–18
		Chromium–vanadium garnet		21.538	
	Sulfide minerals	Arsonite		6.59	2
		Ocher		9.37	
	Water-soluble salts				2
	Adsorption state	Vanadium cation			2
		Vanadium chrome anion			10
Guangshiya, Hubei	Vanadium hydromatite		12.55	3.97	89.29
	Carbonaceous matter		53.02		6.7
	Quartz		17.16	0.006	2.1
	Feldspar		2.13		
	Nephrite, calcite, dolomite		6.81	0.005	0.05
	Pyrite		3.33	0.003	0.02
Zhiji, Zhejiang	Roscoelite		17.0	5.66	89.9
	Vanadium-containing kaolinite		12	6.5	7.4
	Garnet with vanadium		0.5	3.6	1.7
Anren, Zhejiang	Roscoelite		1.0	12.85	16.4
	Clay		20.4	3.23	83.6
Tangwu, Zhejiang	Illite		41.2	3.34	100
Zhejiang	Broken weathering powder	Pascoite	1	59.6	20.8
		Calcium – vanadium garnet	0.5	24.86	4.5
		Schreyerite	2.9	49.5	50.3
		Clay	33.7	1.48	17.4
	Lump stone-like coal	Calcium – vanadium garnet	2.1	24.86	57.7
		Schreyerite	1.0	33.66	37.1
		Clay	31.2	0.61	5.4

(Continued)

Table 3.3: (Continued)

Region	Vanadium mineral	Mineral content of ore (%)	V_2O_5 content (%)	V_2O_5 rate of distribution (%)
Yueyang, Hunan	Kaolinite-based silica–aluminate			70
	Free oxide			10–20
	Carbonaceous			a little
Fangshankou, Gansu	Roscoelite	10	5.5	79.6
	V-bearing kaolinite	2	2.79	10.9
	V-containing ferric oxide	6	0.48	
	Vanadite	1	8.42	9.5
Guida, Jiangxi	Roscoelite	0.78	37.5	33.91
	Vanadinite	8.53	4.41	43.52
	Vanadinite	17.01	0.65	12.85
	Phosphorus diaspore	0.73	0.90	0.81
	Limonite	5.21	0.26	1.62
	Quartz	51.65	0.0075	0.46
	Carbonaceous	5.22	0.65	3.94
Sichuan province	Clay mineral			52.38
	Carbonaceous			27.62
	Sulfide			12.38
	Silicate minerals			7.62

3.3.4 Carnotite

Carnotite has the composition $K_2[VO_2]_2[VO_4]_2 \cdot 3H_2O$ and contains 42%–46% uranium.

3.3.5 Roscoelite

Roscoelite in sandstone has the composition $KV_2[AlSi_3O_{10}](OH \cdot F)_2$. The homogeneity of the homoplasmy is Mg, Cr, Fe^{2+}, and Fe^{3+}. Roscoelite in sandstone from Hubei, China contains SiO_2 48.05%, TiO_2 0.38%, Al_2PO_3 15.00%, V_2O_3 14.62%, Cr_2O_3 14.56%, Fe_2O_3 0.56%, MgO 4.32%, CaO 0.34%, BaO 1.28%, and K_2O 6.19%, giving a total of 98.33%. Roscoelite in sandstone occurs as a monoclinic crystal.

3.3.6 Patronite

The main components of patronite are VS_4 and VS_4, which appear as a monoclinic crystal containing 28.4% V. It is usually dense or powdery, black, or dark gray with streaks of black, with a metallic luster, hardness of 2, and density of 2.98 g/cm^3. The mineral is produced from bituminous shale with acid vein rock intrusions and used as a source of vanadium.

3.3.7 Organovanadium ore

In organovanadium ore, vanadium occurs as a complex with porphyrin, phenol, and other complex organic compounds that are found in high sulfur oil, asphaltene shale, and asphaltene.

3.4 Distribution of vanadium resources

3.4.1 Type and distribution

Vanadium resources are distributed throughout the five continents, where Europe, Asia, and Africa have relatively rich vanadium resources. Vanadium is mainly found in China, Russia, South Africa, Australia, and New Zealand, while which occur in vanadium containing titanoferrous-magnetite and occur in oil deposit in Venezuela, Canada, the Middle East, and Australia, while vanadium ore and clay minerals are found in the United States. Of the known global vanadium reserves, 98% occur as V-bearing titanoferrous magnetite. The V_2O_5 content in V-bearing titanoferrous magnetite can reach 1.8%, and this material is the most widely used at present. Vanadium reserves are also abundant in oil deposits, which are deposited in phosphate rock ore, uranium sandstone, siltstone, bauxite, carbonaceous crude oil, coal, oil shale, and tar sands. The V_2O_5 deposits in North America and Queensland,

Australia are oil shale or bituminous sandstone, with contents of 0.45%. Aluminum clay minerals in Arkansas and uranium mines in Colorado may also be sources of vanadium. Canada extracts vanadium from tar sands in Alberta.

3.4.2 V-bearing titanoferrous magnetite

The global reserves of V-bearing titanoferrous magnetite are large, although concentrated in a few countries and regions, including the Commonwealth of Independent States, the United States, China, South Africa, Norway, Sweden, Finland, Canada, and Australia, where most of the reserves are in South Africa, North America, and the Asia-Pacific region. In South Africa, vanadium is usually produced from vanadium magnetite deposits with an average grade of 1.5%. Chinese V-bearing titanoferrous magnetite ore deposits are either widely distributed or concentrated. Vanadium is also extracted from iron ore deposits. China has proven reserves of 9.83 billion tons, with possible reserves of 30 billion tons, mainly distributed in Sichuan Panzhihua, Hebei Chengde, Shaanxi Hanzhong, Hubei Yunyang, Xiangyang District, Guangdong Xingning and Dai counties of Shanxi Province, and other regions. The Panzhihua region is China's main metallogenic belt, containing V-bearing titanoferrous magnetite that is one of the most important deposits in the world. The deposit runs north to south and is about 300 km long. There are seven large and superlarge proven deposits and six medium-sized deposits. The total V_2O_5 reserve is 2596 million tons, ranking third in the world. Vanadium is mainly mined from magmatic V-bearing titanoferrous magnetite deposits.

3.4.3 Stone-like coal resources

As an independent deposit, vanadium ore mainly occurs as black shale produced during the Cambrian period. The distribution of vanadium is relatively extensive. China's vanadium mineral resources have proved reserves in 19 provinces, where the reserves in Sichuan are the richest, accounting for 49% of the total reserves, followed by Hunan, Anhui, Guangxi, Hubei, Gansu, and other provinces.

3.4.4 Global distribution of vanadium resources

Table 3.4 summarizes the global distribution of vanadium resources.

Table 3.4: Global distribution of vanadium resources.

Country	Main region	Mineral categories	Ore reserves (Mt)	Ore grade V_2O_5 (%)	Reserves V_2O_5 (kt)
South Africa	Highveld	Titanoferrous Magnetite		1.4–1.5	
Uganda		Magnetite	13	1.1	143
Namibia		Vanadate ore	2	0.5	10
Russia	Uralsk, Kusinski, Ilsinsk, and Samchikan	Titanoferrous Magnetite	430	0.6	
Finland	Mustavaala	Titanoferrous Magnetite	40	0.36	144
Sweden, Norway	Otanmaki	Titanoferrous Magnetite			
Britain, France, Germany, Belgium, Luxembourg		Titanoferrous Magnetite Ilmenite Carbonate ore	15	0.43	57.5
China	Panzhihua–Xichang	Titanoferrous Magnetite	168	0.25	29,430
	Chengde	Titanoferrous Magnetite		0.28	10,000
	Manashan	Magnetite		0.22	420
	Southern China	Stone-like coal			3600
India		Ilmenite Bauxite ore		0.24–0.40 0.05–1.0	
Canada	Quebec	Asphaltite Titanoferrous Magnetite		0.19	1900
United States	Colorado	Magnetite Laterite Phosphate ore		0.5–0.7 0.75–1.25 0.14	2–2.8 11–19 0.161
	Idaho	Uranovanite		0.85–1.4	
Peru	Minaragra	Patronite		32	(201.6)
Venezuela		Oil			112
Brazil		Magnetite		1.3	45
Australia	West Australia	Ilmenite		0.75	168
New Zealand		Ilmenite		0.3–9.5	832

3.4.5 Global vanadium reserves

Table 3.5 shows the global reserves and basic reserves of V_2O_5. The total global vanadium reserve is 63 million tons (the same amount as V_2O_5), with a reserve of 10.2 million tons and a base reserve of 31.1 million tons. China is the world's leading source of vanadium, with a total of 19.6% and 9.6% of the global reserves.

Table 3.5: Global reserves and basic reserves of vanadium (V_2O_5, 10,000 t).

Country	Reserves	Basic reserves
Russia	498.8	699.6
South Africa	299.9	1250.0
China	199.9	298.5
Australia	16.3	239.4
United States	–	401.1
Other	5.1	220.8
Total	1020.0	3109.4

Note: China's stone-like coal vanadium resources are not included.

# CHAPTER 4

Extraction of vanadium from titanoferrous magnetite: mineral processing and enrichment methods

Chapter outline
4.1 V-bearing titanoferrous magnetite 60
    4.1.1 Ore structure 60
    4.1.2 Ore-bearing rock types 61
    4.1.3 Mineral structure 61
    4.1.4 Main mineral composition 62
4.2 Main mineral characteristics of V-bearing titanoferrous magnetite 62
    4.2.1 Titanomagnetite $(Fe,Ti)_3O_4$ 63
    4.2.2 Hematite (Fe_2O_3) 63
    4.2.3 Limonite $(Fe_2O_3 \cdot nH_2O)$ 63
    4.2.4 White titanium ore 64
    4.2.5 Ilmenite 64
    4.2.6 Pyrrhotite 64
    4.2.7 Calcite $(CaCO_3)$ 64
    4.2.8 Quartz (SiO_2) 64
    4.2.9 Kaolinite clay rock 64
4.3 Typical mineral processing methods 65
    4.3.1 Ore from Panzhihua, China 68
    4.3.2 Ore from Chengde, China 73
4.4 V-containing titanoferrous magnetite outside of China 77
    4.4.1 Deposits in Russia 77
    4.4.2 Deposits in the United States 78
    4.4.3 Deposits in Finland 80
    4.4.4 Deposits in South Africa 80
    4.4.5 Deposits in Canada 80
    4.4.6 Deposits in Norway 82
    4.4.7 Deposits in New Zealand 82
4.5 Main equipment 82
    4.5.1 Crushing equipment 83
    4.5.2 Grinding equipment 84
    4.5.3 Screening and grading equipment 84
    4.5.4 Magnetic separation equipment 87

The Earth's crust includes the outermost layers of the solid surface structure and is composed of various types of rock, which are aggregates containing one or more minerals that were formed under different geological conditions. The minerals contain naturally occurring elements and inorganic compounds. The chemical compositions and physical properties are relatively homogenous and fixed, and the minerals are generally crystalline. V-bearing titanoferrous magnetite is a valuable ore, even though it has a low vanadium concentration, and is the main raw material for vanadium extraction. V-containing magnetite can occur as either the titanoferrous material or Ti-free magnetite. V-containing titanoferrous magnetite generally has low Ti and Fe contents, and enrichment to produce V-containing iron concentrate is performed to meet the requirements of ironmaking in blast furnaces. In addition, direct reduction or direct processing of mineral-grade vanadium can be performed [14–16,19,39,40].

4.1 V-bearing titanoferrous magnetite

V-bearing titanoferrous magnetite is a sequential crystallization deposit of magmatic rock. The crystallization of magnetite and ilmenite is generally late but varies depending on the rock mass. The magnetite crystals in China's V-bearing titanoferrous magnetite deposits were formed later than pyroxene and plagioclase and have the structural features of late magmatic sedimentary deposit and vein-like intrusions. In the Bushveld deposit in South Africa, magnetite crystals are posterior to the olivine and orthopyroxene, roughly the plagioclase. The layered ore body is mainly distributed in the bottom of the upper rock zone (main anorthosite) and the top of the main rock and has structural features of an early magmatic sedimentary deposit. When it is seen at the bottom of the main ore zone and in various rock phases, the ore may have been formed by slurry penetration into the main rock. The Catchikara deposit of V-bearing titanoferrous magnetite in Russia consists of two independent pyroxenite ore bodies, the Gusevo and Catchikara deposits, where the former is a huge disseminated pyroxenite ore body.

4.1.1 Ore structure

V-bearing titanoferrous magnetite is a typical magmatic ore, where the spatial and genetic properties are consistent with basic and ultrabasic ore-bearing rocks, usually iron-bearing ultrabasic rock and iron rock. According to the formation and occurrence of the deposit, V-bearing titanoferrous magnetite can be divided into late magmatic exogenic and late magmatic injection types. An injection ore body, like a vein, is formed when molten lava containing V-bearing titanoferrous magnetite is injected into the fissure zone of the upper part of a submerged body by external forces. In this case, intergranular extrusion occurs, and the ore structure is dense. There is no external dynamic action to produce disseminated structures and mineral bodies with sponge-siderite structures. Injection-type deposits of

V-bearing titanoferrous magnetite permeate the ore-bearing rock mass and the boundary between the ore and surrounding rock is clear. Often there are surrounding rock breccia and xenoliths in the ore body. The ore bodies are usually arranged in echelon form of lentil body and vein body, and branched structures are common, sometimes accompanied by obvious alteration of surrounding rocks. The amount of magma-differentiated V-bearing titanoferrous magnetite increases gradually from top to bottom with a rhythmic variation, and is related to dark lithofacies, which are formed by differentiation in the rock mass and ore body, and has a high degree of basic, more for layered, stratified, lenticular, with characteristics consistent with the occurrence of the rock mass, where there is usually a gradual transition between the ore and rock.

4.1.2 Ore-bearing rock types

The rocks containing V-bearing titanoferrous magnetite ore include (1) magnesium-ultramafic layered intrusions consisting of dunite, harzburgite, peridotite, pyroxenite, norite, gabbro, and anorthosite; (2) layered intrusions containing troctolite, norite, anorthosite, and iron diorite; (3) layered intrusions containing peridotite, pyroxenite, and gabbro; and (4) composite pluton containing plagioclase and gabbro, where the rock is polymorphic. The most important ore bodies are embedded parallel to igneous accumulations, distributed in layered petrographic-zone rock mass containing gabbro and plagioclase, where the surrounding rock is gabbro, anorthosite, and gabbronorite. The ledge is in abrupt contact with the surrounding rock at the bottom, while there is usually a compositional gradient with the upper rock. Vein and tubular ore bodies are rare and injected into the petrographic zone; they are in abrupt contact with the surrounding rocks. The ore minerals are mainly coulsonite, titanomagnetite, ilmenite, and sulfides. The main gangue mineral are plagioclase, pyroxene, and olivine.

4.1.3 Mineral structure

Metallogenic rock is a natural structure containing one or more minerals formed under geological actions. The rocks are classified into three groups based on their genetic type: magmatic, sedimentary, and metamorphic. Iron ore structures can be disseminated, stockwork disseminated, banded, dense lump, brecciated, cyprinoid, pisolitic, kidney-like, cellular, powdery, or earthy. The structures of V-bearing titanoferrous magnetite are generally mineralized in the late magmatic stage and can have sideronitic, granular mosaic, reticular, reaction rim, graphic, nodular, or solid solution separation structures. Titanomagnetite is a magnetite containing ilmenite and other solid solution decomposition products. During late magmatic differentiation, magnetite precipitated along with ilmenite to form titanomagnetite, where the associated vanadium is an isomorphism.

4.1.4 Main mineral composition

The different origins and mineralization conditions forming the V-bearing titanoferrous magnetite deposits around the world resulted in similar mineral compositions, but different mineral structures. This directly influences the selection of extractive metallurgy processes. The contents of the main useful mineral compositions in V-bearing titanoferrous magnetite are shown in Table 4.1.

Table 4.1: Contents of the main useful minerals in the different mineral structures of V-bearing titanoferrous magnetite.

	Structure of ores				
Mineral	Dense bulk ore (rich ore)	Medium thick disseminated ore (medium ore)	Sparsely disseminated ore (lean ore)	Scattered disseminated ore (off-balance ore)	Rock
TFe	47.44	39.48	25.14	17.47	14.04
TiO_2	14.88	12.8	9.15	7.30	7.00
V_2O_5	0.42	0.38	0.20	0.17	0.09
Cr_2O_3	0.11	0.13	0.15	0.13	0.012

Disseminated deposits of titanomagnetite mainly fill the interstices of diopside, amphibole, and olivine (iron meteorite structures) with allotriomorphic particles. Depending on the titanomagnetite particle size, the disseminated ore body is divided into common disseminated ore (0.074–0.2 mm), fine-grained disseminated ore (0.2–1 mm), and medium-grained disseminated ore (1–3 mm). Table 4.2 gives the V contents of the different ore structures.

Table 4.2: Vanadium contents in different ores (%).

Ore structure type	Fe	V	Ti	Reserve level ratio
Diffuse disseminated	16.8	0.071	0.68	5.0
Particle disseminated	17.2	0.078	0.79	23.7
Fine-grained disseminated	17.8	0.084	0.82	44.3
Medium-grained disseminated	18.1	0.088	0.86	17.0
Coarse-grained disseminated	17.4	0.089	0.87	10.0

The main mineral composition of V-bearing titanoferrous magnetite is shown in Table 4.3.

4.2 Main mineral characteristics of V-bearing titanoferrous magnetite

The structure of V-bearing titanoferrous magnetite is complex and dense symbiosis of major minerals, consisting of metal and gangue minerals. According to the gangue mineral density, they are divided into titania (density >3 g/cm^3) and plagioclase (<3 g/cm^3).

Table 4.3: Main mineral composition of V-bearing titanoferrous magnetite.

Mineral categories	Metallic mineral			Nonmetallic mineral	
	Essential minerals	Auxiliary minerals	Associated impurity minerals		
Mineral name	Titanomagnetite Ilmenite	Hematite Specularite Rutile Chromite	Pyrrhotite Pyrite Chalcopyrite Pentlandite Linnaeite Niccolite Sphalerite Arsenoplatinite	Pyroxene Olivine Anorthose Hornblende Phosphorite Spinel Chiltonite Quartz	Chlorite Serpentine Uralite Biotite Calcite, etc.

4.2.1 Titanomagnetite $(Fe,Ti)_3O_4$

Titanomagnetite structures include titanomagnetite [$Fe_{0.23}(Fe_{1.95}Ti_{0.42})O_4$, $Fe_2O_3 \cdot FeTiO_3$], ilmenite ($FeTiO_3$), and magnetite (Fe_3O_4). Titanomagnetite is generally produced as an automorphism, semiautomorphism, or other granular structure. The grain size is large and the material can be easily crushed. Some particles are lamella covered by titania, which are difficult to disintegrate. Titanomagnetite is often a composite mineral phase containing magnetite, ulvite, magnesium aluminate spinel, and a small amount of ilmenite. The ulvite has a lamella microstructure with a thickness <0.5 μm and length of 20 μm.

4.2.2 Hematite (Fe_2O_3)

Hematite is the most common mineral and the main component of primary iron ore. It is brown, magenta, or red in color, or sometimes reddish-brown with fuchsia stripes. Hematite is nonmagnetic.

4.2.3 Limonite $(Fe_2O_3 \cdot nH_2O)$

Lignite is usually yellow-brown or brown with yellow-brown streaks. It has a semimetallic luster and is blocky and powdery. The hardness is different from the mineral form and it is nonmagnetic. It is found in earthy, schistose, kidney-like, oolitic, and block structures. Limonite is the secondary mineral formed under oxidizing conditions. Although its distribution in the ore body is wide, it usually occurs in thin layers.

4.2.4 White titanium ore

White titanium ore occurs in the ore layer at low concentrations. It appears in the form of bright granules with a glassy luster and is a decomposition product of titanium magnetite $(Fe,Ti)_3O_4$.

4.2.5 Ilmenite

Granular ilmenite is the main mineral used for the recovery of titanium. Granular ilmenite is often associated with titanomagnetite, or distributed in silicate mineral grains, and forms via automorphism and semiautomorphism processes. It has coarse granules and is easily broken and dissociated. Granular ilmenite is about 90% of total ilmenite reserves. In the granules, there are small amounts of magnesium aluminate spinel wafers and fine-veined hematite, which are distributed along fractures, where the length of arteries and veins is 2−2.12−2.1 μm. Sometimes granular ilmenite contains small amounts of sulfides as emulsion droplets or fine veins, which affect the quality of the Ti concentrate.

4.2.6 Pyrrhotite

The chemical formula of pyrrhotite is $Fe_{(1-x)}S$, which belongs to class of hexagonal or monoclinic singlesulfide minerals. It is generally hexagonal, columnar, or barrel-shaped, but rarely seen, and usually forms dense clumps. Pyrrhotite is the most common sulfide mineral, accounting for more than 90% of the total sulfide minerals.

4.2.7 Calcite ($CaCO_3$)

Calcite is mainly found with a grain structure of spath calcite, which is colorless and transparent. It is produced by recrystallization of micrite calcite, which fills cracks in the mineral. Calcite is rare in ores and mainly occurred in limestone and dolomite.

4.2.8 Quartz (SiO_2)

Quartz is colorless and transparent and occurs as grains or granular aggregates in irregular, spherical, or strip-like structures. It is rarely found in ores, and mainly in limestone or dolomite.

4.2.9 Kaolinite clay rock

Kaolinite clay rock is found in soil and is a dense, gray−white clay mineral by the weathering of feldspar. The density, hardness, and magnetism of iron minerals in V-bearing titanoferrous magnetite are shown in Table 4.4.

Table 4.4: Density, hardness, and magnetism of iron minerals in V-bearing titanoferrous magnetite.

Mineral	Density (g/cm^3)	Hardness (HM)	Coefficient of magnetization (10^{-6} cm^3/g)	Specific resistance (Ω)
Titanomagnetite	4.59	6	10,000	1.38×10^6
Ilmenite	4.62	6	240	1.75×10^5
Pyrrhotite	4.52	4		1.25×10^4
Titania	3.25	7	100	3.13×10^{13}
Plagioclase	2.67	6	14.0	$>10^{14}$
Olivine	3.26	7	84	

4.3 Typical mineral processing methods

V-bearing titanoferrous magnetite is used to recover Fe and Ti concentrates and pyrite, while some of the associated minerals are used to recover phosphorite. The weak magnetism of magnetite is widely exploited during ore dressing. According to the ore properties and characteristics of the extraction method, weak magnetism can be used along with either (1) continuous grinding or (2) batch grinding. Method 1 is most appropriate for ores with coarse particle size and a high Fe content, where one or two stages of continuous grinding can be used, followed by magnetic separation after the separation requirements are met. Generally, when the particle size is larger than 0.2–0.3 mm, only one stage of grinding and magnetic separation is performed. Method 2 is suitable for ores with a fine particle size and low-Fe content, where magnetic separation is performed after one grinding stage. Some of the classified tailings are discarded, while the coarser concentrate is subjected to a further grinding step. The separation process involving grinding and dressing steps is shown in Fig. 4.1. The ore product with a particle size of 12–0 mm obtained after crushing is ground in the first stage. Materials with a particle size <0.074 mm account for 30%–40% of the total after classification. After concentration by desilting in a magnetic dewatering tank and magnetic separation, the classified tailings are discarded. After the second stage of grinding and classification, the coarse concentrate is obtained, where 75%–85% of the material has a particle size <0.074 mm. The batch grinding and classification process can maximize tailing disposal and reduce subsequent grinding efforts. After finishing the grinding the classified tailings are discharged via a magnetic de-flume concentration deslimation process and magnetic separation. In order to prevent magnetic agglomeration during the second classification, demagnetization is performed on the concentrate produced by magnetic separation. In order to prevent the loss of fine iron ore, the overflow from the secondary grading is generally premagnetized.

The fine grinding and screening steps increase the grade of the iron concentrate. The material is graded according to the decreasing ratio of a screw classifier and hydrocyclone, where the overflow is separated to obtain some of the coarse-grained poor continuous ore body. The main component of the concentrate after magnetic separation is quartz and iron ore; the quartz reduces

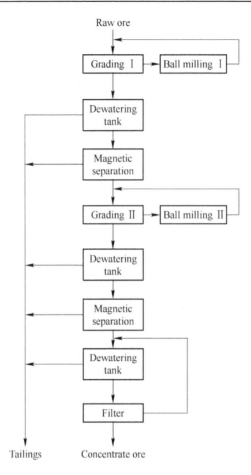

Figure 4.1
Separation process including batch grinding and dressing steps.

the quality of the iron concentrate, especially the iron grade. The grade of magnetic concentrate was significantly improved by screening and removing coarse particles. The typical process flow of batch grinding and concentration, magnetic separation, and fine screening is shown in Fig. 4.2. After three stages of grinding and magnetic separation and four stages of fine screening, the ores are ground to 0.3–0.4 mm, and some gangue and iron ore is dissociated. Then, magnetic separation is performed and part of the final tailings is discarded. The crude concentrate is then ground to a size of ∼0.1 mm via a second process, and then desliming and magnetic separation are performed, followed by additional tailings being discarded. The magnetic concentrate is finely screened, where the finer part is dehydrated and filtered to obtain the concentrate. The coarser fraction is subjected to magnetic separation and a third stage of grinding, followed by the third and fourth stages of fine screening, where the particles captured by the screen are returned to the magnetic separation process and ground again. The final sieved

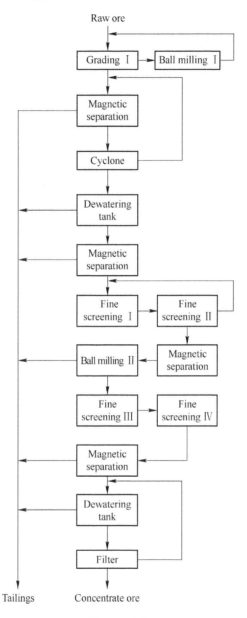

Figure 4.2
Typical process flow of batch grinding and concentration, magnetic separation, and fine screening.

product enters the last stage of magnetic separation to obtain the concentrate and tailings. The concentrates and tailings from each separation step are combined to form the final materials.

Beneficiation production lines are equipped with jaw crushers, hammer crushers, ball mills, classifiers, magnetic separators, flotation machines, thickeners, dryers, and other major equipment, along with feeders, hoists, and transfer equipment. The order of processing the original ore is as follows: vibrating feeder (groove feeder), PE jaw crusher, PEX jaw crusher, vibrating screen, roller dry separator, bin, pendulum feeder (electromagnetic vibrating feeder), ball mill, spiral classifier (high-frequency sieve), magnetic separator, wet concentrate, thickener (separation), and dry powder.

4.3.1 Ore from Panzhihua, China

4.3.1.1 Structure of the deposit

The Panzhihua (China) deposit of V-bearing titanoferrous magnetite is a late magmatic deposit of Hercynian gabbro. The ore body is produced and stratified in the rock mass, and the ore-bearing rock with the V-bearing titanoferrous magnetite is intermittently distributed along two deep faults of the Anning River and Panzhihua region. Most of the ore is immersed in Sinian dolomite, or the nonconformal plane between the Sinian and pre-Sinian structures. The rock body is composed of gabbro, olivine gabbro, and troctolite. It is late-Hercynian-rich iron ore containing basic−ultrabasic rock with high-calcium and low-silicon contents and is slightly alkaline. The deposit is typical of late magmatic crystallization, as it is well differentiated and layered. Due to structural damage and gully cutting, the ore-bearing rock of the Panzhihua deposit of V-bearing titanoferrous magnetite is separated along the direction from northeast to south and divided into six ore bodies (Zhujiabaobao, Lanjia volcano, Jianbaobao, Daomakan, Gongshan, and Naraqin). The Panzhihua gabbro body is about 19 km long and 2 km wide and contains large deposits at Panzhihua, Taihe, Baima, and Hongge. The ore body occurs below the rhythmic layer, produced in lenticular and layered forms that are parallel to one another. A single layer of the mineral is more than 1 km long, with a thickness between a few inches and several hundred meters.

4.3.1.2 Mineral composition

The typical content of the main minerals within the V-bearing titanoferrous magnetite in Panzhihua is shown in Table 4.5, where the multielement mineral contents in V-bearing titanoferrous magnetite from Panzhihua are shown in Table 4.6. The main metallic minerals of V-bearing titanoferrous magnetite from Panzhihua are titanomagnetite and ilmenite. Titanomagnetite is a complex mineral, composed of magnetite, ulvite, aluminum spinel, and lamella of ilmenite, with smaller concentrations of magnetite, limonite, goethite, and

Table 4.5: Main mineral content of V-bearing titanoferrous magnetite from Panzhihua.

Mineral	Titanomagnetite	Ilmenite	Sulfide	Titania	Plagioclase	Total
Content	43−44	7.5−8.5	1.0−2.0	16.50−29.0	10−19.50	100.0

Table 4.6: Typical multielement minerals in V-bearing titanoferrous magnetite from Panzhihua (%).

Component	TFe	FeO	V_2O_5	SiO_2	Al_2O_3	CaO	MgO	S
	30.55	22.82	0.30	22.36	7.90	6.80	6.35	0.64
Component	P_2O_5	TiO_2	Cr_2O_3	Co	Ga	Ni	Cu	MnO
	0.08	10.42	0.029	0.017	0.0044	0.014	0.022	0.294

secondary pyrite. The sulfides are dominated by pyrrhotite, while hengleinite, linnaeite, niccolite, cinerite, chalcopyrite, pyrite, and vallerite are also found.

The gangue mineral is dominated by titania and plagioclase, while titanhornblende, olivine, chlorite, serpentine, iddingsite, tremolite, sphene, sericite, epidote, prehnite, biotite, toposhite, calcite, and apatite also occur.

Titanomagnetite is the main component of iron ore in V-bearing titanoferrous magnetite. It is composed of magnetite, ulvite, aluminum spinel, and lamella of ilmenite. Most of the vanadium in the ore is isomorphic with iron minerals and enters the iron concentrate during ore dressing processes.

Table 4.7 shows the occurrence of Fe, V, and Ti in different deposits in the Panzhihua region (China). Manganese and chromium are also found in the titanomagnetite, along with selenium, tellurium, and platinum-group elements as sulfides, while Ta and Nb are present in ilmenite.

The Panzhihua-region ore bodies in China are divided into four large deposits, named Panzhihua, Baima, Hongge, and Taihe. The Panzhihua deposit has Fe content of 31%–31%, TiO_2 content of 8.98%–8.98%, V_2O_5 content of 0.28%–0.34%, Co content of 0.014%–0.014%, and Ni content of 0.008%–0.015%. The Taihe deposit has a similar composition to the Panzhihua one and is also a high-Ti iron ore. The Baima deposit is a low-Ti ore (TiO_2 content of 5.98%–8.17%), with high iron content of 28.99%, V_2O_5 content of 0.28%, Co content of 0.016%, and Ni content of 0.025%. The Hongge deposit is a low-Fe/high-Ti-type ore with a TiO_2 content of 9.12%–14.04%, average Fe content of 36.39%, V_2O_5 content of 0.33%, with a high average Ni content of 0.27%.

The three mining areas of Panzhihua, Baima, and Taihe have similar ore bodies with slightly different mineral contents. With increasing iron grade, the contents of TiO_2, V_2O_5, Co, and NiO increase, while the contents of SiO_2, Al_2O_3, and CaO decrease. The content of MgO in the Panzhihua and Taihe deposits decreases with increasing iron grade, while the opposite is true for the Baima deposit.

4.3.1.3 Mineral processing

The mineral processing procedure for V-bearing titanoferrous magnetite in Panzhihua is a three-stage closed-circuit process involving crushing, stage grinding, and stage

Table 4.7: Content of Fe, V, and Ti in different ore bodies in Panzhihua (%).

Region	Mineral	Productivity	Fe Grade	Fe Distribution rates	TiO$_2$ Grade	TiO$_2$ Distribution rates	V$_2$O$_5$ Grade	V$_2$O$_5$ Distribution rates
Lanjia volcano Jianbaobao	TitanomAgnetite	43.5	56.7	79.3	13.38	56.0	0.6	94.1
	Ilmenite	8.0	33.03	8.5	50.29	38.7	0.045	1.3
	Sulfide	1.5	57.77	2.8				
	Titanaugite	28.5	9.98	9.1	1.85	5.1	0.045	4.6
	Plagioclase	18.5	0.39	0.2	0.097	0.2		
	Total	100.0	31.08	99.9	10.42	100.0	0.13	100.0
Zhujiabaobao	Titanomagnetite	44.9	55.99	80.04	13.37	52.1	0.54	97.8
	Ilmenite	10.4	32.81	10.9	49.65	44.8	0.05	2.2
	Sulfide	1.5	51.48	2.5				
	Titanaugite	26.7	6.64	5.7	13.4	3.1		
	Plagioclase	16.3	0.88	0.5				
	Total	100.0	31.24	100.0	11.52	100.0	0.247	100.0

concentration. Fig. 4.3 shows the typical iron dressing process of iron concentrate in Panzhihua. The slag composition of these ores is mainly CaO and SiO_2, followed by MgO and Al_2O_3. The CaO, SiO_2, MgO, and Al_2O_3 in the iron concentrate after the separation do

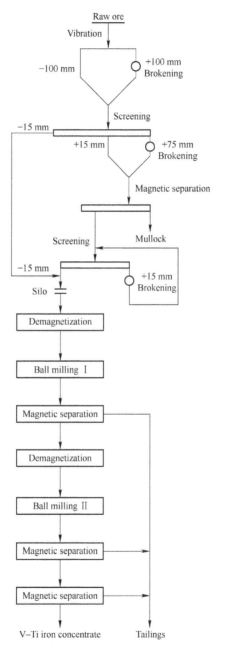

Figure 4.3
Typical iron dressing process for iron concentrate from Panzhihua.

not exceed the standard. There are magnetite, titanite, spinel, and laminite minerals in Panzhihua ore, which contain Fe, V, and Ti. Study of the major basic properties showed that titanomagnetite can be considered as a whole mineral phase to have enrichment and forms V-containing iron concentrate. The proposed enrichment process to produce V-containing iron concentrate includes closed-circuit grinding to 0.4 μm after crushing, two magnetic separation steps, and a scavenging step. The recovery of granular ilmenite from magnetic separation tailings involves three separation processes: (1) spiral dressing, flotation process (for sulfide minerals) and electric separation; (2) strong magnetic separation, spiral concentration, flotation (for sulfide minerals), and electric separation; and (3) spiral concentration, flotation (for sulfide minerals), and electric separation.

Based on the mineral processing flow of V-bearing titanoferrous magnetite from Panzhihua, a treatment plant with an output of 13.5 million Titanium concentrate and concentrating mill of 0.3 million Titanium concentrate were constructed and are currently under operation. A three-stage crushing and single closed-circuit process are used for crushing, while a closed-circuit and three open-circuit processes are used for grinding, along with closed-circuit grinding and three magnetic selection steps to obtain V-containing iron concentrate. A process, including spiral dressing, flotation (for sulfide minerals), and electric separation, has been used in the recovery of granular ilmenite from magnetic separation tailings, resulting in ilmenite and CoS concentrates. Table 4.8 shows the results of multielement analysis of V-containing iron concentrate from Panzhihua, where the TFe varies from 51.0% to 60.6%, TFe has a stable value of ~54%, and the TiO_2 content is in the range of 4.0%–14%.

Table 4.8: Multielement analysis of V-containing iron concentrate from Panzhihua (%).

Component	TFe	FeO	V_2O_5	SiO_2	Al_2O_3	CaO	MgO	S
1	53.56	30.51	0.564	4.64	4.69	1.57	3.91	0.532
2	59.75	26.89	0.767	1.01		<0.1		<0.01
3	59.03	27.30	0.67	2.11	3.29	1.02	1.66	0.32
Component	P	TiO_2	Cr_2O_3	Co	Ga	Ni	Cu	MnO
1	0.0045	12.73	0.032	0.02	0.0044	0.013	0.02	0.33
2	0.007	9.88						
3	0.0023	9.21						

Table 4.9 shows the results of multielement analysis of magnetic separation tailings from the Panzhihua ore.

Table 4.9: Multielement analysis of magnetic separation tailings from Panzhihua.

Component	TFe	V_2O_5	SiO_2	Al_2O_3	CaO	MgO	S
	18.32	0.065	34.4	11.06	11.21	7.66	0.609
Component	P	TiO_2	Co	Ga	Ni	Cu	
	0.034	8.63	0.016	0.0044	0.01	0.019	

4.3.2 Ore from Chengde, China

The Damiao deposit of Chengde V-bearing titanoferrous magnetite is located in the Xuanhua−Chengde−Beipiao deep fault zone of the eastern end of the Inner Mongolia Earth axis, where basic−ultrabasic rock intruded into the pre-Sinian stratum. Residual slurry separated from the late ore-bearing slurry was formed by penetrating structural fractures. More than 50 titanium magnetite ore bodies were formed with lenticular, vein, or cystic structures from anorthosite or anorthosite contact parts of the fracture zone, which have clear boundaries with the surrounding rock. The ore body in gabbro rock is usually disseminated or veined and has a gradual relationship with the surrounding rocks. The ore body is generally 10−360 m long, with a depth of 10−300 m. The ore is either in dense-block form or disseminated.

4.3.2.1 Structural characteristics

The Damiao deposit of Chengde V-bearing titanoferrous magnetite is located in the eastern end of the Inner Mongolia Earth axis and occurs in the basic−ultrabasic rock belt controlled by the Xuanhua−Chengde−Beipiao deep fault zone in an east−west direction. It contains metamorphic rock from the early Sinian Period and is widely distributed, showing mainly amphibole plagiogneiss, hornblende gneiss, biotite oblique gneiss, and the migmatitic granite. The upper part is covered with jura-white archaic sedimentary rocks and volcanic rocks, generally associated with apatite. The Damiao deposit mainly contains anorthosite complex rock, where V-bearing titanoferrous magnetite occurs in plutonic and basic igneous rock bodies, which consist of plagioclase and gabbro. Most have suffered from secondary changes, accompanied by sodium tetrahedrite and chlorolization. Plagioclase is the most widely distributed rock in the mined area, which is the main basic rock in the periphery of the ore body. Gabbro is mostly distributed on the east side of the mining area, and in the northeast extension, which generally contain V-bearing titanoferrous magnetite. The ore content is greater in this region and has commercial value. The ore and rock follow a gradient relationship, changing from dense ore to densely disseminated, to sparsely disseminated, and finally to ore-disseminated gabbro. The ore body produced in the plagioclase has an obvious contact boundary. There are more basic and acidic vein rocks, which are considered a postmetallogenic product. They are interspersed with early rocks. The ore-bearing rock is irregular, where vein-like, lenticular, or block-like ore bodies occur, most above the surface. There are great deep changes to the ore-bearing rock mass, which shows expansion, shrinkage, sprouting, composites, and other behavior.

4.3.2.2 Mineral composition

The main metallic minerals in the V-bearing titanoferrous magnetite are titanium magnetite, ilmenite, and pyrite, while the main gangue minerals are chlorite and amphibole. The structure of the Damiao deposit of Chengde V-bearing titanoferrous magnetite is either dense or disseminated, where the boundary with the surrounding rock is obvious. The main useful minerals include magnetite, ilmenite, pyrite, chalcopyrite, secondary hematite, and limonite. Gangue minerals include chlorite, amphibole, plagioclase, siderophyllite, and apatite. The typical mineral composition of the Damiao deposit in Chengde is shown in Table 4.10.

Table 4.10: Typical mineral composition of the Damiao deposit of V-bearing titanoferrous magnetite from Chengde.

Mineral	Magnetite	Ilmenite	Chalcopyrite	Mannite	Nicopyrite
Content	44.59	9.84	0.123	0.0121	0.014
Mineral	Capillary pyrite	Pyrrhotite	Silicate	Pyrite	
Content	0.00016	0.00004	43.85	1.47	

The typical mineral composition of the Damiao deposit of Chengde V-bearing titanoferrous magnetite is shown in Table 4.11.

Table 4.11: Typical mineral composition of the Damiao deposit of Chengde V-bearing titanoferrous magnetite (%).

Component	TFe	V_2O_5	TiO_2	S	P	Co	Ni
Content	29.76	0.288	7.57	0.364	0.36	0.01	0.019

The Black Mountain deposit of V-bearing titanoferrous magnetite in Chengde is located in the Yanshan section of Chicheng—Pingquan running east—west; this deposit belongs to the Tianshan—Yinshan complex tectonic belt running in the east—west direction and contains a basic rock body in a zone of basic or ultrabasic rock. The basic rock mass occurs in an east—west fault that is 40 km in length and intruded into a pre-Sinian mixed rock, where the rock mass is dominated by plagioclase. The large ore body contains Damiao deposit, Black Mountain deposit and Tougou deposit from east to west, which is referred to as the "big black basic rock mass."

The Black Mountain deposit is a late magmatic rock mass. The ore body morphology was determined by the flow structure of magmatic rocks and primary joints fissures. The rock mass and morphology is relatively complex. Volatile components during mineralization promoted the magma layers; differentiation did not occur perfectly, so different sizes and types of massive rocks are nonhomogeneously mixed with disseminated ores. There is an irregular distribution of the ore in the massive rock and rock. After liquid separation, some of them experienced a period of movement and appear in plagioclase, forming high-quality large-scale ore bodies. Iron phosphate mineralization occurred in the northeast direction of the ore body or the upper part. As the apatite is relatively rich, the phosphorus reaches the industrial grade. Typical elemental compositions of the Black Mountain V-bearing titanoferrous magnetite from Chengde are shown in Table 4.12.

Table 4.12: Typical multielement analysis results of the Black Mountain deposit of V-bearing titanoferrous magnetite from Chengde (%).

Component	TFe	FeO	Fe_2O_3	TiO_2	V_2O_3	SiO_2
Content	14.68	13.12	6.4	8.63	0.102	33.53
Component	Al_2O_3	CaO	MgO	P_2O_5	S	Co
Content	16.27	6.44	4.18	0.19	0.52	0.024

The surrounding rock alteration is not well developed and is dominated by chloritization, micronization, and carbonation. The Black Mountain deposit has more than 60 ore bodies and outcrops of different sizes. The ore bodies are usually produced in lenticular, tubular, or irregular shapes. This deposit is divided into dense and disseminated ores. The main minerals are titanium magnetite, ilmenite, and small amounts of rutile, and pyrite containing cobalt and nickel. Vanadium occurs in magnetite as an isomorphism, where the crystals contain both magnetite and ilmenite, in which magnetite is separated into a solid solution. Cobalt and nickel are found in pyrite and magnetite. The mineral density and magnetism of the primary minerals found in the Black Mountain deposit are shown in Table 4.13.

Table 4.13: Mineral density and magnetism of the primary mineral in the Black Mountain deposit of V-bearing titanoferrous magnetite from Chengde.

Mineral	Density (g/cm^3)	Magnetism (cm^3/g)
Titanomagnetite	4.815	7300×10^{-6}
Ilmenite	4.560	113×10^{-6}
Sulfide	4.830	$< 16 \times 10^{-6}$
Chlorite	3.187	$(50 \times 10^{-6})-(300 \times 10^{-6})$
Anorthose	2.635	10×10^{-6}

Valuable elements, including copper, chromium, phosphorus, gold, silver, and platinum-group metals, are found in the Black Mountain deposit of V-bearing titanoferrous magnetite from Chengde. The typical compositions of the main minerals in this deposit of are shown in Table 4.14.

Table 4.14: Typical composition of the main minerals in the Black Mountain deposit of V-bearing titanoferrous magnetite from Chengde.

Mineral	Ilmenite	Titanomagnetite	Hematite	Sulfide	Chlorite	Plagioclase	Pyroxene	Other
Content	15.6	4.3	3.6	1	24.5	35.6	11.6	3.8

4.3.2.3 Mineral processing technology

The Shuangtashan ore dressing plant is located in Chengde city, Hebei province, China. It started operation in 1959 and was the first plant to process V-bearing titanoferrous magnetite in China. It was designed to process 650 kt of ore annually. The ore comes from mines in the great temple area. The main minerals are titanomagnetite, ilmenite, and pyrite, while the main gangue minerals are chlorite, amphibole, and plagioclase. The original ore contains Fe 30%, TiO_2 8%, and V_2O_5 0.4%. Two-stage grinding and magnetic separation are used to produce V-containing iron concentrate, which contains Fe 61%, V_2O_5 0.7%–0.9%, and TiO_2 9%, for an iron recovery of 70%.

The Black Mountain ore dressing plant is located in Chengde city, Hebei province, China. It started operation in 1985 and was designed to process 900 kt of ore annually. The main metallic minerals in the ore are titanomagnetite, ilmenite, and pyrite, and the main gangue minerals are chlorite and amphibole. The V-containing iron concentrate is produced using a two-stage grinding and magnetic separation process. The original ore contains Fe 33.80%, TiO_2 8.47%, and V_2O_5 0.353%, while the V-containing iron concentrate has a Fe content of 60%, with a recovery rate of ~70%. Table 4.15 gives the typical chemical composition (%) of V-containing iron concentrate from Chengde ore. The multielement analysis results and mineral composition of the magnetic tailings after the beneficiation are shown in Table 4.16 and 4.17, respectively.

Table 4.15: Typical chemical composition of V-containing iron concentrate from Chengde (%).

Species	TFe	FeO	CaO	SiO_2	MgO	Al_2O_3	TiO_2	V_2O_5
V-containing iron concentrate	64.09	26.16	0.78	3.00	1.86	1.60	3.75	0.55
V-containing iron concentrate	60.79	29.38	0.21	2.05	1.94	3.26	7.20	0.77

Vanadium does not exist as a single mineral, but usually in an isomorphic form with magnetite. The vanadium grade in the ore is positively correlated with the iron grade. The metallic minerals in the magnetic tailings are mainly ilmenite and titanomagnetite, with a

Table 4.16: Multielement analysis results of the magnetic tailings (%).

Component	TFe	FeO	Fe$_2$O$_3$	TiO$_2$	V$_2$O$_3$	SiO$_2$
Content	14.68	13.12	6.4	8.63	0.102	33.53
Component	Al$_2$O$_3$	CaO	MgO	P$_2$O$_5$	S	Co
Content	16.27	6.44	4.18	0.19	0.52	0.024

Table 4.17: Main mineral composition of magnetic tailings (mass%).

Mineral	Ilmenite	Titanomagnetite	Hematite	Sulfide	Chlorite	Plagioclase	Pyroxene	Other
Content	15.6	4.3	3.6	1	24.5	35.6	11.6	3.8

small amount of rutile, anatase, leucoxene, limonite, hematite, and sulfide ore. The gangue mineral is dominated by chlorite and plagioclase, with a small amount of biotite, quartz, calcite, and apatite.

4.4 V-containing titanoferrous magnetite outside of China

V-bearing titanoferrous magnetite outside of China generally appears in large ore deposits, such as those in Catchikara, Russia, and Bushveld, South Africa. Many of the V-bearing titanoferrous magnetite ores outside of China are richer than the Chinese ones, such as South Africa's Bushveld raw ore with a V$_2$O$_5$ content of 1.8% and Finland's Otamatti raw ore with a V$_2$O$_5$ content of 0.4%. Stable deposits exist, such as those in Catchikara, Russia, and Bushveld, South Africa.

4.4.1 Deposits in Russia

The main metallic minerals in the V-bearing titanoferrous magnetite of Russia include titanomagnetite, followed by ilmenite, and small amounts of sulfides. The gangue minerals include plagioclase, olivine, amphibole, plagioclase, epiphyllite, and serpentine. The grain size of titanium magnetite is 0.9–1 mm. The V-bearing titanoferrous magnetite has low grade, with average values of Fe 15%–17%, TiO$_2$ 0.43%–1.88%, and V$_2$O$_5$ 0.13%. The Catchikara concentrator is located in the city of Catchikara, Sverdlov province, Russia, where mineral processing is based on the iron concentrate.

There are 29 production steps in this concentrator process: 3 crushing steps, 14 medium and fine crushing steps, and 1–20 series are two-stage grinding and four-stage magnetic separation (including dry magnetic separation). The 21–29 series is a three-stage grinding and four-stage magnetic separation process. Two products of coarse iron concentrate and fine iron concentrate are obtained. The mined ore contains an Fe content of 16%–18%,

Table 4.18: Typical multielement analysis of Catchikara V-containing iron concentrate.

Item	Chemical composition (%)								
	TFe	TiO_2	V_2O_5	FeO	Fe_2O_3	SiO_2	Al_2O_3	CaO	MgO
Raw ore	15.9	1.5	0.13	5.5	16.6	47.37	10.08	14.1	8.9
Concentrate-1	60.3	0.66		27.2	55.91	5.10	2.62	1.73	2.57
Concentrate-2	62.5			28.75	57.33	3.96	2.44	1.35	2.14
Wet magnetic tailings	6.55			4.2	4.96	46.86	5.90	15.29	13.05

resulting in an Fe content of 60.59% in the coarse iron concentrate, 62.5% in the fine iron concentrate, and a total iron recovery of 65%–66%. Table 4.18 shows the results of typical multielement analysis of Catchikara iron concentrate.

The Kyiu deposit of Calley-cora is a submerged ore, mined by digging and land cutting, producing ore blocks. The main rocks are gabbro, anorthosite, alkaline, and ultramafic rocks. The ore grade is TFe 29%–45%, TiO_2 5%–10%, and V_2O_5 0.15%–0.75%. The iron ore concentrate after beneficiation is TFe 66%–69% and V_2O_5 1.1%–1.4%. The Afrikaner deposit of Calley-cora is a disseminated rock mass, occurring in layers, blocks, and lenticular forms, and containing mainly alkaline, mafic, and ultramafic rocks. The ore grade is TFe 36%, TiO_2 7%, and V_2O_5 0.26%, while the V-containing iron concentrate contains TFe 58% and V_2O_5 0.5%–0.6%. The Jarrett Lake deposit of Calley-cora is mined by digging and land cutting, producing ore blocks. The ore grade is TFe 13%–37%, TiO_2 8%–26%, and V_2O_5 0.13%, while the iron ore concentrate after beneficiation contains 58% TFe and 0.6% V_2O_5. The Prdozhgask deposit of Calley-cora is mined by digging and land cutting, producing ore blocks. The main rocks are gabbro, anorthosite, and alkaline, mafic, and ultramafic rocks. The raw ore contains 13%–37% TFe, 8%–26% TiO_2, and 0%/13% V_2O_5. After beneficiation the iron concentrate contains 58% TFe and 0.6% V_2O_5.

4.4.2 Deposits in the United States

The Macintyre ore dressing plant is located in the Adirondack Mountains of Essex County, northeast of New York. It started operation in 1942 and processes 10,600 kt/day of ore. The main useful minerals in the ore are ilmenite and magnetite. The main gangue minerals are labradorite, amphibole, pyroxene, garnet, and biotite. The iron concentrate is produced by a two-stage process of continuous grinding and magnetic separation. The raw ore contains 33% Fe and 16% TiO_2, while the iron concentrate contains 63% Fe. The process flow of recovering ilmenite in iron tailings includes both magnetic separation and fine particle floatation. The titanium concentrate contains 45% TiO_2. The mineral deposits of V-bearing titanoferrous magnetite are very rich in the United States and are distributed in Alaska, New York, Wyoming, and Minnesota, although most have not yet been exploited. There are also deposits in Sanford Lake, New York, which is related to the Adirondack complex of Precambrian gabbro and plagioclase, with an ore body 1600 m long. The footwall rocks are dense coarse anorthosite, where the hanging side contains disseminated or dense ore in fine-to-medium-grain gabbro rock, which are parallel to each other or at an angle of 45 degrees. The average iron content of the ore is 34% Fe, 18%–20% TiO_2, 0.45% V_2O_5, where the iron concentrate after beneficiation contains 59% Fe, 9%–10% TiO_2, and 0.7% V_2O_5. The Dianna comprehensive ore body in New York is mined in layers and produces block and lenticular forms. The main rock types are gabbro, anorthosite, alkaline, and mafic rocks. The ore contains 20% TFe, 7% TiO_2, and 0.05% V_2O_5, while the iron concentrate after beneficiation contains 65% TFe, 4% TiO_2, and 0.15% V_2O_5. The Hager mine in New Jersey is a disseminated ore type, mined in layers, and produces block and lenticular forms. The main rocks are gabbro, anorthosite, alkaline, mafic, and ultramafic rocks, containing 60% TFe, 6% TiO_2, and 0.4% V_2O_5. New Jersey typically has disseminated ore types, with layered lenticular ore, mined by digging and land cutting. The main rocks are gabbro, anorthosite, alkaline, mafic, and ultramafic rocks with an ore containing 50% TFe, 10%–15% TiO_2, and 0.5% V_2O_5.

The Piedmont mine in North Carolina is disseminated ore, produced in blocks, where the main rocks are gabbro, anorthosite, alkaline rock, and ultramafic rocks, with an ore grade of 40%–65% TFe, 12% TiO_2, 0.13%–0.38% V_2O_5. The iron concentrate after beneficiation contains 0.6% V_2O_5. The Appalachian ore in North Carolina and Tennessee is disseminated and occurs in blocks, where the main rocks are gabbro, anorthosite, alkaline rock, and ultramafic rocks. The ore contains 40%–60% TFe and 5%–7% TiO_2, where the iron concentrate after beneficiation contains 0.2%–0.4% V_2O_5. The Iron Mountain deposit in Wyoming occurs in lamellar and lenticular forms, where the main rock is gabbro, alkaline, and ultramafic iron rock, with an ore composition of 17%–45% TFe, 10%–20% TiO_2, and 0.17%–0.64% V_2O_5. The ore from the Euwen Lake mine in Wyoming is block type, where the main rock is gabbro, alkaline, and ultramafic iron rock. The raw ore contains 29% TFe, 5% TiO_2, and 0.2% V_2O_5. The Iron Mountain deposit in Colorado is in the form of blocks within plagioclase, alkaline, mafic, and ultramafic iron rocks. The ore contains 40%–50% Fe, 14% TiO_2, and 0.41%–0.45% V_2O_5. California's San Gabriel ore deposit is in the form of lamellar, lenticular, and mass blocks within alkaline, mafic, and ultramafic iron rocks. The raw ore contains 46% TFe, 20% TiO_2, and 0.53% V_2O_5. The Sneam ore in Alaska is layered and lenticular and is mined by digging and land cutting. The main rocks are gabbro, anorthosite, alkaline, and ultramafic rocks. The ore contains 19% TFe, 2.6% TiO_2, and 0.09% V_2O_5. The iron concentrate after beneficiation contains 64% TFe, 3.5% TiO_2, and 0.7% V_2O_5. The Kliuke deposit in Alaska is mined by digging and land cutting, and the main rocks are gabbro, anorthosite, alkaline, and ultramafic rocks. The ore contains 15%–20% TFe,

2.0% TiO$_2$, and 0.05% V$_2$O$_5$. The iron concentrate after beneficiation contains 62%–64% TFe, 2.4% TiO$_2$, and 0.3%–0.4% V$_2$O$_5$. The Lake Erie deposit in Alaska is layered and lenticular and mined by digging and land cutting. The main rocks are gabbro, anorthosite, alkaline, and ultramafic rocks. The ore contains 12%–19% TFe, 1.3% TiO$_2$, and 0.02% V$_2$O$_5$. The iron concentrate after beneficiation contains 40%–60% TFe, 3.1% TiO$_2$, and 0.3%–0.5% V$_2$O$_5$.

The main useful minerals in the Adirondack deposits of the northeastern part of New York City are ilmenite and magnetite. The main gangue minerals are labradorite, amphibole, pyroxene, garnet, and biotite.

4.4.3 Deposits in Finland

The main metallic minerals in the V-bearing titanoferrous magnetite in Finland are magnetite, ilmenite, and pyrite, while the main gangue minerals are chlorite, amphibole, and plagioclase. The Otamatti dressing plant is located in Central Finland, where the deposit contains an average of 35%–40% Fe, 35% TiO$_2$, and 0.38% V$_2$O$_5$. The coarse-grained tailing is ground to 0.2 mm. After magnetic separation and flotation, the iron concentrate contains 69% Fe, 2.5% TiO$_2$, and 1.07% V$_2$O$_5$.

4.4.4 Deposits in South Africa

There are titanoferrous magnetite deposits containing high vanadium contents in South Africa. Minerals with a particle size over 6 mm are used as a raw material for direct reduction, while those below 6 mm are coarsely ground, screened, finely ground, and then magnetically separated to remove the nonmagnetic parts. The typical composition of iron concentrate after beneficiation is 53%–57% TFe, 1.4%–1.9% V$_2$O$_5$, 12%–15% TiO$_2$, 1.0%–1.8% SiO$_2$, 2.5%–3.5% Al$_2$O$_3$, and 0.15%–0.6% Cr$_2$O$_3$.

4.4.5 Deposits in Canada

The Steel Mountain mine in Newfoundland, Canada, uses ore that is disseminated, layered, and lenticular and occurs in blocks. The main rocks are gabbro, alkaline, mafic, and ultramafic rocks with an ore content of 50%–55% TFe, 10% TiO$_2$, and 0.4% V$_2$O$_5$. The iron concentrate after beneficiation contains 64.5% TFe, 8% TiO$_2$, and 0.75% V$_2$O$_5$. The Indian Hurd mine in Newfoundland is mined by digging and land cutting. The main rocks are gabbro, alkaline, mafic, and ultramafic rocks. The ore contains 64% TFe, 2%–6% TiO$_2$, and 0.2%–0.7% V. The iron ore concentrate after beneficiation contains 1.8% V$_2$O$_5$. The Matawa mine, Ontario, contains disseminated ore that is mined by digging and land cutting to produce blocks. The main rocks are alkaline, mafic, and ultramafic rocks, and the ore contains 38% TFe, 8% TiO$_2$, and 0.76% V$_2$O$_5$. The iron concentrate after beneficiation contains 60% TFe and 1.4% V_2O_5. The Lake Cross mine in Manitoba is disseminated ore mined by digging and land cutting to produce blocks. The main rock is anorthosite, alkaline, mafic, and ultramafic rocks. The ore contains 28%−60% TFe, 3%−10% TiO_2, and 0.02%−0.5% V_2O_5. The iron concentrate after beneficiation contains 69% TFe, 0.11% TiO_2, and 1.6% V_2O_5. The Banks Island deposit in British Columbia is mined by digging and land cutting to produce blocks. The main rock is anorthosite, alkaline, mafic, and ultramafic rocks. The ore contains 20%−50% TFe, 1%−3% TiO_2, and 0.07%−0.55% V_2O_5. The iron concentrate after beneficiation contains 60% TFe, 1.5%−5% TiO_2, and 0.9%−1.8% V_2O_5. The Poacher Island mine in British Columbia uses digging and land cutting methods to produce blocks. The main rocks are gabbro, anorthosite, alkaline, and ultramafic rocks. The ore contains 25% TFe, 2% TiO_2, and 0.2%−0.35% V_2O_5. The iron concentrate after beneficiation contains 60% TFe, 0.3%−1.5% TiO_2, and 0.3%−0.9% V_2O_5.

The Cevennes deposit in Quebec has layered and lenticular ore that is mined by digging and land cutting to produce blocks, mainly of alkaline, mafic, and ultramafic rocks. The ore contains 11%−42% TFe and 3%−16% TiO_2. The iron concentrate after beneficiation contains 60% TFe, 14%−18% TiO_2, and 0.25%−0.8% V_2O_5. The Magpie Mountain mine in Quebec has layered and lenticular ores mined by digging and land cutting. The main rock is gabbro, mafic, and ultramafic rocks, where the ore contains 43% TFe, 10% TiO_2, and 0.2%−0.35% V_2O_5, where the iron concentrate after beneficiation contains 49% TFe, 16% TiO_2, and 0.2% V_2O_5. The Saint-Uparin mine, Quebec, has block-like ore, where the main rocks are gabbro, alkaline, mafic, and ultramafic rocks. The ore contains 30%−40% TFe, 38%−45% TiO_2, and 0.17%−0.34% V_2O_5. The iron ore concentrate after beneficiation contains 67% TFe, 2% TiO_2, and 0.30%−0.6% V_2O_5. The Maureen mine in Quebec uses digging and land cutting methods to , the rocks are the gabbro, alkaline, mafic, and ultramafic rocks. The ore contains 25%−43% TFe, 19% TiO_2, and 0.05%−0.34% V_2O_5. The iron concentrate after beneficiation contains 60%−65% TFe, 1%−4% TiO_2, and 0.4% V_2O_5. The Lake Dore mine in Quebec contains disseminated ore that is mined by digging and land cutting to produce blocks. The main rocks are gabbro, anorthosite, alkaline, and mafic rocks. The ore contains 28%−53% TFe, 5%−8% TiO_2, and 0.3%−1.0% V_2O_5. The iron concentrate after beneficiation contains 62%−64% TFe, 4%−5% TiO_2, and 1.0%−2.5% V_2O_5.

Lac tio mine in Allard, Canada, is ilmenite covered by hematite, which occurs as blocks of hematite and ilmenite with a $TiO_2 \cdot FeO$ and Fe_2O_3 ratio of ∼2:1. The gangue is mainly plagioclase, with small amounts of pyroxite, biotite, pyrite, and magnetite, occurring in lamellar and lenticular forms, which is mined by digging and land cutting to produce block. The raw ore contains 0.27% V_2O_5. Lake Michekamo mine in Labrador province is mined by digging and land cutting to produce blocks. The main rocks are gabbro, alkaline, mafic, and ultramafic rocks. The iron concentrate after beneficiation contains 58% TFe, 9% TiO_2, and 1.9% V_2O_5.

4.4.6 Deposits in Norway

The terness deposit in Norway is the largest titanium mine in Europe. The ore is the impregnated lenticular type. The main rocks are gabbro, plagioclase, alkaline, mafic, and ultramafic iron rocks. The ore reserves are about 300 million tons. The raw ore contains 20% TFe and 17%–18% TiO_2. There is a deposit of low-Ti magnetite in the Rhodes district, which is disseminated, and mined by digging and land cutting to produce blocks. The main rocks are gabbro, anorthosite, alkaline, and ultramafic rocks. The ore contains 30% TFe, 4% TiO_2, and 0.30% V_2O_5. The ore from the Selvag mine in Lovertown is disseminated and mined by digging and land cutting to produce blocks. The main rocks are anorthosite, alkaline, mafic, and ultramafic rocks. The ore contains 35% TFe, 4.0% TiO_2, and 0.4% V_2O_5. The iron concentrate after beneficiation contains 60% TFe, 5% TiO_2, and 0.7% V_2O_5. The Rhodes Sade mine in Morley contains disseminated ore that is mined by digging and land cutting to produce blocks. The main rocks are gabbro, anorthosite, alkaline, and ultramafic rocks. The ore contains 35% TFe, 6.0% TiO_2, and 0.5% V_2O_5. The iron concentrate after beneficiation contains 62% TFe, 2% TiO_2, and 0.9% V_2O_5. The Zog mine in Moray uses digging and land cutting methods to produce blocks. The main rocks are alkaline, mafic, and ultramafic iron rocks. The ore contains 10%–30% TFe, 5%–50% TiO_2, and 0.1%–1% V_2O_5. The Oslo mine in Morey has disseminated ore that is mined by digging and land cutting. The main rocks are anorthosite, alkaline, mafic, and ultramafic iron rocks. The raw ore contains 10%–30% TFe, 5.0%–50% TiO_2, and 0.1%–1.0% V_2O_5. The Stolgongan mine in Egersund has disseminated layered and lenticular ore that is produced as blocks. The main rocks are gabbro, alkaline, mafic, and ultramafic rocks. The ore contains 5% TFe, 17.0% TiO_2, and 0.14% V_2O_5. The iron ore concentrate after enrichment contains 65% TFe, 5% TiO_2, and 0.73% V_2O_5.

4.4.7 Deposits in New Zealand

The V-bearing titanoferrous magnetite in New Zealand contains 58% iron, 8%–10% TiO_2, and 0.6% V_2O_5.

4.5 Main equipment

Ore dressing is the process of selecting useful minerals from the mined raw ore according to their physical and chemical properties and overall performance differences. According to the dressing process, the ore dressing machinery is divided into crushing, grinding, vibrating screens, sorting, and dewatering machines. Crushing machines include jaw crushers, rotary crushers, cone crushers, roller crushers, and counterattack crushers. The most widely used grinding machinery includes cylindrical mills, rod mills, ball mills, pebble mills, and autogenic mills. Vibrating screens and resonant screens are commonly used in screening machines. Hydraulic and mechanical classifiers are widely used for wet grading. According to the separation mechanism, separation machines are divided into gravity dressing machines, magnetic separators, flotation devices, and special concentrators. Mineral processing equipment mainly includes crushing, fine grinding, and ore dressing devices. It is necessary that the power configuration is high, the process is connected smoothly, and the overall system meets the requirements of high capacity and efficiency with minimal energy use.

4.5.1 Crushing equipment

Crushing equipment is mainly used for rough processing of the raw ore. Crushers generally handle larger chunks of the ore and produce quite coarse product, usually larger than 8 mm. Crushers have a certain gap between the crushing faces, which do not touch each other. Crushers are classified into coarse, medium, and fine crushers. Common crushers include jaw crushers, cone crushers, impact crushers (e.g., sand-making machines), hammer crushers, and roll crushers.

The main parts of a jaw crusher are the frame, eccentric shaft, belt groove wheel, flywheel, movable jaw, elbow plate, fixed jaw plate, and movable jaw plate. Through the periodic motion of the moving jaw, the ore material is crushed between the two jaw plates (fixed and movable jaws). At the start of the process, the pressure is low, which compacts the materials. When the pressure is increased above the strength of the material, it is then crushed. Conversely, when the movable jaw is wiggled to the opposite direction of jaw, the material moves downward under gravity. When it comes in contact with the jaws, it experiences periodic motion, and to discharge the distance, after several cycles, has been broken, which is discharged from outlet. A motor continuously runs the jaw crusher and batches of material are crushed and then the product is discharged. Jaw crushers are characterized by a large crushing ratio, uniform product size, simple structure, reliable operation, simple maintenance, and low operation cost.

The main parts of a cone crusher are the frame, belt pulley, horizontal shaft, eccentric sleeve, upper crushing wall (fixed cone), lower crushing wall (dynamic cone), hydraulic coupling, lubrication system, and control system. During operation of the cone crusher, a motor provides rotation via a belt wheel or linkage shaft, which moves a drive shaft attached to the cone. Crushing is achieved via the pressure exerted by the eccentric sleeve around the fixed point of the rotation pendulum motion. As a result, the crushing cone sometimes comes close to and sometimes leaves the surface of the mortar wall fixed on the adjusting sleeve, so that the ore is continuously impacted in the crushing chamber to process the material.

4.5.2 Grinding equipment

Grinding equipment is mainly used for fine grinding of crude V-bearing titanoferrous magnetite minerals. Fine grinding mainly consists of grinding and impact processes, which grind the crushed product to a size of 10–300 mm. The product granularity depends on the disseminated grain size and selection method used during processing. Commonly used grinding equipment includes rod mills, ball mills, autogenous mills, and semiautomatic mills. Dry or wet ball milling is chosen depending on the ores and drainage. In general, ball mills include parts for feeding, discharging, rotation, transmission (reduction box, small transmission gear, motor, and electric control), and other main parts. The ball mill barrel is made of cast steel and the inner lining can be changed. The rotary gear is machined with the casting gear, and the shell has a lining with good wear resistance. The ball mill is usually a horizontal cylinder that is rotated via an external transmission, two cantant-lattice type. The ore material is filled into the first chamber of the mill using the feeding device through a hollow shaft screw. There are ladder-like or corrugated linings inside the barrel, along with steel balls of different specifications. Centrifugal force is generated when the barrel rotates, which lifts the steel balls to a certain height, which then drop and crush the material. After the material has been crushed in the first barrel, it enters a second one through a single-layer compartment. The second chamber is lined with a flat liner and also contains steel balls for further grinding of the material. The resulting powder is discharged through the discharge grate to complete grinding.

Rod mills are driven by a motor, either directly by a low-speed synchronous motor or indirectly through a large peripheral gear, which controls the rotation of the barrel body. The barrel body contains appropriate grinding medium, such as steel rods. Under the action of centrifugal and frictional forces, the grinding medium is lifted to a certain height and then falls in a state of throwing or leakage. A characteristic of rod mills is that during grinding, the grinding medium is in contact with the ore, so it has a certain selective grinding effect. When used for coarse grinding, rod mills have greater capacity than ball mills of the same specification.

4.5.3 Screening and grading equipment

Sieving and grading methods are used to classify the materials into appropriate size ranges after they have been crushed. Screen classification is mainly used for the classification and grading of fine V-bearing titanoferrous magnetite minerals. The size of the mesh screen determines the classification of the material into different particle size ranges; this method is often used for coarse-grained materials. According to the different sedimentation velocity in a medium (usually water), ore material can be divided into different particle size grades; this method is more suitable for materials with smaller grain sizes. Screening machinery includes fixed bar screens, vibrating screens, and wet fine screens. Bar screens have steel bars, and channel steel and rail, which are welded. It can be designed for use as either a screen or grizzly and can be installed horizontally at the top of a silo in a coarse crushing machine to prevent large rocks entering the crusher. They can also be mounted on an angle for prescreening. Vibrating screens contain a screen box, screen mesh, vibrator, and vibration damping. The vibrator is installed on the side plate of the sieve box, and the motor is driven by a triangular belt to generate the centrifugal force and the power required to vibrate the box. The sieving side plate is made of high-quality steel plate and is connected with the beam and pedestal of the vibrator with high-strength bolts or ring-groove rivets. The vibrator is installed on the side plate of the screen box, which is driven by the motor via a coupling, generating centrifugal force and the power to vibrate the screen. In the process of extracting iron, circular-motion vibrating screens are often used, such as single-axis inertia vibrating screens, self-fixed center vibrating screens, and heavy vibrating screens. The wet fine screen device has various layers of sieve boxes, with feeding boxes for each layer. Through the feeding box, the slurry is distributed over the full width of the screen surface, which can avoid counter-enrichment by cyclone methods.

Classifier include thickener, hydrocyclones, and spiral classifiers. Classifiers are widely applicable for grading ore pulp in the process of metal dressing and can also be used to remove mud and dehydrate the slurry during washing operations. They are often used to form a closed-circuit process with a ball mill. Commonly used classifiers include spiral classifiers, hydrocyclones, and thickeners. In the case of spiral classifiers, spiral blades are connected to a hollow shaft, which is supported by bearings at both ends. A gear is installed at the upper end, and the motor drives the spiral via a bevel gear drive shaft. The bottom bearing is installed at the bottom of the lifting mechanism, which moves the shaft up or down. The lifting mechanism is driven by a motor through a reducer and a pair of bevel gears, so that the lower end of the screw rises and falls. In the process of mineral gradation, due to the different particle sizes and size distributions, and hence, settling velocities in the fluid, the fine particles float in the water and exit via an overflow from the upper part, while the coarse grains sink on the bottom. Filtering of the grinding mill powder is then performed; the coarse material is moved using the spiral screw to the mill inlet, while the fine material is discharge from the overflow pipe.

Hydrocyclones are efficient devices for ore grading via desilting by rotary flow. They are also used for concentration, dehydration, and separation processes. The hydrocyclone consists of a hollow cylinder in the upper part and an invertebrate connected with the cylinder at the bottom, which constitute the working barrel of the hydrocyclone. This system also has a feed pipe, overflow pipe, and sediment orifice. Hydrocyclones are mainly used for grading, separation, concentration, and desilting in mineral processing. When the hydrocyclone is used for grading, it is mainly used as the grinding classification system within the grinding unit. In the case of desliming, it can be used in reprocessing plants. In the case of concentration and dehydration, it is used to concentrate tailings before they are sent to fill underground mining tunnels. External and internal vortices are the main dynamic action within the hydrocyclone; they rotate in the same direction but move in the opposite direction. The external cyclone carries thick and heavy solid material, which is discharged from the sediment orifice as the sediment product. The internal cyclone carries the small and light solid material that is discharged via the overflow outlet as the overflow product.

A hydrocyclone uses a sand pump (or a height difference) to rotate the slurry in the cylinder along the tangential direction at a certain pressure (usually 0.5–2.5 kg/cm) and flow rate (about 5–12 m/s). The slurry then rotates along the wall of the cylinder at a rapid speed, generating a centrifugal force. The coarse and heavy ore grains are moved to the outside of the chamber by the effect of the centrifugal force and gravity.

Thickener systems are solid–liquid separation devices based on gravity separation. They are usually made of concrete, wood, or welded metal plates to form a cylindrical shallow trough with a cone-shaped bottom. Slurry with a solid weight of 10%–20% can be concentrated into an underflow slurry with a solid content of 45%–55% via gravity sedimentation. The thickened underflow slurry is discharged via an underflow port at the bottom of the thickener by the action of a slow operation (1/3–1/5 rpm) mounted in the thickener. The thickener produces a relatively clean clarified liquid as the overflow, which is discharged by a circular chute at the top. There are three main types of transmission modes in thickener devices, where the first two are the most common: (1) central transmission thickeners have a small diameter, usually <24 m; (2) peripheral roller-drive thickeners are relatively common in large- and medium-sized systems, which are run using a car transmission, from which it gets its name. They usually have a diameter of ~53 m but can be up to 100 m; (3) peripheral rack-drive-type thickeners.

The main features of thickeners include (1) a degassing tank to avoid solid particles becoming attached to bubbles, like the "parachute" settling phenomenon; (2) the ore pipe is located below the liquid level in order to prevent gas being brought into the mine; (3) the mine sleeve is lowered and the feeding plate is raised to allow the slurry to be brought into the mine in an even and steady flow, effectively preventing turning flowers caused by residual pressure; (4) the internal overflow weir ensures that the material can flow adequately and prevent "short circuit" phenomenon; (5) the overflow weir has a zigzag structure to improve suction of local drainage; (6) the rake teeth design can be changed from a slash to curve shape, to make pulp center rake, and gave it back to a "backlog" force at the center, to row ore underflow concentration is high, thus increasing the processability.

4.5.4 Magnetic separation equipment

Magnetic separation systems are mainly used in magnetic concentration of minerals, such as finely ground V-bearing titanoferrous magnetite. Magnetic separation equipment consist of a magnetic system and selected boxes. Depending on the magnetic properties of the targeted mineral particles, alternating magnetic fields can be used to separate magnetite and classify magnetic minerals. Common magnetic systems include several magnetic poles, each of which consists of a permanent magnet and magnetic guide plate. The polarity of the magnetic pole is usually changed along the circumference direction, while that in the axial direction stays constant. The boxes is made of ordinary steel or hard plastic plates, and a nonconductive material is used for the parts near the magnetic system. The lower part of the boxes is the feed area, and the bottom plate is generally open with rectangular holes to discharge the tailings. The gap between the bottom plate and magnetic roll barrel can be adjusted between 30 and 40 cm. Strongly magnetic minerals (e.g., magnetite and pyrite) are separated using a weak magnetic field, while weakly magnetic minerals (e.g., hematite, siderite, ilmenite, and black tungsten ore) require the use of a strong magnetic field. The magnetic separator using a weak magnetic field consists of an open magnetic field provided by a permanent magnet. Weakly magnetic iron ore can also be magnetized to become a strong magnetic mineral in order to be separated using a system with a weak magnetic field. Magnetic separators with a strong magnetic field use a closed-circuit magnet system with multipurpose electromagnets. The main types of magnetic separator are cylindrical, belt, rotating ring, disk, and induction roller types. A magnetic pulley is used to preselect massive magnetic ore particles. Fig. 4.4 shows a schematic diagram of the magnetic extraction process.

Separation occurs within the magnetic field of the magnetic separator via both magnetic and mechanical forces. Particles with different magnetism move along different trajectories,

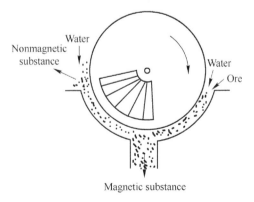

Figure 4.4
Schematic diagram of the magnetic separation process.

thus separating them into various mineral products. According to the type of magnet used, magnetic separators are classified into permanent magnet and electromagnetic iron systems. Based on the use of either dry or wet ore, magnetic separators are divided into dry iron or wet iron removal systems, respectively. Magnetic separators for sorting mineral dry allow the particles to move freely, otherwise magnetic separation can be degraded or severely affected to the point of separation not occurring. Wet magnetic separation uses high-quality ferrite materials or rare-earth magnetic steel. Magnetic rollers are commonly used, where the magnetic ore is selected by sorption processes. When the ore layer is thick, some magnetic particles in the upper part of the magnetic ore can enter the tailings as the magnetic force is small, which increases the grade of the tailings. The electromagnetic system at the top of the sorting point is mainly used to select magnetic materials. Permanent magnet systems are used alongside electromagnetic systems, mainly to keep the magnetic material on the cylinder surface, where rotation of the cylinder is carried to the weak magnetic field. This system is mainly used for dry selection of fine magnetic ore, and also for removing magnetic impurities from powdery materials.

CHAPTER 5

Extraction of vanadium from V-bearing titanoferrous magnetite by smelting to produce V-containing iron

Chapter Outline

5.1 Blast furnace smelting   90
    5.1.1 Sintering process   90
    5.1.2 Sintering in practice   93
    5.1.3 Pellet production   96
    5.1.4 Blast furnace smelting   98
5.2 Ironmaking by direct reduction   108
    5.2.1 Process characteristics   108
    5.2.2 Chemical reactions during direct reduction   109
    5.2.3 Direct reduction in practice   110
5.3 Main production equipment   116
    5.3.1 Blast furnace   116
    5.3.2 Direct reduction system   117

The selection of either pyrogenic or wet processes to extract vanadium from V-bearing titanoferrous magnetite depends on the V content of the ore (which varies because of regional differences) and the developmental stage of the extraction. Vanadium extraction by pyrogenic processes can handle raw materials with low-V grades, which are then enriched, processed, and recycled; this is an indirect process. Wet processes are direct methods of vanadium extraction, which have the advantages of short processing time and high recovery; however, the raw materials need to have a relatively high vanadium content.

Vanadium extraction by pyrogenic processes first involves smelting and reduction of vanadium from V-bearing titanoferrous magnetite to produce V-containing iron. This process includes smelting in a blast furnace, direct reduction, and then smelting reduction processes. The ore is processed in a specific atmosphere formed by reductants such as CO, H_2, and C, where reduced pig iron is obtained by a physicochemical reaction. Producing V-containing iron requires mineral sintering or pelletizing to produce a sinter or pellet, respectively, which are reduced in a blast furnace, or directly reduced to form metallic pellets, which are then smelted, separated, and thoroughly reduced in an electric furnace to produce V-containing iron [14–17,23–25,28,31,37,52,56,81].

5.1 Blast furnace smelting

Smelting in a blast furnace is a continuous process of reducing V-bearing titanoferrous magnetite into iron. Raw materials, including iron concentrate, coke, and flux, are mixed according to the desired proportion, then sintered, and pelletized. The pellets are fed into the top of the blast furnace via a charging device, where the throat of the furnace is maintained at a certain height. Air is heated to 900°C–1200°C using a stove at the bottom of the tuyere drum and blown into the blast furnaces using a blower. High-temperature combustion of the coke results in a series of physical and chemical changes in the vanadium and iron materials. The coke and ore form alternating layers in the furnace. The ore is reduced and melted gradually in the process of settling in the hearth to form V-containing iron and slag. The resulting liquid gathers at the bottom of the furnace and is released regularly from a taphole.

5.1.1 Sintering process

Sintering is the process of cementing mineral particles into lumps. Before sintering, various powdered iron materials are mixed with appropriate amounts of fuel, flux, and water. This mixture is sintered to produce a series of physical and chemical changes in the material.

5.1.1.1 Sintering characteristics

The iron concentrate formed from V-bearing titanoferrous magnetite is more stable than that of ordinary iron ore, with a less-variable iron content, and low metallogenic grade of iron, where separation of gangue minerals is more difficult. The iron concentrate of V-bearing titanoferrous magnetite from Panzhihua, China, is characterized by high ferrous, titanium, sulfur, and aluminum contents, and a low silicon content, where (CaO + MgO style)/(SiO$_2$ + Al$_2$O$_3$) is generally greater than 0.5. Because of high content of TiO$_2$ in the iron concentrate from Panzhihua ore, the probability of perovskite formation increases with increasing TiO$_2$ content. The CaO·TiO$_2$ perovskite has a high melting temperature, low surface tension, and low compressive strength. The perovskite negatively affects bonding of the ore during sintering as it weakens interactions between titanomagnetite and titanium hematite. The perovskite content is often 25% for compound ferrite (SFCA), competed calcium with silicate. This results in insufficient amounts of liquid phase being produced during sintering, which is not conducive to good consolidation. Hence, the resulting sintered material is brittle, with low strength and poor low-temperature reduction behavior. The perovskite is precipitated preferentially during sintering, filling the gaps between the grains of silicates, titanium magnetite, and hematite, resulting in poor sinter strength, and high degradation during low-temperature reduction. This is accompanied by a high porosity, low strength, and high return rate of the sinter, and high energy use during the process. Perovskites are much more likely to form than calcium ferrite in the sintering temperature range used. Stable calcium ferrite is not easily formed at high temperatures ($>1250°C$). The relationship between calcium ferrite and perovskite is reciprocal. With increasing calcium ferrite content in the iron concentrate, the perovskite content in the sinter increases, while increasing the perovskite content limits the formation of calcium ferrite and reduces the sinter strength.

In the case of V-bearing titanoferrous magnetite, the strength of the sintered ore is generally lower than that of ordinary sinter; its drum index is generally 81%–82%, while that of ordinary sinter can reach 83%–85%. After cooling, the drum index of the sinter is 6%–7% higher than before cooling, indicating that the sinter ore is more brittle when hot and less than normal sinter. In addition, the low-temperature pulverization index of the sintered ore is much higher than that of ordinary sinter, which is generally greater than 60%, and up to 80%–85%. The iron concentrate of V-bearing titanoferrous magnetite generally has a poor grain size, with weak hydrophilicity. Hence, sintering and pelletizing can be challenging.

The properties of the sinter are very sensitive to the amount of fuel used in sintering the iron concentrate of V-bearing titanoferrous magnetite, which is usually lower than that used for ordinary iron concentrate. The main reason for this is that the SiO_2 content of vanadium and titanium magnetite concentrate is low ($<6.18\%$), and few silicate minerals are produced during sintering, which consumes less heat. The FeO content in the iron concentrate of V-bearing titanoferrous magnetite is high (30.25%), and it is oxidized via an exothermic reaction during sintering. For example, the sulfur content in the iron concentrate of V-bearing titanoferrous magnetite from Panzhihua is high (0.8%), where the S is oxidized exothermically during sintering because of the particularities of smelting V-bearing titanoferrous magnetite, it must be combined in specific ratios with ordinary iron ore to control the TiO_2 content in the blast furnace slag. During sintering of the iron concentrate of V-bearing titanoferrous magnetite, a titanium hematite ($Ti-Fe_2O_3$) solid solution is formed, which is laminated with a high degree of oxidation and low FeO content in the sinter. Because sulfur in the iron concentrate of V-bearing titanoferrous magnetite is in the form of magnetite and pyrite, it is easily removed. The carbon content in the mixture is relatively low, and highly oxidizing conditions are present during sintering, which are beneficial for desulfurization. The basicity and FeO content of the sinter affect the low-temperature pulverization rate of the sinter, which is generally higher than that of the common ore. After the sinter exits the drum, the particles have a diameter distribution of 65.44% <3.15 mm and 29.31% <0.5 mm. The sinter has good storage properties; after storage for 72 h, the particle size variation is smaller, with only 1.56%–2.27% <5 mm.

5.1.1.2 Chemical reactions during sintering

5.1.1.2.1 Combustion of solid carbon

The solid carbon combustion reaction generates CO and CO_2, while some oxygen remains, providing REDOX gas and heat for other reactions. The reactions are as follows:

$$C + \frac{1}{2}O_2 = CO \tag{5.1}$$

$$CO + \frac{1}{2}O_2 = CO_2 \tag{5.2}$$

$$CO_2 + C = 2CO \tag{5.3}$$

The composition of the combustion gases depends on the sintering materials, the amount of fuel used, the degree of reduction and oxidation reactions, and the gas composition pumped through the combustion layer.

5.1.1.2.2 Decomposition and mineralization of carbonate

The carbonates in the sinter include mainly $CaCO_3$, with some $MgCO_3$, $FeCO_3$, and $MnCO_3$. Under the sintering conditions, decomposition of $CaCO_3$ begins around 720°C, and chemical boiling starts at 880°C; other carbonates have lower decomposition temperatures.

$$CaCO_3 = CaO + CO_2 \tag{5.4}$$

CaO is a decomposition product of calcium carbonate during the mineralization reaction, where CaO can react with other minerals in the sinter to form new compounds. The reactions are as follows:

$$CaCO_3 + SiO_2 = CaSiO_3 + CO_2 \tag{5.5}$$

$$CaCO_3 + Fe_2O_3 = CaO \: Fe_2O_3 + CO_2 \tag{5.6}$$

$$CaCO_3 + TiO_2 = CaO \cdot TiO_2 + CO_2 \tag{5.7}$$

If mineralization is incomplete, there will be residual free CaO, which will absorb atmospheric water during storage.

$$CaO + H_2O = Ca(OH)_2 \tag{5.8}$$

The volume of the sinter increases as it expands and becomes pulverized.

5.1.1.2.3 Decomposition, reduction, and oxidation of iron and manganese oxides

Fe_2O_3 can be decomposed under the sintering conditions when the temperature is higher than 1300°C. The dissociation pressure of Fe_3O_4 is very small under the sintering conditions, but it can also be decomposed in the presence of SiO_2 above 1300°C. As the sinter is the main raw material fed into the blast furnace, it needs to have good reducibility and high-temperature properties to improve the slag composition in the sinter, that is, reduce its SiO_2 content.

$$3Fe_2O_3 + CO = 2Fe_3O_4 + CO_2 \tag{5.9}$$

$$Fe_3O_4 + CO = 3FeO + CO_2 \tag{5.10}$$

$$4FeO + O_2 = 2Fe_2O_3 \tag{5.11}$$

$$FeS_2 = FeS + S \tag{5.12}$$

$$4FeS + 7O_2 = Fe_2O_3 + 4SO_2 \tag{5.13}$$

$$S + O_2 = SO_2 \tag{5.14}$$

5.1.1.3 Sintering operation and control

The average volume of the sinter is 3200 m³ for a sintering area of (70–90)m³/(cm² min). The vacuum level is determined by the fan capacity, exhaust system resistance, permeability of the material layer, and air loss via leakage. The thickness of the feed layer should be optimized to ensure high yield and high quality. An appropriate machine speed (generally 1.5–4 m/min) ensures that the sintering material will be thoroughly combusted at the end of the predetermined sintering cycle. To control the end point of sintering, the position of the platform is controlled. The end points of small and medium sintering systems are generally controlled in the second-to-last bellows, while that of large sintering systems is controlled in the third-to-last bellows.

After ignition, the combustion of fuel in the sintering mixture releases a large amount of heat, which melts the minerals in the charge layer. As the combustion layer moves down and cold air passes through it, the resulting molten liquid phase is cooled and recrystallized (1000°C–1100°C) and then solidified into a mesh structure of sinter. Along with the crystallization and precipitation of new minerals, the cold air flow is preheated, while the sinter is cooled, and the poorly oxidized material may be reoxidized when exposed to air. After the ignition starts, a sinter layer, combustion layer, preheating layer, drying layer, and overwetting layer appear. Finally, the last four layers disappear, and only the sinter layer remains.

5.1.2 Sintering in practice

V-bearing titanoferrous magnetite from several countries and regions is smelted in normal blast furnaces. Different additive materials and sintering control technologies need to be selected to suit the different Ti contents in the V-bearing titanoferrous magnetite. The iron concentrate of V-bearing titanoferrous magnetite from Panzhihua, China, has ~10% TiO_2. For the equivalent material produced from ore from Chengde, China, the TiO_2 content is 3.5%–7.0%, while that from Russia has ~3.5%. After sintering in the blast furnace, different burden structures are obtained.

5.1.2.1 Sintering in China

In the sintering batching process, ratio selection is that, Panzhihua iron concentrate is 41.76% Panzhihua V-bearing titanoferrous magnetite, 10% Australian iron ore, 26.03% Chinese high-grade iron ore, 2.21% rich brown powder, 4% bag filter dust, 8.5% lime, 3.88% limestone, 2% steel slag, 4.5% coke powder, and 22% returning charge. The thickness of the feed layer is 679 mm, machine speed is 1.8 m/min, sintering velocity is 20.9 mm/min, negative pressure of main circuit is 1534 kPa, +3 mm particle size in mixture 67.12%, the solid fuel consumption is 46.99 kg/t, and the drum index is 72.79%, resulting in a yield of 74.59%. The sinter ore contains 48.53% TFe, 7.94% FeO, 5.69% SiO_2, and 0.39% V_2O_5. The chemical compositions of the raw materials used in sintering processes in Pan steel, China are shown in Table 5.1.

The typical chemical composition of sinter from Pansteel, China, is shown in Table 5.2.

The typical chemical composition of the raw materials used for sintering by Chengsteel, China, is shown in Table 5.3. In accordance with the desired composition, the mixture contains 39% V-containing iron concentrate, 10% marine sand, 5% ordinary iron concentrate, 7.5% pellet return fines, 16% returns of the sintering machine, 4% steel slag,

Table 5.1: Chemical composition of materials sintered in Pan steel, China (%).

Species	TFe	FeO	CaO	SiO_2	MgO	Al_2O_3	TiO_2	V_2O_5	S	P
Panzhihua Concentrate	53.98	31.87	1.15	3.52	2.54	4.01	12.72	0.56	0.63	0.0029
Australian iron ore	61.41	0.56	0.25	3.82	0.33	2.26	0	0	0.04	0.09
Ordinary concentrate	58.59	20.32	3.22	6.67	1.97	2.30	0	0	0.67	0.065
Limonite	48.22	1.91	2.15	16.82	0.53	4.23	0	0	0.18	0.069
Steel slag	34.52	30.92	31.78	6.49	7.99	2.34	1.26	1.34		
Bag filters dust	33.86	6.10	6.34	7.20	1.72	3.28	3.16	0.19	0.35	0.036
Limestone	0.82	0	53.25	1.26	1.36	0.62	0	0	0	0
Quick lime	0.63	0.13	90.03	1.82	0.53	0.51	0	0	0	0
Active lime	0.38	0.21	86.15	1.12	0.31	0.32	0	0	0	0
Coke powder			3.01	5.22	0.97	1.58	0.045			

Table 5.2: Typical chemical composition of sinter from Pansteel, China (%).

Item	TFe	FeO	CaO	SiO_2	MgO	Al_2O_3	TiO_2	V_2O_5	CaO/SiO_2
	48.56	8.05	13.11	5.48	1.98	2.95	5.87	0.36	2.39

10% iron concentrate from Black Mountain, 4.8% lime, 3.4% light burned dolomite, and 5% coke powder. Sintering is carried out in a sintering pot, which has a size of Φ 300 mm × 600 mm. After the uniform and artificial processing of the material, the sintering raw materials are fed into the cylinder mixer for pelletizing for 10 min.

The water content in the mixture is 8%. Around 4.0 kg of finished sinter ore is placed in the bottom of the sintering pot as the base material, which is with a height of larger than 10 mm. The thickness of the feed layer is 600 mm, and the sintering negative pressure control is 1.2 kPa. Liquefied petroleum gas is used for combustion. The ignition temperature, time, and negative pressure are 1150°C, 1.0 min, and 8 kPa, respectively, where the end of sintering is defined as the time when the sintering exhaust temperature begins to decrease. Table 5.4 shows the typical mineral composition of sinter from Chengsteel, China, where the corresponding chemical composition of the sintering ore is shown in Table 5.5.

The iron concentrate from V-bearing titanoferrous magnetite in Catchikara, Russia, is usually smelted in ordinary blast furnaces and contains 60% TFe, 0.66%−3.5% TiO_2, and

Table 5.3: Chemical composition of sintered raw materials from Chengsteel, China (%).

Species	TFe	FeO	CaO	SiO_2	MgO	Al_2O_3	TiO_2	V_2O_5
Chengde iron concentrate	64.09	26.16	0.78	3.00	1.86	1.60	3.75	0.55
Black Mountain iron concentrate	60.79	29.38	0.21	2.05	1.94	3.26	7.20	0.77
Marine sand	56.58	28.30	1.77	4.07	3.81	2.61	7.50	0.56
Ordinary iron concentrate	66.78	28.03	0.68	4.05	0.86	2.19	1.71	0.16
Pellet return fines	62.57	15.10	1.35	4.55	1.12	0.90	3.24	0.40
Returns of sintering machine	53.68	8.88	9.14	4.06	4.25	2.20	3.72	0.48
Steel slag	34.52	30.92	31.78	6.49	7.99	2.34	1.26	1.34
Quick lime			39.00	1.58	27.28	0.66	0.075	
Active lime			44.76	2.48	31.35	0.55	0.050	
Coke powder			3.01	5.22	0.97	1.58	0.045	

Table 5.4: Typical mineral composition of sinter from Chengsteel, China (vol.%).

Metallic phase				Binding phase		
Magnetite	Hematite	Perovskite	Porosity	Calcium ferrite	Dicalcium silicate	Vitric
35−40	30−35	8−10	35−40	10−12	A little	8−10

Table 5.5: Typical chemical composition of sintering ore from Chengsteel, China (%).

No.	TFe	FeO	CaO	SiO_2	MgO	Al_2O_3	TiO_2	V_2O_5	CaO/SiO_2
X-2	55.68	13.14	8.93	4.64	3.88	1.85	3.58	0.44	1.92

Table 5.6: Typical chemical composition of sinter produced in Catchikara, Russia.

Composition	TFe	FeO	CaO	SiO$_2$	TiO$_2$	V$_2$O$_5$
1	54.16	11.11	10.56	4.88		0.510
2	54.31	10.73	10.48	4.76		0.509

0.95%–3.3% V$_2$O$_5$. Including the auxiliary materials added during sintering, the typical chemical composition of the sinter produced in Catchikara, Russia is shown in Table 5.6.

The typical chemical composition of the sinter produced in Chousoff, Russia, is shown in Table 5.7.

5.1.3 Pellet production

Pellets of iron concentrate from V-bearing titanoferrous magnetite are produced by high-temperature oxidation and roasting, which are then used as raw materials for ironmaking in blast furnaces. Pellet production includes two main processes: molding and consolidation during roasting. Molding involves the use of a disk pelletizer or drum pelletizer to form green pellets from the iron ore concentrate, which has a certain density and strength. Roasting consolidation is the most complex step in the production process as many physical and chemical reactions occur that affect the metallurgical properties of the pellets, such as the strength, porosity, and degree of reduction. The sintering pellets are exposed to high-temperature metallurgical processes to prepare raw materials with various chemical compositions and metallurgical properties. Controlling the blast furnace conditions results in particles with a certain size and strength, chemical stability, and good metallurgical properties.

5.1.3.1 Pellet characteristics

There are three main types of equipment used to roast pellets: shaft furnaces, belt calciners, and chain-grate rotary kilns. All of these systems include drying, preheating, soaking, and cooling processes. Unlike the sintering process, pellet production has the following characteristics: (1) strict raw material requirements, where the raw material should be relatively pure and finely ground, with a specific surface area of 1500–1900 cm^2/g. The moisture should be below the critical limit, and the SiO$_2$ content should not be too high. (2) Pellets that are too dense and have high moisture content can rupture or burst at high temperature, so a drying and preheating process is required before high-temperature roasting. (3) The pellets should have a uniform size and shape, and good air permeability.

Table 5.7: Typical chemical composition of sinter from Chousoff, Russia.

Composition	TFe	FeO	CaO	SiO$_2$	TiO$_2$	V$_2$O$_5$	MnO	Cr$_2$O$_3$	P	C
1	50.98	30.11	9.83	7.29	2.08	0.91	2.14	1.75	0.024	0.30
2	50.1	27.21	10.22	7.68	2.18	0.91	2.21	1.70	0.026	0.40

It is common to use a low-negative-pressure fan when producing pellets in a rotary kiln with a belt-type or grate-chain-type roasting machine. (4) There is no solid fuel in most of the pellets, so roasting requires a large amount of heat, which is provided by liquid or gas fuel from burning coal. To maximize the heat utilization rate, hot waste gas from the bed after heating, or recycled hot flue gas can be fed into the pellet material layer.

5.1.3.2 Control of pellet production

The pellets are made from iron concentrate from V-bearing titanoferrous magnetite and magnetite concentrate in proportions of 80% and 20%, respectively. Bentonite is used as a binder in the pellets. The fuel is usually anthracite and gas. The addition of 1%–2% water to the pelleting process is common to optimize the moisture content of the mixture. The temperature of the drying process is generally 200°C–400°C, which is mainly for evaporating moisture, and ensuring that crystallized water is expelled from the pellets. The temperature of the preheating process is 900°C–1000°C. The small amounts of water not eliminated in the drying process are removed in this step. The main reactions in this process are the oxidation of magnetite to hematite, the decomposition of carbonate minerals, the decomposition and oxidation of sulfides, and some solid-state reactions. The temperature of the calcining zone is generally 1200°C–1300°C. Reactions that have not yet been completed in the preheating process, such as decomposition, oxidation, desulfurization, and solid-phase reaction, should also continue in this process, where the chemical reactions are similar to those occurring during sintering.

The major reactions are iron oxide crystallization and recrystallization, grain growth, and solid-phase reactions in which low-melting-point compounds dissolve and some liquid phase forms. Therefore the pellets shrink and densify. The furnace temperature should be slightly below the roasting temperature. The main purpose of this process is to allow complete growth of crystals within the pellet to homogenize the mineral composition and minimize internal stresses. The cooling phase reduces the temperature of the pellets from above 1000°C to lower temperatures that the transportation belt can withstand. Air is used as the cooling medium, which has a high oxidizing potential; hence, any nonoxidized magnetite inside the pellet will be oxidized during cooling. Table 5.8 shows the typical chemical composition of pellets from Chengde and Panzhihua, China.

Common equipment for pellet roasting includes vertical furnaces, belt roasters, and chain-grate rotary systems. Table 5.9 shows the typical chemical composition of pellets from Catchikara, Russia.

Table 5.8: Typical chemical components of pellets in Chengde and Panzhihua.

Origin	TFe	FeO	CaO	SiO_2	MgO	Al_2O_3	TiO_2	V_2O_5
Chengde	62.57	15.10	1.35	4.55	1.12	0.90	3.24	0.40
Panzhihua	53.96	3.23	3.63	5.01	3.32	0.78	9.02	0.58

Table 5.9: Typical chemical composition of pellets from Catchikara, Russia.

Sample	TFe	FeO	CaO	SiO$_2$	TiO$_2$	V$_2$O$_5$
1	60.45	2.62	1.25	4.62		0.567
2	60.39	2.70	1.20	4.31		0.566

5.1.4 Blast furnace smelting

5.1.4.1 Smelting characteristics

The raw material used for blast furnace smelting is mainly composed of iron ore, fuel (coke), and flux (limestone). Generally, the smelting of 1 t of pig iron requires 1.5–2.0 t of iron ore, 0.4–0.6 t of coke, 0.2–0.4 t of flux, giving a total of 2–3 t of raw materials. If the sinter grade is increased by 1%, the coke ratio can be reduced by 2%, resulting in a 3% increase in yield from the blast furnace. After granulation, the sinter output can be increased by 5.5%, while the coke ratio is reduced by 3.3%, the sinter basicity drops by 0.1, the coke ratio of the blast furnace increases, and the output of pig iron is reduced by 3.5%. When using sinter powder with a size <5 mm, a 1% increase can reduce blast furnace production by 6%–8%, with a higher coke ratio. Decreasing the FeO content of the sinter by 1%, the change in intensity is not big, reduces the coke ratio by 1%. Sintering with the addition of 3% MgO can improve the slagging system and stabilize blast furnace operation.

The rate of gas reducing iron ore in blast furnace is affected by many factors, including the properties of the ore (e.g., particle size and porosity), along with the gas composition, gas velocity, and reduction temperature. The gas–solid reduction process includes the following basic steps: (1) the reduced gas diffuses to the ore surface via a gas film on the ore grain surface; (2) the reduced gas diffuses into the ore through the reduced metal layer; (3) chemical reactions occur at the interface between metallic iron and wüstite; (4) the reduced gas product diffuses outward through the reduced metal layer; and (5) the reduced gas diffuses outwards through the gas film on the surface.

The main ingredients of blast furnace slag are derived from the gangue minerals in the raw ore. A quaternary (CaO–MgO–SiO$_2$, Al$_2$O$_3$) slag system is formed in the process of smelting ordinary iron ore, while smelting V-bearing titanoferrous magnetite concentrate is a pentabasic (CaO–MgO–SiO$_2$–Al$_2$O$_3$–TiO$_2$) slag system. Compared with quaternary slag, pentabasic slag is characterized by a higher melting temperature, formation of foam slag, thickening of the slag, and difficulties in removing sulfur. The melting temperature of pentabasic slag with low-Ti content is similar to quaternary slag from smelting ordinary iron ore, while the formation of foamy slag in the smelting of pentabasic slag with high-Ti content is enhanced. The thickening of the furnace slag occurs during the reduction process in the blast furnace. Some TiO_2 in the slag is reduced to produce titanium carbides and nitrides. The melting points of TiC and TiN are 3140°C ± 90°C and 2950°C ± 50°C, respectively, which are much higher than the maximum temperature obtained in the furnace. The desulfurization ability of high-Ti slag is much lower than that of ordinary slag, with an Ls value of only 5–9.

Adhesion of some compounds on the sides of the reaction vessel occurs as the molten iron contains oxides of vanadium and titanium with higher melting points than the tapping temperature. The layer of adhered material cannot be melted by the molten iron in the subsequent processing batch, resulting in an increasingly thick layer. Hence, the volume of the reaction vessel rapidly decreases, and it can only be used a few times before the adhesion layer seriously affects normal use and circulation. This is a challenge for the design and operation of blast furnaces.

5.1.4.2 Chemical reactions during smelting

Fig. 5.1 shows the Gibbs free energy diagram of the oxide. Reduction of iron oxide in the blast furnace is mainly because of CO reducing the iron oxide to generate CO_2 and low-valence iron, or the process of generating H_2O and low-valence iron by hydrogen reduction. The reduction sequence is Fe_2O_3, Fe_3O_4, FeO, and Fe. Below 570°C, FeO is unstable, and the reduction sequence is Fe_2O_3–Fe_3O_4–Fe. The main reduction reactions of iron oxide are as follows:

$$3Fe_2O_3 + CO \rightarrow 2\ Fe_3O_4 + CO_2 \tag{5.15}$$

$$Fe_3O_4 + CO \rightarrow 3FeO + CO_2 \tag{5.16}$$

$$FeO + CO \rightarrow Fe + CO_2 \tag{5.17}$$

$$3Fe_2O_3 + H_2 \rightarrow 2\ Fe_3O_4 + H_2O \tag{5.18}$$

$$Fe_3O_4 + H_2 \rightarrow 3FeO + H_2O \tag{5.19}$$

$$FeO + H_2 \rightarrow Fe + H_2O \tag{5.20}$$

When H_2 and CO are used as reducing agents, they are restricted by water and gas reactions:

$$H_2 + CO_2 \rightarrow H_2O + CO \tag{5.21}$$

Direct reduction: When the material is first exposed to a high temperature of 850°C, because of the presence of a large amount of coke, the generated CO_2 and H_2O react and are immediately converted into CO and H_2:

$$CO_2 + C \rightarrow 2CO \tag{5.22}$$

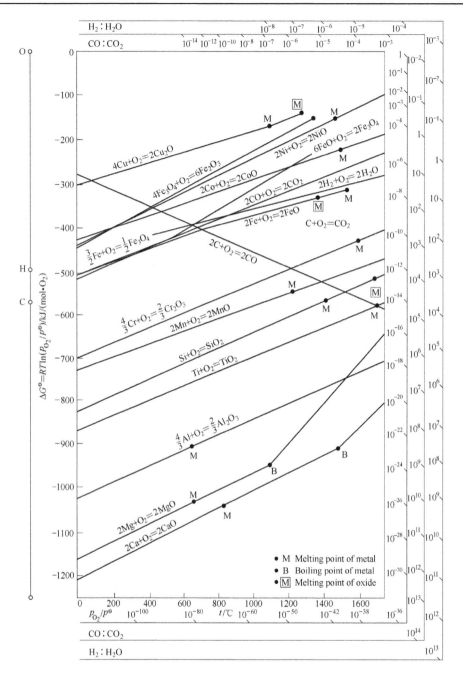

Figure 5.1
Gibbs free energy diagram of the oxide.

$$H_2O + C \rightarrow H_2 + CO \tag{5.23}$$

The carbon in the molten iron also participates in reduction and carburization reactions in the blast furnace:

$$3Fe + CO \rightarrow Fe_3C + CO_2 \tag{5.24}$$

$$FeO(MnO,\ SiO_2) + C \rightarrow Fe(Mn,\ Si) + CO \tag{5.25}$$

$$3Fe + C \rightarrow Fe_3C \tag{5.26}$$

The gangue, coke ash, and ore entering the blast furnace contain other oxides (e.g., SiO_2, Al_2O_3, CaO, and MgO), sulfides (FeS_2), and phosphate [$Ca(PO_4)_2$]. Some symbiosis iron ore also contains manganese, titanium, chromium, vanadium, copper, cobalt, nickel, niobium, arsenic, potassium, sodium, and other oxygenated compounds, and small amounts of sulfides. The various oxides have different chemical stability, which determines whether they are fully or partly reduced in the blast furnace, where those that are not reduced at all enter the slag. Silicon is more difficult to reduce than iron, and can only be reduced by carbon at high temperature and enters the liquid phase in the molten iron.

$$(SiO_2) + 2[C] \rightarrow [Si] + 2CO \tag{5.27}$$

Most Si in the cast iron is from reduction of SiO_2 from the coke ash or slag, which passes through the tuyere at high temperature (above 1700°C), where the reaction occurs to first generate gaseous SiO, which is reduced to silicon on the rise again and enters the molten iron.

Manganese compounds in the ore, such as MnO_2, Mn_3O_4, Mn_2O_3, and $MnCO_3$, are easily reduced by CO to form MnO, which is directly reduced by carbon in the furnace slag and enters the molten iron:

$$(MnO) + [C] \rightarrow [Mn] + CO \tag{5.28}$$

Phosphorus enters the blast furnace in the form of $3CaO \cdot P_2O_5$ or $3FeO \cdot P_2O_5$, while other metals, such as copper, nickel, cobalt, arsenic, and lead exist as oxides or sulfides and are all reduced. Generally, 75%–80% of the oxides of vanadium, niobium, and chromium can be reduced.

$$(TiO_2) + [C] \rightarrow [TiC] + CO \tag{5.29}$$

Any elements that can form carbides and dissolve in the molten iron, such as manganese, vanadium, chromium, and niobium, can increase the carbon content of the molten iron. Elements such as silicon, phosphorous, and sulfur, which induce the decomposition of carbides in molten iron, will prevent its carburization. Ordinary pig iron contains about 4%

carbon. When the carbides dissolved in the molten iron reach saturation, the remaining carbides remain in the slag, such as SiC when high-Si pig iron is smelted, and TiC formed when the ore contains TiO_2. The melting temperatures of carbides are generally high (melting point of TiC $>$ 3290°C, and SiC $>$ 2700°C). Hence, mixing carbides with the solid phase in the slag makes the slag liquidity difficult to refine.

The raw material for smelting V-bearing titanoferrous magnetite in a blast furnace is the sinter of V-bearing titanoferrous magnetite, which contains haplotypite, titanomagnetite, perovskite, and titanosilicate phases, as well as a small amount of calcium ferrite, pseudobrookite, and residual ilmenite. In the process of the sinter falling from the throat to the belly of the blast furnace, the basic reactions and changes in phase composition of the smelt are achieved sequentially in the different temperature zones.

The compositions of V-containing pig iron from Panzhihua and Chengde are shown in Table 5.10, where the two compositions of the molten iron are similar.

The chemical compositions of vanadium pig iron from the Nizhny Tagil and Chousoff plants in Russia are shown in Tables 5.11 and 5.12, respectively.

5.1.4.3 Reduction of vanadium

In sinter containing vanadium and titanium, vanadium enters the iron concentrate as the vanadium spinel ($FeO \cdot V_2O_3$). The reduction reaction of vanadium is as follows:

Table 5.10: Composition of V-containing pig iron from Panzhihua and Chengde (%).

	C	V	Si	Mn	Ti	S	P	Cr
Panzhihua	4.25	0.32	0.18	1.11	0.12	0.065	0.055	<0.10
Chengde	3.8–4.5	0.25–0.5	0.20–0.80	0.10–0.70	0.10–0.50	≤0.08	≤0.15	<0.10

Table 5.11: Chemical composition of vanadium pig iron from the Nizhny Tagil, Russia.

C	V	Si	Mn	Ti	S	P	Cr
4.52	0.462	0.12	0.30	0.18	0.022	0.06	<0.10
4.55	0.465	0.10	0.33	0.16	0.021	0.07	<0.10
4.61	0.451	0.25	0.27	0.16	0.030	0.045	0.08

Table 5.12: Chemical composition of vanadium pig iron from Chousoff, Russia.

V	Si	Mn	Ti	S	P	Cr
0.482	0.264	0.387	0.22	0.033	0.036	0.289
0.481	0.290	0.360	0.26	0.033	0.039	0.240

$$FeO \cdot V_2O_3 + C \rightarrow Fe + 2VO + CO\uparrow \quad (5.30)$$

In the drop zone, because of the formation of liquid iron phase, the reduction of vanadium occurs via the following reaction:

$$FeO \cdot V_2O_3 + 4C \rightarrow 2[V] + Fe + 4CO\uparrow \quad (5.31)$$

$$VO + C \rightarrow [V] + CO\uparrow \quad (5.32)$$

$$V_2O_3 + 3C \rightarrow 2[V] + 3\ CO\uparrow \quad (5.33)$$

The reduction rate of vanadium is mainly limited by the thermodynamics, which is controlled by the furnace temperature and depends on the basicity of the slag. During smelting the reduction rate of V is mainly determined by the furnace temperature after the selected five-element slag system with different TiO_2 content. Second, the reduction rate of V decreases with increasing TiO_2 content in the slag. The actual rate of vanadium recovery from smelting is in the case of low-titanium slag smelting. During smelting of titanium slag, low-titanium, medium-titanium, and high-titanium slag give η_v values of 80%, 75%, and 70%, respectively.

5.1.4.4 Reduction of titanium

The blast furnace can be divided into three temperature zones from the throat to the furnace body, which vary over the temperature range of 650°C–900°C. In addition to the indirect reduction of Fe_2O_3, Fe_3O_4, FeO, and calcium ferrite, haplotypite, titanium magnetite, and hematite begin to lose oxygen, as shown in the following chemical reactions:

$$MFe_3O_4 + n(Fe_2TiO_4) \rightarrow mFe_3O_4 \cdot nFe_2TiO_4 \quad (5.34)$$

$$Fe_2O_3 \cdot TiO_2 + CO \rightarrow 2FeO \cdot TiO_2 + CO_2\uparrow \quad (5.35)$$

$$FeO \cdot TiO_2 + FeO \rightarrow 2FeO \cdot TiO_2 \quad (5.36)$$

The reaction product involves magnetite, wüstite, and a small amount of fine iron grains. In the middle of the furnace, with a temperature range of 900°C–1150°C, magnetite is reduced, where the main chemical reactions are as follows:

$$mFe_3O_4 \cdot N[2(Fe,\ Mg,\ Mn)O \cdot TiO_2 \cdot FeO \cdot V_2O_3] + 3mCO \rightarrow 3mFeO \\ + n[2(Fe,\ Mg,\ Mn) \cdot TiO_2 \cdot FeO \cdot V_2O_3] + 3mCO_2\uparrow \quad (5.37)$$

$$FeO + CO \rightarrow Fe + CO_2\uparrow \quad (5.38)$$

$$FeO \cdot TiO_2 + FeO \rightarrow 2FeO \cdot TiO_2 \quad (5.39)$$

After the reaction the solid solution of pontoid and ulvite with some wüstite is reduced to metal iron. The temperature range of 1150°C–1250°C in the furnace body allows reduction decomposition of ulvite to occur, where the main chemical reactions are as follows:

$$FeO + CO \rightarrow Fe + CO_2 \uparrow \tag{5.40}$$

$$2FeO \cdot TiO_2 + CO \rightarrow Fe + FeO \cdot TiO_2 + CO_2 \uparrow \tag{5.41}$$

$$FeO \cdot TiO_2 + CO \rightarrow Fe + TiO_2 + CO_2 \uparrow \tag{5.42}$$

$$TiO_2 + CaO \rightarrow CaO \cdot TiO_2 \tag{5.43}$$

The phase formed after the reaction is composed of metallic iron, ulvite, and small amounts of wüstite, ilmenite, brookite solid solution, and perovskite.

The soft-melting zone (1250°C–1350°C) occurs from the bosh to the bottom of the furnace body. In this region, direct reduction begins, where the formation of the soft-melting sinter is characterized by the bond content. The initial slag in the lower part of the soft-melting zone begins to form, and the iron grains are polymerized. The main chemical reactions are as follows:

$$FeO \cdot 2TiO_2 + C \rightarrow Fe + 2TiO_2 + CO \uparrow \tag{5.44}$$

$$MgO \cdot TiO_2 + TiO_2 \rightarrow MgO \cdot 2TiO_2 \tag{5.45}$$

$$CaO \cdot FeO \cdot SiO_2 + C \rightarrow CaO \cdot SiO_2 + Fe + CO \uparrow \tag{5.46}$$

$$Al_2O_3 + TiO_2 \rightarrow Al_2O_3 \cdot TiO_2 \tag{5.47}$$

$$Al_2O_3 + MgO \rightarrow MgO \cdot Al_2O_3 \tag{5.48}$$

$MgO \cdot 2TiO_2$ and $Al_2O_3 \cdot TiO_2$ produced by the reaction and silicate from the original sinter should generate low-melting furnace slag phases, such as titania, where the metal iron diffuses and crystallizes into large iron beads in the slag. Dripping with reaction from the bosh to the tuyere area occurs above 1350°C, where metallic iron is carburized and early slag melting temperature drop, slag and iron melt dripping. Smelting of V-containing iron concentrate and titanium in the drop zone is characterized by reduction of titanium oxide and vanadium oxide by carbon. According to the change of the free energy of reaction, the main reactions occurring at the slag coke interface are as follows:

$$2TiO_2 + C \rightarrow Ti_2O_3 + CO \uparrow \tag{5.49}$$

$$TiO_2 + C \rightarrow TiO + CO \uparrow \tag{5.50}$$

$$TiO_2 + 3C \rightarrow TiC + 2CO \uparrow \tag{5.51}$$

$$TiO_2 + \frac{1}{2}N_2 + 2C \rightarrow TiN + 2CO \uparrow \tag{5.52}$$

$$TiO_2 + C \rightarrow [Ti] + CO\uparrow \tag{5.53}$$

The TiC and TiN generated by the reaction are dispersed in the slag in the form of a solid solution. The TiC/TiN ratio in the Ti(C, N) solid solution is related to the temperature at the time of formation and the partial pressure of nitrogen. In general rule, with increasing temperature, the partial pressure of nitrogen increases, and the proportion of TiN in the Ti(C, N) solid solution increases, followed by the formation of Ti(C, N) solid solution with different tones. Then, [Ti] and [V] enter the molten iron phase. The reactions at the slag iron interface mainly occur between saturated carbon from the molten iron and TiO_2 in the slag:

$$TiO_2 + 3[C] \rightarrow TiC + 2CO\uparrow \tag{5.54}$$

$$TiO_2 + \frac{1}{2}N_2 + 2[C] \rightarrow TiN + 2CO\uparrow \tag{5.55}$$

Ti(C, N) generated after the reaction and Ti(C, N) dispersed in the slag are adsorbed on the iron beads in the slag to form a thin Ti(C, N) shell, which impedes their aggregation and growth; this is a major reason for the high iron loss of titanium slag. During the dropping process, the molten iron phase gradually forms melted iron that is over-saturated with carbon and nitrogen as the temperature decreases in the furnace cylinder. Because of the solubility, the following precipitation reactions for Ti(C, N) can occur:

$$[Ti] + [C] \rightarrow TiC \tag{5.56}$$

$$[Ti] + [N] \rightarrow TiN \tag{5.57}$$

The Ti(C, N) solid solution generated by the precipitation reaction and repeated deslagging and tapping of the iron result in a Ti(C, N) slag deposit at the bottom of the furnace, which has a high melting point and different tones, which is beneficial for furnace protection. If not controlled, the bottom will rise and the hearth will pile up.

5.1.4.5 Characteristics of blast furnace smelting

The raw fuel requirements are more stringent for vanadium production than for ordinary iron ore smelting. Coke strength of M_{40} is required to reach 76%, while that of M_{10} is less than 9%. The sulfur content should be below 0.50% to ensure that the sulfur load is less than 4 kg/t pig iron. The TiO_2 content in the V-containing concentrate is controlled, with a maximum of 13% and minimum of 0.50%. The sinter must be screened before entering the furnace.

Various problems are encountered during smelting, such as the thickening of the slag in the blast furnace. China and Russia use blast furnaces to smelt V-containing iron concentrate. The Nizhny Tagil and Chousoff factories in Russia restrict the TiO_2 content in the smelting slag to less than 10%. The Maanshan iron and steel companies in China restrict the TiO_2

content in the smelting slag to less than 8%. Typical components of the low-titanium slag from Russia's Nizhny Tagil and Chousoff factories are shown in Table 5.13. Small amounts of V-containing concentrates are used in the Chongqing, Shuicheng, and Kunming iron and steel companies in China, where the TiO_2 content of the blast furnace slag is ~3.5%. The typical composition of high-Ti blast furnace slag used by the Panzhihua Iron and Steel Company is shown in Table 5.14. The TiO_2 content in the slag of a Chengde steel plant is 16%–18%. Limiting the amount of Ti in iron during the smelting process can restrict the production of Ti(C, N), which inhibits the formation of foam slag and prevents the furnace slag from thickening.

The selected slag in the process of smelting V-containing iron concentrate in blast furnaces is required for homogeneous slag and effective desulfurization, which is favorable for reduction of V-containing ore to produce iron. Similar to the smelting of common ore slag, alkaline substances (such as CaO and MgO style) can reduce the activity of TiO_2 and SiO_2, increase the fluidity of slag, react with TiO_2 to form perovskites and lower the activity of TiO_2 in the slag, and inhibit Ti compounds formed at a lower valence, which is good for desulfurization and reduction of vanadium into the iron. The degree of slag basicity (CaO/SiO_2) with medium-titanium content is 0.9–1.0 and the melting temperature is below 1325°C, which is favorable for slag and iron production, although the desulfurization ability is poor. To achieve desulfurization and favorable reduction of vanadium, it is necessary to increase the degree of basicity of the slag; however, when the slag basicity (CaO/SiO_2) increases to 1.33, the melting temperature increases to 1350°C. In smelting of V-containing iron concentrate in blast furnaces, the key problem is the low desulfurization ability, where a particular challenge is thickening of the Ti-containing slag and high iron loss. The key technology of smelting Ti-bearing magnetite is controlling the excess titanium and inhibiting the production of Ti(C, N) to prevent thickening of the slag. Higher TiO_2 contents in the metallurgical slag result in enhanced thickening effect. In general, the slags

Table 5.13: Typical components of the low-titanium slag from Russia's Nizhny Tagil and Chousoff factories (%).

Chemical composition	TiO_2	CaO	SiO_2	Al_2O_3	MgO	FeO	S	V_2O_5	Ratio of slag to iron
	7.9	32.8	28.5	15.3	11.7	0.75	0.63	0.13	0.440 t slag/t iron

Table 5.14: The typical composition of high-Ti blast furnace slag used by the Panzhihua Iron and Steel Company (%).

Chemical composition	TiO_2	CaO	SiO_2	Al_2O_3	MgO	TFe	MnO	Sc_2O_3	V_2O_5	Ratio of slag to iron
	23.9	26.10	23.4	13.1	8.10	2.20	1.0	20 (g/t)	0.35	0.750 t slag/t iron

are classified by their TiO_2 contents as low-Ti slag (no more than 10% TiO_2), medium-Ti slag (10%−20% TiO_2), and high-Ti slag (no less than 20% TiO_2). In blast furnace ironmaking, the furnace temperature is generally discussed relative to the [Si]% content in the iron. Similarly for smelting V-containing iron concentrate in blast furnaces, as TiO_2 and SiO_2 have similar properties, the furnace temperature is expressed as the content of [Si + Ti]% in the iron.

Low-Ti slag will thicken slowly with increasing furnace temperature during smelting. Under inappropriate operating conditions, the fluidity of the slag will become poor, even to the point that the slag is difficult to separate, resulting in ineffective smelting. When smelting low-titanium slag in Russia, the [Ti] and [Si] contents in the iron are limited to around 0.30%, which can prevent thickening. Combined aeration and natural gas blasts can be used to limit the [Si] and [Ti] content to about 0.20%, which improves the technical and economic indexes of low-titanium slag smelting and achieves performance close to that of normal smelting under the same conditions.

When smelting low-Ti slag in China, appropriate control of the furnace temperature is performed to prevent the titanium slag from thickening, control iron loss, and enhance desulfurization ability. The suitable temperature range to be controlled is Σ [Si + Ti] = 0.63%−0.75%, [Si] = 0.38%−0.45%, and [Ti] = 0.25%−0.30% to protect the blast furnace and extend its life, the vanadium titanium ferro mineral containing TiO_2 was added to the blast furnace. TiO_2 in the blast furnace slag reached 1.5%−3.0%, forming a high-strength carbonitride consolidation layer in the hearth and bottom of the furnace, which adhered to the furnace wall.

The main problem of smelting medium-Ti slag is that this slag thickens faster than low-Ti slag. The proper range of furnace temperatures for smelting 13%−15% TiO_2 slag in China is to ensure [Si] = 0.25%−0.31%, [Ti] = 0.20%−0.25%, and [Si + Ti] = 0.45%−0.56%. If [Ti] in iron reaches 0.30%, the iron loss will be increased, even if the slag does not obviously thicken. If [Ti] is greater than 0.30%, the slag will thicken. For smelting titanium slag in a blast furnace with a volume of 1000 m^3 or above, dedensification measures should also be taken.

The main problem when smelting high-Ti slag is that the slag thickens fast, iron loss is high, the slag desulfurization capacity is low, and slag foaming can occur. In general, high-titanium slag and a suitable melting temperature are selected. The high-titanium slag has a high melting point, short slag, and strong crystallization. For a range of MgO and Al_2O_3 concentrations, the SiO_2 content remains unchanged, and the melting temperature increases with increasing TiO_2 content and increasing CaO/SiO_2 ratio. To be suitable for smelting, it is necessary to control the two ratios SiO_2/TiO_2 and CaO/SiO_2, to ensure that the melting temperature of the slag is in the low-melting zone. The typical SiO_2/TiO_2 content used in Chinese Panzhihua steel production is 1, and CaO/SiO_2 is 1.07−1.13. The melting temperature is about 1380°C−1400°C, which is within the lower melting zone of the high-titanium five-element slag system. In addition, a reasonable distribution of dispensing gas flow and active furnace cylinder work is a very important factor for high-titanium slag smelting and prevention of thickening. The control process is the same as that used for ordinary ore. According to the softening temperature of high-V titanium sinter, it is suitable for gas flow to maintain bosh area edge, to maintain the soft-melting zone and enhance fusion ability, and prevent the central junction thick stroke disorder caused by blast furnace smelting. Smelting applications shows that the CO_2 content at the edge should be 2%−3% higher than at the center, and that the furnace is active and can maintain long-term stability. Finally, the choice of a stable furnace temperature range is the key to preventing foaming and thickening of the furnace slag. The relationship between iron [Ti] and [Si] is [Ti] = 1.353[Si] + 0.012. Under normal smelting conditions the content of [Ti] is greater than the [Si] content; hence, high-Ti slag smelting is mainly dependent on the changes in iron [Ti] content. In the case of [Ti] < [Si], this characteristic is an important factor influencing the accumulation of high-Ti furnace slag at the furnace cylinder center.

5.2 Ironmaking by direct reduction

Direct reduction of iron ore is used to reduce iron ore into iron through solid reduction at temperatures below the melting point of the ore. This iron preserves a large amount of micropores formed when oxygen is lost and appears like a sponge under the microscope; hence, it is also known as sponge iron. Sponges in the form of pellets are also called metallized pellets. Similar to steel, directly reduced iron is characterized by low carbon and silicon contents. This material is also used in the steelmaking process, which is usually referred to as the process of direct reduction of ironmaking.

5.2.1 Process characteristics

The main components of V-containing iron concentrate are TiO_2 and FeO, while the remainder is SiO_2, CaO, and MgO, Al_2O_3, and V_2O_5. Ironmaking by direct reduction is via the use of carbon pellets. Under high temperature reducing conditions, iron oxide reacts with carbon, titanium slag, and metallic pellets formed via solid-state reactions. Reduction smelting is carried out in electric furnaces by melting separation or depth, where vanadium steel and low-grade titanium slag are obtained. Because of differences in the concentration and melting point, effective separation of titanium slag and metal iron is achieved.

Currently, there are three kinds of direct reduction processes, including the tunnel kiln, rotary kiln, and rotary hearth furnace processes, where all use coal-based reduction methods. V-containing iron concentrate is mixed with the coking coal as reductant, then a binder is added to prepare coal-based reduction pellets that are reduced at high temperature under a reducing atmosphere. The oxides of iron, vanadium, titanium, and silicon in magnetite are reduced and metallized pellets containing iron are formed. According to the toward of vanadium, there are two main electric furnace processes. First, melting and phase separation are performed, in which the V and Ti enter the slag, which is used as the material for extracting vanadium and titanium. Second, melting and deep reduction in the electric furnace, where the vanadium is reduced and enters the molten iron, and the obtained titanium slag is used as the raw material for titanium extraction. Alternatively, metallized pellets are finely ground to extract metallic iron, where the extracted iron slag is melted and separated to obtain molten V-containing iron. The grinding waste for carbon and recycling can be used as a low-grade titanium raw material. Processes, including chain gas machine, rotary kiln prereduction, cooling magnetic separation, thermal reduction of the ore-smelting electric furnace, oxygen blowing vanadium of converter, and semisteel steelmaking, are the most mature production processes used in practice. A typical process flow is shown in Fig. 5.2.

5.2.2 Chemical reactions during direct reduction

The solid-state reduction of V-containing iron concentrate is a stepwise process, following $Fe_2O_3 \rightarrow Fe_3O_4 \rightarrow FeO \rightarrow Fe$, which is accompanied by the partial reduction of Ti^{4+}, while the mineral components MgO and MnO dissolve in M_3O_5 solutes (M = Fe, Ti, Mg, Mn). The reduction of iron oxides is retarded, and the final product of the direct reduction is a mixture of metal iron, reducible rutile, and M_3O_5 solid solutes. The main reduction reactions involving iron oxides are as follows:

$$3Fe_2O_3 + CO \rightarrow 2\,Fe_3O_4 + CO_2 \tag{5.58}$$

$$Fe_2O_3 + C = 2FeO + CO \tag{5.59}$$

$$Fe_3O_4 + CO \rightarrow 3FeO + CO_2 \tag{5.60}$$

$$FeO + CO \rightarrow Fe + CO_2 \tag{5.61}$$

$$3Fe_2O_3 + H_2 \rightarrow 2Fe_3O_4 + H_2O \tag{5.62}$$

$$Fe_3O_4 + H_2 \rightarrow 3FeO + H_2O \tag{5.63}$$

$$FeO + H_2 \rightarrow Fe + H_2O \tag{5.64}$$

$$CO_2 + C \rightarrow 2CO \tag{5.65}$$

$$H_2O + C \rightarrow H_2 + CO \tag{5.66}$$

$$FeO(MnO,\ SiO_2) + C \rightarrow Fe(Mn,\ Si) + CO \tag{5.67}$$

$$V_2O_3 + 3C = V + 3CO \tag{5.68}$$

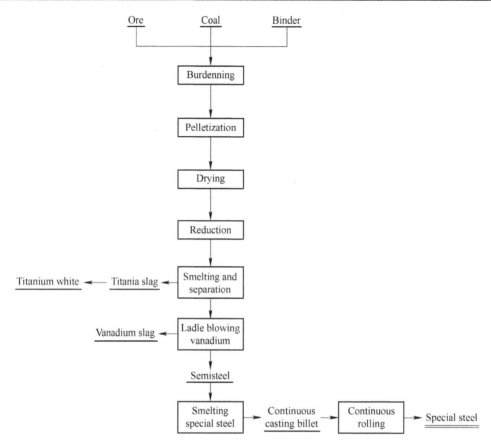

Figure 5.2
Typical process flow of ironmaking by direct reduction.

$$V_2O_5 + 3C = V + 5CO \tag{5.69}$$

$$MO + CO \rightarrow M_3O_5 \tag{5.70}$$

Here, M denotes the metals Fe, Ti, Mg, or Mn. The mineral-phase reactions occurring during direct reduction of V-containing iron concentrate are shown in Table 5.15.

5.2.3 Direct reduction in practice

Direct reduction is commonly applied in South Africa, while other such plants are experimental and limited in size. The development of direct reduction in Panzhihua, China, is unbalanced. Different direct reduction systems are used depending on the scale. Pilot-scale devices include fluidized furnaces, while scaled-up equipment includes rotary

Table 5.15: Mineral-phase reactions during direct reduction of V-containing iron concentrate.

No.	Chemical reaction	Details
1	$3Fe_2O_3 + CO \rightarrow 2Fe_3O_4 + CO_2$	Hematite is reduced to magnetite
2	$Fe_3O_4 + CO \rightarrow 3FeO + CO_2$	Magnetite is reduced to wüstite
3	$XFeO + yCO \rightarrow xFe + yCO_2$	Wüstite is reduced to metallic iron
	$XFeO + (x-y)FeO \cdot TiO_2 \rightarrow (x-y)Fe_2TiO_4$	Some wüstite and ilmenite react to form titanium ferrite
4	$[mFe, nMg]TiO_4 + qCO \rightarrow [m-q]Fe, nMg]_2 TiO_4 + qFe + 1/2qTiO_2 + qCO_2$	Part of the iron in the MgO-containing titanium ferrite is reduced, and N_{MgO} is increased to form Mg-rich titanium ferrite, while some TiO_2 is precipitated
5	$[(m-q)Fe, nMg]_2 TiO_4 + q'CO + qTiO_2 \rightarrow (1+q)[(m-q-q')Fe, Mg]TiO_3 + q'Fe + CO_2$	FeO in Mg-rich titanium ferrite continues to be reduced, when $(m-q-q') = 1+q$, it is converted into magnesium ilmenite
6	$[(m-q-q')Fe, Mg] \cdot TiO_3 + \tilde{q}CO \rightarrow 1/2[(M-q-q'-\tilde{q})Fe, nMg]T_2O_5 + q'Fe + \tilde{q}CO_2$	FeO in Mg-rich ilmenite continues to be reduced, when $(m-q-q'-\tilde{q}) + n = 0.5$, it is converted into iron containing anosovite

and tunnel kilns, rotary hearth furnaces, and vertical furnaces. Although the test product is different, various parameters are being optimized.

5.2.3.1 South African rotary kiln reduction production line

Highveld, South Africa, performed a test from April 1963 to May 1964 using a 15 t/day semiindustrial test facility; after 10 months of research, iron, steel, and vanadium products were produced simultaneously. This plant is equipped with 13 sets of rotary reduction kilns with a diameter of 4 m and length of 61 m, and rotational speed of 0.40–1.25 rpm. Seven submerged arc furnaces were used for smelting, two with a power of 45 MVA, and the other two of 33 MVA, with a diameter of 14 m, smelting period of 3.5–4 h, and furnace capacity of 70 t of iron/furnace. Another 63-MVA furnace has a diameter of 15.6 m, a smelting period of 3.5–4 h, and capacity of 80 t iron/furnace.

The composition of V-containing iron concentrate in South Africa is as follows: 53%–57% TFe, 1.4%–1.9% V_2O_5, 12%–15% TiO_2, 1.0%–1.8% SiO_2, 2.5%–3.5% Al_2O_3, and 0.15%–0.6% Cr_2O_3. V-containing iron concentrate was mixed with coal, silica, and dolomite; then fed into a rotary kiln; and reduced at 1100°C. This mixture is then delivered to an electric furnace where the slag and molten metal were separated. The composition of the molten metal is 3.95% C, 1.22% V, 0.24%Si, 0.22%Ti, 0.22% Mn, 0.08% P, 0.037% S, 0.29% Cr, 0.04% Cu, and 0.11% Ni. The corresponding low-grade titanium slag composition is 32% TiO_2, 22% SiO_2, 17% CaO, 15% MgO, 14% Al_2O_3, 0.9% V_2O_5, and 0.17% S. Fig. 5.3 shows the direct reduction process flowchart of South Africa's Highveld company.

5.2.3.2 Direct reduction production line in New Zealand

Granular coarse sand (72% > 0.1 mm) is used as a raw material in New Zealand, which does not need to be crushed; the iron concentrate is simply obtained by screening, magnetic separation, and gravity separation. The mineral composition is as follows: 57% TFe, 8% TiO_2, and 0.4% V_2O_5. The iron concentrate is mixed with coal, then roasted, and preheated in multihearth furnaces around 650°C to remove the volatile components. The ore is then placed in a rotary kiln for prereduction, where the metallization rate is controlled to 80%, and the temperature is 900°C–1000°C. A mixture of 90% reduced raw material and 10% carbocoal is transferred to the top of a rectangular furnace for melting and separation. The obtained metal melt composition is as follows: 3.5% C, 0.2% Si, 0.2% Ti, and 0.45% V, which is used to extract vanadium slag using a ladle. The vanadium slag has a composition of: 16%–20% V_2O_5, 16% SiO_2, 34% TFe, 1% CaO, 13% MgO, and 0.1% P. The semisteel contains 3% C, suitable for converter steelmaking, where the vanadium recovery is about 60%.

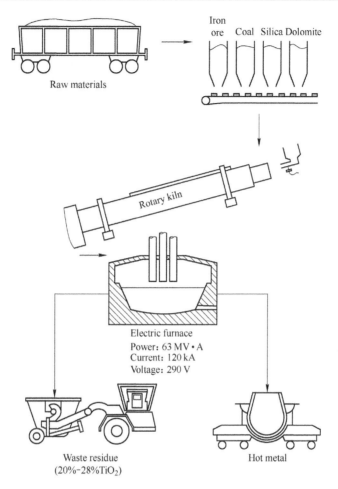

Figure 5.3
Process flow of ironmaking used by the Highveld company.

5.2.3.3 Rotary kiln reduction test lines in Panzhihua Steel, China

Within the national sixth and seventh Five-Year Plans, the Chinese government launched and subsidized a program named "new process for Panzhihua V-bearing titanoferrous magnetite." Management and experts from national scientific research institutes, universities, and relevant enterprises have been involved in the program, sharing the work and cooperating with one another. The work has been divided into processes for the north and south regions. New processing experiments with Panzhihua V-bearing titanoferrous magnetite are based on the flow of vanadium during melting in an electric furnace. One process allows vanadium to enter the slag phase, and then vanadium and titanium are extracted from the slag. In another process the vanadium is reduced and enters the molten iron to produce Ti slag as the raw material for Ti extraction. Panzhihua V-containing iron concentrate is prereduced in a chain-grate rotary kiln system. After cooling and magnetic separation, reduction is carried out in an ore-smelting electric furnace, and an oxygen converter is used for blowing the vanadium. The semisteel is used for steelmaking. Panzhihua V-containing iron concentrate used as material contains 54% TFe and 12% TiO_2. The highest temperature of prereduction is in the range of 1000°C−1050°C, while the metallization rate is kept at 65%−75%. The prereduced metallized pellets are fed into an ore-smelting electric furnace (with coke powder as the reductant) for melting and separation, resulting in a titanium slag with 60%−65% TiO_2, molten iron containing 2.01%C, 0.47% Si, 0.43% V, and 0.12% Ti.

The molten iron produced by the two different melting and deep reduction methods varies greatly. The typical components of vanadium slag produced from molten iron and deep-reduced molten iron are shown in Table 5.16.

When treated by the new process of "direct reduction electric furnace melting," the titanium slag can be recycled and used in the form of molten titanium slag, where the TiO_2 content can be enriched to more than 50%.

5.2.3.4 Direct reduction by rotary hearth furnace in Panzhihua, China

Direct reduction by rotary hearth furnace uses a mixture of iron ore concentrate, coal powder, and binder, which is pelletized, dried, and then fed into the rotary hearth furnace. During rotation of the furnace, the material experiences reactions in the preheating, reduction, and neutral zones. When the reactions are complete, the pellets are discharged into a heat transfer tank which is lined internally with a refractory material, where rapid cooling occur the hot metal components shown in Table 5.17. The direct reduction process flow of the rotary hearth furnace system is shown in Fig. 5.4.

Table 5.18 shows the chemical composition of the molten iron produced by the rotary hearth furnace, where Table 5.19 shows the chemical composition of the reduced TiO_2 slag produced by this process.

Table 5.16: Typical components of vanadium slag from molten iron and deep-reduced molten iron.

Item	V_2O_5	Al_2O_3	CaO	MgO	MnO	SiO_2	TiO_2	TFe
Further reduced V_2O_5 slag	8.72	2.59	1.50	4.50	-	24.63	2.11	35.80
Melted and separated V_2O_5 slag	1.92	10.92	5.27	12.33	1.01	11.12	41.44	13.1

Table 5.17: Composition of molten iron.

Item	C	Si	Mn	V	Ti	P
Range	1.5−2.5	0.40−1.47	0.20−0.45	0.40−0.60	0.20−0.38	0.05−0.15
Average	1.92	0.78	0.27	0.55	0.16	0.08

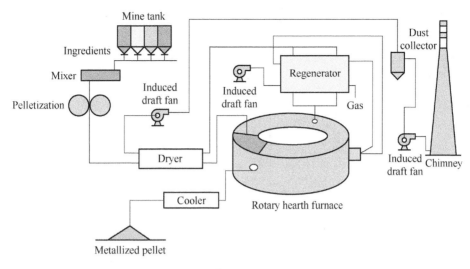

Figure 5.4
Direct reduction process flow of the rotary hearth furnace.

Table 5.18: Composition of molten iron produced by the rotary hearth furnace /%.

No.	C	S	V	Si	Ti
1	2.94	0.076	0.517	0.53	0.252
2	2.81		0.505	0.54	
3	2.36	0.10	0.59	0.50	0.096
4	2.41	0.064	0.515	0.705	0.26
5	3.31	0.078	0.507	0.484	0.178
6	2.73	0.060	0.57	0.615	0.348
7	2.69	0.073	0.509	0.450	0.202
Average	2.75	0.073	0.53	0.55	0.223

The typical V-containing iron concentrate produced in Panzhihua, China, has 54% TFe and 12% TiO_2. When used as a raw material, which is mixed with coal powder and binder for pelletizing. This mixture is reduced in a rotary hearth furnace at high temperature (1100°C) under a reducing atmosphere to produce metallized pellets with iron. These pellets are further reduced, melted, and separated using an electric furnace to produce a molten metal containing vanadium and reduced TiO_2 slag.

Panzhihua companies have adopted ways to use the Panzhihua V-containing iron concentrate with a composition of 56% TFe and $TiO_2 \leq 12\%$ as a raw material, which is mixed with the coking coal and binder to produce coal-based reduction pellets. Then, the

Table 5.19: Chemical composition of reduced TiO_2 slag produced by a rotary hearth furnace (%).

Item	CaO	FeO	SiO_2	TiO_2	V_2O_5	MgO
1	7.51		9.72	44.92	0.17	
2	8.76		8.48	53.00	0.29	
3	11.11		7.35	53.21	0.38	
4	11.55	1.45	9.85	53.85	0.184	
5	13.35	0.86		50.26	0.516	10.34
6	12.96	0.31		50.69	0.182	10.21
7	13.60	0.37		52.31	0.233	9.56
Average	11.26	0.478	8.85	51.18	0.279	10.04

pulverized coal is placed in an enclosed reduction pot, which is placed in a tunnel kiln. Gas is combusted to maintain a high temperature of 1100°C and a reducing atmosphere, which results in a mixture of iron and slag. After gravity separation, reduced iron and coal are obtained; the coal dust is recycled and the reduced iron is remelted to obtain hot metal containing around 0.35% V.

5.3 Main production equipment

The common systems are classified as blast furnace or direct reduction systems.

5.3.1 Blast furnace

5.3.1.1 Sintering system

Sintering systems are suitable for sintering in large ferrous metallurgy plants and are the main equipment used in the process of convulsions sintering. Sintering systems are used for sintering blocks of V-containing iron concentrate and mineral-rich powders with various compositions and granularities. Sintering partially removes sulfur, phosphorus, and other harmful impurities from the ore. Sintering systems with different sizes are characterized by their different lengths and widths. The main parts of the sintering process line include a quick lime crushing chamber, coal (coke) powder crushing room, raw material store, primary mixing room, secondary mixing room, sintering room, feeder, and hot crushing and screening room. In the raw lime crushing chamber, large calcareous lime chunks are crushed using a jaw crusher and roller mill and then screened. This material is then tested to ensure that fine lime is obtained (85% of the particles <3 mm), which is then stored in a flux bin. The calcareous powder material is fed based on its weight. In the coal (coke) powder crushing chamber, the tested and qualified coal (coke) powder is ground in a four-roll crusher and then screened. The fine coal (coke) powder (85% of the particles <3 mm) is stored in a fuel bin and then weighed before use in further processes. The stored raw materials are used for ore mixing, lime drying, and sieving. The store has six bins of reserve material for sinter production. The mineral powder and fuel are weighed using an electronic scale, while the flux is transferred and weighed by a screw feeding scale (according to the required ratio of ingredients) and supplied to the mixer chamber. Here, water is added to achieve a certain humidity and mixed well during primary mixing. In the second mixing step, the mixture is pelletized and heated. In the sintering chamber the mixture, mineral powder, flux, and bulk are mixed with a certain proportion of water. A reciprocal feeder is used to distribute the mixture along the length direction of the mixer. Using a combination of round roll and multiroll feeders, the mixture is evenly distributed over a sintering width of 1.5 m. The sintering material is moved into the igniter via rotation of a broad plate. After hot air preheating the sintering material is heated by high-temperature combustion of flue gas. Under the action of negative-pressure ventilation, the sintering material is moved through the process and simultaneously combusted. All physical and chemical changes are completed at the end of the process. The hot sinter is crushed to a size of ϕ 1.5x2.0 m using a single-roll crusher and then sieved through a screening system. The resulting sinter has a high iron content.

5.3.1.2 System installation

The blast furnace shell is made of steel plate. The blast furnace has five main parts: the furnace throat, body, waist, bosh, and hearth. The main products of smelting in blast furnaces are pig iron, blast furnace slag (as by-product), and blast furnace gas. When the blast furnace is operated, it is loaded with sinter, iron ore, coke, and slagging (limestone) from the top of the furnace and is blown with preheated air from the tuyere at the bottom of the furnace. At high temperature the carbon in the coke (some blast furnaces use auxiliary fuels, such as coal, heavy oil, and natural gas) is reacted with oxygen in the injected air and burned to produce carbon monoxide and hydrogen. These gases rise in the furnace and remove oxygen from the iron ore, producing iron. The molten iron is then tapped from the taphole. Unreduced impurities in the iron ore are combined with the flux of the limestone to produce the slag, which is tapped from another taphole. The generated gas is removed from the top of the furnace and used as a fuel for the hot blast furnace to heat the furnace, coke oven, and boiler. Blast furnaces for smelting V-bearing titanoferrous magnetite generally have a volume of 300–2000 m^3, where the slag and coke ratios are higher than used in normal blast furnaces. V bearing titanomagnetite sinter and pellets of V-bearing titanoferrous magnetite are reduced in the blast furnace to obtain V-containing iron and blast furnace slag.

5.3.2 Direct reduction system

5.3.2.1 Rotary kiln

Rotary kilns are classified according to their shape and can be divided into fixed-size and variable-diameter rotary kilns. The fixed-size rotary kiln is used for extracting vanadium pentoxide. The rotary kiln body is composed of a kiln head, body, and end. The kiln head has a larger diameter than the kiln body, and is where the material is discharged. The sealing is achieved by stainless steel sheets and the kiln body, which consists of the access port, nozzles, cars, and observation hole. The kiln body is the main part of the rotary kiln, which is usually 30–150 m long, cylindrical, and has 3–5 hoops. For normal operation, rotary kilns are lined with brick. The kiln end section is also an important part of the rotary kiln and is similar to the cover of a rotary kiln at the feed end, which mainly achieves feeding and sealing.

Rotary kilns are a thermal system composed of gas flow, fuel combustion, material movement, and heat transfer parts. Correct operation ensures that the fuel burns adequately, and heat can effectively be transferred to the material, which undergoes a series of physical and chemical changes during heating to produce the final clinker product. The working area of the rotary kiln has three parts: a drying section, heating section, and roasting section. Ventilation is one of the two material conditions of the combustion reaction, where an oxygen flow enables fuel combustion to release heat energy and maintain the desired temperature. Under normal operation, when the ventilation volume is low, the oxygen supply is insufficient, and the combustion speed is low, resulting in high heat consumption. For higher gas flows the air volume increases, and the time of combustion and decomposition decreases, resulting in an increase in generated combustion gas, and high heat loss, which affects normal operation of the kiln. Hence, accurate and timely control of the kiln ventilation is important. Under normal circumstances a fan at the tail of the system runs stably to extract the air. Ventilation in the kiln should be stable, although there are some factors that affect the flow, such as resistance to change the whole system, the air temperature in the kiln, air leakages, the fan inlet flow being mixed with cold air, dust accumulation, blockages in the pipes, uniformity and distribution of the material, and kiln interference factors, such as two systems.

Rotary kilns often use the latest wireless communication technology to transfer the temperature data from the thermocouple to the operation room. The kiln temperature transmitter is battery operated and can collect multiple thermocouple signals simultaneously. It is installed in the kiln body and can withstand the cylinder body rotation and temperatures above 300°C because of thermal insulation measures. The cylinder is required for radiates heat, antirain, antisun, and antivibration. The kiln temperature receiver is installed in the operation room, which directly displays the temperature in the kiln, and has 4–20 mA of output, which can be sent to a computer or other instrument display.

V-containing iron concentrate (56% TFe and 12% TiO_2) is mixed with noncoking coal and binder to make carbon-coal-based reduction pellets, which are treated at 1100°C under reducing conditions to obtain metallized pellets.

5.3.2.2 Rotary hearth furnace

A common structure of the rotary hearth furnace is a ring heating furnace of rolling steel. Typically, the top and wall of the furnace do not move, while the hearth of the furnace rotates and receives the raw material sent to the furnace for heating. The main process steps include: V-containing iron concentrate (56% TFe and 12% TiO_2) is mixed with noncoking coal and binder to prepare carbon-coal-based reduction pellets, which are dried and fed into the furnace. With the rotation of the hearth of the furnace, the reduction pellets pass through a preheating zone, reduction zone, and neutral zone. After the reactions are complete, metallized pellets are obtained, which are discharged into a heat transport tank lined with a refractory material for rapid cooling. Fig. 5.5 shows the profile and cross-sectional views of the rotary hearth furnace.

V-containing iron concentrate (56% TFe and 12% TiO_2) is mixed with noncoking coal and binder to make carbon-coal-based reduction pellets, which are reduced at 1100°C under reducing conditions to obtain metallized pellets.

5.3.2.3 Shaft furnace

In shaft furnace systems the furnace body is upright and the furnace cover has a shaft. The furnace gas rises within the furnace and heat is exchanged between the hot gas and the material during counterflow. The furnace burden in most shaft furnaces is in direct contact with the fuel. In the case of internally cooled shaft blast furnaces, the cooling and roasting is completed in the same furnace, and the combustion chamber surrounds the rectangular roasting chamber. Hence, it is easy to supply gas to the center of the furnace burden using fire holes on both sides. In the case of long furnaces, long cooling belts are used, which effectively cool the pellets. However, the discharge temperature is still 427°C–540°C, which requires that the water outside the furnace is cooled, which affects the quality of the

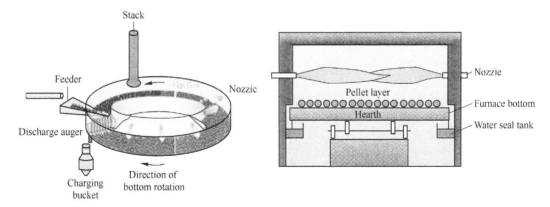

Figure 5.5
Profile and cross-sectional views of the rotary hearth furnace.

finished pellets. In medium-sized externally cooled shaft furnaces, roasting is completed in the furnace body, and then the pellets are cooled outside the shaft furnace. A heat recycling system is employed to efficiently use heat from the shaft furnace. The finished pellets are effectively cooled and the discharge temperature can be under 100°C. In the smelting of V-containing titanoferrous magnetite, shaft furnaces are mainly used to produce oxidized pellets and reduced metallized pellets but can also be used directly to produce V-containing iron.

5.3.2.4 Tunnel kilns

Tunnel kilns are long straight tunnels with fixed walls and arch apexes on both sides. Kiln cars run along tracks laid at the bottom of the kiln. Combustion equipment in the tunnel kiln is arranged in the central area, which constitutes the fixed thermal belt and calcining zone. High-temperature flue gas from combustion reactions in the tunnel kiln exits via a chimney at the kiln head or under the action of a fan-induced draft. Meanwhile, the material is gradually preheated in the preheating zone of the tunnel kiln. Cold air is blown from the kiln tail. After a period of cooling in the cooling zone of the tunnel kiln, the air flow is heated and then sent to a dryer as the heat source for drying raw materials. The combustion fuel used in tunnel kilns is mainly heavy oil, light diesel oil, natural gas, and coal gas. Combustion of oil requires storage tanks and combustion nozzles. Because of the high cost of oil, natural gas and coal gas are still used, where natural gas is transported to the kiln by pipelines and then sprayed into the kiln for combustion using a special natural gas burner. If this gas cannot be transported to the kiln by a gas pipeline, a gas generator is installed to produce gas and transport it through a pipeline to the tunnel kiln.

V-containing iron concentrate (56% TFe and 12% TiO_2) is mixed with coking coal and binder to prepare coal-based reduction pellets containing carbon to facilitate reduction reactions. These pellets are loaded in a reduction pot, where gas combustion provides the high temperature (1100°C) and reducing atmosphere for reduction in the tunnel kiln to produce an iron-slag mixture. After gravity separation, reduced iron and coal are produced. The coal dust is recycled and the reduced iron is melted to produce molten V-containing iron with a V content of $\sim 0.35\%$.

5.3.2.5 Electric furnaces

Ore-smelting electric furnaces are industrial furnaces that use an electrical system to supply heat. Such furnaces are one of the many systems used in the overall production line. The furnace body is the main part of the furnace, where the electrode is one of the most important parts. Other parts include auxiliary electromechanical equipment, such as the transformer, short net, electrode holder, lifting device, and feeding and discharging systems. Related equipment includes the substation, water supply system, and dust-removal device. The furnace body is composed of brick masonry, steel structures (steel frame, furnace shell, or coaming) and a foundation pier. Because of the high temperature at the bottom of the furnace, an overhead structure is generally used, which is cooled by natural wind or forced air blasts. Direct reduction of the metallized pellets is performed, or magnetic separation of the iron in the electric furnace, which can be melted, separated, and reduced to produce molten iron containing carbon and vanadium.

CHAPTER 6

Extraction of vanadium from V-containing titanoferrous magnetite: enrichment of vanadium slag

Chapter Outline
6.1 Enrichment of vanadium slag   124
    6.1.1 Vanadium grade of molten iron   126
    6.1.2 Blowing temperature   126
    6.1.3 Blowing time   129
    6.1.4 Blowing method   129
    6.1.5 Selection of coolant   129
    6.1.6 Slagging agents   130
    6.1.7 Variable-valence elements   130
    6.1.8 Vanadium slag quality   130
    6.1.9 Semisteel quality   134
6.2 Production of V-containing steel slag from molten iron by steelmaking   136
    6.2.1 Reactions occurring during slag production   136
    6.2.2 Main slag blowing equipment   136
6.3 Pretreatment of V-containing molten iron for slag blowing   137
    6.3.1 Blowing vanadium slag   137
    6.3.2 Main slag blowing equipment   138
6.4 Enriched vanadium slag from V-containing molten iron by sodium treatment and oxidation   140
6.5 Comparison of various vanadium extraction methods   141
    6.5.1 Tank furnace method   141
    6.5.2 Molten iron tank method   142
    6.5.3 Atomization furnace method   142
    6.5.4 Converter process   142
    6.5.5 Direct-reduction—electric furnace—shaking ladle—vanadium slag method   143
6.6 Main equipment   145
    6.6.1 Vanadium recovery converter   145
    6.6.2 Atomization furnace for extracting vanadium   146

Vanadium slag is a material rich in vanadium. Vanadium is reduced when producing pig iron to obtain molten iron containing about 0.25% V. Vanadium-containing iron is the main raw material for the production of vanadium dioxide, accounting for 50%−60% of vanadium produced globally. In the process of making steel with pig iron, vanadium is oxidized and is effectively enriched by entering the slag to form a V-containing slag. There are three main processes for extracting vanadium from V-containing iron: vanadium slag blowing, sodium-treated vanadium slag blowing, and steel-slag/V-slag blowing. The V-containing molten iron is poured into the converter for steelmaking, where V is oxidized and enriched in the V-containing steel slag. The steel for steelmaking and vanadium slag for extracting vanadium pentoxide are processed simultaneously using the steelmaking and vanadium slag blowing processes, respectively. The thermal efficiency of this process is high, but the steel slag and vanadium slag are mixed, resulting in a V-containing steel slag that generally has a low V_2O_5 content, which affects later extraction of vanadium pentoxide. First, vanadium in the molten iron is blown and oxidized to obtain vanadium slag, where the vanadium is highly enriched in the slag. The melt is drossed to obtain the vanadium slag, while the semisteel is used for steelmaking. Steelmaking and vanadium slag blowing processes are used for their respective functions. The thermal efficiency of this process is low, but a high V_2O_5 grade in the slag is achieved, which is advantageous for extracting V_2O_5. The V-containing molten iron can be treated with sodium and then oxidized to form sodium-treated vanadium slag. It can dispense roasting when extracting V_2O_5 from the vanadium slag. The sodium-treated vanadium slag is finely ground directly into the leaching process. It is generally accepted that the circulation of the process of sodium salt, but heavy loaded, the thermal efficiency is low, and the operating environment is poor, resulting in a decrease in the temperature of the iron. Typical production plants that extract vanadium by pyrolysis exist in China, Russia, and South Africa [24,36−39,49−51,58,59,64−67,69−71].

6.1 Enrichment of vanadium slag

The enrichment of vanadium slag is carried out in molten iron, which is a typical oxidation and separation process that aims to balance the vanadium slag and semisteel. As the oxidation process continues, a slag−iron interface is formed, and alternating reduction and oxidation reactions occur. Fig. 6.1 shows a schematic diagram of slag−metal−gas transfer in the vanadium slag blowing process. As the blowing process continues, the slag surface changes as the slag volume increases, and the heat transfer and mass transfer process are fast.

The vanadium−oxygen phase diagram is shown in Fig. 6.2. The oxidation of vanadium has a characteristic subarea, phase separation, and the layered, at different temperatures. The

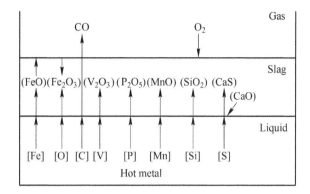

Figure 6.1
Schematic diagram of the slag–metal–gas mass transfer process.

amount of vanadium oxide increases with increasing blowing (oxidation) time. Various oxides of vanadium are present, where different oxidation reactions occur and various compound oxides are produced. For all valence states, the state and phase of the compounds are changing. The process flow of producing V-containing molten iron and vanadium slag from V-containing titanoferrous magnetite is shown in Fig. 6.3.

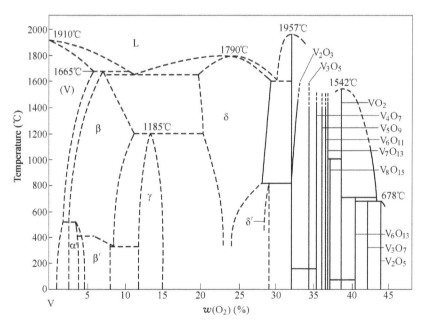

Figure 6.2
Vanadium–oxygen phase diagram.

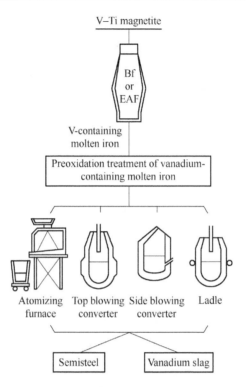

Figure 6.3

Process flow of V-containing molten iron and vanadium slag from V-containing titanoferrous magnetite.

6.1.1 Vanadium grade of molten iron

The content of vanadium in iron is related to the properties of the iron ore used. Higher contents of vanadium in V-containing iron concentrate result in higher V contents in the molten iron, where a higher grade of V_2O_5 in the vanadium slag can be obtained by blowing. V-containing molten iron can be divided into four grades: V02 with vanadium content $\geq 0.2\%$; V03 with vanadium content $\geq 0.3\%$; V04 with vanadium content $\geq 0.4\%$; and V05 with vanadium content $\geq 0.5\%$. It is necessary that the vanadium content of V-containing molten iron is $\geq 0.15\%$ to obtain vanadium slag. The V-containing molten iron must have a vanadium content of $\geq 0.1\%$ to produce V-containing steel slag. The V_2O_5 content in the vanadium slag increases with increasing initial vanadium content in the V-containing molten iron and decreases with increasing contents of Si, Ti, Mn, and residual vanadium in the V-containing semisteel.

6.1.2 Blowing temperature

Vanadium oxidation involves complex multiphase exothermic reactions with very large changes in the heat of combustion. The standard free energies of the oxidation reactions between each element in the process of vanadium slag blowing with respect to temperature are shown in Fig. 6.4.

In general, studies showed that the temperature is relatively low (1300°C) in the early stage of vanadium slag blowing of V-containing molten iron in the pool. At the start of blowing, Ti, Si, Cr, V, and Mn in the V-containing molten iron are preferentially oxidized compared to carbon, releasing a large amount of heat that rapidly increases the temperature of the molten pool. When the temperature exceeds 1400°C (1673K), the affinity of carbon and oxygen is greater than that of vanadium and oxygen ($\triangle G^\theta V > \triangle G^\theta C$). Therefore the carbon in the steel begins to rapidly oxidize and inhibit the oxidation of vanadium, which impedes vanadium recovery. At the same time the rapid oxidation of carbon reduces the carbon content in the semisteel,

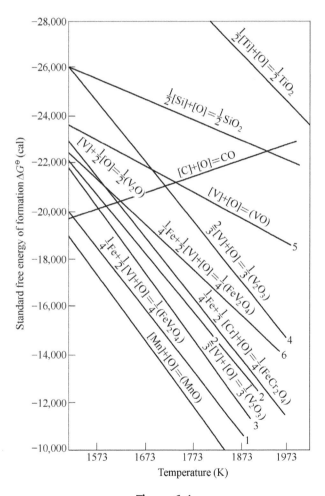

Figure 6.4
Standard free energy of oxidation for each element in the process of vanadium-slag blowing from V-containing molten iron as a function of temperature.

Table 6.1: Vanadium and titanium phases present in vanadium slag with silica (%).

SiO_2 (wt.%)	V-containing phase	Si-containing phase
15	$Fe_xV_{3-x}O_4$ $Mn_xV_{3-x}O_4$	Mg_2SiO_4 Fe_2SiO_4
25	$Fe_xV_{3-x}O_4$ $Mn_xV_{3-x}O_4$	(Fe, Mg, Ca)SiO_3 $CaFeSi_2O_6$

which degrades the steelmaking process. To limit carbon oxidation and extract vanadium in the vanadium slag blowing process, the temperature of the molten pool need to be strictly controlled to below 1400°C. Higher contents of SiO_2, TiO_2, and V_2O_3 (and lower FeO contents) in the final slag result in higher critical temperatures for selective oxidation of carbon and vanadium. As the TiO_2 and SiO_2 contents in the vanadium slag increase and the molten pool temperature increases, the distribution ratio of vanadium in the steel and slag decreases. As the FeO content in vanadium slag decreases, the distribution ratio of vanadium in the steel and slag decreases. Experimental results showed that higher V_2O_3 contents in the vanadium slag resulted in the distribution ratio of vanadium in the steel and slag first increasing and then decreasing. The MnO content in the vanadium slag has only a minor effect on the V distribution in the slag and steel.

The temperature range at which the vanadium slag increases most rapidly, and the spinel grain size is the largest, is 1200°C–1250°C. At 1250°C and a holding time less than 60 min, the spinel phase grain grew with a high crystallization rate, while holding times above 60 min resulted in slower grain growth.

Table 6.1 shows the vanadium and titanium phases occurring in vanadium slag in the presence of silica, while Table 6.2 shows the vanadium and titanium phases occurring in the slag in the presence of titania. When the SiO_2 content of the vanadium slag is low, SiO_2 mainly occurs as olivine, while when the SiO_2 content is high, a small amount of pyroxene appears in the vanadium slag. When the TiO_2 content of vanadium slag is low, Ti mainly occurs as $Mg_xTi_{3-x}O_4$ and Fe_2TiO_4, while when the TiO_2 content is high, a small amount of $MgTi_2O_5$ appears in the vanadium slag.

Table 6.2: Vanadium and titanium phases present in vanadium slag with titania (%).

TiO_2 (wt.%)	V-containing phase	Ti-containing phase
15	$Fe_xV_{3-x}O_4$ $Mg_xV_{3-x}O_4$	$Mg_xTi_{3-x}O_4$ Fe_2TiO_4
25	$Fe_xV_{3-x}O_4$ $Mg_xV_{3-x}O_4$	$Mg_xTi_{3-x}O_4$ Fe_2TiO_4 $MgTi_2O_5$

6.1.3 Blowing time

The blowing time is generally short to ensure the efficient vanadium recovery and to balance the quality of the vanadium slag and semisteel. The blowing time is closely related to the intensity of oxygen gas, where a coolant is used to meet the requirements of residual vanadium in the semisteel. The blowing time is determined by the quantity of V-containing molten iron, vanadium content, equipment structure, oxidant and coolant types, gas pressure, and the method for blowing the vanadium slag. The blowing of vanadium slag in a converter consists of an early stage of blowing and a late reblow. The oxygen blowing time is generally limited to 6–12 min. After blowing, if the residual vanadium content in the semisteel is >0.04%, a short reblow period is used to control the oxidation rate of vanadium. Blowing vanadium slag by atomization is only related to the flow of molten iron. After separation of the vanadium slag and semisteel, the thermal insulation time and mineralization time significantly affect the quality of the vanadium slag.

The total content of the V_2O_5 and FeO in the vanadium slag is constant, where $(V_2O_5) +$ (TFeO) = 60%. If the vanadium slag is limited in the allowed range, measures for decreasing (FeO) to increase the (V_2O_5) content in the vanadium slag are effective. Comprehensive control measures can avoid overblowing and high residual vanadium contents in the semisteel. Hence, the blowing time and rhythm are very important in the production of vanadium slag.

6.1.4 Blowing method

The blowing method can be divided into atomization and blowing steps, where the blowing process uses either compressed air or oxygen. According to the requirements of the converting equipment, blowing can be applied at the top, bottom, side, or a mixture of these. The different blowing methods for vanadium slag require different auxiliary facilities and achieve various vanadium extraction efficiencies. The major component of vanadium slag is iron (35%–40%); hence, reducing the iron content effectively increases the relative V content. Even a decrease in the content of total iron in the slag of 5% is beneficial to the vanadium content in the slag. The total iron content in the vanadium slag depends on the intensity of the oxygen blowing and position of the oxygen gun.

6.1.5 Selection of coolant

The coolant benefits the slag-blowing process by increasing the supply of vanadium and increasing the supply of oxygen. Its main function is to absorb process heat during slag blowing, while not affecting the quality of the vanadium slag and semisteel. The coolant is added to the molten iron, usually in the form of a suitable amount of V-containing iron block or scrap steel ($\sim$20%–30%) and iron scale ($\sim$5%–7%). Commonly used coolants are iron oxide, water, scrap steel, and iron ore.

6.1.6 Slagging agents

During blowing of vanadium slag, SiO_2, FeO, and MnO react in the initial slag to produce silicate phases, such as iron and manganese olivine. According to the $FeO-SiO_2$ phase diagram, the melting point of the silicate phase is 1205°C. A mixed silicate phase of Fe_2SiO_4 and FeO is formed (FeO: 76%, SiO_2:76%), which is a fusible material with a melting point of 1177°C. The formation of silicate phases is beneficial for decreasing the early slag melting point and vanadium slag viscosity, which helps increase the liquidity of the slag. As the fluidity of the initial vanadium slag is increased, the oxidation of vanadium is enhanced and the content of the vanadium−iron spinel increases. An increase in the fraction of silicate phase results in the formation of a bonding phase during solidification, ensuring that the vanadium slag is not too dry or thick at the end of the blowing process. According to the $FeO-V_2O_3$ phase diagram, at the end of blowing, the viscosity of the vanadium slag increases quickly, and the final slag changes from a half-frozen state to a granular or half paste, resulting in low TFe and MFe contents in the converter vanadium slag. There are significant advantages in using a converter to recover vanadium. SiO_2 is used as a slagging agent, which can achieve lower (TFe) and higher (V_2O_5) contents in the vanadium slag.

Table 6.3 compares the effect of using SiO_2 as a slagging agent. The addition of SiO_2 and TiO_2 is beneficial for the diffusion of ions in the slag, and the precipitation and growth of the main phase. SiO_2 and TiO_2 will form olivine and titanium spinel, respectively, which affect the subsequent vanadium content and extraction. Hence, formation of these phases should be limited.

6.1.7 Variable-valence elements

Generally, the presence of variable-valence elements in the formed vanadium−iron spinel, such as Fe, V, Mn, and Cr, increases the porosity of the spinel grains.

6.1.8 Vanadium slag quality

The quality of the vanadium slag is characterized by its crystal structure, chemical composition, and vanadium and iron content. The structure is controlled by the crystallization and mineralization time after the separation of the vanadium slag and semisteel.

Table 6.3: Effect of using SiO_2 as a slagging agent.

	[C] $_{semisteel}$(%)	[V] $_{semisteel}$(%)	$t_{semisteel}$ (°C)	(V_2O_5) (%)	(TFe) (%)	(SiO_2) (%)	Statistics/furnace
No slagging	3.362.16−4.12	0.040.01−0.116	1376,1325−1431	16.51,13.4−27.8	35.32,20.6−42.6	15.6,12.4−20.6	2130
Slagging	3.622.79−4.09	0.0350.01−0.10	1387,1335−1420	20.09,11.7−31.2	28.3,17.6−42.9	18.01,14.2−20.7	1983

6.1.8.1 Oxidation of vanadium slag

Fig. 6.5 shows the relationship between the distribution coefficient of vanadium in the vanadium slag and semisteel (V)/[V] and the ratio (TFe)/(SiO₂) in the vanadium slag at the critical temperature of slag blowing. Blowing vanadium slag from V-containing molten iron is performed under a suitable molten pool temperature. The distribution coefficient of vanadium in the vanadium slag and semisteel (V)/[V] and the ratio of the iron oxide and silica contents in the vanadium slag (TFe)/(SiO₂) have a clear linear relationship, as shown in Fig. 6.5. Using single-furnace slag basicity operation or duplexing acid operation methods, the distribution coefficient of vanadium in vanadium slag and semisteel (V)/[V] dramatically increases with increasing (TFe)/(SiO₂) ratio.

To increase the oxidation of vanadium and its content in the vanadium slag as much as possible, strong oxidizing conditions throughout the process are required. During the single-furnace slag basicity process (CaO/SiO₂ = 1.0−2.0), (V)/[V] is expressed as follows:

$$(V)/[V] = -44.3 + 237.7\ TFe/Si, correlation\ coefficient\ of\ Y = 0.96 \quad (6.1)$$

In the duplexing acid operation, (V)/[V] is expressed as follows:

$$(V)/[V] = -23.5 + 74.7\ TFe/SiO_2 \\ \gamma = 0.91 \quad (6.2)$$

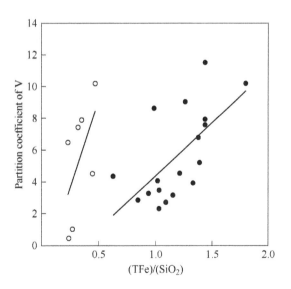

Figure 6.5
Relationship between (V)/[V] and (TFe)/(SiO₂) in the vanadium slag at the critical temperature of blowing vanadium slag: (1) single-furnace slag basicity method (CaO/SiO₂ = 1.00 − 2.00); (2) duplexing acid method.

In the equations, (V) is the vanadium content in vanadium slag (%); [V] is the residual vanadium content in the semisteel (%); TFe is the content of FeO in the vanadium slag (%); and SiO_2 is the content of SiO_2 in vanadium slag (%).

The content of iron oxide in the vanadium slag is very high, generally over 30%. If this content is too high, the V_2O_5 content will be reduced, which increases the iron consumption when blowing vanadium slag and complicates extraction of V_2O_5 from the vanadium slag. In the case of a high oxidation rate of vanadium the iron oxide content in the vanadium slag is optimal and does not need to be increased.

6.1.8.2 Basicity of vanadium slag

When the ratio of TFe/SiO_2 is certain, and the temperature is suitable for blowing the vanadium slag, the distribution coefficient of vanadium in the slag and semisteel increases with increasing basicity of the vanadium slag. When the acid slag R is <1.0 or ≥ 2.0, the oxidation rate of vanadium increases by $\sim 30\%$ for the same carbon oxidation rate. To obtain alkaline vanadium slag, a certain amount of alkaline substances, such as CaO, are generally added during vanadium slag blowing. This can increase the activity of iron oxide, a(FeO), in the $FeO-SiO_2$ system, which is advantageous for the formation and separation of vanadium oxide. However, the amount of vanadium slag is increased, and the content of V_2O_5 in the vanadium slag decreases, while the addition of CaO produces insoluble calcium vanadate and reduces the recovery of vanadium during the extraction process. This results in a need to increase the acid leaching process, which complicates the overall production process. Furthermore, CaO and P_2O_5 in the vanadium slag react to produce calcium phosphate, which greatly reduces the slag quality.

The quality of the vanadium slag refers to the V_2O_5 content in the slag, where a high V_2O_5 content is desirable, with impurity contents as low as possible. There are two main factors affecting the quality of the vanadium slag. The first factor is the composition of the molten iron. The relationship between V_2O_5 in the vanadium slag and the elements in the V-containing molten iron, the original vanadium content, and the vanadium content of the semisteel is expressed by the following multiple regression equation:

$$(V_2O_5) = 6.22 + 31.916[V] - 10.556[Si] - 8.964[V] - 2.314[Ti] - 1.8555[Mn] \qquad (6.3)$$

The equation shows that the content of V_2O_5 in the slag (V_2O_5) increases with increasing initial vanadium content in the molten iron, and decreases with increasing Si, Ti, and Mn contents in the molten iron and increasing residual vanadium content in the semisteel. The second factor is the content of FeO in the slag. The typical FeO content in vanadium slag is

15%–50% for the typical blowing methods, which is represented by FeO. The relationship between FeO and (V₂O₅) is expressed as follows:

$$V_2O_5 = -1.94(TFe) + 64.18 \qquad (6.4)$$

Calculating the sum of (V₂O₅) and (FeO) contents in the vanadium slag gives a constant (V₂O₅) + (TFeO) = 60%, where both of these are main ingredients in the vanadium slag. Any measure for lowering the (FeO) content in the vanadium slag effectively increases the (V₂O₅) content.

6.1.8.3 Quality of vanadium slag

The vanadium slag is an acid slag, where some of the slag is dark gray to black. The structure varies from relatively loose porous foamed slag to quite dense rock-like slag. The mineral composition of vanadium slag is mainly vanadium–iron spinel, iron olivine, quartz, and metallic Fe, as shown in Table 6.4. In general, the presence of variable-valence elements in the structure of the vanadium–iron spinel, such as Fe, V, Mn, and Cr, result in increased porosity of the spinel material.

The main chemical components are shown in Table 6.5.

Depending on the V₂O₅ content, China's national standards of vanadium slag define six grades: vanadium slag 11 (V₂O₅ 10.0%–12.0%); vanadium slag 13 (V₂O₅ 12.0%–14.0%); vanadium slag 15 (V₂O₅ 14.0%–16.0%); vanadium slag 17 (V₂O₅ 16.0%–18.0%);

Table 6.4: Mineral composition of vanadium slag (%).

Vanadium–iron spinel			Fayalite volume fraction (%)	Quartz volume fraction (%)	Metallic iron volume fraction (%)
Volume fraction (%)	Crystalline form	Grain size (mm)			
~40	Relative integrity	~10	35	25	Few
~60	Relative integrity	20–30	35	5	5
70–75	Relative integrity	~30	25–30	No	15–20

Table 6.5: Main components of vanadium slag (%).

Vanadium slag source	Method	V₂O₅	SiO₂	FeO	CaO	TiO₂	MnO
Pansteel	Atomization	15–19.5	13.05	40	0.78	11.51	5.8
Pansteel	Converter	15–19.5	13.05	40	0.78	11.51	5.8
Chengde-steel	Converter	15–18	16.5	32.5	1.32	6.00	2.5
Ma-steel	Converter	12.78	31.2	39.48	2.52	5.77	4.37
Russia	Converter	16–19	17–20				
South Africa	Ladle	27.8					
New Zealand		15.48	20.30	30.30	1.13	15.00	14.7

vanadium slag 19 (V_2O_5 18.0%−20.0%); and vanadium slag 21 (V_2O_5 > 20.0%). The contents of P, CaO, SiO_2, and Fe in the vanadium slags are restricted.

The contents of V_2O_5 and other components in the vanadium slag are determined by the composition of the molten iron. The Si and V contents in molten iron are the primary factors directly influencing the vanadium slag quality. Low Si and V contents in the molten iron result in low SiO_2 and V_2O_5 contents in the vanadium slag. On the contrary, a low content of SiO_2 and high content of V_2O_5; that is, SiO_2 and V_2O_5 in the vanadium slag are negatively correlated. Therefore, to obtain the required grade of vanadium slag, there are strict limits on the Si content of V-containing molten iron.

6.1.9 Semisteel quality

Semisteel should have high carbon content, a certain degree of superheat, and low residual vanadium content. The carbon and residual vanadium contents of the semisteel are controlled by the final temperature of the molten pool and total oxygen blowing. High final temperatures result in semisteel with low carbon content and high vanadium content. Experimental results from the recovery of vanadium using an oxygen converter showed a relationship between ΔC (decarburization quantity) and T(end) (final temperature of the molten pool), as follows:

$$\Delta C = 56.67 T(end) - 7.03 \tag{6.5}$$

This equation shows that the temperature of the molten pool should be strictly controlled to ensure high carbon content in the semisteel. Production practices have shown that for molten pool temperatures <1400°C, the semisteel has high carbon content and sufficient degree of superheat for effective subsequent steelmaking.

In the case of blowing vanadium slag in a furnace with oxygen or compressed air, elements such as Si and Ti in the molten iron are first oxidized into the slag, followed by oxidation of V:

$$\frac{4}{3}[V] + O_2 = \frac{2}{3}[V_2O_3] \tag{6.6}$$

The V_2O_5 produced by oxidation is enriched in the slag and separated from the molten iron. In addition to the requirement of sufficient oxidation of the vanadium, oxidation of carbon should be limited to retain enough carbon in the molten iron for subsequent steelmaking. The end temperature must be strictly controlled in the range of 1623K−1693K for optimal oxidation and conversion of vanadium. During blowing, the temperature of the vanadium slag is controlled by feeding sufficient amounts of V-containing pig iron, iron, sinter, and pellets as coolant materials, or chopped ferrosilicon and coke breeze to increase the temperature.

The main technical and economic indicators from vanadium slag production companies in both China and around the world are shown in Table 6.6.

Table 6.6: Comparison of main technical and economic indicators from vanadium slag production companies in China and around the world.

Producer	Semisteel C (%)	Semisteel, temp (°C)	Vanadium slag (%)					Oxidation (%)	Recovery (%)	Capacity (10 kt)
			V_2O_5	TFe	MFe	P	CaO			
Highveld	–	1270–1400	23–25	29	–	–	3	>93	82	18
Nizhni Tagil	3.20	1380	16	32	9–12	0.04	1.5	≥90	82–84	17
Chengsteel	3.17	1400	12	34	21	0.07	0.8	87.8	77.6	>6
Pansteel	3.57	1375	19.40	28–33	9–12	<0.1	2.0	91.42	82.0	17

6.2 Production of V-containing steel slag from molten iron by steelmaking

V-containing molten iron is directly used for steelmaking. With the addition of oxygen, excess carbon and other impurities in the molten iron are oxidized and removed simultaneously via the gas or slag phases. Suitable amounts of slagging materials (such as lime) and deoxidizers are added. Oxides such as FeO, V_2O_3, TiO_2, SiO_2, and MnO react with lime (slagging material) to form V-containing steel slag.

6.2.1 Reactions occurring during slag production

V-containing molten iron is directly used for steelmaking, where the oxidation of Fe, V, Ti, Si, and Mn occurs. The reaction between the oxides and slagging agents occur as follows:

$$\frac{4}{3}[V] + O_2 = \frac{2}{3}[V_2O_3] \tag{6.7}$$

$$[Ti] + O_2 = [TiO_2] \tag{6.8}$$

$$[Si] + O_2 = [SiO_2] \tag{6.9}$$

$$2[Fe] + O_2 = 2[FeO] \tag{6.10}$$

$$4[FeO] + O_2 = 2Fe_2O_3 \tag{6.11}$$

$$2[V] + 3[Fe_2O_3] = [V_2O_3] + 6[FeO] \tag{6.12}$$

$$5[P] + O_2 = P_2O_5 \tag{6.13}$$

$$2FeO + Si = SiO_2 + 2Fe \tag{6.14}$$

$$FeO + Mn = Fe + MnO \tag{6.15}$$

$$FeO + C = CO + Fe \tag{6.16}$$

Silicon dioxide and manganese oxide produced by oxidation during oxygen blowing react with the slagging agent (lime) to produce V-containing steel slag. The desulfurization reaction and production of phosphorus slag are as follows:

$$FeS + CaO = FeO + CaS \tag{6.17}$$

$$2P + 5FeO + 3CaO = 5Fe + Ca_3(PO_4)_2 \tag{6.18}$$

6.2.2 Main slag blowing equipment

V-containing steel slag is produced via steelmaking processes using V-containing molten iron, resulted from no special disposition of vanadium recovering steelmaking. The vanadium content of V-containing steel slag is commonly 2%–4% (as V_2O_5), while the remainder is iron, calcium, magnesium, and silicon and aluminum oxides, into which vanadium is dispersed as a variety of mineral phases. It is difficult to directly separate vanadium by mineral sorting and smelting; usually direct alloying is used, followed by additional sintering for comprehensive recycling and extraction of the vanadium.

Depending on the different steelmaking equipment, the V-containing steel slag can be produced in the steelmaking process using a converter, open hearth, or electric arc furnace. The V-containing steel slag produced by CAP Mills (Chile) using oxygen blowing in an alkaline converter has the following composition: 5.7% V_2O_5, 47.0% CaO, 2.5% MgO, 11.0% SiO_2, 3.2% P_2O_5, 4.0% MnO, 15.1% Fe, and 1.2% Al_2O_3, because of the high content of CaO and P in the slag, vanadium mainly consisted of $3CaO \cdot P_2O_5 \cdot V_2O_5$, $CaO \cdot 3V_2O_5$, $CaO \cdot V_2O_5$, and $3CaO \cdot V_2O_5$.

6.3 Pretreatment of V-containing molten iron for slag blowing

Oxygen or compressed air is generally used for the pretreatment of V-containing molten iron to dissolve the vanadium oxide and enrich the slag phase of the molten iron with V, which is combined with other components and metallic iron to form vanadium slag. After separation of the vanadium slag, the molten iron is called semisteel, which is used for steelmaking. The V_2O_5 content of the vanadium slag is generally 10%–30%.

6.3.1 Blowing vanadium slag

With oxygen or compressed air blowing of V-containing molten iron in a furnace, Fe, Si, and Ti in the molten iron are first oxidized and enter the slag. The oxidation reaction between vanadium, oxygen, and ferric oxide occurs, followed by the scorification reaction and slagging of the components of the vanadium slag. The relevant chemical reactions are as follows:

$$\frac{4}{3}[V] + O_2 = \frac{2}{3}[V_2O_3] \tag{6.19}$$

$$[Ti] + O_2 = [TiO_2] \tag{6.20}$$

$$[Si] + O_2 = [SiO_2] \tag{6.21}$$

$$2[Fe] + O_2 = 2[FeO] \tag{6.22}$$

$$2[V] + 3[Fe_2O_3] = [V_2O_3] + 6[FeO] \tag{6.23}$$

$$C + [O] = CO \tag{6.24}$$

$$[V] + 1.5[O] = (VO_{1.5}) \tag{6.25}$$

$$[Si] + 2[O] = (SiO_2) \quad (6.26)$$

$$[Mn] + [O] = (MnO) \quad (6.27)$$

$$[P] + 2.5[O] = (PO_{2.5}) \quad (6.28)$$

$$Fe + [O] = (FeO) \quad (6.29)$$

$$[S] + (CaO) = (CaS) + [O] \quad (6.30)$$

The main slagging reactions are as follows:

$$FeO + V_2O_3 = FeV_2O_4 \quad (6.31)$$

$$MnO + SiO_2 = MnSiO_3 \quad (6.32)$$

$$FeO + SiO_2 = FeSiO_3 \quad (6.33)$$

It is necessary to have adequate oxidation of vanadium in V-containing molten iron during blowing of vanadium slag, while the oxidation of carbon should be restricted to ensure sufficient carbon in the semisteel for subsequent steelmaking processes. Hence, the blowing temperature should be limited to 1623–1693K as this temperature range is optimal for oxidation and conversion of vanadium. Appropriate amounts of V-containing pig iron, iron, sinter, and pellets are added as coolants, while ferrosilicon and broken coke block are added to increase the temperature, allowing the process temperature to be controlled.

6.3.2 Main slag blowing equipment

Different equipment and methods are adopted by different countries, where the main iron-making equipment used in Russia and China is a blast furnace. The main blowing equipment and conversion methods are divided into atomization furnace blowing, air-side converter blowing, top- or bottom-converter blowing, and ladle (vibrating tank) blowing. The atomization furnace and converter are used when the supply of molten iron is from a large-scale blast furnace, while ladle (vibrating tank) blowing is used in conjunction with small blast furnaces.

China uses top-blown converters, air-side converters, and atomization furnace blowing to produce vanadium slag. Russia uses top- or bottom-blown converters with oxygen to produce vanadium slag. (1) Atomization furnace blowing. China invented vanadium extraction by atomization in Panzhihua using an atomization furnace design (as shown in Fig. 6.6), which has been successfully applied to industrial production. In this process, the V-containing molten iron carried by iron pots is mixed in the tundish, in which the flow is controlled. The V-containing molten iron is atomized via a nozzle at the bottom of the tundish by injection of a high-speed airflow from the atomizer nozzle, where droplets are

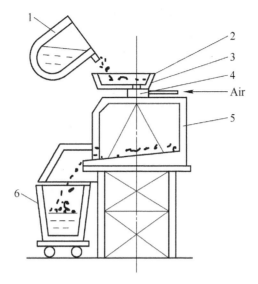

Figure 6.6
Device diagram of the atomization furnace.

formed with a size less than 2 mm. When the droplets fall in the atomizing chamber and the semisteel tank, an oxidation reaction occurs with the oxygen in the airflow, forming crude vanadium slag. The refined vanadium slag is obtained by crushing and magnetic separation and is then used as the main raw material for chemical extraction of vanadium. (2) Air-side converter blowing. China's Chengde Iron and Steel plant (in the early 1960s), and Maanshan Iron and Steel Company (in the early 1970s), adopted the process of air-side converter blowing to produce vanadium slag, where semisteel for steelmaking was prepared using oxygen top blowing, that is, double production of vanadium slag. Compressed air or oxygen-rich air is injected from the side air nozzle of the converter, which stirs the V-containing molten iron; Si, Ti, V, and other metallic elements are oxidized, producing the vanadium slag. The semisteel is used as the raw material for steelmaking. When the vanadium content of molten iron is low, pig iron containing a high V content is used as a coolant during slag blowing, where the vanadium grade of the slag increases by 16%–20%. (3) Vanadium slag blowing in Russia uses top- or bottom-blown oxygen converters to produce vanadium slag. The air is injected into the top- or bottom-blown converters, where oxygen-rich air is used to stir the V-containing molten iron, resulting in oxidation of Si, Ti, and V in the V-containing molten iron to produce the vanadium slag.

- Vanadium extraction from bottom-blown converters: Russia's Chusov joint company places V-containing molten iron into a bottom-blown converter for blowing, where the vanadium slag is produced during a steelmaking process. The total recovery of vanadium slag from V-containing molten iron is about 90%.

- Double vanadium extraction by top-blown converters: top-blown converters are used to blow V-containing molten iron and separate it into semisteel and vanadium slag in the Nizhny Tagil steel plant in Russia. The recovery of vanadium slag from V-containing molten iron is 92%–94%. This method is also used to produce vanadium slag in Chengsteel, Masteel, and Panzhihua Steel companies in China.
- Atomization process for vanadium extraction from molten iron from a blast furnace. Before 1997 the Panzhihua Iron and Steel Corporation, China poured V-containing molten iron into intermediate tanks, which then supplied the atomizers. After atomization, vanadium was oxidized from V_2O_3 to V_2O_5, where a mixture of V_2O_4 and V_2O_3 sunk in the semisteel tank and entered the semisteel, while vanadium slag was formed on the surface. The atomized vanadium extraction process was invented by Panzhihua Iron and Steel Co., Ltd., and successfully applied to industrial production. It was the main method of vanadium slag production in China in the 1970s and 1980s. Now it has been replaced by converter vanadium extraction processes, where the oxidation rate of vanadium is generally up to 85%–90%, the recovery rate of vanadium is $\sim$73.6%, and the recovery rate of semisteel is 93.9%. The main advantages of the atomization vanadium extraction process include a long furnace life, large processing capacity, semicontinuous production, simple equipment, and easy operation.

Vanadium extraction using slot-type furnaces has been used by Masteel in China. The capacity of the slot-type furnace is 70 t/h, where the oxidation rate of vanadium is 88.5%–95.2%, the recovery of vanadium is 81.3%–90.49%, and the rate of semisteel recovery is 90.20%–94.1%. Advantages include continuous production, simple equipment, and low production cost, while the disadvantages are a high iron content in the vanadium slag and low recovery rate of vanadium.

6.4 Enriched vanadium slag from V-containing molten iron by sodium treatment and oxidation

The high sulfur and phosphorous contents of V-containing molten iron result in processing difficulties. V-containing molten iron is treated with sodium for desulfurization, dephosphorization and the recovery of vanadium slag. The traditional oxidation roasting process used for vanadium extraction is omitted, where the aim is to combine the purification outside the furnace and vanadium recovery and steelmaking processes, practiced less slag steelmaking, and no slag steelmaking. The sodium is injected into the V-containing molten iron via an airflow, where vanadium in the molten iron is oxidized to vanadium oxide, which combines with sodium carbonate at high temperature to form sodium vanadate. The vanadium extraction process takes advantage of sodium's strong reactivity with sulfur and phosphorus to achieve desulfurization and dephosphorization, respectively, to obtain high-quality semisteel and water-soluble sodium-treated vanadium slag. Trials were performed in a steel plant of Panzhihua Steel with a test scale of

20–85 t of sodium-treated vanadium slag, where sodium carbonate was blown into the molten iron with compressed air. A gas medium was used to produce sodium carbonate in the molten iron via a diffusion reaction, where a simultaneous oxygen flow oxidized dissolved vanadium. The sulfur and phosphorus in the molten iron reacted with sodium carbonate to form polymers with the vanadium slag. Vanadium mainly existed in the form of sodium vanadate. The chemical reactions involving sodium-treated vanadium slag are very similar to those occurring during blowing of vanadium slag. The chemical reactions for sodium treatment, desulfurization, and dephosphorization are as follows:

$$[S] + Na_2CO_3 = Na_2SO_4 + CO_2 \quad (6.34)$$

$$[P] + Na_2CO_3 = NaPO_3 + CO_2 \quad (6.35)$$

$$V_2O_3 + Na_2CO_3 = Na_2O \cdot V_2O_3 + CO_2 \quad (6.36)$$

The preparation of sodium-treated vanadium slag belongs to the experimental requirements, where the main part of the equipment is the tundish. Because of the particularity of sodium salt, the sodium-treatment process is decomposed at high temperature. There is a large loss of sodium and salt during the process of gas injection and blowing, where dust is dispersed in the operation space and surrounding area of the furnace platform. Hence, the operating environment is extremely poor. If the vanadium slag is excessively treated with sodium, more than 50% of the slag weight can be directly dissolved in water, resulting in aqueous solutions with high basicity, ionic strength, and impurity contents, which can damage the reductant, and after hot leaching may produce secondary precipitation. The treatment of a ton of molten iron requires 48 kg of sodium salt, which produces 30 kg of sodium-treated vanadium slag.

6.5 Comparison of various vanadium extraction methods

The main methods for extracting vanadium from V-containing molten iron include use of converters, ladles, and atomizers for vanadium recovery. These three methods are mainstream processes in modern processing. They all use special equipment, where the process is well controlled, and the technical and economic indicators are stable.

6.5.1 Tank furnace method

A flow of V-containing molten iron enters the tank furnace at a constant velocity, while being aerated with air, oxygen, or oxygen-rich gas via nozzles installed at specific locations along the furnace. The gas flow has the same velocity as the V-containing molten iron. Si, Mn, and V are oxidized and separated from the molten iron. The tank furnace is generally composed of three sections: (1) control section, where a front chamber or intermediate pot is used to ensure a constant flow of the molten iron into the next section;

(2) blowing section, where the air, oxygen, or oxygen-rich gas are blown into the molten iron at the side, bottom, top, or a combination of these, and an appropriate coolant is added to control the temperature of the molten pool; (3) separation section, where semisteel and vanadium slag are mechanically separated by removing the vanadium slag from the surface of the semisteel. The main process parameters for the tank furnace method are the gas pressure and flow rate, nozzle height, melting pool depth, and blowing time. It is also necessary to limit the molten pool temperature to below 1400°C. The advantage of this method is that it does not require a large plant or heavy lifting equipment, although the process control is poor, and corrosion damage of the refractory material is serious.

6.5.2 Molten iron tank method

The V-containing iron is contained in the molten iron tank, where a movable injection gun can be inserted into the molten iron to blow the vanadium slag. The injection gun is generally a tube with its exterior protected by a refractory material. The gun penetration depth, pressure and flow rate of the blowing gas, time of blowing, and temperature of the molten pool are the important process parameters for this method. The disadvantages of this method include poor process control, splashing, and low yield.

The molten iron tank method can also be used with oxygen or air as a carrier gas to inject Na_2CO_3 into the molten iron. In this case, Si, Mn, and V in the molten iron are oxidized, along with S and P, and enter the vanadium slag simultaneously. Advantages of this process are a low residual vanadium content in the semisteel, and low contents of S, P, and other impurity elements, which is particularly conducive to steelmaking. However, Na_2CO_3 is volatile at high temperature, pollutes the environment, and also seriously damages equipment and refractory materials.

6.5.3 Atomization furnace method

The molten iron contained within iron pots is supplied to the tundish under controlled-flow conditions, and then atomized via a nozzle at the bottom of the tundish, where the high-speed airflow produces droplets with a size less than 2 mm. When the droplets fall in the atomizing chamber and semisteel tank, they are oxidized via contact with the oxygen in the airflow to produce crude vanadium slag, which is separated by crushing and magnetic separation to produce the refined vanadium slag.

6.5.4 Converter process

China's Chengde Iron and Steel plant (in the early 1960s), and Maanshan Iron and Steel Company (in the early 1970s), adopted the use of air-side blowing converters to produce vanadium slag and top-blown oxygen converters for producing semisteel for steelmaking.

This is the so-called double technique for production of vanadium slag. The converter method is currently the main technique for blowing vanadium slag. Compressed air or oxygen-rich air is injected from an air nozzle at the side of the converter to stir the V-containing molten iron, resulting in Si, Ti, V, and other elements being oxidized to produce vanadium slag. The semisteel is used as a raw material for steelmaking. When the vanadium content of molten iron is low, pig iron with a high vanadium content is used as a coolant when blowing vanadium slag, resulting in the vanadium grade of the slag increasing by 16%−20%. Table 6.7 compares the main technical and economic indicators of the main vanadium production processes in China and around the world.

6.5.5 Direct-reduction−electric furnace−shaking ladle−vanadium slag method

In direct-reduction−electric-furnace−shaking-ladle−vanadium-slag processing of V-containing titanoferrous magnetite the V-containing iron concentrate is pelletized with carbon, and reduced in a rotary kiln (or rotary hearth furnace, shaft furnace, or tunnel kiln). A temperature range of 1000°C−1050°C is maintained to prepare the metallic pellets, which are smelted and further reduced in an electric furnace. Then, separation of V-containing molten iron and titanium slag is performed. The titanium slag is further processed by the titanium industry, while V-containing molten iron is used for extracting vanadium slag by shaking ladle or vibrating pot methods. These extraction methods have low productivity, where the main consideration is to adapt to the electric furnace cycle. South Africa uses rotary kiln−electric furnace processes, which are combined with a shaking ladle or vibrating tank to produce vanadium slag, where the average content of vanadium in the slag produced by the plant is ∼23%.

The ladle is placed on a cradle and shaken in an eccentric manner 30 times/min. The amounts of oxygen and coolant were calculated according to the composition and temperature of the molten iron. Iron block and scrap steel coolants are added before blowing with oxygen, where the height of the gun is 750 mm, the oxygen flow is 28−42 m^3/min, and the oxygen pressure is 0.15−0.25 MPa. When the amount of oxygen blown reaches the predetermined value, the gun is stopped, then the shaft continues for 5 min to reduce the iron oxide content in the slag and increase the grade of vanadium slag. When blowing of the vanadium slag is finished, the semisteel is poured into the converter and the vanadium slag is transported to the slag field for cooling. The process of blowing vanadium slag with oxygen in a shaking ladle involves eccentric shaking of the ladle, which results in good agitation of the molten iron. Hence, the oxygen can be passed into the molten pool under low pressure, and high recovery of vanadium is achieved, while adhesion of the gun can be prevented.

The technological parameters and main indicators for a 60-t shaking process performed by the Highweld Vanadium and Steel Co., Ltd were as follows. The charge weight of molten

Table 6.7: Comparison of main technical and economic indicators from vanadium slag production companies in China and around the world.

Producer	C content of semisteel (%)	Semisteel temperature (°C)	Vanadium slag (%)					Oxidation (%)	Recovery (%)	Capacity (10 kt)
			V_2O_5	TFe	MFe	P	CaO			
Highveld	–	1270–1400	23–25	29	–	–	3	>93	82	18
Nizhny Tagil	3.20	1380	16	32	9–12	0.04	1.5	≥90	82–84	17
Chengde Steel	3.17	1400	12	34	21	0.07	0.8	87.8	77.6	>6
Pan steel	3.57	1375	19.40	28–33	9–12	<0.1	2.0	91.42	82.0	17

iron was 66.8 t, weight of added iron ore was 1.5 t, charge weight of iron block was 6.0 t, amount of added river sand was 0.19 t, and oxygen consumption was 21.54 m^3/t. The amount of semisteel produced was 68.2 t (93% yield) and output of vanadium slag was 5.85 t (93.4% yield). The main technical indicators were: oxidation of vanadium was 93.4%, recovery of vanadium was 91.6%, recovery of semisteel was 93%, total blowing time was 52 min, vibrating time was 59 min, total cycle time was 90 min/furnace, temperature of molten iron before oxygen blowing was 1180°C, metal melting temperature was 1270°C, oxygen nozzle diameter was 2 in., height of oxygen blowing tube from static pool surface was 76.2 cm, normal oxygen flow rate was 28.3 Nm^3/min, oxygen flow rate was 42.5 Nm^3/min, and oxygen blowing tube pressure (under normal flow) was 160 kPa.

6.6 Main equipment

6.6.1 Vanadium recovery converter

The converter is composed of a furnace shell and lining. The furnace shell is made of welded steel plate, and the lining consists of three parts: a working layer, permanent layer, and filling layer. The working layer is in direct contact with the liquid metal, slag, and furnace gas and can be easily eroded. The permanent layer is fixed to the furnace shell to protect the shell steel plate and is not removed when the furnace is repaired. The filling layer between the permanent layer and working layer is composed of tar magnesia or tar dolomite, which is used to reduce the pressure because of thermal expansion of the working layer on the furnace shell. Hence, it is convenient to dismantle the furnace.

Generally, the furnace can be rotated and belongs to the class of metallurgical furnaces used for blowing steel or blowing mattes. Converter furnaces are manufactured from steel plate and refractory linings and are generally cylindrical. The chemical reaction heat is generated by the blowing process when smelting, and an additional heat source is not used. The converter is one of the most important steelmaking components, which can also be used for Cu and Ni smelting. According to the refractory properties, the converter can be divided into alkaline converters (using magnesite or dolomite as lining) and acid converters (lined with siliceous material). The converter can also be classified by the mode of gas delivery: bottom-blown converter, top-blown converter, and side-blown converter. Depending on the gas blown into the furnace, converters are classified as air or oxygen converters. Steelmaking in converters is the main intensive processing of molten iron. In this process, heat is required, which is provided by physical heat and reaction heat generated by the chemical reaction between melt components (such as carbon, manganese, silicon, and phosphorus), and the injected oxygen, which allow the steelmaking process to achieve the compositional and temperature requirements. The furnace charge mainly consists of molten iron and slag (such as lime, quartz, and fluorite). To adjust the temperature, scrap steel and a small amount of cold pig iron and ore are added. In the process of converter steelmaking, carbon in the molten iron can produce a mixture of carbon monoxide and a small amount of carbon dioxide, which is used as the converter gas. The amount of converter gas produced in a smelting process is not steady, and components also change. The gas from the converter smelting process is collected, usually by passing it through cooling and dust removal systems, where it is accumulated and stored in a gas vessel and is in mixed equilibrium before delivery to the user.

6.6.2 Atomization furnace for extracting vanadium

The main equipment includes the atomizer, atomizing chamber, and tundish. There is a rectangular furnace in the atomization chamber where atomization and oxidation reactions of the molten iron occur. The side and top walls are composed of water-cooled parts, while the bottom of the furnace and the lower furnace wall are built of firebrick. The chamber is designed according to the angle α at which two rows of gas inlets meet in the atomizer and the amount of the molten iron to be treated. The volume−capacity ratio is 0.7−0.85 m^3/t. The best lining material is magnesium refractory, which is both erosion resistance and does not pollute the vanadium slag. The slope of the bottom is 10%. The tundish is a rectangular container for welding that is lined with refractory materials. to avoid the blast furnace slag entering the atomization chamber, the middle section has a wall of slag and the bottom has a rectangular nozzle of 220 mm × 20 mm to control the flow and shape of the iron flow. With high-speed airflow, the molten iron is atomized into small droplets, and the vanadium in the molten iron is oxidized to vanadium slag. The process focuses on separation of vanadium and all other elements present, such as silicon and titanium oxides, which have an activity coefficient close to that of vanadium. Some manganese, phosphorus, and iron are oxidized and separated, while sulfur is generally not oxidized significantly.

The molten iron is poured into a rectangular tundish and flows across the rectangle nozzle at the bottom of the tundish into the atomizing chamber. As the flow of molten iron meets the high-speed airflow from the atomizer in the atomization chamber, tiny iron droplets are produced. This rapid increase in surface area results in good oxidation kinetics of the elements, enhancing their oxidation and slag formation. The semisteel and slag pass through the chute into the tundish, where they are separated. First, the semisteel is poured out for steelmaking, and then the vanadium slag is removed.

Process characteristics: as the molten iron is atomized into tiny iron droplets, each one behaves as a small independent pool. After the vanadium atoms at the surface are oxidized, internal vanadium atoms diffuse to the surface, while other elements at the droplet surface are oxidized. The iron droplets fall into the bottom of the atomization chamber and are collected in the semisteel tank, where oxygen is provided by the FeO to allow the oxidation reaction to continue as $(FeO) = [Fe] + [O]$, $2[V] + 3[O] = (V_2O_3)$, $[C] + [O] = CO$. Around 60% of the vanadium is oxidized in the furnace and about 40% in the tank. Therefore the atomization of vanadium can meet the vanadium–carbon balance requirements and obtain high oxidation (over 90%), while the resulting semisteel meets the requirements for steelmaking.

CHAPTER 7

Extraction of vanadium from V-containing titanoferrous magnetite: extraction of V_2O_5 from vanadium slag

Chapter Outline
7.1 Sodium salt roasting   151
    7.1.1 Main process   155
    7.1.2 Main reactions   155
    7.1.3 Preparation of raw materials   158
    7.1.4 Main equipment   160
    7.1.5 Roasting of vanadium slag   161
    7.1.6 Leaching of roasted vanadium slag   164
    7.1.7 Purification of vanadium leaching solution   165
    7.1.8 Precipitation of vanadate   166
    7.1.9 Melting process   171
    7.1.10 Main equipment and process parameters   172
    7.1.11 Calcining and sodium roasting with phosphate   174
7.2 Extracting vanadium pentoxide from vanadium slag by calcium salt roasting   175
    7.2.1 Extraction process   175
    7.2.2 Extraction reactions   175
    7.2.3 Extraction characteristics   178
    7.2.4 Calcium salt roasting of vanadium slag   179
    7.2.5 Leaching of Ca salt−roasted vanadium slag   180
    7.2.6 Precipitation of vanadate   180
    7.2.7 Roasting and leaching equipment   181
7.3 Extracting vanadium pentoxide from Na-treated vanadium slag   181
7.4 Extracting vanadium from V-containing steel slag by acid leaching   182
    7.4.1 Extraction process   182
    7.4.2 Extraction reactions   183
    7.4.3 Main equipment and process parameters   184
7.5 Extracting V_2O_5 from vanadium slag by sulfide addition   184
7.6 Main raw materials   184
    7.6.1 Industrial raw materials   184
    7.6.2 Industrial fuel   189
7.7 Main vanadium extraction equipment   193
    7.7.1 Rotary kilns   193

7.7.2 Multihearth roasting furnace   195
7.7.3 Stove   196
7.7.4 Ball mills   197
7.7.5 Thickener   197
7.7.6 Magnetic separators   198
7.7.7 Filter   198
7.7.8 Melting furnace   199

The technological process for extracting vanadium was designed for standard vanadium pentoxide products and includes the entire process from receiving the V-containing raw materials to producing the final standard vanadium pentoxide (V_2O_5) products. The technical parameters and processing capacity of the equipment are designed based on the desired downstream standard vanadium pentoxide products. Vanadium extraction processes based on the available raw materials include the main extraction process, where the main purpose is to extract vanadium by enriching all V-containing materials, and subextraction processes where vanadium is extracted via the enrichment of other by-products. The design of vanadium extraction processes also considers the initial chemical additives for either alkali or acid treatments. Alkali treatments are classified into sodium and calcium salt methods. The ore can also be calcined without additives to achieve conversion reactions and extract vanadium. The conversion of vanadium is achieved via alkali, acid, or hot-water leaching [13,22,25,32,33,41–48,57,60,61,69,70,71,73–79]. For the treatment of low-grade vanadium slag, it is necessary to develop a simple and feasible process that is optimized for local conditions and can be applied to other industries. Then, vanadium extraction is performed after enrichment and transformation.

The process of extracting vanadium from vanadium slag generally follows: calcination, leaching, purification, precipitation, and additional calcining steps. This produces the final V_2O_5 product, where the key technologies are the roasting and transformation steps. Roasting vanadium slag is an oxidation process, where the vanadium slag is heated and oxidized at high temperature; trivalent vanadium is oxidized to tetravalent and pentavalent vanadium. This process degrades the mineral structure of the vanadium slag. To facilitate oxidation of vanadium and convert it into soluble vanadium salt, additive materials must be used, where sodium salt and calcium salt are the two main types of additives. High-Ca vanadium slag is calcined with calcium phosphate, while high-Ti vanadium slag is treated with sulfide roasting to extract vanadium.

In the 1990s salt-free oxidizing roasting was attempted in vanadium extraction processes. Oxygen in an airflow is used to oxidize vanadium slag at high temperature without any additives, where low-valence vanadium oxide is directly converted to high-valence vanadium oxide. This method reduces environmental pollution, although the roasting effect is not ideal. At the beginning of this century, high-temperature roasting with various additives was developed for extracting vanadium from vanadium slag. This method is suitable for extracting vanadium from V-bearing titanoferrous magnetite and increases the efficiency of vanadium transformation. The most mature roasting process is still the sodium salt method, although it causes serious environmental pollution. Calcium salt roasting for extracting vanadium is a promising process, because it both eliminates environmental pollution and the roasting effect is improved. However, there exist problems with this method; specifically, the required industrial production technology is not mature, the raw and auxiliary materials are not standard, and the control of the product quality is challenging. Calcium salt roasting for extracting vanadium has been used successfully in Russia to solve the problem of ammoniacal nitrogen in wastewater, which has been enhanced by the development of the vanadium industry. However, the product quality is still low. In China, calcium salt roasting for extracting vanadium is undergoing industrial trials.

The extraction of V_2O_5 from vanadium slag is a system selection process. To a certain extent, it is the typical mainstream process that represents an era, a new industry, and a product of the development of the advanced level and its characteristics. It is necessary to reflect the core ideas of vanadium extraction by systematic theoretical analysis, along with consideration of production experience. This will enhance the valuable industrial products and resources. Depending on the technical means and parameter selection, the core idea runs through all-round runs through the whole process. Comprehensive consideration of the availability of raw materials should be first addressed, followed by the preparation of a complete set of equipment satisfying the requirements of advanced technology, economic viability, and controllability. Finally, the process must satisfy the market demand for high-end products while meeting the environmental and safety standards of the industry [41–45,50,51].

7.1 Sodium salt roasting

The extraction of V_2O_5 from vanadium slag is a system selection process. External controllable measures must be taken to achieve the required changes in the structure of the vanadium slag to produce a water-soluble material suitable for alkali or acid leaching and maximize the separation of vanadium and impurities in the vanadium slag. Fig. 7.1 shows the thermodynamic equilibrium diagram of sodium vanadate, sodium silicate, and sodium aluminate. The standard free energy of sodium vanadate is less than those of sodium silicate and sodium aluminate. Hence, during sodium roasting, sodium vanadate is prior to the sodium silicate and sodium aluminate. Fig. 7.2 shows the phase diagram of the V–O–H system, where different systems containing vanadium ions occur at different pH ranges. The Al_2O_3 content in vanadium slag is very low (in the microlevel), and it does not generally produce significant amounts of sodium aluminosilicate [$1/2Na_2O \cdot 1/2Al_2O_3 \cdot 2SiO_2$] during the extraction of V_2O_5 from vanadium slag. The standard free energies of the formed single silicates and aluminates are higher, resulting in a higher temperature of salification. It is less likely for sodium aluminosilicate [$1/2Na_2O \cdot 1/2Al_2O_3 \cdot 2SiO_2$] to salify at the salification temperature of vanadate and will not affect the extraction of V_2O_5 from vanadium slag.

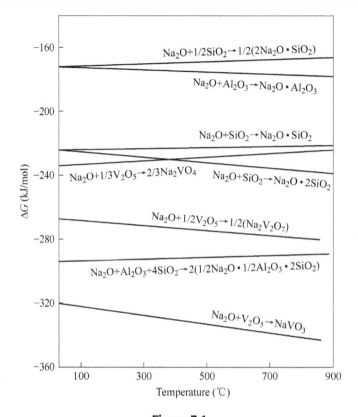

Figure 7.1
Thermodynamic equilibrium diagram of sodium vanadate, sodium silicate, and sodium aluminate.

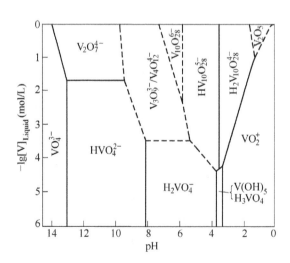

Figure 7.2
Phase diagram of the V–O–H system.

There are many restrictions on the formation of vanadate. Appropriate parameters must be optimized and selected to overcome the coionic effect and adapt to the actual separation requirements. Vanadates have different salt-forming temperatures and can be converted into each other at high temperatures. The melting points of the major vanadates are shown in Table 7.1; there are larger differences in the melting points of the same series of vanadates, which can be used to obtain the desired vanadate. By optimizing the temperature, a water-soluble system suitable for alkaline leaching or acid leaching can be achieved, which maximizes the separation of vanadium and impurities in the vanadium slag.

Table 7.1: Melting points of major vanadate compounds.

Vanadate	Molecular formula	Melting point (°C)
Sodium orthovanadate	Na_3VO_4	850–866
Sodium pyrovanadate	$Na_4V_2O_7$	625–668
Sodium metavanadate	$NaVO_3$	605–630
Calcium metavanadate	$CaO \cdot V_2O_5$	778
Calcium pyrovanadate	$2CaO \cdot V_2O_5$	1015
Calcium orthovanadate	$3CaO \cdot V_2O_5$	1380
Iron orthovanadate	$FeVO_4$	870–880

In the process of extracting vanadium from vanadium slag by sodium salt roasting, the aim is the formation and transformation of pentavalent vanadium, where the vanadates in the system are mainly sodium vanadate, sodium orthovanadate, and sodium pyrovanadate, where sodium vanadate is dominant. Fig. 7.3 shows the solubility diagram of the $Na_2O-V_2O_5-H_2O$ system. All vanadates, including sodium vanadate, sodium orthovanadate, and sodium pyrovanadate, are water soluble, and low vanadate has acid solubility.

Sodium metavanadate can be dissolved in water, although the increase in solubility with increasing temperature is limited as it easily forms a hydrate in the presence of water. When the vanadium concentration is high, an anhydrous crystal is produced from sodium metavanadate at temperatures above 35°C, while hydrated sodium metavanadate is generated at temperatures below 35°C. Vanadium slag containing high-calcium content can be treated using calcium phosphate to reduce the calcium and perform sodium salt roasting. The slag is ground in a ball mill, mixed with a certain amount of sodium phosphate and sodium carbonate, and then roasted at high temperature for a certain time to convert calcium ions into calcium phosphate. Vanadium reacts with sodium carbonate to produce water-soluble sodium vanadate, which can be water leached. Roasting at 1373K for 2 h showed that the recovery of vanadium increases with increasing phosphate content. For a phosphate ratio of 300%, the recovery rate of vanadium reached 99%. Theoretically, when the temperature reaches 1473K, 100% of the vanadium can be recovered.

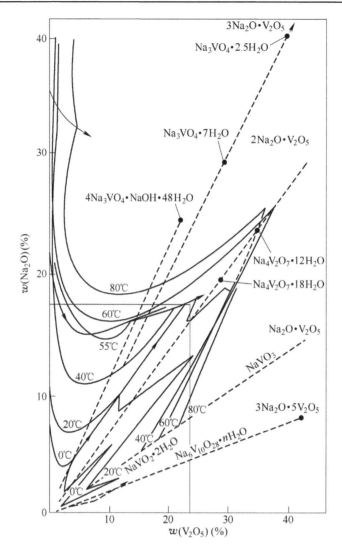

Figure 7.3
Solubility diagram of the $Na_2O-V_2O_5-H_2O$ system.

High-Ti vanadium slag can also be treated with sulfur and magnesium oxide as additives. The mixture is then heated to 1873K, so that the slag can be fully cured, followed by cooling. Vanadium is mainly concentrated as oxides and sulfides. Studies showed that the ratio of V/VO_3 increases with decreasing FeO content in the slag. The slag pellets were then ground to 200 mesh, and SO_2 and O_2 gases were injected at a temperature of 723K. The vanadium was converted into $VOSO_4$ that is highly soluble in water and can be water leached at room temperature. Longer water-leaching times resulted in better vanadium recovery, with a recovery rate up to 88.2%. The Russian Tula factory can leach more than 90% of the vanadium into solution through roasting, pulping, and acid leaching. This method is simple and easy to perform but results in a low grade of V_2O_5. Plants in former West Germany used to mix slag and lime for roasting a certain amount of time after alkali leaching, and obtained 85%−90% recovery.

7.1.1 Main process

In the process of extracting V_2O_5 from vanadium slag by oxidizing and sodium salt roasting, vanadium slag is crushed and ground in a ball mill. Then, iron or iron grains are removed via magnetic separation and air separation. A mixture containing vanadium slag (grain size <0.1 mm) and sodium salt (Na_2CO_3, NaCl, and Na_2SO_4) is oxidized and roasted at high temperature, where the vanadium in the slag is converted into a water-soluble sodium vanadate, which is then leached with hot water and dilute sulfuric acid. Fig. 7.4 shows a process flowchart of extracting V_2O_5 from vanadium slag by oxidizing and sodium salt roasting. After the aqueous solution containing vanadium is purified, acid is used to adjust the pH of the solution, and ammonium polyvanadate (APV) is precipitated in the presence of NH_4^+ and NH_3. Finally, V_2O_5 is obtained by calcination of high temperature.

7.1.2 Main reactions

Vanadium in the slag exists as an iron−vanadium spinel ($FeO \cdot V_2O_3$), which is not water soluble. There is a phase transition during high-temperature roasting, where the phase structure of the iron−vanadium spinel is damaged, and trivalent vanadium (V^{3+}) is oxidized and converted into pentavalent vanadium (V^{5+}) under the oxidizing atmosphere. The V^{5+} ions react with alkalis (Na_2CO_3 or NaCl and Na_2SO_4) to produce soluble vanadium compounds.

In the process of sodium roasting, there are two kinds of important chemical reactions occurring in the vanadium slag. The first is the oxidation reaction, where low-valence vanadium oxide in the slag is completely oxidized to high-valence vanadium oxide. The second is the sodium reaction, where high-valence vanadium oxide (V_2O_5) reacts with sodium salt (Na_2CO_3, NaCl) to produce water-soluble sodium vanadate ($NaVO_3$).

7.1.2.1 Oxidation of metallic iron

Oxidation of metallic iron starts at about 300°C, where the reactions are as follows:

$$Fe + \frac{1}{2}O_2 \rightarrow FeO \tag{7.1}$$

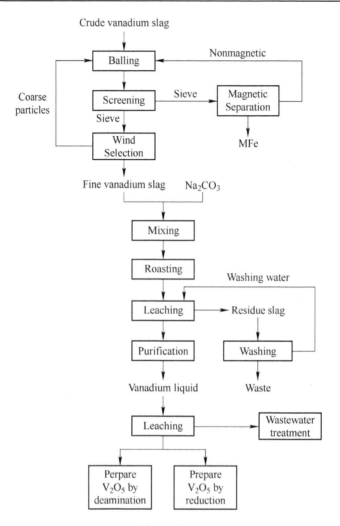

Figure 7.4
Process flowchart for extracting V_2O_5 from vanadium slag by oxidizing sodium salt roasting.

$$2FeO + \frac{1}{2}O_2 \rightarrow Fe_2O_3 \quad (7.2)$$

7.1.2.2 Iron olivine decomposition by oxidation

The oxidative decomposition of iron olivine occurs at a temperature range of 500°C–600°C, where the reactions are as follows:

$$2FeO \cdot SiO_2 + \frac{1}{2}O_2 \rightarrow Fe_2O_3 + SiO_2 \quad (7.3)$$

7.1.2.3 Vanadium spinel decomposition by oxidation

The oxidative decomposition of vanadium spinel occurs at a temperature range of 600°C–700°C, where the reactions are as follows:

$$FeO \cdot V_2O_3 + \frac{1}{2}O_2 \rightarrow Fe_2O_3 \cdot V_2O_3 \quad (7.4)$$

$$Fe_2O_3 \cdot V_2O_3 + \frac{1}{2}O_2 \rightarrow Fe_2O_3 \cdot V_2O_4 \quad (7.5)$$

$$Fe_2O_3 \cdot V_2O_4 + \frac{1}{2}O_2 \rightarrow Fe_2O_3 \cdot V_2O_5 \quad (7.6)$$

$$Fe_2O_3 \cdot V_2O_5 \rightarrow Fe_2O_3 + V_2O_5 \quad (7.7)$$

7.1.2.4 Nitrification of vanadium pentoxide

After low-valence vanadium oxide in the vanadium slag is completely oxidized to high-valence vanadium oxide, the latter begins to react with sodium at 600°C–700°C. The vanadium pentoxide reacts with sodium salt to produce sodium vanadate, which is soluble in water. The reactions are as follows:

$$V_2O_5 + Na_2CO_3 \rightarrow 2NaVO_3 + CO_2 \uparrow \quad (7.8)$$

$$V_2O_5 + 2NaCl + H_2O \rightarrow 2NaVO_3 + 2HCl \uparrow \quad (7.9)$$

$$V_2O_5 + 2NaCl + \frac{1}{2}O_2 \rightarrow 2NaVO_3 + Cl_2 \uparrow \quad (7.10)$$

$$4FeO \cdot V_2O_3 + 4Na_2CO_3 + O_2 = 4Na_2O \cdot V_2O_3 + 2Fe_2O_3 + 4CO_2 \uparrow \quad (7.11)$$

$$2NaCl + \frac{1}{2}O_2 = Na_2O + Cl_2 \uparrow \quad (7.12)$$

$$Na_2O + V_2O_3 + O_2 = 2NaVO_3 \quad (7.13)$$

$$Na_2SO_4 + V_2O_3 + \frac{3}{2}O_2 = 2NaVO_3 + SO_2 \uparrow + O_2 \uparrow \quad (7.14)$$

The edulcoration reactions of desiliconization and dephosphorization are as follows:

$$3(Ca, Mg)Cl_2 + 2PO_4^{3-} = 6Cl^- + (Ca, Mg)_3(PO_4)_2 \downarrow \quad (7.15)$$

$$(Ca, Mg)Cl_2 + SiO_3^{2-} = 2Cl^- + (Ca, Mg)SiO_3 \downarrow \quad (7.16)$$

At the same temperature, vanadium pentoxide reacts with oxides of iron, manganese, and calcium to produce vanadates that are insoluble in water but soluble in acid. The reactions are as follows:

$$V_2O_5 + CaO \rightarrow Ca(VO_3)_2 \tag{7.17}$$

$$V_2O_5 + MnO \rightarrow Mn(VO_3)_2 \tag{7.18}$$

$$3V_2O_5 + Fe_2O_3 \rightarrow 2Fe(VO_3)_3 \tag{7.19}$$

7.1.2.5 Crystallization and deoxidization of sodium metavanadate

If the roasted material slowly cools during the cooling process in the kiln, sodium metavanadate can be crystallized and deoxidized to produce vanadium bronze, which is insoluble in water and inevitably reduces the leaching efficiency of vanadium. If the sodium metavanadate content is maintained by keeping the temperature above the melting point of sodium metavanadate (550°C), or quickly removing the material from the kiln and quenching it, crystallization and deoxidization of the sodium metavanadate can be avoided or reduced. The chemical reaction of vanadium precipitation is as follows:

$$Na_2H_2V_{10}O_{28} + 2NH_4^+ = (NH_4)_2H_2V_{10}O_{28} \downarrow + 2Na^+ \tag{7.20}$$

$$Na_2V_{12}O_{31} + 2NH_4^+ = (NH_4)_2V_{12}O_{31} \downarrow + 2Na^+ \tag{7.21}$$

The chemical reaction of deamination and molten pieces is as follows:

$$(NH_4)_2V_6O_{16} \rightarrow 2NH_3 \uparrow + 3V_2O_5 \tag{7.22}$$

$$V_2O_5 \text{(reduction in ammonia atmosphere)} = V_2O_4 + \frac{1}{2}O_2 \uparrow \tag{7.23}$$

$$Na_2SO_4 + V_2O_5 = 2Na_2VO_3 + SO_2 \uparrow + \frac{1}{2}O_2 \tag{7.24}$$

7.1.3 Preparation of raw materials

The first step in extracting vanadium is preparation of the raw materials. First, jaw crushers are used to crush the massive vanadium slag pieces in a coarse crushing process where a jaw plate is rotated around a fixed spindle. The massive vanadium slag is broken via bending deformation and fracture. Commonly used jaw crusher models include the PEE250 × 400 and PEE600 × 400, which have a processing capacity of about 6–8 t/h. Ball mills are used for fine-grinding processes, where the movement of the balls grinds and compresses the material, which accelerates dissociation of the ore. To achieve the required finely ground vanadium materials, commonly used ball mill models include 1500 × 5700, which has a processing capacity of 3−4 t/h of slag. In addition, magnetic separation equipment is used to remove iron from the vanadium slag. This process includes (1) magnetic separation of metallic iron, (2) sieving to remove large iron particles, and (3) and gravity separation methods that take advantage of the fact that the specific gravity of iron is greater than those of other components, for example, air separation of metallic iron commonly uses equipment, including a dry magnetic separation machine, vibration sieve, air separator, and cyclone classifier. The metallic iron content in the vanadium slag is limited to 5%−10%.

During the preparation of the raw materials, the vanadium slag is crushed and ground in a ball mill, and the iron is removed using magnetic separation and air separation. Vanadium slag with a grain size <0.1 mm is obtained, which ensures that the spinel grains have a very high degree of monomer dissociation, and the specific surface area of the vanadium slag is high. Good dynamic conditions are created for dissociation and the reactions between sodium and oxygen that can prevent burden adhesion and overheating during roasting because of oxidation of iron, which can lead to sintering of the material, influence of vanadium oxidation and sodium, and avoids Fe wastage. To achieve the desired amount of sodium salt (Na_2CO_3, NaCl, and Na_2SO_4), a standard sodium salt product is used after simple processing, including a screening process to remove inclusions or lumps and get rid of deteriorated or failed parts.

The prepared raw materials are separated into raw material storage bins, and metering equipment is installed for bulk ingredients. According to the desired intermittent or continuous flow of the ingredients, weight control is used for continuous weighing of batch ingredients. The different materials are loaded into the bins, the flow is calibrated, and the calculated flow is controlled. The feeding device types include disk feeders, electromagnetic vibration feeders, and screw feeders, where the screw feeder disk is installed under the tremie pipe, and powdery or granular materials are discharged from the bin and pile up in the distribution cone. The movable sleeve covers the discharge pipe of the bin, where the feeding rate is regulated by adjusting the position of a scraper, lifting activities, or the disk speed. The material moisture content needs to be <12%. Electromagnetic vibrating feeders use an electromagnetic vibrator to produce high-frequency vibration and drive a conveyor, where the material is fed in a parabolic jumping movement. Screw feeders adjust the feeding rate via the rotational speed of the screw. The movement of the materials needs to be taken into consideration. In general, screw, scraper, belt, bucket, and pneumatic conveyer types are used. According to the required composition, a proper mixture of the ingredients is selected. To reduce mixing losses and avoid dust in the environment, some moisture is required, but limited to <5%. Wet material is beneficial for increasing the contact area of different materials and roasting efficiency.

The second step is roasting and leaching. During high-temperature roasting the vanadium in the slag undergoes a phase transition, where the spinel phase structure is degraded and V^{3+} is oxidized and converted into V^{5+} under the oxidizing atmosphere. The V^{5+} ions react with the alkali (NaCl or Na_2SO_4), converting the vanadium into water soluble sodium vanadate, which is leached with hot water, resulting in the vanadium components being separated from most of the insoluble impurities. The raw material used in the roasting process is covered the corresponding transformation additive of sodium chloride, sodium carbonate, or sodium sulfate. The oxidants are mainly oxygen-rich compressed air, potassium permanganate, and soft manganese ore. Hot water and dilute sulfuric acid are needed for leaching. Hot water can be pure hot water, washing water, or dilute solutions. The third step is to purify the vanadium aqueous solution, where calcium chloride is used for phosphorous removal. In the fourth step the pH of the vanadium solution is adjusted with acid and APV and ammonium bivanadate are precipitated in the presence of ammonium (NH^{4+}, NH_3). The main ammonium salts used for vanadium precipitation are ammonium chloride and ammonium sulfate. The fifth step is calcination of the APV and ammonium bivanadate to obtain V_2O_5.

In the process of extracting vanadium by sodium salt roasting, sodium vanadate is leached as the vanadium aqueous solution that is converted into the liquid phase. The vanadium components are separated from most of the nonvanadium impurities that are insoluble in water. After purification of the aqueous solution of sodium vanadate and vanadium, APV and ammonium bivanadate are precipitated in the presence of NH^{4+} and NH_3, which is calcined to form V_2O_5.

Depending on the choice of materials and processes, materials involved in extracting vanadium include V-containing minerals, such as vanadium slag, and additives such as sodium salt (Na_2CO_3, NaCl, and Na_2SO_4), sulfuric acid, ammonium salt (NH_4Cl and $(NH_4)_2SO_4$), dephosphorization agent ($CaCl_2$), exchange resin, and extracting agents.

7.1.4 Main equipment

In the preparation of the raw material, ball mills are used for grinding, while magnetic separation equipment is used for separating iron. The roasting process uses multihearth furnaces, rotary kilns, stoves, smoldering furnaces, and fluidized furnaces. The various hearth furnaces and rotary kilns are large-scale equipment. These systems have several advantages, such as stable operation and good electrical performance. Selecting similar equipment can be advantageous to ensure commonality and practicality of the process that helps meet the requirements of large-scale production. Equipment such as the stove, smoldering furnace, and fluidized furnace are smaller-scale equipment, which are more suitable for manual control and small-scale resource development. The roasting furnace is divided into a preheating zone, preoxidation zone, reaction zone, and cooling zone. The calcining reaction temperature is generally 1123K–1223K; the actual temperature is determined by the properties of the vanadium slag, the type of sodium salt, and the amount of additives.

7.1.5 Roasting of vanadium slag

During roasting of the vanadium slag, it is oxidized and reacts with sodium to form a salt. The structure of the vanadium slag is broken down and dissociated during oxidation, where high-valence vanadium oxide and sodium salt react to form a salt. The roasting conversion rate of vanadium refers to the soluble vanadium content in the roasted slag.

7.1.5.1 Quality of vanadium slag

There are many factors affecting the roasting conversion rate of vanadium slag. First, the quality (i.e., structure and chemical composition) of the vanadium slag; vanadium in the slag mainly exists as the vanadium–iron spinel phase [$FeO \cdot V_2O_3$]. In addition to the vanadium phase in the slag, silicate (iron olivine) and metallic iron phases are dominant. The main composition of iron olivine is Me_2SiO_4 (Me is mainly Fe^{2+}, Mn^{2+}, and other metallic ions), where its chemical composition is characterized by a high FeO content. In the vanadium slag, iron olivine is deposited between the spinel granules making up the matrix of vanadium slag. The specific chemical composition of vanadium slag is as follows: 15%–21% V_2O_5, 37%–45% FeO, 13%–23% SiO_2, 0.5%–2.3% CaO, 0.9%–1.4% MnO, 9%–11% TiO_2, 1.2%–1.7% Cr_2O_3, 0.015%–0.04% P, and 12%–30% metallic Fe.

Control of the vanadium slag quality is generally required to lower the metallic Fe and P contents as much as possible. Metallic Fe can be removed by magnetic separation. When the Fe content of vanadium slag is high, rapid oxidation of Fe will occur, leading to local overheating of the furnace and even material sintering, which affects the roasting process. The particles of V–Fe spinel in the slag are very small. The vanadium slag must be finely ground to ensure that these spinel particles have sufficient area for exposure to oxygen and sodium salt to enable the oxidation and salt reactions, respectively. The particle size is controlled to 120–160 μm during production.

7.1.5.2 Roasting temperature

The roasting temperature is also very important. The roasting materials in the reaction furnace experienced a gradual change from low to high temperature and then high to low temperature. During this process, dissociation oxidation reactions occur at low temperature, while reactions with sodium occur at high temperature. After cooling, classified clinker is obtained. The temperature in the furnace can be divided into the oxidation zone (under 600°C), sodium zone (600°C–800°C), and cooling zone. Furnace cooling is generally from the highest temperature to 600°C, where the material is discharged. As the melting point of sodium metavanadate is 550°C, controlling the temperature of different types of roasting equipment is different. Multihearth furnaces discharge at the bottom, where the temperature is controlled to about 700°C. The discharge temperature of rotary kilns is controlled above 550°C. As the roasting temperature increases, the rates of vanadium oxidation and the salification reaction increase, along with the equilibrium constant of the reactions. In the range of 800°C–900°C, vanadium conversion is generally 90.2%–90.2%. Above 900°C the vanadium conversion rate falls sharply as sintering of the material begins and vitrification occurs. In addition, sodium carbonate reacts with SiO_2 to generate insoluble vitreous $Na_2O \cdot V_2O_5 \cdot SiO_2$, which affects the roasting process.

7.1.5.3 Roasting additives

The roasting additives used with vanadium slag are mainly sodium salts (Na_2CO_3, NaCl, and Na_2SO_4), where NaCl is an anaerobic acid salt. In the process of vanadium slag roasting, sodium salt decomposes to produce Na_2O that reacts with V_2O_5 to produce $NaVO_3$. NaCl is a metal halide salt with low melting point, which is volatile, easily decomposes, and has large losses during the roasting process. When NaCl is used as a single additive, as it is highly volatile, the NaCl content in the burden decreases, and inadequate amounts can remain. Increasing the NaCl content may lead to sintering of the charge material, kiln bonding ring, and contamination with volatile matter; as this affects the roasting temperature, NaCl cannot be used as a single additive. Na_2CO_3 is an aerobic acid salt that decompose and reacts with V_2O_5 to produce $NaVO_3$ and CO_2. Na_2SO_4 is an aerobic acid salt that decomposes and reacts with high-Ca vanadium slag to produce $NaVO_3$ and SO_2. The CaO vanadium slag reacts with the generated SO_2 to produce insoluble $CaSO_4$, which acts as fixed calcium. Na_2CO_3 is the main additive for roasting vanadium slag, supplemented with a small amount of NaCl. The role of NaCl is to nucleate in the early reaction, and increasing the porosity and reaction surface area of the material.

The additive ratio is referred to as the soda ratio that is associated with the vanadium content and impurities in the vanadium slag. The soda ratio reflects the overall sodium salt content that includes soda and NaCl salt. The soda ratio should be 1.2, where the amount of NaCl salt is 30% of that of the soda. The soda ratio is the amount of sodium carbonate to V_2O_5 in the vanadium slag.

Sodium carbonate is commonly known as soda. It is a stable compound in the form of a white powder, with a molecular formula of Na_2CO_3, melting point of 850°C, and decomposition temperature around 2000°C at atmospheric pressure. The decomposition temperature can be reduced in the presence of acidic oxides, which can dissolve in water and release heat. There are three kinds of sodium carbonate: sodium carbonate monohydrate, sodium carbonate pentahydrate, and sodium carbonate decahydrate, where anhydrous sodium carbonate is routinely used in industry.

Sodium chloride is commonly known as salt. It is a stable compound in the form of a white powder, with a molecular formula NaCl and melting point of 800°C. It deliquesces easily in air and is soluble in water. The forms include sodium chloride and sodium chloride dihydrate, where anhydrous sodium chloride is generally used in industry.

Sodium sulfate is commonly known as Glauber's salt. It is a stable compound that is difficult to decompose. It is a white powder with a molecular formula of Na_2SO_4, melting point of 884°C, and is soluble in water. The decomposition temperature of the pure substance is about 3177°C at atmospheric pressure, which can be significantly reduced in the presence of acid oxides. For example, in the presence of vanadium pentoxide, the decomposition temperature drops to 740°C. This property is successfully applied in the process of extracting vanadium. There are three kinds of sodium sulfate: sodium sulfate monohydrate, sodium sulfate pentahydrate, and sodium sulfate decahydrate, where anhydrous sodium sulfate is used industrially.

7.1.5.4 Oxidizing atmosphere

When roasting vanadium slag with the addition of sodium salt, the V−Fe spinel is decomposed, and low-valence vanadium oxide is oxidized and reacted with sodium salt. These reactions require the participation of oxygen. As the oxygen concentration of the gas increases, the conversion rate of vanadium increases. In general, it is required that the oxygen concentration in the tail gas discharged from the roasting furnace is >5%, which is achieved by adjusting the gas displacement; as the displacement increases, the oxygen content increases. If the oxygen charging exceeds a certain limit, the internal calcining zone of the kiln can shift backward, shortening the preheating zone. All physical and chemical changes in the material need to occur in the furnace, as they cannot take place in the preheating zone. Hence, the process cannot finish on time or advance to the high-temperature belt. If the sodium salt additive is fused in advance to produce a low-melting-point vitreous material, sintering of the material in the furnace occurs, which is not conducive to the production of soluble sodium metavanadate. The cooling zone is extended at the same time, resulting in a lower cooling speed. The release of soluble oxygen in the metavanadate may lead to the production of insoluble vanadium bronze [NaV_6O_{15} and $Na_8V_{24}O_{63}$], which reduces the leaching efficiency of vanadium. Fig. 7.5 shows the relationship between vanadium oxides of different valence.

7.1.5.5 Slag recycling

The roasting of high-grade vanadium slag is required to adjust the total vanadium content, which avoids excessive low-melting-point substances that aid sintering of materials in the furnace, and the need to add returning slag. The returning slag is usually vanadium tailings from leaching and washing or cleaning charge and roasting fault material, etc. The total quantity of low-melting-point sodium salt needs to be controlled while optimizing the second roasting to increase vanadium recovery. The addition of returning slag is determined by the quality of the vanadium slag. To accelerate oxidation, some manufacturers introduce oxygen-enriched gas and oxidants (such as oxygen, soft manganese, and potassium permanganate) during roasting.

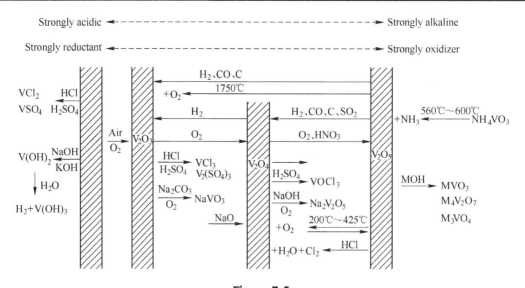

Figure 7.5
Relationship between different valence states of vanadium oxide.

7.1.5.6 Roasting in practice

The practice of sodium roasting requires (1) increasing the mixing time to ensure uniform mixtures; (2) reduction of the rotating speed of the rotary kiln, thus increasing the residence time of the materials; (3) a stable combustion system; (4) fixed quantity of material to reduce fluctuations in the material conversion rate; and (5) an increase in the soda ratio from ~1.2 to ~1.4.

7.1.6 Leaching of roasted vanadium slag

Fig. 7.6 shows the relationship between the vanadate, vanadium concentration, and pH of the aqueous solution at 25°C. There is state of variable valence hydration at pH values. During leaching the soda-roasted vanadium slag is finely ground using wet ball milling, leached in hot water, and then filtered using a rubber belt filter that recovers most of the vanadium in the soluble sodium vanadate salt solution. The leaching residue after filtering is washed and removed for secondary pickling, which mainly leaches slightly water soluble and slightly acid soluble vanadates, such as $Ca(VO_3)_2$, $Mn(VO_3)_2$, $Fe(VO_3)_2$, $Fe(VO_3)_3$, and composite salts. After leaching, all vanadium occurs as acid-soluble vanadate salts. Sulfuric acid or hydrochloric acid is used as a solvent. In general, sulfuric acid is used, where low-valence vanadium reacts with sulfuric acid to produce soluble vanadyl sulfate. To increase the leaching efficiency of vanadium, the roasting process needs to be optimized by ensuring good conversion and rapid cooling. On the other hand, to strengthen the leaching effect,

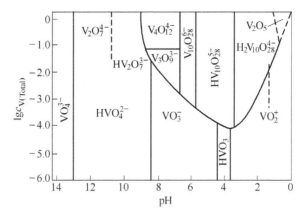

Figure 7.6
Relationship between the vanadate and vanadium concentrations and pH of the aqueous solution at 25°C.

especially secondary leaching, highly concentrated acid is used to dissolve the maximum amount of vanadium in the solution while preventing leaching of Fe^{2+}, Fe^{3+}, Mn^{2+}, Cr^{3+}, and AlO_2^-. To avoid complications during the purification of the vanadium solution, optimization of the roasting and leaching effects is required, along with appropriate selection of the pH.

7.1.7 Purification of vanadium leaching solution

There are metal impurity cations in the vanadium leaching solution, mainly Fe^{2+}, Mn^{2+}, and Mg^{2+}. The Fe^{2+} in solution is oxidized and converted into Fe^{3+} by reaction with oxygen. Hydrolysis of Fe^{3+} occurs at pH values of 1.5–2.3, where precipitation of $Fe(OH)_3$ occurs. In addition, Cr^{6+} is reduced to obtain Cr^{3+}, where hydrolysis of Cr^{3+} occurs at pH 4.0–4.9, and hydrolysis precipitation of $Cr(OH)_3$ is observed. The hydrolysis of Fe^{2+} occurs at pH 6.5–7.5, resulting in $Fe(OH)_2$ precipitation. The hydrolysis of Mn^{2+} occurs at pH 7.8–8.8, where $Mn(OH)_2$ is precipitated. All of the Mn^{2+} and Mg^{2+} ions are removed by hydrolysis and precipitation under alkaline conditions. Anions in the vanadium leaching solution, such as CrO_4^{3-}, SiO_3^{2-}, and PO_4^{3-}, can be removed by addition of an ion-precipitating agent at 90°C. When a precipitating agent containing Mg^{2+} is added, the hydrolysis of CrO_4^{3-} occurs at pH 9–10 to create precipitation of $MgCrO_4$, the hydrolysis of SiO_3^{2-} occurs at pH 6.5–7.5 to precipitate $MgSiO_3$. Hydrolysis of PO_4^{3-}, NH_4^+, and Mg^{2+} occurs at pH 8–9 to precipitate $MgNH_4PO_4$. Hydrolysis of PO_4^{3-} and Ca^{2+} occurs at pH 8–9 to precipitate $Ca_3(PO_4)_2$. In addition, calcium chloride is added to the vanadium solution for dephosphorization.

7.1.8 Precipitation of vanadate

The methods for vanadate precipitation from purified vanadium solution include precipitation by ammonium salt, calcium salt, or hydrolysis. The precipitation of the vanadate leaching solution by ammonium salt is divided into acid precipitation, weak acid precipitation, and alkaline precipitation by ammonium salt. The intermediate products of vanadate precipitation are different, but the target product is always V_2O_5. Comparison of vanadium precipitation methods is shown in Table 7.2.

7.1.8.1 Precipitation of vanadate by hydrolysis

After purification, vanadium mainly exists as V^{5+} in solution, with a small amount of V^{4+}. The vanadate exists in different agglomeration states during hydrolysis that depends on the pH, temperature, concentration, and impurities. With increasing solution acidity, vanadate is precipitated in a stepwise manner, and a red wine–like color gradually develops, which is known as red cake vanadate. The vanadium solution needs to boil. The vanadium concentration of the solution is kept at 5–7 g/L; if the concentration is too high, nucleation will be too fast and impurity adsorption will occur, resulting in the red cake vanadate being loose with a poor product quality. In particular, when Fe^{3+}, Al^{3+}, and P^{5+} ions are present, $FePO_4$ and $AlPO_4$ are produced, which contaminate the product. Fig. 7.7 shows the relationship between the state of V^{5+}, vanadium concentration, and pH of the aqueous solution, while the pH–Eh potential diagram of vanadium ions in the aqueous solution is shown in Fig. 7.8.

7.1.8.2 Precipitation of vanadate by ammonium salt

Under weak alkali conditions the vanadium solution is a $Na_2O-V_2O_5-H_2O$ system after purification. At pH 8–9 the vanadium is present as $V_4O_{12}^{4-}$, where the addition of NH^{4+} leads to the production and crystallization of $((NH_4)VO_3)$ or $2NH_3-V_2O_5-H_2O$, where the crystallization temperature is 20°C–30°C.

Under weak acid conditions the vanadium solution is a $Na_2O-V_2O_5-H_2O$ system after purification. At pH 4–6, vanadium exists as $V_{10}O_{28}^{6-}$, where the addition of NH^{4+} leads to decavanadate precipitation as the Na^+ ions in the vanadium solution result in precipitation of $(NH_4)_{6-x} \cdot Na_xV_{10}O_{28} \cdot 10H_2O$ ($x = 0-2$). To reduce the precipitation of Na, recrystallization is conducted at pH 2 to obtain ammonium hexavanadate $[(NH_4)_2V_6O_{16}]$. Residual vanadium in solution after crystallization can be reduced to 0.05–0.5 g/L.

Under acidic conditions the vanadium solution is a $Na_2O-V_2O_5-H_2O$ system after purification. At pH 2–3 the addition of NH^{4+} produces ammonium hexavanadate $[2NH_3 \cdot 3V_2O_5 \cdot H_2O]$. There are comprehensive advantages of this process, such as fast vanadate precipitation, high product purity, high efficiency of vanadate precipitation, low residual vanadium content in solution, and low ammonium consumption.

Table 7.2: Comparison of vanadium precipitation methods.

Item	Precipitation of vanadate by hydrolysis(V^{5+})	Precipitation of vanadate by hydrolysis (V^{4+})	Acidic precipitation of vanadate leaching solution by ammonium salt	Weak acidic precipitation of vanadate leaching solution by ammonium salt	Basic precipitation of vanadate leaching solution by ammonium salt	Use of calcium vanadate	Use of iron vanadate
Sediment	Red cake, sodium hexapolyvanadate	Tetravalent vanadium acid	Ammonium hexapolyvanadate	Ammonium decavanadate	Ammonium metavanadate	Calcium pyrovanadate	Iron vanadate
Chemical formula	$Na_2O \cdot 3V_2O_5 \cdot H_2O$	$VO_2 \cdot nH_2O$	$2NH_3 \cdot 3V_2O_5 \cdot H_2O$	$(NH_4)_{6-x} \cdot Na_xV_{10}O_{28} \cdot 10H_2O$	$2NH_3 \cdot V_2O_5 \cdot H_2O$	$2CaO \cdot V_2O_5$	$xFe_2O_3 \cdot yV_2O_5 \cdot zH_2O$
Precipitation pH of vanadate	1.5–3	3.5–7	2–3	4–6	8–9	5–11	<7
Vanadium solution concentration (g/L)	5–8	20	Low or high	High	High	Low or high	Low or high
Acid consumption	Large	Large	Large	Minor	Minor		
Ammonium consumption	No	Minor	Minor	Large	Large		
Precipitation temperature of vanadate (°C)	90	Room temperature	75–90	Room temperature	Room temperature	Room temperature	Room temperature
The main waste	Effluent	Effluent	Effluent, waste gas	Effluent, waste gas	Effluent, waste gas	Effluent	Effluent
Production cycle	Short	Short	Short	Long	Long	Short	Short
Precipitation efficiency of vanadate (%)	~98	99.5	>98	>98	>98	97–99.5	99–100
V_2O_5 purity (%)	80–90		>99	>99	>99	Low	Low

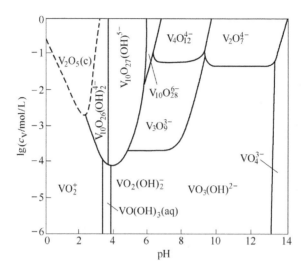

Figure 7.7
Relationship between the state of V^{5+}, vanadium concentration, and pH in aqueous solution.

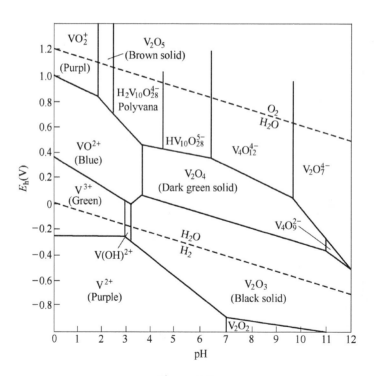

Figure 7.8
pH–Eh potential diagram of vanadium ions in aqueous solution [vanadium concentration 10^{-2} mol/L, 25°C, 100 kPa].

7.1.8.3 Precipitation of vanadate by calcium salt

After purification the vanadium solution is a $Na_2O-V_2O_5-H_2O$ system. Under the condition of strong mixing, the addition of [$Ca(OH)_2$, $CaCl_2$] results in the production of calcium vanadate. With increasing solution alkalinity, calcium metavanadate, calcium orthovanadate, and calcium pyrovanadate are obtained in turn. Calcium metavanadate has a high vanadium content and solubility, which easily results in low efficiency of vanadate precipitation. With increasing pH a silica gel precipitation occurs in the presence of SiO_3^{2-}, while Ca^{2+} and PO_4^{3-} are combined and precipitated as impurities with calcium vanadate. The relevant chemical reactions are as follows:

$$Ca(OH)_2 + 2NaVO_3 = CaO \cdot V_2O_5 + 2NaOH \qquad (7.25)$$

$$CaCl_2 + 2NaVO_3 = CaO \cdot V_2O_5 + 2NaCl \qquad (7.26)$$

If calcium vanadate is an intermediate, which can be dissolved with sulfuric acid, the reaction conditions are pH 2 and room temperature. Calcium is precipitated as $CaSO_4$ that can be removed by filtering. For a vanadium solution concentration of 25 g/L V_2O_5, sodium salt is added and the solution is heated, producing red cake vanadium at 90°C. A sodium bicarbonate solution can also be used, with reaction conditions of pH 9.4 and a temperature of 90°C, where calcium is precipitated as $CaCO_3$ and removed by filtering. For a vanadium solution concentration of 120 g/L V_2O_5, NH_4^+ is added, and reaction conditions of pH 8.3–8.6, and temperature of 25°C are used. The reaction time is controlled to 6 h, and ammonium metavanadate is precipitated.

7.1.8.4 Precipitation of vanadate by iron salt

After purification the vanadium solution is a $Na_2O-V_2O_5-H_2O$ system. Under the condition of strong mixing, Fe^{2+} and Fe^{3+} agents are added. Under a weakly acidic reaction condition, different colored precipitates are produced upon heating. The precipitate is green when ferrous compounds are added, while it is yellow when ferric compounds are added. V^{5+} ions are reduced to V^{4+} by ferrous compounds during this process, where ferrous ions oxidize into ferric ions. The vanadium product is precipitated as $Fe(VO_3)_2$, $Fe(VO_3)_3$, $Fe(OH)_3$, and $VO_2 \cdot H_2O$. The ferric irons are introduced during the precipitation of vanadates, where the vanadium product is precipitated as $xFe_2O_3 \cdot yV_2O_5 \cdot zH_2O$.

7.1.8.5 Precipitation of vanadate in industry

Before the start of vanadate precipitation, the precipitation tank must be calibrated to determine the amount and concentration of the standard vanadium solution, where the required amount of ammonium sulfate is calculated. The measured vanadium solution is transferred into the vanadium precipitation tank after preheating. The vanadium solution is heated to a certain temperature, and then acid is added to adjust the pH. Then, the solution is heated again, and excess ammonium sulfate is added. The solution is stirred well, and the pH value is adjusted again using acid. The classified product in the vanadium precipitation tank is transported to an APV collecting tank for temporary storage and then sent to a chamber filter press for filtration, washing, and blowing, where an APV cake containing about 40% water is obtained. The precipitated supernatant is pumped to a tank for gravity settling. The vanadium material at the bottom is recovered, and the supernatant is discharged into a wastewater pipe and sent to a water-treatment system. The main reaction involving vanadium is as follows:

$$6NaVO_3 + (NH_4)_2SO_4 + 2H_2SO_4 = (NH_4)_2V_6O_{16} + 3Na_2SO_4 + 2H_2O \quad (7.27)$$

The vanadium solution is adjusted to a pH of 2.0–2.5 and then stirred and heated to begin precipitation of vanadate. The equation for calculating the amount of sulfuric acid is as follows:

$$V_{H_2SO_4} \frac{2.88 \times C_{\text{vanadium solution}} \times V_{\text{vanadium solution}} K_{NH_3}}{D_{\text{sulfuric acid}} \times C_{\text{sulfuric acid}} \times 1000} (m^3) \quad (7.28)$$

During precipitation of the vanadate, all metrics should be within the required limits to ensure normal operation. The typical production indexes of the vanadate precipitation process include classified liquid V 23.81 g/L, classified liquid P 0.011 g/L, and superclear liquid V 0.074 g/L.

7.1.8.6 Factors influencing vanadate precipitation

The vanadate precipitation efficiency is influenced by the Si content in the vanadium solution. For a SiO_2/V_2O_5 molar ratio >0.89, the vanadate precipitation efficiency is <99%, while the vanadate precipitation time increases and filtration becomes difficult. The product quality decreases with increasing Si content in the vanadium solution. In industrial production the SiO_2/V_2O_5 moddlar ratio is ~0.05 that has little effect on the precipitation efficiency of vanadate or the product quality.

Phosphorus has a major influence on acidic precipitation of vanadate leaching solution by ammonium salt, because a complex clathrate (heteropolyacid) is formed by a reaction between phosphorus and ammonium. If the complex clathrate content reaches a critical value, the vanadate precipitation efficiency is dramatically affected. When the pH of vanadium solution is limited to ~2.5 during vanadate precipitation, and $[P]/[V_2O_5] > 0.0056$, the vanadate precipitation efficiency drops below 99% and then decreases dramatically. In industrial production, calcium chloride is used to remove phosphorus so that $[P]/[V_2O_5]$ of the vanadium solution is < 0.003 that has no effect on vanadate precipitation.

Manganese has little effect on the vanadate precipitation efficiency and only a slight effect on the product quality. During production the solution is usually alkaline when the water is leached, and the Mn is easily hydrolyzed and left in the slag. The content of Mn in the solution is very low that does not affect precipitation of vanadate. However, in the process of acid leaching, Mn is the main impurity in the product.

The effect of Fe on precipitation of vanadate is significant. When [Fe]/[V_2O_5] in the vanadium solution is >0.053, the vanadate precipitation efficiency is below 99%. When water leaching is used to extract vanadium, Fe is easily hydrolyzed under alkaline conditions. In the vanadium solution, [Fe]/[V_2O_5] is typically below 0.002, which has a small effect on vanadate precipitation but a significant effect on the product quality.

The effect of aluminum on vanadate precipitation is great. For a vanadium solution pH of ~2.5 and [Al_2O_3]/[V_2O_5] above 0.002, the effect of Al on the vanadate precipitation efficiency is significantly reduced. When the pH of the vanadium solution is controlled to ~2.0, the influence of Al on the vanadate precipitation efficiency is reduced, and the precipitation efficiency is greatly increased. The influence of Al is because of the production of heteropolyacid from reactions between aluminum, ammonium, and vanadium, which limits the precipitation of APV. Under real production conditions, [Al_2O_3]/[V_2O_5] of the vanadium solution is limited to 0.001–0.0015 that has little effect on vanadate precipitation.

The effect of Ca on precipitation of vanadate is negligible. Under actual production conditions, Ca is added to the vanadium solution as $CaCl_2$ to purify and clarify the solution. The amount of $CaCl_2$ added in production is controlled to 1–1.5 g/L, and the effect on vanadate precipitation is very small. When excessive $CaCl_2$ is added, although the vanadium solution will be clarified quickly, it will produce calcium vanadate that is insoluble in water and results in loss of vanadium. Because magnesium vanadate is similar to sodium vanadate, it is also soluble. Ammonium can be substituted for magnesium ions in the vanadium solution, where magnesium has little effect on vanadate precipitation. When the pH of the vanadium solution is around 2.5, Cr has a certain effect on the precipitation of vanadate. Proper reduction of the pH can avoid this effect, which is beneficial for separation of V and Cr; Cr has no effect on the vanadium product quality.

7.1.9 Melting process

The moisture content of APV cake obtained from vanadate precipitation is high; usually, the APV is heated to 500°C–550°C and then dried, resulting in the crystallized and adsorbed water being removed and powdery vanadium pentoxide being obtained. To produce vanadium pentoxide flakes, the APV can be heated above the melting point of vanadium pentoxide (800°C–900°C), then dehydration and melting are performed to produce the flakes. The main equipment used in the melting process is a reverberatory furnace. The main reactions during the melting stage are as follows:

$$\text{Below } 670°C: (NH_4)_2V_6O_{16} = 3V_2O_5 + H_2O + 2NH_3\uparrow \quad (7.29)$$

$$\text{Above } 670°C: 3V_2O_5 + 2NH_3 = 3V_2O_4 + 3H_2O + N_2\uparrow \quad (7.30)$$

With the consumption of ammonia gas, the partial pressure of ammonia in the system decreases, and the partial pressure of oxygen gradually increases, resulting in low-valence vanadium oxide being oxidized to V_2O_5, as follows:

$$2V_2O_4 + O_2 = 2V_2O_5 \tag{7.31}$$

The melting process also includes a desulfurization reaction, namely, a reaction with Na_2SO_4, which is carried by V_2O_5 from the upper layer of the vanadium precipitation supernatant, to produce vanadate. The reaction is as follows:

$$Na_2SO_4 + V_2O_5 = 2NaVO_3 + SO_2\uparrow + \frac{1}{2}O_2 \tag{7.32}$$

7.1.10 Main equipment and process parameters

A rotary kiln and multihearth furnace are usually used for roasting. The multihearth furnace is a vertical rotary kiln. The general roasting furnace is divided into a preheating/preoxidation zone, reaction zone, and cooling zone. The roasting reaction temperature is generally 1123K–1223K, where the exact temperature is mainly dependent on the characteristics of the vanadium slag, and the type and quantity of additive. The aim of roasting is to produce sodium vanadate, which is soluble in water and is separated from the water-insoluble impurities.

Vanadium pentoxide is extracted from vanadium slag by sodium roasting, where rotary kilns are used for this process in most vanadium plants in China. The Pansteel Group Co., Ltd. in China used sodium roasting processes with a multihearth furnace in 1989. In addition, the German electric metallurgy company Nuremberg Vanadium Co., Ltd. employed a sodium roasting process using a multihearth furnace. The leaching residue from a single roasting has a V_2O_5 content below 0.6%, while the recovery of vanadium is 85%–90%. The process of sodium roasting oxidation of vanadium slag in a rotary kiln results in a leaching residue with 1.2%–2% V_2O_5 after two-stage leaching, which requires that the vanadium slag is returned for a second sodium roasting cycle to reduce the vanadium content in the residue to <0.8% and achieve vanadium recovery of 80%–88%. Fig. 7.9 shows the roasting process with a rotary kiln.

Fig. 7.10 shows a schematic diagram of roasting in a multihearth furnace. There is vertical shaft with a rake arm and rake teeth in the center of the multihearth furnace. Rotating the vertical shaft turns the rake that moves the material in a set direction. Coal gas or natural gas is used as a heat source for roasting to maintain the furnace temperature at 1473K.

In the process of extracting vanadium from vanadium slag in the Chusov vanadium plant in Russia, the slag is coarsely crushed to 60–80 mm, and then ground in a ball mill to

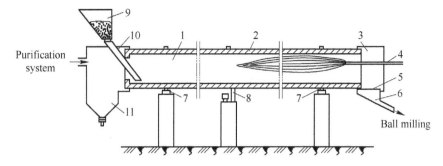

Figure 7.9
Roasting using a rotary kiln. 1—kiln body; 2—firebrick lining; 3—kiln head; 4—burner; 5—bashan; 6—row hopper; 7—roller; 8—transmission gear; 9—bin; 10—blanking; 11—gray box FIG.

<10 mm, followed by magnetic separation to ensure that the iron residue is no more than 6%. The slag is ground in a rod mill to below 0.15 mm and then mixed with sodium salt to produce a uniform material. With a screw feeder the mixture is fed into a rotary kiln with an internal diameter of 2500 mm, external diameter of 3000 mm, and length of 42,000 mm.

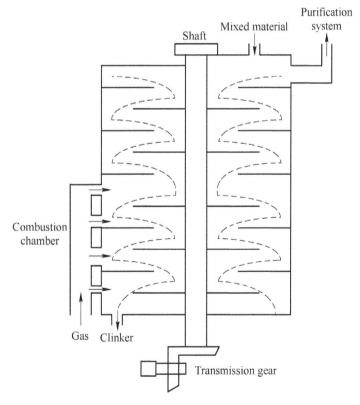

Figure 7.10
Diagram of furnace roasting process.

Heavy oil is used as a fuel for sodium and oxidizing roasting, where the efficiency of vanadium conversion is 85%–92%.

Table 7.3 compares the recovery of extracting vanadium oxide from vanadium slag in China and other countries. After wet grinding of Na-roasted vanadium slag, it is added to the leaching tank intermittently to give a liquid-to-solid ratio is 3.5:1. The solution is stirred and leached at a temperature of 313K–323K and then filtered. The filtered liquor contains 15 g/L of V_2O_5, while the filtered residue contains 0.6% V_2O_5. Acid leaching is used to recover the low-valence vanadium and vanadium salt compounds. The roasted material is mixed with the filter residue and placed in an acid solution for leaching, resulting in a V-containing acidic solution. The two leaching solutions are mixed, and then ammonium is added to produce APV, which is deaminated and fused. Flake V_2O_5 is obtained after fusion in a melting and casting furnace, which is used for smelting of ferrovanadium.

Table 7.3: Comparison of the recovery of vanadium oxide extracted from vanadium slag in China and other countries.

Process	Chinese advanced one-stage roasting	Chinese advanced two-stage roasting	Global advanced roasting
Recovery of material (%)	98	98	99
Recovery of conversion and leaching (%)	87–88	90	91
Recovery of precipitation (%)	99	99	99
Recovery of melting (%)	96	97	99
Total recovery (%)	81.31–81.96	84.70	88.29
Consumption of vanadium slag (t/t)	12.30–12.20	11.80	11.33

7.1.11 Calcining and sodium roasting with phosphate

This process is suitable for slag with relatively high-CaO content (>50%), and the CaO content is between 15% and 20%. The vanadium slag is ground using a ball mill, and then the fine vanadium slag is mixed with a certain amount of sodium phosphate and sodium carbonate and then roasted at high temperature for a certain time, where the phosphate radicals and calcium ions combine to form calcium phosphate. Vanadium in the slag reacts with sodium carbonate to produce water-soluble sodium metavanadate, which is water leached to extract vanadium. For a temperature of 1373K and roasting time of 2 h, the recovery of vanadium increases with increasing phosphate content. For a phosphate ratio of 300% the recovery of vanadium reaches 99%. Theoretically, when the temperature reaches 1473K, 100% of the vanadium can be recovered.

7.2 Extracting vanadium pentoxide from vanadium slag by calcium salt roasting

Calcium carbonate and vanadium slag are ground in a ball mill, and then iron is removed from the slag by magnetic separation. The mixture is roasted at high temperature and then leached in dilute sulfuric acid. The phosphorus-containing vanadium solution is treated with a phosphorus removal agent, and then concentrated sulfuric acid is added to precipitate vanadate. The vanadate is dried, washed, filtered, and then melted to obtain flake vanadium pentoxide with a purity of 97%–98.5%.

7.2.1 Extraction process

Vanadium slag is crushed and then ground using a ball mill. The iron is removed from the vanadium slag by magnetic and air separation. Vanadium slag (with a granularity <0.1 mm) is mixed with limestone ($CaCO_3$), and then calcium roasting is conducted at high temperature, where vanadium in the slag is oxidized and converted into soluble calcium vanadate and calcium metavanadate by the addition of dilute acid. The roasted material is leached by dilute sulfuric acid. The resulting vanadium solution after purification treatment is mixed with a special additive to precipitate vanadate and obtain V_2O_5. Fig. 7.11 shows a process flowchart of extracting vanadium pentoxide from vanadium slag by calcium salt roasting.

7.2.2 Extraction reactions

The vanadium in the vanadium slag exists as vanadium−iron spinel ($FeO \cdot V_2O_3$) that is insoluble in water. The precipitation sequence of several typical spinels in vanadium slag is $FeCr_2O_4$, FeV_2O_4, and Fe_2TiO_4. Fig. 7.12 shows the V_2O_5-CaO phase diagram. Fig. 7.13 shows the solubility diagram of calcium vanadate salt. When the pH is between 2.5 and 3.0, calcium pyrovanadate is the best choice. During high-temperature roasting, phase transitions occur, and the vanadium−iron spinel structure is damaged. During the calcining roasting process, $Ca_{0.17}V_2O_5$ forms at 400°C, V_2O is oxidized to V_2O_5 at 500°C, while above 600°C, calcium vanadate is produced. As the temperature increases, CaV_2O_6 is converted to $Ca_2V_2O_7$, which is converted into $Ca_3(VO_4)_2$. V^{3+} is oxidized to V^{5+} in the oxidizing atmosphere. The reaction between V^{5+} and limestone ($CaCO_3$) is as follows:

$$4FeO \cdot V_2O_3 + 12CaCO_3 + 5O_2 = 4Ca_3(VO_4)_2 + 2Fe_2O_3 + 12CO_2 \uparrow \quad (7.33)$$

Calcium pyrovanadate has the highest leaching efficiency. Hence, the weight ratio of CaO/V_2O_5 is controlled in the range of 0.5–0.6, where the best roasting time is 1.5–2.5 h, and the best roasting temperature is 890°C–920°C. The optimum cooling time for the roasted vanadium slag is 40–60 min, and the optimum cooling end temperature is 400°C–600°C.

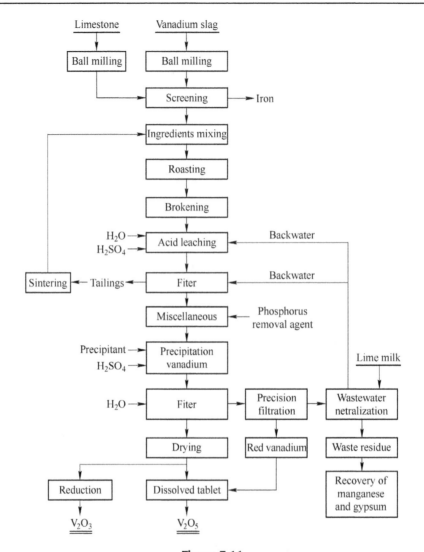

Figure 7.11
Process flow of extracting vanadium pentoxide from vanadium slag by calcium salt roasting.

$$2FeO \cdot V_2O_3 + 4CaCO_3 + \frac{5}{2}O_2 = 2Ca_2V_2O_7 + Fe_2O_3 + 4CO_2\uparrow \tag{7.34}$$

$$Ca_3(VO_4)_2 + 4H_2SO_4 = 3CaSO_4 + (VO_2)_2SO_4 + 4H_2O \tag{7.35}$$

$$Ca_2V_2O_7 + 3H_2SO_4 = 2CaSO_4 + (VO_2)_2SO_4 + 3H_2O \tag{7.36}$$

$$2VO_2^+ + 2H_2O = V_2O_5 \cdot H_2O + 2H^+ \tag{7.37}$$

$$V_2O_5 \cdot H_2O = V_2O_5 + H_2O \tag{7.38}$$

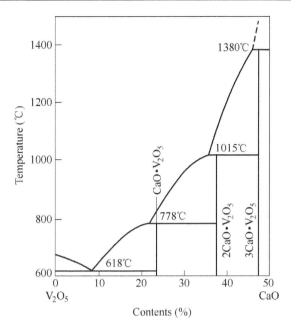

Figure 7.12
V_2O_5–CaO phase diagram.

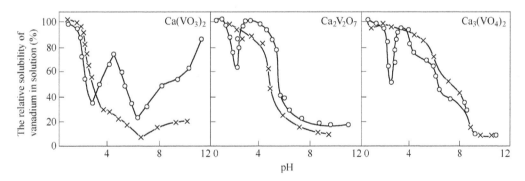

Figure 7.13
Solubility diagram of calcium vanadate salt.

In the range of 300°C–700°C the olivine and spinel are gradually degraded. At 400°C, $Ca_{0.17}V_2O_5$, CaV_2O_6, CaV_2O_5, and V_nO_{2n-1} ($2 \leq n \leq 8$) are obtained. Below 500°C, FeO_x ($4/3 < x < 3/2$) appears. At 500°C, olivines are decomposed completely. At 600°C, Fe_2O_3 is formed, where the content increases with increasing temperature. At 800°C the spinel phase disappears and Fe_2TiO_5 forms.

In the process flow of extracting vanadium pentoxide from vanadium slag by calcium salt roasting, low-valence iron begins to oxidize to ferric iron oxide at 500°C, which is characterized by weight gain of the vanadium slag. The low-valence vanadium in the spinel starts to oxidize to V^{5+} when the temperature reaches 650°C, which is observed as an increase of vanadium slag. At higher temperatures the oxidation speed is faster. The calcium carbonate begins to decompose and release CO_2 after the temperature reaches 750°C, until decomposition is complete, showing that the material is weightless.

7.2.3 Extraction characteristics

During extraction of vanadium pentoxide from vanadium slag by calcium salt roasting, the vanadium slag and limestone are weighed and mixed, and then combined with a certain amount of returned slag. These materials are transported to the stock bin at the top of the rotary kiln, from where they fed into the kiln for roasting. The clinker from the rotary kiln after roasting enters a coarse-material bin and is then cooled by a water-cooled screw conveyor. The rotary kiln clinker is ground using a rod mill to make it finer. The classified clinker is stored in a bin for fine material and then moved to a leaching tank after weighing. The pH is adjusted and the leaching reaction is performed. The resulting mixture of water-soluble vanadium and slag is transferred to a belt-type vacuum filter to be washed and filtered. The leaching residue is separated, where one part is returned after desulfurization for roasting ingredients, while most continues to the following sintering step. The leaching liquor is purified to remove impurities. Then, acid and a precipitating agent are added to the vanadium solution to precipitate the vanadate. The vanadate recovered from the precipitation tank is sent to the red vanadium collection tank and then transferred to a plate-and-frame filter press for filtering, washing, blowing, and drying. Intermediate vanadium pentoxide products are obtained, which contain about 25% water. After pneumatic drying, red vanadium is obtained from the plate-and-frame filter press, most of which is reduced in the kiln to produce V_2O_3, while some is used to produce flake V_2O_5. The wastewater produced by precipitation and filtration is filtered and recycled by a leaf filter and then sent to a wastewater treatment station for recycling. The red vanadium recovered from the leaf filter is sent to the melting furnace for V_2O_5 production.

Lime or limestone is used as an additive in the rotary kiln during oxidation and roasting to produce calcium vanadate. This can avoid charge adhesion problems at high temperature that arise when using traditional soda additives. In addition, it prevents pollution of the environment with harmful gas caused by release and decomposition of sodium salt (in particular NaCl and Na_2SO_4) during extraction of V_2O_5 from vanadium slag by sodium salt roasting. The production efficiency of the roasting equipment is greatly liberated and the oxidation efficiency of vanadium increases. Strict limits on the content of calcium oxide in vanadium slag are lifted. Vanadium slag and additives (lime or limestone) are ground in a wet ball mill, followed by wet magnetic separation to reduce dust pollution to the environment and facilitate contact between the additives and vanadium slag. The calcined clinker is crushed and ground to 0.074 mm, then water is added for mixing. The liquid–solid ratio is controlled in 4–5:1, and dilute sulfuric acid (5%–10% H_2SO_4) is used to adjust the pH to 2.5–3.2 under stirring. The leaching temperature is in the range of 50°C–70°C, where more than 90% of vanadium in the clinker is dissolved in the solution, along with Mn and Fe. Vanadate is precipitated by the traditional hydrolysis method. The purity of the product is higher than that of the sodium roasting process, while the purity of V_2O_5 is at least 92%, and the phosphorus content is 0.010%–0.015%. The impurities in the product are mainly Mn and Fe. The recovery of vanadium is about 2% higher than that of the traditional sodium roasting method.

A vanadium slag with high calcium oxide (where CaO/V_2O_5 in the vanadium slag is ~0.6) can be directly produced, which is called "calcium vanadium slag" and used for direct roasting after ball milling. The roasting temperature is 900°C–930°C. After oxidizing and roasting the vanadium product is calcium vanadate. The roasted clinker is continuously leached with dilute sulfuric acid, and red vanadium is produced by hydrolysis.

7.2.4 Calcium salt roasting of vanadium slag

7.2.4.1 Roasting characteristics

The calcium salt roasting of vanadium slag has three main stages that overlap. In the range of 300°C–500°C the Fe-containing olivine phase (Fe_2SiO_4) in the vanadium slag is oxidized and decomposed, some free FeO is oxidized, SiO_2 is dissociated, and showed an increasing trend. The vanadium–iron spinel [FeV_2O_4] gradually loses the surrounding [Fe_2SiO_4] phase. In the range of 500°C–600°C the FeV_2O_4 is oxidized and decomposed, the FeO in the FeV_2O_4 is oxidized to Fe_2O_3, and V_2O_3 is oxidized to V_2O_5 by reaction with surface-adsorbed oxygen. Some V_2O_3 and Fe_2O_3 combine to produce a solid solution [R_2O_3]. As the temperature increases from 600°C to 900°C, the oxidation of V_2O_3 is accelerated, low-valence vanadium is oxidized to high-valence vanadium, and some V_2O_5 and Fe_2O_3 react to produce iron vanadate, while the rest of the V_2O_5 and generated $FeVO_4$ react with CaO to form calcium vanadate. Both CaV_2O_6 and $Ca_3V_2O_8$ are present, where VO_2 reacts with CaO to produce CaV_3O_7. When the temperature exceeds 900°C, $FeSiO_3$ decomposes into Fe_2O_3 and SiO_2, which hinder the diffusion of oxygen. The generated SiO_2 from decomposition reacts with CaO to generate Ca_3SiO_5 with a high melting point, which is detrimental to calcium salt roasting of vanadium slag.

7.2.4.2 Influence of CaO/V_2O_3 ratio

When the ratio of CaO/V_2O_3 increases, the contact area between CaO, V_2O_3, and V_2O_5 increases that favors calcium vanadate production. When $CaO/V_2O_3 > 1.125$, the conversion and leaching efficiency of vanadium is reduced, where V_2O_5 reacts with excessive CaO to produce CaV_2O_6 and $Ca_3V_2O_8$ in the presence of SiO_2, which have a high melting point and crystallize late, shaped in the space of limitations, self-shaped error. It usually forms as irregular granules between other minerals and is encased. Silica gel is produced during acid leaching that hinders the reaction between the vanadium-containing phase and acid, resulting in vanadium loss.

7.2.4.3 Influence of roasting temperature

During calcium salt roasting the production of calcium vanadate starts from 600°C, which intensifies at 700°C and tends to be complete at 800°C. Above 800°C, vanadium oxide is oxidized completely, but the reaction temperature is too high, resulting in sintering and a slower reaction process.

7.2.4.4 Influence of roasting time

During calcium salt roasting the roasting time includes the time of oxidation, roasting decomposition, calcification, and salification, which is controlled to 1.5–2.0 h. Increasing the roasting time increases energy consumption and processing costs and reduces process efficiency, while disproportionation reactions occur and the structure of the sintering material changes, which affects the leaching efficiency.

7.2.5 Leaching of Ca salt–roasted vanadium slag

In general, Ca salt–roasted vanadium slag is cooled and leached by sulfuric acid to produce vanadium sulfate and vanadyl sulfate, which are ground in a wet ball mill, degraded, and leached by alkali liquor. The chemical reactions are as follows:

$$Ca(VO_3)_2 + Na_2CO_3 = CaCO_3 + 2NaVO_3 \tag{7.39}$$

$$Ca(VO_3)_2 + 2NaHCO_3 = CaCO_3 + 2NaVO_3 + CO_2 + H_2O \tag{7.40}$$

Because of the low solubility of $CaCO_3$, continuous intake of CO_2 gas can accelerate the reaction process.

7.2.6 Precipitation of vanadate

The hydrolysis of vanadium solution obtained by sulfuric acid leaching produces vanadium hydrate that is filtered and washed. Powdery V_2O_5 is produced by calcination of vanadium hydrate. The alkaline leaching solution is the same as that used for sodium salt roasting.

7.2.7 Roasting and leaching equipment

The process of calcium salt roasting and leaching of vanadium slag involves oxidation of the low-valence vanadium and its conversion into water-soluble vanadate. Various roasting equipment can realize calcium roasting and oxidation. Equipment commonly used in Ca salt roasting of vanadium slag and leaching of Ca salt−roasted vanadium slag include rotary kilns, leaching systems, purification equipment, and melting equipment. These systems are similar to those used for extracting V_2O_5 by sodium salt roasting; however, there are differences in the tank arrangements, where some technical parameters are also adjusted correspondingly. There is no wastewater treatment equipment, and the vanadium slag remains unchanged. The amount of slag is increased by precipitation of vanadate via neutralization of the solution, where the temporary slag field is added. The V_2O_5 product purity is 97%−98.5%, and the recovery of V_2O_5 is >83%.

7.3 Extracting vanadium pentoxide from Na-treated vanadium slag

Molten iron containing Na-treated vanadium is produced by injecting airborne soda into V-containing molten iron. Vanadium in the molten iron oxidizes to form vanadium oxide, while sodium carbonate reacts with vanadium oxide to produce sodium vanadate at high temperature. The soda-treated vanadium extraction process uses the strong reactivity between sodium and sulfur (as well as phosphorus), to remove S and P from the V-containing molten iron and produce high-quality semisteel and water-soluble soda-treated vanadium slag. Soda-treated vanadium slag is affected by moisture and wet milling. Large pieces of slag are broken into coarse particles and directly transferred to the ball mill for grinding and leaching. Under the joint action of the water and grinding balls, the slag is quickly converted into slurry, which can fully contact state of slurry. The soda-treated vanadium slag is used as the raw material for vanadium extraction, where the alkali is recovered and recycled. Aluminum sulfate and calcium chloride are added to the soda-treated vanadium slag leaching liquid for purification, where the S and P are removed. CO_2 is injected into the purified vanadium liquid to recover sodium carbonate via mineralization. In addition, hydrogen peroxide and ammonium persulfate are added to the vanadium solution to remove sodium carbonate for oxidation. Finally, ammonium vanadate is precipitated after deamination by addition of ammonium chloride to the purified vanadium liquid. The ammonia is recovered from the vanadium slag by evaporation.

Depending on the characteristics of the soda-treated vanadium, which contains Na, P, S, and high Si contents, it is managed according to the carbonation. Precipitation of ammonium vanadate via solution neutralization and recovery of the sodium salt is performed. Fine V_2O_5 is obtained with a purity of >99%. Sodium carbonate is recovered as a high-purity sodium salt and is recycled as a treatment agent for the molten iron. Based on laboratory tests, commercial production processes have been developed, giving a V recovery of 92.6% and Na recovery of 42.8%.

7.4 Extracting vanadium from V-containing steel slag by acid leaching

In general, V-containing steel slag has high calcium oxide, S, and P contents that complicate extraction of V_2O_5 in the sodium salt and calcium salt systems. An acid system needs to be used to minimize the influence of the calcium oxide, S, and P. The V_2O_5 extracted by acid leaching is mainly used for V-containing steel slag, which is ground in a ball mill, followed by removal of iron. Acid is used to leach the V-containing steel slag, and the vanadium is dissolved into the solution.

7.4.1 Extraction process

In the first stage of leaching, the pH is controlled to ~ 4, and the V-containing steel slag is leached by hot steam and insulation. The vanadium is present as calcium vanadate and is retained in the residue. Large amounts of Fe, Cr, Mn, S, P, and other impurities are leached as ions into the supernatant. Fe occurs as ferrous sulfate, and high-valence Cr is reduced by the ferrous sulfate to produce Cr^{5+} ions that dissolve in the supernatant after solid−liquid separation. The large amounts of Fe, Cr, Mn, S, P, and other impurities are separated and remain in the liquid. Calcium vanadate enters in the form of bottom flow in the second period of leaching. Iron is removed from the supernatant by addition of yellow ammonium iron vitriol; after solid−liquid separation, this material is used as a raw material for ironmaking, and the clear liquid is returned to the first leaching stage. In the second stage of leaching, the pH is controlled to ~ 1, and V-containing steel slag is leached with hot steam and insulation, where vanadium is present as V^{3+} and V^{5+} in the leach liquor. The residual Fe and Cr from the first stage of leaching are dissolved into the solution because of the high acidity. After solid−liquid separation the liquid is purified and treated, and then the residue is sent to a special storage area.

An oxidizing agent is added to the leaching liquid from the second stage of leaching, converting the V^{3+} to V^{5+}. Various oxidizing agents can be used, such as hydrogen peroxide, sodium hypochlorite, and ozone. The pH is adjusted with ammonia solution to around 1.5. After system processing of N235 (trifatty tertiary amine) and TBP (tributyl phosphate), and sulfonated kerosene extraction, vanadium enters the organic phase, while Fe is removed by washing. The V-containing organic phase is extracted by soda, where the extract liquor is processed and recycled. Vanadium pentoxide is extracted from the reverse extract liquid using the traditional method of extracting vanadium by sodium salt roasting.

7.4.2 Extraction reactions

There are five steps in extracting V_2O_5 by acid leaching of V-containing steel slag. The first is to remove impurities from V-containing steel slag by acid leaching; the second is high-temperature vanadium leaching, followed by oxidation of vanadium in the liquid; the fourth is purification of the solution; and, finally, the vanadate is precipitated using the traditional process of sodium salt roasting for extracting V_2O_5 from the vanadium slag. The chemical reactions occurring during the first stage of leaching are as follows:

$$Fe + H_2SO_4 = FeSO_4 + H_2 \quad (7.41)$$

$$FeO + H_2SO_4 = FeSO_4 + H_2O \quad (7.42)$$

$$MnO + H_2SO_4 = MnSO_4 + H_2O \quad (7.43)$$

$$3Fe^{2+} + Cr^{6+} = 3Fe^{3+} + Cr^{3+} \quad (7.44)$$

$$S + 2H_2SO_4(\text{concentrated}) = 3SO_2 \uparrow + 2H_2O \quad (7.45)$$

$$2P + 5H_2SO_4(\text{concentrated}) = 5SO_2 + 2H_2 + 2H_3PO_4 \quad (7.46)$$

$$Cr_2O_3 + 3H_2SO_4 = Cr_2(SO_4)_3 + 3H_2O \quad (7.47)$$

$$CaO + FeS + 2O_2 = FeO + CaSO_4 \quad (7.48)$$

The chemical reactions occurring during the second stage of leaching are as follows:

$$Fe_2(SO_4)_3 + 2H_2O = 2Fe(OH)SO_4 + H_2SO_4 \quad (7.49)$$

$$2Fe(OH)SO_4 + 2H_2O = Fe_2(OH)_4SO_4 + H_2SO_4 \quad (7.50)$$

$$Fe(OH)SO_4 + Fe_2(OH)_4SO_4 + 2NH_4OH \rightarrow (NH_4)_2Fe_6(SO_4)_4(OH)_{12} \quad (7.51)$$

$$V_2O_5 + H_2SO_4 \rightarrow (VO_2)_2SO_4 + H_2O \quad (7.52)$$

$$Ca(VO_3)_2 + 2H_2SO_4 \rightarrow (VO_2)_2SO_4 + CaSO_4 + 2H_2 \quad (7.53)$$

In the third stage the oxidation of vanadium in the liquid occurs as follows:

$$V^{3+} + H_2O_2 = V^{5+} + H_2O \quad (7.54)$$

The purification reactions in the vanadium solution during the fourth stage are as follows:

$$VO_3^- + [R]^+ = VO_3[R] \quad (7.55)$$

$$VO_3[R] = VO_3^+ + [R]^- \quad (7.56)$$

$$[R]OH + HCl \rightarrow [R]Cl + H_2O \quad (7.57)$$

The precipitation reactions occurring in the vanadium solution during the fifth stage are as follows:

- Precipitation reaction of vanadate. $NaVO_3$ is reacted to generate NH_4VO_3 with excessive NH^{4+} ions.

$$NaVO_3 + NH_4^+ \rightarrow NH_4VO_3 + Na^+ \tag{7.58}$$

- Chemical reaction. NH_4VO_3 is decomposed at high temperature to produce V_2O_5:

$$NH_4VO_3(s) \rightarrow V_2O_5 + 2NH_3\uparrow + H_2O\uparrow \tag{7.59}$$

7.4.3 Main equipment and process parameters

The V-containing steel slag produced in the Chile CAP steelworks using a converter with a basic lining had the following composition: 5.7% V_2O_5, 47.0% CaO, 2.5% MgO, 11.0% SiO_2, 3.2% P_2O_5, 4.0% MnO, 15.1% Fe, and 1.2% Al_2O_3. Because of the high contents of CaO and P in the V-containing steel slag, the vanadium occurred as $3CaO \cdot P_2O_5 \cdot V_2O_5$, $CaO \cdot 3V_2O_5$, $CaO \cdot V_2O_5$, and $3CaO \cdot V_2O_5$. To reduce the acid consumption during leaching, CaO in the slag must be converted into calcium sulfate. Hence, additives are added to reduce the calcium content. When pyrite was added during the roasting process, the total V recovery from steel slag to red cake (81%–82% V_2O_5) was ~80%.

7.5 Extracting V_2O_5 from vanadium slag by sulfide addition

Extracting V_2O_5 from vanadium slag by sulfide addition is mainly used for high-Ti vanadium slag. The sulfide is added to high-Ti vanadium slag with sulfur and magnesium oxide, and heated to 1873K, where the slag is fully cured, and then cooled. Vanadium is mainly concentrated as oxides and sulfides. With decreasing FeO content in the slag, the ratio of V/VO_3 increases gradually. Then, the slag is ground in a ball mill to below 200 mesh and then treated with SO_2 and O_2 gas at a temperature of 723K. Vanadium is easily converted to water-soluble $(VO_2)_2SO_4$, which is leached by water at room temperature. Longer water leaching times result in better V recovery rates, up to 88.2%.

7.6 Main raw materials

7.6.1 Industrial raw materials

The main raw materials include industrial sodium chloride, industrial sodium carbonate, industrial sulfuric acid, ammonium sulfate, and ammonium chloride.

7.6.1.1 Industrial soda

The pure alkali is a white crystalline powder with a density of 2.53 g/cm³. Depending on its packing density, it is classified as light pure alkali and heavy quality soda. Soda has the formula Na_2CO_3, molecular weight of 106, and melting point of 845°C–852°C. It is soluble in water and alkaline aqueous solutions, where the maximum solubility is at 36°C. Table 7.4 shows the Chinese national standard for industrial soda (GB210.1-2004). In the process of extracting vanadium from vanadium slag, sodium carbonate is mainly used as a calcining additive, alkaline leaching agent, and solution pH regulator. It is also used as an absorbent of SO_2 during roasting and calcining.

Table 7.4: Industrial soda GB210.1-2004.

Index items	I	II		
	Superior product	Superior product	Top quality goods	Classified product
Total alkali content (%) (calculated from the mass fraction of dry base Na_2CO_3)	99.6	99.6	98.8	98.0
Sodium chloride (in terms of the mass fraction of dried NaCl) (% ≤)	0.30	0.70	0.90	1.20
Mass fraction of iron (dry basis) (% ≤)	0.003	0.0035	0.006	0.010
Sulfate content (with respect to dry SO_4 mass) (% ≤)	0.03	0.03	–	–
Water-insoluble fraction (% ≤)	0.02	0.03	0.10	0.15
Bulk density (g/mL ≥)	0.85	0.90	0.90	0.90
Particle size (180 μm screen, % ≥ 1.18 mm ≤)	75.0	70.0	65.0	60.0
	2.0	–		–

7.6.1.2 Industrial sodium chloride

Industrial salt (sodium chloride) has the molecular formula NaCl and molecular weight of 58.44. GB/T5462-2003 sodium chloride is widely used in industry and is one of the basic raw materials for the chemical industry. Sodium chloride is a colorless or white cubic crystal, which is soluble in water and glycerin, slightly soluble in ethanol and liquid ammonia, but insoluble in hydrochloric acid. There is moisture in the air. The typical chemical analysis of NaCl used in industrial extraction of vanadium is shown in Table 7.5.

In the process of extracting vanadium from vanadium slag, sodium chloride is mainly used as a roasting additive.

7.6.1.3 Industrial ammonium sulfate

Ammonium sulfate has the molecular formula $(NH_4)_2SO_4$, with a molecular weight of 132.13. The pure product is a colorless or white crystalline powder, which is soluble in

Table 7.5: Typical chemical analysis of sodium chloride for industrial vanadium extraction.

Executive standard GB/T5462-2003			
Analysis item		Technical index	Analysis result
Appearance		White or yellow-gray crystal, with no obvious impurities	
Sodium chloride	% ≥	99.1	99.65
Water	% ≤	0.30	0.01
Water-insoluble materials	% ≤	0.05	0.01
Magnesium ions (Ca^{2+}\Mg^{2+})	% ≤	0.25	Not detected
Sulfate ion(SO_4^{2-})	% ≤	0.30	0.052
Antitackiness (mg/kg)	% ≤	10	5.51
The product conforms to the GB/t5462-2003 standard of industrial excellence			

water, insoluble in alcohol and acetone, and forms an acidic in aqueous solutions. It easily absorbs moisture and agglomerates and has strong corrosion and permeability. The quality standard of ammonium sulfate for vanadium extraction is shown in Table 7.6.

Ammonium sulfate is mainly used in precipitation of vanadate to supply ammonium ions.

Table 7.6: Ammonium sulfate industry standard (GB535-83) (%).

Item	N (dry basis)	Water	Free acid	Fe	Heavy metal (Pb)	Water insoluble
Index	≥ 21.0	≤ 0.5	≤ 0.05	≤ 0.007	≤ 0.005	≤ 0.05
Index(1)	≥ 21.0	≤ 0.5	≤ 0.08	≤ 0.007		
Index(2)	≥ 20.8	≤ 1.0	≤ 0.20	≤ 0.007		
Product	20.89	1.4	<0.03	0.016	<0.005	0.554

7.6.1.4 Industrial ammonium chloride

Ammonium chloride has the chemical formula NH_4Cl and is a colorless or white cubic crystal. Its taste is salty and cool and slightly bitter. This material is soluble in water and liquid ammonia, slightly soluble in alcohol but insoluble in acetone and ether. The aqueous solution is weak acidic, and the acidity increases when heated. It is corrosive to ferrous metals and other metals, especially to copper, but has no corrosive effect on pig iron. Industrial ammonium chloride is a white powder or granular crystals, and it is odorless, salty, and cool. It easily absorbs moisture and agglomerates. It is soluble in water, glycerol, and ammonia but insoluble in ethanol, acetone, and ethyl fan. At 350°C, sublimation occurs in weak acid solutions.

The Chinese national standard GB-2946-92 includes the following details:

1. Appearance: white crystal.
2. Content of ammonium chloride (dry basis) is 99.5%.
3. Water content: 0.4%.

4. Content of sodium chloride (dry basis): 0.2%.
5. Iron content is 0.001%.
6. Content of heavy metals (Pb): 0.0005%.
7. Content of water-insoluble components: 0.02%.
8. Content of sulfates (SO_4^{2-} meter): 0.02%.
9. pH value: 4.0–5.8.
10. Ignition residue: 0.4%.

Ammonium chloride is mainly used in the precipitation of vanadate to provide ammonium ions.

7.6.1.5 Industrial sulfuric acid

Industrial sulfuric acid has the molecular formula H_2SO_4, relative molecular weight of 98.08, and specific gravity of 1.83. Industrial sulfuric acid should meet the requirements of the GBT534-2002 industrial standard. Sulfuric acid is mainly used for the dissolution and leaching of vanadate, as the acid solution medium for high-calcium vanadium slag, for acid washing of residue, and for pH adjustment of the vanadium solution. Table 7.7 shows details of the GBT534-2002 standard for industrial sulfuric acid.

Table 7.7: GBT534-2002 standard for industrial sulfuric acid.

Item	Index					
	Concentrated sulfuric acid			Fuming sulfuric acid		
	Superior product	Top quality goods	Classified product	Superior product	Top quality goods	Classified product
Mass fraction of H_2SO_4 (% ≥)	92.5 or 98.0	92.5 or 98.0	92.5 or 98.0	—	—	—
Mass fraction of free SO_4^{2-} (% ≥)	—	—	—	20.0 or 25.0	20.0 or 25.0	20.0 or 25.0
Mass fraction of ash (% ≤)	0.02	0.03	0.10	0.02	0.03	0.10
Mass fraction of iron (% ≤)	0.005	0.010	—	0.005	0.010	0.030
Mass fraction of arsenic (As) (% ≤)	0.0001	0.005	—	0.0001	0.0001	—
Mercury (Hg) mass fraction (% ≤)	0.001	0.01	—	—	—	—
Mass fraction of lead Pb (% ≤)	0.005	0.02	—	0.005	—	—
Transparency (mm ≥)	80	50	—	—	—	—
Transparency (/mL ≤)	2.0	2.0	—	—	—	—

Note: "—" indicates that there is no such item in the technical requirements of the category product.

7.6.1.6 Sodium sulfate

Sodium sulfate is an inorganic compound. Sodium sulfate decahydrate is known as Glauber's salt, sodium dihydrate, and sodium sulfate, which occur in the form of white crystals or powder, which are odorless, bitter, and hygroscopic. The material is colorless, transparent, and in the form of large crystals or granules. The chemical formula for sodium sulfate with internal water is $Na_2SO_4 \cdot 10H_2O$ (10 hydrate) or $Na_2SO_4 \cdot 7H_2O$ (7 hydrate). Sodium sulfate is a monoclinic system with short columnar crystals. The aggregates are compact, and massive or crustose. Sodium sulfate is mainly used as a roasting additive during the extraction of vanadium from vanadium slag.

7.6.1.7 Limestone

Limestone is mainly composed of calcium carbonate ($CaCO_3$). Limestone can be directly processed into building stone and fired to produce lime. Limestone can be processed into quicklime and lime. The main component of lime is CaO that is generally lumpy, pure white, with light gray or yellowish impurities. The physical and chemical properties of lime are shown in Table 7.8. Typical limestone block/powder contains 40.79% burned loss, 4.62% Si, 1.21% Al, 0.52% Fe, 50.16% Ca, and 1.10% Mg. Calcium carbonate limestone is mainly used as a calcining additive during the extraction of vanadium.

Table 7.8: Physical and chemical properties of lime (YB/ T042-2004).

Category	Grade	Chemical composition (%)						Activity degree, 4 mol/ml HCl, 40°C ± 1°C, 10 min
		aO	CaO + MgO	MgO	SiO_2	S	gloss	
		Not less than		No greater than				Not less than
Ordinary metallurgical lime	Lass four	80	—	5	0	0.100	9	180

7.6.1.8 Sulfur

Sulfur is a light yellow, brittle, crystal, or powder, which has a unique odor. The specific density of sulfur is 2 (water = 1), and it has a melting point of 119°C and boiling point of 444.6°C. Sulfur is insoluble in water and contains ≥99% S. Most sulfur products are derived from petroleum refining. The chemical composition is shown in Table 7.9. The sulfur of vanadium residue is mainly used in the roasting of high-Ti vanadium slag.

Table 7.9: Chemical composition of sulfur.

Content (%)	Unit	Content
Sulfur (S)	%	≥ 99.5 (dry basis)
Carbon (C)	%	≤ 0.1 (dry basis)
Acidity (calculated using H_2SO_4)	%	≤ 0.005
Iron	%	≤ 0.005
Organics	%	≤ 0.3
Moisture	%	≤ 0.50

7.6.1.9 Sodium phosphate

Sodium phosphate is also called trisodium phosphate. It has a molecular formula of Na_3PO_4, molecular weight of 163.94, chemical formula of $Na_3PO_4 \cdot 12H_2O$, density of 1.62 g/cm^3, melting point of 73.4°C, and pH of 11.5−12.5. Trisodium phosphate is a colorless or white crystal, which dissolves in water to form a strongly alkaline solution. It is insoluble in ethanol and carbon disulfide. It is important to have trisodium phosphate (12-hydrate) and anhydrous trisodium phosphate. Anhydrous trisodium phosphate is a white crystal with a density 2.536 g/cm^3 and melting point of 1340°C. Trisodium phosphate (12-hydrate) is a colorless cubic crystal or white powder with a density of 1.62 g/cm^3, melting temperature of 73.3°C, and decomposition temperature of 76.7°C. When it is heated to 100°C, the 12 water molecules are lost to form the anhydrous material. It can be weathered in dry air. Anhydrous trisodium phosphate is soluble in water and its aqueous solution is strongly alkaline, although it is insoluble in carbon disulfide and ethanol. Trisodium phosphate is prepared by neutralization of a phosphoric acid and sodium carbonate solution to control the pH to 8−8.4. Then, the filter cake residue is filtered, the filtrate enriched, and liquid caustic soda is added to achieve a Na/P ratio of 3.24−3.24. Trisodium phosphate is obtained after cooling and crystallization, followed by solid−liquid separation and drying. Anhydrous trisodium phosphate is prepared by dissolving trisodium phosphate (12-hydrate) in water. The material with a moisture content of 10%−15% is heated to 85°C−90°C for dehydration. Sodium phosphate is mainly used in the phosphorylation of high-Ca vanadium slag.

7.6.2 Industrial fuel

Industrial gases include coke oven gas and natural gas, while industrial coal is mainly used as a solid fuel. These materials are mainly used as fuels and reducing agents, for example, as the heating fuel for vanadium slag roasting in rotary or multichamber kilns, the fuel and reducing agent for the preparation of vanadium trioxide, and the heat source for producing hot water.

7.6.2.1 Gas fuel

The main composition of natural gas is paraffin that accounts for the vast majority of methane, along with small amounts of ethane, propane, and butane. In addition, the gas contains hydrogen sulfide, carbon dioxide, nitrogen, and water vapor and small amounts of carbon monoxide and trace gases, such as He and Ar. In their standard state, methane to butane exists in the gaseous state, while pentane and above are liquids. Methane is the shortest and lightest hydrocarbon molecule. A typical natural gas composition is shown in Table 7.10.

Table 7.10: Typical natural gas composition (%).

CH_4	C_2H_6	C_3H_8	C_4H_{10}	$CO_2 + H_2S$	CO	H_2	N_2	Unsaturated hydrocarbon	Low calorific value (kJ/m^3)
96.67	0.63	0.26		1.64		0.13	0.07	1.30	35421

Coal gas is a gas prepared from coal that contains combustible components. Depending on the processing method, the properties and uses of the gas are classified into coal gasification results in a product containing water gas and semiwater gas, air gas, or producer gas. The calorific value of the gas is low, so it is also referred to as a low-calorific value gas. Gas produced by coking using the coal distillation method is called coke oven gas or blast furnace gas and is a medium-heat value gas. The composition and gas contents of common blast furnace and coke oven gases are shown in Tables 7.11 and 7.12.

Table 7.11: Composition and gas contents of blast furnace gas.

Gas	CO	CO_2	H_2	N_2	O_2	CH_4
Volume (%)	25.2	16.1	1.0	57.3	0.2	0.2

Table 7.12: Composition and gas contents of coke oven gas.

Gas	CO	CO_2	H_2	N_2	O_2	CH_4	C_3H_8
Volume (%)	8.6	2.0	59.2	3.6	1.2	23.4	2.0

The properties of the hazardous chemical components in the gas are shown in Table 7.13.

7.6.2.2 Solid fuel

Coal is a solid combustible mineral that has undergone complex biochemical and physical−chemical changes underground. It is a solid combustible organic rock, which is mainly composed of plant remains that are then transformed by geological processes. Coal is the most widely distributed fossil energy resource on Earth and is classified into four main categories: bituminous coal, anthracite, subbituminous coal, and lignite. The specific heat and thermal conductivity of common coal types are shown in Table 7.14.

The organic coal compounds contain mainly carbon, hydrogen, oxygen, nitrogen, and sulfur. In addition, there are also small amounts of phosphorus, fluorine, chlorine, and arsenic. Carbon, hydrogen, and oxygen are the main elements in organic coal, accounting

Table 7.13: Properties of hazardous chemicals in the gas.

Item	Explosion limit (%)	Flash point (°C)	Light (°C)	Heat of combustion (KJ/mol)	Hazardous characteristics	Health hazard
Phenol	1.7–8.6	79	715	3050.6	High heat of combustion in case of open fire.	Strongly corrosive to skin and mucous membranes. Can damage central nervous system, liver, and kidney function.
Hydrogen sulfide	4.0–46.0	≤50	260	—	Forms explosive mixture with air. In contact with open fire, the high heat energy causes explosion. At high temperature, pressure in the container increases, with a risk of cracking and explosion.	Strong nerve poison with a strong stimulation effect on the mucous membrane. High concentrations can directly inhibit the respiratory system, causing rapid asphyxia and death.
Hydrogen cyanide	5.6–40.0	−17.8	—	—	Its vapor and air form an explosive mixture. In contact with an open fire, the high heat energy causes explosion. At high temperature, pressure in the container increases, with a risk of cracking and explosion.	Toxicity is so rapid that the tissue cannot use oxygen and produces intracellular asphyxiation.
Carbon monoxide	12.5–74.2	≤50	—	—	Flammable and explosive gas. Forms explosive mixture with air. In contact with an open fire, the high heat energy causes explosion.	Tissue-forming carbon monoxide binds to hemoglobin in the blood, leading to oxygen deficiency.
Hydrogen	4.1–74.1	≤50	400	241.0	Forms explosive mixture with air, which will explode when heated or exposed to fire. This gas is lighter than air and a spark can ignite an explosion.	At high concentrations, asphyxiation results from a decrease in oxygen partial pressure in the air. Anesthesia may be present at very high partial pressures.
Ammonia	16–25	—	651.1	—	Forms explosive mixture with air. In contact with open fire, the high heat energy causes explosion. At high temperature, pressure in the container increases, with a risk of cracking and explosion.	Low concentration of ammonia can stimulate the mucosa, while high concentrations can cause tissue dissolution and necrosis.

(Continued)

Table 7.13: (Continued)

Item	Explosion limit (%)	Flash point (°C)	Light (°C)	Heat of combustion (KJ/mol)	Hazardous characteristics	Health hazard
Methane	5.3–15	−188	538	889.5	It can easily come into contact with bromine, chlorine gas, hypochlorous acid, nitrogen trifluoride, liquid oxygen, oxygen difluoride, and other strong oxidants and can form an explosive mixture when mixed with air.	Nontoxic to humans, but when the concentration is too high, the oxygen content in the air is significantly reduced, resulting in suffocation. Contact with liquid methane can cause frostbite on the skin.

Table 7.14: Specific heat and thermal conductivity of common coal types.

Item	Specific heat (kJ/kg/C)	Thermal conductivity coefficient (W/m/C)
Anthracite, lean coal	1.09–7.17	0.19–0.65
Bituminous coal	1.25–1.50	0.19–0.65
Lignite	1.67–1.88	0.029–0.174
Coal ash	~0.84	0.22–0.29

for more than 95%. The deeper the coal was formed, the higher the carbon content, and the lower the hydrogen and oxygen contents. Carbon and hydrogen are the elements that generate heat during the burning of coal, where oxygen is the combustion element. When coal is burned, nitrogen does not produce heat but is converted to nitrogen oxides and ammonia at high temperatures and is then removed. Coal is mainly used for gas generation during the process of extracting vanadium.

7.7 Main vanadium extraction equipment

In the process of extracting vanadium pentoxide from vanadium slag, vanadium slag is used as the main raw material, which is ground and then separated from metallic iron using magnetic separation. Related equipment includes ball mill, magnetic separation, and screening devices. The roasting equipment includes rotary kilns and multihearth furnaces. The leaching apparatus includes a ball mill and thickener.

7.7.1 Rotary kilns

Rotary kilns are classified according to their shape and can be divided into fixed-size and variable-diameter rotary kilns. The fixed-size rotary kiln is used for extracting vanadium pentoxide. The rotary kiln body is composed of a kiln head, body, and end. The kiln head has a larger diameter than the kiln body and is where the material is discharged. The sealing is achieved by stainless-steel sheets and the kiln body, which consists of the access port, nozzles, cars, and observation hole. The kiln body is the main part of the rotary kiln, which is usually 30–150 m long, cylindrical and has 3–5 hoops. For normal operation, rotary kilns are lined with brick. The kiln end section is also an important part of the rotary kiln and is similar to the cover of a rotary kiln at the feed end that mainly achieves feeding and sealing.

Rotary kilns are a thermal system composed of gas flow, fuel combustion, material movement, and heat transfer parts. Correct operation ensures that the fuel burns adequately, and heat can effectively be transferred to the material that undergoes a series of physical and chemical changes during heating to produce the final clinker product. The working area of the rotary kiln has three parts, a drying section, heating section, and roasting section. Ventilation is one of the two material conditions of the combustion reaction, where an oxygen flow enables fuel combustion to release heat energy and maintain the desired temperature. Under normal operation, when the ventilation volume is low, the oxygen supply is insufficient, and the combustion speed is low, resulting in high heat consumption. For higher gas flows the air volume increases, and the time of combustion and decomposition decreases, resulting in an increase in generated combustion gas and high heat loss, which affects normal operation of the kiln. Hence, accurate and timely control of the kiln ventilation is important. Under normal circumstances, a fan at the tail of the system runs stably to extract the air. Ventilation in the kiln should be stable, although there are some factors that affect the flow, such as resistance to change the whole system, the air temperature in the kiln, air leakages, the fan inlet flow being mixed with cold air, dust accumulation, blockages in the pipes, uniformity and distribution of the material, and kiln interference factors, such as two systems.

Rotary kilns often use the latest wireless communication technology to transfer the temperature data from the thermocouple to the operation room. The kiln temperature transmitter is battery operated and can collect multiple thermocouple signals simultaneously. It is installed in the kiln body and can withstand the cylinder body rotation and temperatures above 300°C because of thermal insulation measures. The cylinder radiates heat, antirain, antisun, and antivibration. The kiln temperature receiver is installed in the operation room, which directly displays the temperature in the kiln and has 4–20 mA of output that can be sent to a computer or other instrument display.

The vanadium slag from the converter is cooled, crushed, and Fe is removed by magnetic separation to obtain coarse vanadium slag, which is placed into a slag bin. The coarse vanadium slag is ground in a ball mill and screened to obtain refined vanadium slag that is transferred and stored in the refined vanadium slag bin. Classification of refined vanadium slag powder is performed using a winnowing machine, where the classified material is added to the ingredients of bunker, while coarse particles are returned to the ball mill. The limestone is transported to the underground bunker by truck, and then ground in the ball mill and transferred into the limestone silo and then into a bin. Vanadium slag, limestone (or sodium carbonate and sodium chloride) are weighed and then mixed with a certain amount of recycled vanadium slag. This mixture is transferred into a bin at the top of the rotary kiln and then fed into the rotary kiln for roasting. The roasted clinker from the rotary kiln is passed to a water-cooled screw conveyor. After cooling in a coarse-cooked bin, it is then finely ground using a rod mill. Classified granular clinker is obtained, which is fed into the mastery of the bunker, and then weighed and added to the leaching tank. Rotary kilns are mainly used in the roasting of vanadium materials.

7.7.2 Multihearth roasting furnace

Multihearth roasting furnaces are vertical cylindrical furnaces equipped with multilayer horizontal furnaces. The round outer shell is made from 7 to 12-mm-thick steel plate, with a 230-mm-thick lining of firebrick. It is usually used to build 6–8 layers of furnace arches at regular intervals along the height of the furnace, where the inner space of the furnace is divided into 5–6 horizontal furnaces. A hole is placed in the center of each layer, in which a rotating spindle is installed across the bottom of the furnace base through the layers of the center round hole, and shaft in each layer is equipped with two grilled arms with teeth. The spindle speed is 0.75–1.5 rpm, which is driven by the grilled arm that slowly moves the calcined material. On the contrary, grilled on the opposite direction from the tooth layer and layer while the roasting vanadium slag moves in the opposite direction. Each layer of the furnace arch is connected to the others, and the charge is fed from the furnace top and successively moves down layer by layer through the feeding holes. The lower layers are located on the periphery of the furnace wall if the upper layer is located near the rotating main shaft. The burden of the furnace is located at the top of the central area that extends to the wall of the furnace in the lower layers, and the roasted vanadium slag is discharged from the bottom of the furnace. The air required for roasting is fed by a blower to the rotating main shaft in each layer to cool the main shaft and grilled arm, which preheats the air before it enters the furnace area. Compared with other roasting furnaces, multihearth roasting furnaces have the advantage of good flexibility to different raw materials. In the process of landing and raking, the furnace is continuously mixed, where the material at the surface, center, and bottom of the furnace are constantly changing position. Hence, all material roasting in the furnace is fully exposed to the air, allowing complete reactions to occur.

The aim of multihearth furnace roasting is to maintain high temperatures, a certain negative pressure, and an oxidizing atmosphere to facilitate the reactions between vanadium slag and additives and the corresponding physical and chemical changes. Vanadium slag, sodium carbonate, and sodium chloride are weighed and then mixed with a certain amount of recycled slag. This mixture is delivered to the bunker at the top of the multihearth furnace and then fed into the furnace for roasting. The roasted clinker from the furnace is passed to a water-cooled screw conveyor. After cooling in a coarse-cooked bin, it is then finely ground using a rod mill. Classified granular clinker is obtained, which is fed into the mastery of the bunker and then weighed and added to the leaching tank. In the process of extracting vanadium pentoxide from vanadium slag, the furnace is mainly used for roasting vanadium-containing materials.

7.7.3 Stove

The stove is mainly composed of a steel cylinder lined with refractory brick or a reinforced concrete shell. The raw material block or pellets are added to the top of the stove and air is supplied at the lower part. If a solid fuel is used, it is added gradually or incorporated into the raw material. If gas or liquid fuel is used, it is sprayed into the stove with the airflow. All raw materials gradually sink in the stove via gravity, and the product is produced by preheating, combustion, and cooling, and then discharged from the bottom of the stove. The stove is characterized by a simple structure and continuous operation. Preheating, calcination, and cooling adopts stratified loading. In general, the material layer is thicker, and the interface between the coal and vanadium material is thin, resulting in a high heat of fuel combustion because of the high temperature and rapid heating. Hence, when V-containing materials are in contact with the fuel, they are easily burned. However, when they are far from the fuel, the temperature is lower and convective heat transfer is slow, resulting in underfiring and uneven heat use. Sometime the roasted clinker has poor quality and high coal consumption. Operating with a concealed fire, the temperature of the smokestack is generally <200°C. The main advantages of this system include its simple structure and convenient masonry with a small volume and hence, small area for heat loss. However, a disadvantage of this process is that when the particle size of the mixture changes, poor uniformity of the material and airflow along the cross section of the kiln body occurs. This results in the "kiln wall effect," where the fuel does not burn adequately, resulting in nodular, risk control system, and poor ventilation. As the gas concentration is low and the raw material in the stove wall enclosed brush, serious deterioration of the stove occurs, which affects its service life.

The inner diameter of the conical stove gradually increases from top to bottom, where the lower part of the kiln is a cone and the upper part is shaped like a horn mouth. This stove shape is beneficial for uniform settling of the materials and can reduce the kiln wall effect. The stove is divided into three parts: a preheating zone, roasting zone, and cooling zone. Mixtures of V-containing material and fuel are delivered to the top of stove by a small feeding car and then fed into the preheating zone of stove through the distribution system. The material gradually progresses and moves downward through the roasting and cooling zones. The roasted clinker is discharged from an ash system. The combustion air is delivered by a blower at the bottom of the stove, which gradually rises through the cooling, roasting, and preheating zones of the stove. The gas produced after calcining is discharged from the top of the stove. In the preheating zone, heat exchange occurs between the cold material in the stove and the rising hot gas, resulting in the material being heated and the gas being cooled. In the roasting zone the heated fuel and rising hot air react in a combustion reaction that releases a lot of heat. The calcining materials absorb this heat and produce the vanadium clinker. In the process of vanadium slag extraction, stoves are mainly used for the roasting of vanadium materials.

7.7.4 Ball mills

Ball mills are composed of a horizontal cylinder, hollow shaft for feeding and discharging, and a grinding unit, cylinder for long cylinder. The cylinder is equipped with the grinding unit and cylinder for steel processing, where the steel lining plate and cylinder are fixed. The grinding medium is generally steel balls, where their diameter and content are varied. Steel balls are fed into the cylinder. Cylpebs are used as the grinding medium, when available. Based on the particle size of the raw material, the grinding medium is selected. The material is fed into the cylinder via the feeding end of the hollow shaft. When the ball mill cylinder rotates, the grinding medium is influenced by inertia, centrifugal forces, and friction. The steel balls are moved by the cylinder, and when they reach a certain height, they fall under gravity and act as a projectile that hits and breaks the material in the cylinder. Ball mills are composed of feeding, discharging, rotary, transmission (reducer, small transmission gear, motor, electric control), and other main parts. The hollow shaft is made of cast steel, where the inner lining can be changed. The rotary gear is machined with the casting gear, and the shell has a wear-resistant lining that has good wear resistance. Dry ball milling or wet grid ball milling can be selected according to the material and drainage requirements. The length-to-diameter ratio of a rod mill should be <5, while the ratio of the length of the rod storehouse to effective mill diameter should be controlled to 1.2–1.5. The length of the rod needs to be shorter than the rod storehouse (100 mm), which is beneficial for parallel arrangement of steel rods and prevents crossing and disordered rods. In the process of vanadium extraction from vanadium slag, ball mills are mainly used for the preparation of V-containing materials and the leaching of roasted clinker.

7.7.5 Thickener

The thickener is a solid–liquid separation device based on gravity settling. It is usually made of concrete, wood, or welded metal plates as structural materials to construct a cylindrical trough with a conical bottom. In this structure, pulp containing a solid fraction of 10%–20% is converted to an underflow pulp containing a solid weight of 45%–55% by gravity settling enrichment. With the help of a rake installed in the thickener that is moving slowly (1/3–1/5 rpm), the underflow pulp is discharged from an outlet at the bottom of the thickener. The thickening process produces a clean fluid (overflow) that is discharged from a circular chute at the top. Thickeners can be classified into three main types according to the transmission modes, where the first two are the most common: (1) central transmission: usually, this kind of thickener has a small diameter, usually below 24 m; (2) peripheral roller drive type: more common for large- and medium-sized thickeners as it is driven by a transmission car, it is usually about 53 m in diameter and 100 m in length; and (3) peripheral rack drive type.

The main features of thickeners include (1) a degassing tank to avoid solid particles becoming attached to bubbles, such as the "parachute" settling phenomenon; (2) the ore pipe is located below the liquid level to prevent gas being brought into the mine; (3) the mine sleeve is lowered and the feeding plate is raised to allow the slurry to be brought into the mine in an even and steady flow, effectively preventing turning flowers caused by residual pressure; (4) the internal overflow weir ensures that the material can flow adequately and prevent "short circuit" phenomenon; (5) the overflow weir has a zigzag structure to improve the suction of local drainage; and (6) the rake teeth design can be changed from a slash to curve shape, to make pulp center rake and gave it back to a "backlog" force at the center, to row ore underflow concentration is high, thus increasing the processability.

7.7.6 Magnetic separators

Magnetic separators are mechanical devices that achieve separation based on the different magnetic properties of materials. The magnetic separation process applies magnetic and mechanical forces to the ore particles. Particles with different magnetism move along different trajectories, thus separating them into various mineral products. The main part of the magnetic separator is a "mountain"-type electromagnet and a rotating suspension induction disk. The disk is a turnbuckle dish with spikes, with a larger diameter than the feeding belt width about half, which uses a worm and worm wheel reduction drive. The polar distance between the electromagnet and disk can be adjusted in the range of 0–20 mm via a handwheel. To prevent blockage a weak magnetic field is applied to the ore cylinder, so strongly magnetic minerals can be separated first. In the process of extracting vanadium pentoxide from vanadium slag, magnetic separators are mainly used for the processing of vanadium slag.

7.7.7 Filter

The filter is a device that uses a porous media to intercept solid particles in a solid–liquid mixture. The filter medium is used to separate a container into upper and lower chambers. The solid–liquid mixture is added to the upper chamber, and under pressure the liquor passed through the filter medium and the filtrate enters the lower chamber. The solid particles are trapped on the surface of the filter medium, producing a filter residue (or filter cake). The filter cake layer gradually thickens during the filtration process, resulting in higher resistance to the liquid passing through the filter layer, and a decrease in the filtration rate. When the filter chamber is full of filter residue or the filtration rate is too low, filtering is stopped and the filter residue is removed, thereby regenerating the filter medium for further filtration cycles.

The liquid must overcome the resistance supplied by the filtrate layer and the filter medium. Hence, a pressure difference between the sides of the filter medium must be applied, which is the driving force for filtration. Increasing the pressure difference can accelerate filtration. The crushed and transformed particles can easily plug the filtration media pores when the pressure is low, resulting in a low filtration rate. The methods for applying pressure as the driving force for filtration are classified into gravity filter, vacuum filter, and pressure filter types. The choice of filter is based on the concentration of the slurry, solid particle size, liquid viscosity, and requirements for the filter quality.

7.7.8 Melting furnace

The furnace for melting vanadium pentoxide is a traditional metallurgical reverberatory furnace. Its main function is dehydration, deamination, and desulfurization of ammonium vanadate and melting of vanadium pentoxide. Heat transfer in the furnace is provided by both combustion reactions and radiative heat transfer from the roof, walls, and hot gas. Reverberatory furnaces are composed of a furnace base, hearth, furnace walls, furnace top, feeding port, product outlet, and flue. Auxiliary equipment includes feeding and blower devices, a smoke exhaust, and waste heat utilization systems. The main body is composed of a combustion chamber, melting chamber, and exhaust flue (chimney). The whole furnace is a rectangular smelting chamber lined with refractory materials.

CHAPTER 8

Extracting vanadium from shale

Chapter Outline
8.1 V-bearing carbon shale 202
    8.1.1 Vanadium shale structure 202
    8.1.2 Classification of V-bearing carbonaceous shale 203
    8.1.3 Demand for V-containing shale to extract vanadium pentoxide 205
8.2 Extracting vanadium pentoxide from stone coal 207
    8.2.1 Sodium salt roasting 208
    8.2.2 Calcium salt roasting 214
    8.2.3 Direct acid leaching 218
    8.2.4 Salt-free roasting 223
8.3 Comparison of vanadium extraction processes from stone coal 224

Shale is a rock formed by the compression of clay during movement of the Earth's crust. It is a type of sedimentary rock that is formed by extrusion, dehydration, recrystallization, and cementation of weak clay. It is called shale because it is stratified and easily delaminated. Shale is generally brown, gray, or black, with low hardness and high fragility and is easy to process. Depending on the silicon, calcium, and carbon contents in different shales, they are classified as siliceous shale, calcareous shale, or carbonaceous shale. Siliceous shale has small deformation and low hygroscopicity and is not easily weathered. Part of the shale contains high contents of K_2O, Na_2O, and CaO. The composition of clay rocks is complex; in addition to clay minerals (such as kaolinite, montmorillonite, hydromica, and biolite), it contains many detrital minerals (such as quartz, feldspar, and mica) and authigenic minerals (such as aluminum, iron, and manganese oxide and hydroxide) with foliate or lamellar bedding.

Stone coal is a type of carbonaceous shale or black shale. It contains a large amount of carbonized organic matter and is commonly found at the top of coal-bearing strata. In addition to carbon, black shale contains various elements, such as V, Fe, Al, Si, Ni, Cu, Mo, and S. Depending on the dominant types of metal in the black shale, it is classified as nickel molybdenum ore, stone coal, or carbon uranium ore.

As in the case of extracting vanadium pentoxide (V_2O_5) from V-bearing titanoferrous magnetite, extraction of V_2O_5 from V-bearing carbonaceous shale can be achieved via sodium salt treatment, calcium salt treatment, roasting leaching, and direct acid leaching. In the sodium salt treatment process, V-bearing carbonaceous shale is generally crushed and finely ground and then roasted with the addition of sodium salt (such as sodium chloride, sodium sulfate, or sodium carbonate) at 850°C. All of the vanadium oxide reacts with sodium salt to produce water-soluble sodium metavanadate ($NaVO_3$), sodium orthovanadate (Na_3VO_4), and sodium pyrovanadate ($Na_4V_2O_7$), which are then leached by water. The vanadium solution is purified by extraction, adsorption, and precipitation to remove impurities. The purified vanadium solution is then adjusted to pH 2–3 by addition of sulfuric acid, where the ammonium vanadate is precipitated in the presence of ammonium. Ammonium vanadate is calcined and melted at 700°C to produce dense purple-black industrial V_2O_5 with a purity >98%.

Since Bleecker published a patent in 1912 describing roasting with sodium salt and water leaching recovery of vanadium, this process has been in continuous use. In the calcium salt treatment process, calcium carbonate is used as an additive for extracting vanadium, which is mixed with finely ground V-bearing carbonaceous shale and then roasted at 850°C. All of the vanadium oxide reacts with calcium oxide to produce water-soluble calcium pyrovanadate ($Ca_2V_2O_7$), calcium metavanadate (CaV_2O_6), and calcium orthovanadate ($Ca_3V_2O_8$), which are leached using dilute acid or alkali. The leaching liquid is purified and then transferred to adjust the acidity. Hydration vanadium is precipitated by the addition of a precipitation agent. The sediment is washed, filtered, dried, and then calcined and melted to produce V_2O_5 flakes. During direct acid leaching, the crushed and ground V-bearing carbonaceous shale is placed in a tank containing dilute acid for leaching. In general, a low pH of 2.0–3.0 is used at room temperature. Precipitation is used to remove impurities, where the pH of the purified vanadium liquid is adjusted and vanadium hydrolysis is precipitated or added cleaning agent to precipitate. Because of the different properties of various vanadium shales, some acid leaching methods need to be coupled with calcination steps to transform the mineral structure into phases more suitable for acid leaching [22,25,30,32–34,46–48,62,63,73–79].

8.1 V-bearing carbon shale

V-bearing stone coal is a fine-grained shale containing vanadium, which also contains 0.4% Zn and 0.005% U_3O_8. The main minerals are quartz, fluorite, clay, calcite, and dolomite.

8.1.1 Vanadium shale structure

Stone coal was formed in ancient time and the early Paleozoic period. It is a type of sedimentary combustible organic rock, which is black or black and gray. Stone coal was formed in the ancient stratum, where biological remains, such as bacteria and algae, were transformed in shallow seas, lagoons, and bays under the conditions of sag and coal. The material composition of V-bearing stone coal is complex, where the content and valence of the vanadium can vary, and the vanadium is dispersed as small particles. Most of the vanadium in V-bearing stone coal occurs as V^{3+} in silicate minerals, such as roscoelite, tourmaline, and garnet, where some of the Si is replaced by V in silicon oxy-tetrahedral multilayers by isomorphism. Similarly, some Al^{3+} is replaced by vanadium in aluminum oxide octahedron monolayers. The vanadium in the V-bearing stone coal occurs in minerals such as V−Ti garnet, Ca−V garnet, and vanadium uranium ore. It can also be found in metallic organic complexes and vanadium leaf moieties. In some cases, adsorbed complex anions occur in iron oxide and clay minerals.

Stone coal is a type of highly metamorphic sapropelic or algal coal, which generally has high ash and sulfur contents, and low heat and hardness. In addition to organic carbon, it also contains silica, calcium oxide, and small amounts of iron oxide, aluminum oxide, and magnesium oxide. It looks like stone and is not easily distinguished from limestone or carbon shale. It is a combustible and kata-metamorphism organic mineral with high ash content (generally greater than 60%). High-quality stone coal with high-carbon content is black, with a semi-bright luster, few impurities, and relative density of 1.7−2.2. Stone coal with a low carbon content is grayish and a little dull and contains significant amounts of pyrite, quartz vein, and phosphorus. The calcium nodules have a relative density of 2.2−2.8. The calorific value of stone coal is not high, between 3.5 and 10.5 MJ/kg and is hence a type of low-calorific fuel.

V-bearing stone coal can be used for extraction of vanadium pentoxide. The V_2O_5 content in the stone coal is low, usually about 1.0%. In V-bearing stone coal, most vanadium occurs as V(III), with some V (IV), and rarely as V(V). As the ionic radii of V(III) and Fe(II) are equal (74 pm), and that of Fe(III) is similar (64 pm), V(III) does not occur as a single-metal mineral but occurs in the silica tetrahedral structure of ferroaluminum minerals such as roscoelite and kaolin by isomorphism.

8.1.2 Classification of V-bearing carbonaceous shale

Stone coal is a type of multimetal complex ore, which has high recycling value of carbon, aluminum, and potassium. Stone coal is not a coal but a type of marine sediment. The material that forms the stone coal includes inorganic salts of mud, Si, Ca, and organic matter composed of low-level organisms, such as phycomycetes. Depending on the composition and structure, stone coal can be classified into siliceous rock and clay carbon types. Depending on the degree of weathering, it can be divided into weathered shale and carbonaceous shale.

There are various types of stone coal. Based on the ash and calorific value, it can be classified into general stone coal and high-quality stone coal. The ash content of general stone coal is 40%−90%, and the calorific value is below 16.7 kJ/g. The ash content of high-quality stone coal is 20%–40%, and the calorific value is 16.7–27.1 kJ/g. Based on the texture and structure, stone coal can be divided into block, granular, flake, and powdery stone coal. Depending on the mineral impurities in the stone coal, it can be divided into siliceous stone coal and calcareous stone coal types.

Vanadium in V-bearing stone coal is adsorbed and occurs in the clay mineral (e.g., kaolin and roscoelite), which corresponds to 66.38%, while the vanadium isomorphism corresponds to 33.62%. Vanadium in V-bearing stone coal is adsorbed and occurs in siliceous rocks, which account for 25.83%, while the vanadium isomorphism in roscoelite accounts for 74.17%.

The mineral composition of V-bearing carbonaceous shale is mainly nonmetallic minerals and less metallic. Nonmetallic minerals are mainly siliceous compounds (e.g., quartz), mud, calcite, and carbon. The metallic minerals are mainly limonite, secondary pyrite, V-containing iron ore, anatase containing Fe and V, and a small amount of roscoelite. The mineral containing vanadium in stone coal is mainly hewettite ($CaO \cdot 3V_2O_5 \cdot 9H_2O$).

Silica: Cryptocrystalline quartz and chalcedony are present in siliceous stone coal ($\sim 80\%$ of the total silica mass), followed by micritic structures with a size of 0.01–0.02 mm ($\sim 20\%$). Quartz is composed of siliceous crystals, and fine silty clastic quartz, where the latter is scattered as particles with a size of 0.01–0.05 mm.

Mud: It is extremely fine and mainly contains kaolinite and roscoelite, where some colloidal limonite impurities are found on the surface of the rocks.

Carbon: Carbon is uniformly distributed in the carbonaceous mudstone, which is fine and dotted in the carbon siliceous rocks, with a particle size generally above 0.003 mm, while some individual grains can reach 0.005 mm. Discontinuous lamellar and lamellar distributions are clustered along the bedding.

Limonite: It is common in rock with a particle size of 0.005–0.1 mm. It is mainly gelatinous and followed by pyrite micrite. It is nonhomogeneously distributed in the lithostratification zone and fissures, as in lenticular and fine veins.

Pyrite: It is a cubic crystal distributed in the rock with a sparse star-dotted with a particle size of 0.005–0.1 mm. Shallow rocks were oxidized to limonite.

Ore structure and structure: The ore structure is mainly a cryptocrystalline and variable residual fine silt structure. The secondary structure is a granular, colloidal, and pseudocrystalline structure. The ore texture mainly consists of microparallel—laminar, block structure, interbedded structure (strip), nodules, and tabular structures.

Depending on the host rocks, the ore type can be divided into three categories: carbonaceous siliceous rock, carbonaceous shale, and siliceous carbonaceous rock. The first two are mainly area natural ore types, while siliceous carbonaceous rock only has a local distribution.

Carbonate-containing silica-type vanadium ore: It mainly occurs in black carbon siliceous rocks with little or no mud interlayer. The V_2O_5 content is generally 0.76%–1.27%, with the highest value of 1.91% and lowest value of 0.68%.

Carbonate-containing mudstone-type vanadium ore: It mainly occurs in mud (shale) stone, with very few thin siliceous layers or strips. The average content of single-sample V_2O_5 is 0.85%–1.16%, with a highest value of 1.69% and lowest value of 0.60%; the average content is 1.06%.

Siliceous carbonaceous vanadium ore: It has characteristics of carbon siliceous rock and mudstone ore, with a V_2O_5 content of generally 0.83%–1.53%, with a highest value of 1.75%, lowest value of 0.59%, and average value of 1.05%.

Vanadium-containing shale in the United States is divided into weathered shale (containing organic carbon below 1%) and carbonaceous shale, where the main compositions are shown in Table 8.1.

Table 8.1: Main compositions of vanadium-containing shale from Nevada, United States.

Shale	V_2O_5	P_2O_5	SiO_2	Al_2O_3	MgO	CaO	Fe	Organic C
Weathered shale	0.93	0.64	54	4.0	6.5	11.2	1.6	<0.1
Weathered shale	1.28	0.55	57	3.2	5.8	11.2	1.4	<0.1
Carbonaceous shale	0.87	0.87	57	4.2	5.6	9.6	2.5	10
Carbonaceous shale	0.80	0.70	53	3.2	4.8	8.4	1.8	10

8.1.3 Demand for V-containing shale to extract vanadium pentoxide

Stone coal deposits containing vanadium are a new metallogenic type, where V-bearing black-shale deposits were formed at marginal sea slope zones. The main vanadium mineral is V-containing illite. The high calorific value of V-bearing stone coal can be used as fuel for thermal power generation after improving combustion technology. Vanadium is enriched in the soot and can be used as a raw material for vanadium extraction. The low-calorific and low-carbon-containing vanadium materials can be used as raw materials for vanadium extraction.

8.1.3.1 Extraction of vanadium pentoxide from shale by pyrogenic processes

Vanadium stone coal occurs with a silicon–oxygen tetrahedron structure of clay minerals with trivalent by isomorphism, which is strongly combined and insoluble in acid and alkali solutions. At high temperature, additives are introduced to transform trivalent vanadium into soluble pentavalent vanadium; meanwhile, carbon in the stone coal is removed. Hence, roasting and transformation of V-bearing stone coal is an indispensable process.

The main steps in V_2O_5 extraction from shale by pyrogenic processes are transformation and roasting of the mineral. Roasting processes include blank roasting and doping roasting. Blank roasting does not use any additives, and high concentrations of acid are required for leaching and decomposition. Additives (such as Na, Ca, Fe, Ba and other salts, and sulfuric acid) are used in doping roasting to produce sodium vanadate, calcium vanadate, and other vanadates that are soluble in water or acid.

In traditional pyrogenic processes for extracting V_2O_5 from shale, NaCl and Na_2CO_3 are used as roasting additives. This results in large amounts of Cl_2, HCl, and other poisonous and harmful gas (e.g., SO_2) being produced, which leads to serious smoke pollution. The salinity of the wastewater is high, and the vanadium recovery is generally only about 50%. In addition, severe resource wastage occurs and the production environment is poor. Many sodium ions are present in the leaching residue, which cannot be scale multipurpose utilization. Calcium roasting does not produce Cl_2, HCl, or poisonous and harmful gases, such as SO_2. However, this roasting process is greatly influenced by the mineral type and its properties, and the roasting temperature, time, atmosphere, and dosage of calcium salt. If there is insufficient process control, calcium roasting can easily produce insoluble silicates, where some vanadium particles are coated with silica, or some of the vanadium reacts with other elements, such as Fe and Ca, to produce insoluble calcium vanadate and iron metavanadate compounds. The calcification processing residue can be large-scale multipurpose utilization.

Blank roasting is mainly used for stone coal decarburization and low-valence vanadium oxidation and has certain limitations related to the mineral structure. However, the roasting equipment is simply traditional shaft kilns and fluidized beds. The production scale is limited, and the vanadium crystal structure is not completely converted during the roasting process, which limits the recovery of vanadium. This process has poor adaptability to stone coal resource utilization. Sulfation roasting can enhance the mineral decomposition process. The sulfation roasting temperature is 200°C–250°C, roasting time is 0.5–1.5 h, and leaching is achieved using an aqueous solution with calcined clinker with pH 1.0–1.5, which results in a high use of sulfuric acid. The boiling point of sulfuric acid is 338°C and its concentration is 98.3%. The gas from roasting is mainly water vapor, so cleaning the gas is easy. Sulfation roasting of stone coal at low temperature is simple without oxidation.

The leaching step during pyrogenic extraction of vanadium pentoxide from shale is classified into three types: water leaching, alkali leaching, and acid leaching. Water leaching is only applicable for sodium vanadate, which is soluble in water, and this process has been widely used with sodium salt roasting. Alkali leaching is suitable for the calcium salt roasting process. It is highly selective and can be used for circulating treatments. It is suitable for the treatment of stone coal with more alkaline gangue. Acid leaching processes are classified into concentrated-acid leaching and weak-acid leaching methods. Concentrated-acid leaching uses large amounts of acid, but can leach more impurities; the acidity of the reacted acid is high, and its recovery is low. The distinctive features of weak-acid leaching include a long reaction time, moderate leaching of impurities, low acidity, and low recovery of the reacted acid. The leaching method can also be divided into powder leaching and pellet leaching. Powder leaching is faster than pellet leaching, although similar leaching efficiencies are obtained.

The purification and enrichment of the leaching solution use adsorption and extraction of a resin, where resin adsorption is only suitable for neutral leaching solutions. Extraction is divided into tetravalent and pentavalent vanadium extraction.

High-carbon stone coal needs to be decarbonized, and the vanadium is enriched as ash, which is used as a raw material for extracting V_2O_5. Some of the enriched ash has high contents of V_2O_5, which is suitable for extracting V_2O_5 in combination with vanadium slag or recycled V-containing materials.

8.1.3.2 Extraction of vanadium pentoxide from shale by wet processes

The extraction of V_2O_5 from shale by wet processes is mainly used for weathered stone coal. To obtain a high efficiency of V_2O_5 leaching, a large amount of H_2SO_4 is consumed. In production the H_2SO_4 dosage is commonly 25%–40% of the ore quantity, resulting in a V_2O_5 leaching efficiency of 65%–65%, where a leaching efficiency above 80% is rare, and the V_2O_5 recovery is generally not more than 70%. The challenge is purification of the acid leach liquor to remove impurities, where Fe(III) reduction and pH adjustment steps consume a large amount of additives, especially ammonia, leading to the production of ammonia nitrogen wastewater that needs to be treated.

8.2 Extracting vanadium pentoxide from stone coal

Extracting vanadium pentoxide from stone coal is similar to extracting it from vanadium slag. In both processes, vanadium components must be converted into soluble substances. Depending on the properties of the mineral, sodium salt roasting or calcium salt roasting can be used as the roasting transformation process, while the leaching process can be acid or alkali leaching. An appropriate mineral decomposition process needs to be selected depending on the composition and structure of the stone coal. Different vanadium grades of the V-bearing stone coal from the various vanadium mines require the selection of different raw materials and processes. Some typical vanadium extraction processes are discussed here, which can be classified into three parts: decomposition of the vanadium ore, enrichment and separation of the vanadium solution, and pure vanadium preparation.

Typical methods for extracting V_2O_5 from stone coal include (1) sodium salt roasting, which involves roasting (smoldering), acid leaching, and ion exchange steps; (2) calcium salt roasting, which involves roasting, acid leaching, and ion exchange steps; (3) salt-free roasting, which includes roasting, acid leaching, and solvent extraction steps; (4) direct acid leaching, which includes leaching, followed by extraction of vanadium pentoxide from an intermediate salt. Table 8.2 shows the melting point and molecular formula of the main vanadates.

Table 8.2: Melting point and molecular formulas of the main vanadates.

Vanadate	Molecular formula	Melting point (°C)
Sodium orthovanadate	Na_3VO_4	850–866
Sodium metavanadate	$Na_4V_2O_7$	625–668
Sodium pyrovanadate	$NaVO_3$	605–630
Calcium metavanadate	$CaO \cdot V_2O_5$	778
Calcium pyrovanadate	$2CaO \cdot V_2O_5$	1015
Calcium orthovanadate	$3CaO \cdot V_2O_5$	1380
Iron orthovanadate	$FeVO_4$	870–880

8.2.1 Sodium salt roasting

Sodium salts (Na_2CO_3, NaCl, or Na_2SO_4) can be used as an additive for vanadium extraction. The vanadium in stone coal is converted into a soluble component by roasting and oxidation and is then separated from other components via a leaching process.

8.2.1.1 Process details

V-bearing stone coal is crushed, ground, and screened. The resulting V-bearing stone coal with a granularity <0.1 mm is mixed with sodium salt (Na_2CO_3, NaCl, and Na_2SO_4) and then roasted and oxidized at high temperature. The vanadium in the stone coal is converted into water-soluble sodium vanadate and sodium metavanadate and then leached with hot water and dilute sulfuric acid, or an alkaline solution. The aqueous solution containing vanadium is enriched after purification and treatment. Then, acid is added to adjust the solution pH, and ammonium vanadate is precipitated in the presence of ammonium (NH^{4+}, NH_3) to obtain V_2O_5 after high-temperature calcination.

During hot-water leaching of clinker from the sodium salt roasting process, the sodium vanadate and sodium metavanadate are dissolved in the hot water, and insoluble impurities are separated from the solution. The pure vanadium compound is produced by purification and separation. Fig. 8.1 shows a typical process flowchart of extracting vanadium pentoxide from stone coal by sodium salt treatment.

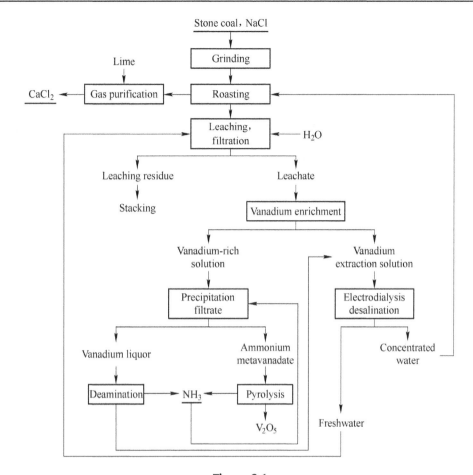

Figure 8.1
Typical process flowchart of extracting vanadium pentoxide from stone coal by sodium salt treatment.

8.2.1.2 Process for high-carbon stone coal

Vanadium-containing carbonaceous shale is decarbonized by combustion in a boiler power generation or fluidized bed system. Vanadium in the carbonaceous shale is enriched in the ash during combustion. Then, the enriched ash is mixed with NaCl or Na_2CO_3, followed by oxidization and roasting. The vanadium compounds are converted into water-soluble $NaVO_3$, NaH_2VO_4, and Na_3VO_4. Vanadium-containing carbonaceous shale can also be mixed with anthracite to increase its calorific value, followed by mixing with NaCl or Na_2CO_3, oxidization and roasting, and conversion into water-soluble $NaVO_3$, NaH_2VO_4, and Na_3VO_4.

8.2.1.3 Process reactions

The metallic oxides in stone coal include oxides of sodium, iron, vanadium, magnesium, and calcium. Sodium oxide and vanadium oxide react during the roasting process to form water-soluble sodium orthovanadate, sodium pyrovanadate, and sodium metavanadate. Calcium oxide and vanadium oxide also react during roasting to produce calcium pyrovanadate, calcium orthovanadate, and calcium metavanadate, which are acid-soluble compounds. Magnesium oxide and vanadium oxide react during roasting to produce magnesium pyrovanadate and magnesium metavanadate, which are selectively dissolved. Iron oxide and vanadium oxide react during roasting to produce ferric orthovanadate, which is also selectively dissolved. Sodium orthovanadate, sodium pyrovanadate, and sodium metavanadate have the highest solubility in water, while calcium vanadate and iron vanadate are less soluble in water, but soluble in dilute acid and alkali. The free energy of vanadium oxide is shown in Fig. 8.2. The carbon combustion energy is less than the Gibbs free energy of vanadium oxidation, so combustion of carbon occurs first during roasting. When the carbon content is low, oxidation of trivalent vanadium begins. Before oxidizing and roasting of the stone coal, the raw ore is usually decarburized.

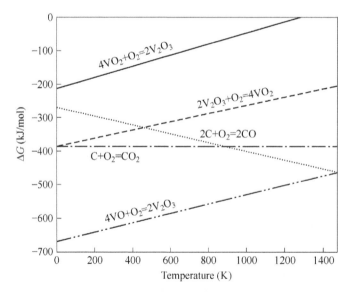

Figure 8.2
Relationship between free energy of vanadium oxide and temperature.

The chemical reactions occurring during sodium salt calcination are as follows:

$$4FeO \cdot V_2O_3 + 4Na_2CO_3 + 5O_2 = 4Na_2O \cdot V_2O_5 + 2Fe_2O_3 + 4CO_2 \quad (8.1)$$

$$2NaCl + \frac{1}{2}O_2 = Na_2O + Cl_2 \quad (8.2)$$

$$V_2O_3 + O_2 = V_2O_5 \tag{8.3}$$

$$xNa_2O + yV_2O_5 = xNa_2O \cdot yV_2O_5 \tag{8.4}$$

$$Na_2O + V_2O_5 = 2NaVO_3 \tag{8.5}$$

During oxidation and roasting the main chemical reactions involving high-carbon stone coal and sodium salt additives are as follows:

$$C + \frac{1}{2}O_2 = CO \tag{8.6}$$

$$CO + \frac{1}{2}O_2 = CO_2 \tag{8.7}$$

$NaVO_3$ and KVO_3 in the roasted clinker are soluble in water. V_2O_5 easily reacts to produce vanadate during the roasting process in the presence of sodium oxide, which is soluble in water and dilute acid, depending on the following chemical reaction:

$$NaVO_3 + H_2O \rightarrow Na^+ + VO_3^- + H_2O \tag{8.8}$$

The leaching liquid from the roasted clinker of V-bearing stone coal is clarified in a tank and then leached using a new resin leaching liquid. The vanadium solution is qualified without purification treatment, and then acid is added for ion exchange adsorption reactions after acid transformation treatment. During adsorption the flow should be controlled to ensure a good recovery. The wastewater from the adsorption step is returned for pulping or leaching processes. After the resin is saturated the NaOH solution with pH of 10–11 is used to obtain a liquid with high vanadium concentration. After analysis the resin is washed with NaOH solution and pure water to give a pH value of 3.5–4. An HCl solution is used for regeneration treatment to restore the adsorption capacity.

Ion exchange is a type of displacement reaction. The resin molecules have lower affinity with Cl^- ions than Na^+ ions. When the resin is in contact with the $NaVO_3$ solution, the following reaction occurs:

$$[R]Cl + NaVO_3 \rightarrow NaCl + [R]VO_3 \tag{8.9}$$

The saturated resin is desorbed by the alkali solution:

$$[R]VO_3 + NaOH \rightarrow NaVO_3 + [R]OH \tag{8.10}$$

After desorption a high-concentration vanadium solution is obtained. The resin is reprocessed with HCl to restore the adsorption capacity:

$$[R]OH + HCl \rightarrow [R]Cl + H_2O \tag{8.11}$$

NH_4VO_3 is precipitated from the $NaVO_3$ solution in the presence of excess NH^{4+} ions:

$$NaVO_3 + NH_4^+ \rightarrow NH_4VO_3 + Na^+ \tag{8.12}$$

NH_4VO_3 is calcined in high temperature to obtain V_2O_5 products:

$$2NH_4VO_3(s) \rightarrow V_2O_5 + 2NH_3\uparrow + H_2O\uparrow \tag{8.13}$$

8.2.1.4 Process operation and technical indexes

The stone coal is screened after two stages of crushing and classification by sieving. The fine material (<15 mm) is transferred to a fluidized bed furnace for decarburization, while the coarse material (>15 mm) is used as an ingredient. Decarburized material mixed with 5% anthracite was calcined for 1.5 h at 820°C, giving a leaching efficiency of 82.08%. For a roasting time of 1 h the leaching efficiency was 81.96%. When the decarburized powder and anthracite mixture is roasted at high temperature, point contacts between the particles facilitate heat transfer and oxidation of vanadium, which accelerates the conversion of vanadium. The addition of anthracite coal to the decarburized material reduced the roasting temperature by 30°C, shortened the roasting time by 0.5 h but did not affect the leaching efficiency, resulting in greatly reduced energy consumption during roasting. After decarburization the stone coal is evenly mixed with recycled ash (particle size below 25 mm) to achieve a calorific value in the range of 400 ± 50 kcal/kg. The mixture is fed into a dry ball mill for grinding to achieve a mineral powder with 0.104 mm grains, which is then pelleted and transferred to a rotary kiln for roasting. The calcination temperature is controlled at 800°C, with a roasting time of 1–1.5 h. The typical chemical composition of high-carbon stone coal is shown in Table 8.3.

Table 8.3: Typical chemical composition of high-carbon stone coal (%).

V_2O_5	SiO_2	Al_2O_3	Fe_2O_3	CaO	MgO	K_2O	Na_2O	C	Volatile
0.82	66.14	6.46	3.49	2.96	1.43	1.65	0.69	9.38	4.59

The roasted clinker is crushed again and mixed into a pulp with a solid–liquid ratio of 1:3 in a pulping tank, which is transferred by a slurry pump to a leaching mixer, where fluidization technology is used for leaching and the pulp is leached by continuous agitation. The leaching acidity is 3–5 vol.% with pH = 7. To accelerate the leaching process and increase the leaching efficiency, the leaching solution is heated to 50°C. The qualified leaching solution (containing vanadium concentration ≥ 2.2 g/L) is pumped into a belt-type vacuum filter, and the filtrate directly enters a solution tank. The filter residue is washed and then filtered, and the filtrate is used in the next leaching cycle or is returned to the pulping step. The leaching residue is transported by a belt conveyor to the dry slag mixer and mixed with 2% lime and an alkali material for neutralization, before being sent to a tailings pond.

The V-containing solution is clarified in the solution tank; purification is not necessary as the V-containing leaching liquid contains the new resin. After the acid transformation process, ion exchange adsorption is performed. During adsorption the flow should be controlled to ensure good recovery. The tailing water from the adsorption step is returned to pulping or leaching processes. After the resin is saturated a NaOH solution with pH 10−11 is used to remove the adsorbent and obtain a liquid with high vanadium concentration. After the removal of the adsorbent the resin is washed with NaOH solution and pure water to achieve a pH of 3.5−4. HCl solution is used to regenerate the resin and restore its adsorption capacity.

Industrially pure NH_4Cl is added to the V-containing solution to precipitate the vanadate. During precipitation, adequate stirring is required to increase the recovery. The precipitate is filtered using a filter box to obtain the intermediate product (ammonium vanadate). The ammonium metavanadate is calcined, resulting in dehydration and recovery of vanadium powder.

The process control parameters are as follows: the ratio of salt is 100−200 kg/t stone coal, depending on the stone coal structure, which can match into a small amount of Na_2SO_4; roasting temperature of 750°C−850°C; and process time of 1−4 h. These conditions result in 50%−65% conversion efficiency, 88%−93% water leaching efficiency, 92%−96% recovery of precipitated vanadate by hydrolysis, 90%−93% refined vanadium yield, and 98.5% V_2O_5 grade product. The recovery of V_2O_5 is $\sim 45\%$ when using a kiln, while it is $\sim 55\%$ when using a fluidized bed. In the process of extracting V_2O_5 from stone coal by sodium salt roasting, the production of 1 t V_2O_5 consumes 20−28 t of sodium chloride, 1−1.5 t of caustic soda, and 1.5−1.8 t of industrial HCl. If the process uses sulfuric acid, 0.8−1.0 t of sulfuric acid and 1.2−2 t of ammonium chloride are required.

V-containing shale from Nevada, United States is used to extract vanadium pentoxide. The shale is crushed to a particle size below 50 mm, and then dried at 383K with 5%−10% salt, ground in an air classifier ball mill to achieve a particle size below 0.42 mm. The powder is roasted for 3 h at 1198K in a rotary kiln. The roasted clinker is cooled and leached with a weak-acid solution. After filtration the pH of the filtrate is adjusted to 2.5−5 to remove Si. The Si-free filtrate is adjusted to pH 3 with sulfuric acid, and vanadium in solution is extracted via an organic phase of 0.075 mol/L DITDA. The vanadium-loaded organic phase is treated with a pure alkali solution, and then NH_4Cl is used to precipitate ammonium vanadate. The sediment is calcined and deaminated to produce V_2O_5.

8.2.2 Calcium salt roasting

The process of extracting V_2O_5 from stone coal by calcium salt roasting involves the addition of lime or limestone during roasting, depending on the research results of the high-temperature reaction of the mineral. V-bearing stone coal is roasted at high temperature, where low-valence vanadium is oxidized and converted into high-valence vanadium. The vanadium in stone coal mainly occurs as calcium–silicon vanadate and calcium–titanium oxide. The mineral is chemically unstable and can dissolve rapidly in weak-acid solutions. Calcium metavanadate compounds are easily dissolved in weak acidic solutions, which are used to separate and extract vanadium from the minerals.

8.2.2.1 Production process

The V-bearing stone coal is crushed to a granularity <0.1 mm and then mixed with calcium salt ($CaCO_3$ limestone or active lime CaO). The mixture is roasted at high temperature to achieve oxidation and calcification. The vanadium in the stone coal is converted into water-soluble calcium pyrovanadate, calcium orthovanadate, and calcium metavanadate compounds, which are leached with dilute sulfuric acid or alkali solutions. The aqueous V-containing solution is purified and then acid is added to adjust the solution pH. Finally, V_2O_5 is obtained after hydrolysis. Fig. 8.3 shows a typical process flowchart of extracting vanadium pentoxide from stone coal by calcium salt roasting.

8.2.2.2 Process reactions

The chemical reactions during calcium salt roasting are as follows:

$$V_2O_3 + O_2 = V_2O_5 \tag{8.14}$$

$$2V_2O_4 + O_2 = 2V_2O_5 \tag{8.15}$$

$$CaCO_3 = CaO + CO_2 \tag{8.16}$$

$$2V_2O_5 + 4CaO = 2Ca_2V_2O_7 \tag{8.17}$$

$$V_2O_5 + 3CaO = Ca_3(VO_4)_2 \tag{8.18}$$

Dilute sulfuric acid leaching occurs as follows:

$$Ca_2V_2O_7 + 3H_2SO_4 = 2CaSO_4 + (VO_2)_2SO_4 + 3H_2O \tag{8.19}$$

$$Ca_3(VO_4)_2 + 4H_2SO_4 = 3CaSO_4 + (VO_2)_2SO_4 + 4H_2O \tag{8.20}$$

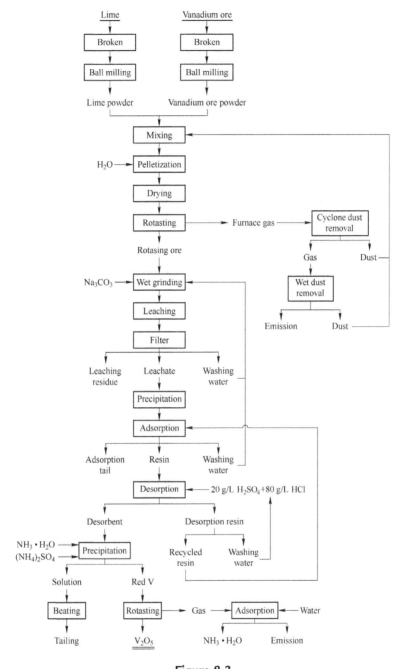

Figure 8.3
Typical process flowchart of extracting vanadium pentoxide from stone coal by calcium salt roasting.

The alkali leaching reactions are:

$$Ca_2V_2O_7 + Na_2CO_3 + H_2O = 2NaVO_3 + CaCO_3\downarrow + Ca(OH)_2 \quad (8.21)$$

$$Ca_3(VO_4)_2 + Na_2CO_3 + 2H_2O = 2NaVO_3 + CaCO_3\downarrow + 2Ca(OH)_2 \quad (8.22)$$

V^{4+} occurs with VO^{2+} in acidic solutions of sulfate and chloride (containing 0.01–0.1 mol of acid). The vanadium is solvent extracted using D2EHPA kerosene, where polymerization occurs in the extraction compound. The organic phase is 6% D2EHPA kerosene solution and 3% TBP. Six stages of mixed extraction are performed, and ammonia is used to control the pH to 1.9. The reverse extraction liquid is 140 g/L of sulfuric acid solution, the extraction temperature is 38°C–49°C, and quaternary reverse extraction is performed.

The leach liquor is purified by extraction, N-236 is used as the extractant, *sec*-octyl alcohols as the synergistic extractant, and sulfonated kerosene as the solvent. The extraction parameters are 15% N-236; 3% 2-octyl alcohol; 82% sulfonated kerosene; phase ratio of extraction of 1:2–3; 2–3 stages of countercurrent extraction; organic phase saturated concentration of 30 g/L; pH of 2–9; and temperature of 5°C–45°C. These conditions result in an extraction efficiency of 99.55%, where the vanadium content in the raffinate is 0.0114 g/L. In the case of N-235 extraction with a pH of ~2.5 the extraction efficiency is 99.62%. The saturated organic phase is reverse extracted with 1 M NH_4OH + 4 M NaCl; two stages of countercurrent reverse extraction are performed to obtain sodium polyvanadate, with a concentration of ~30 g/L.

After reverse extraction the vanadium solution is heated to 70°C to precipitate the vanadate. The red cake precipitate is roasted to produce 98% V_2O_5.

$$2VO_2^+ + 2H_2O = V_2O_5 \cdot H_2O + 2H^+ \quad (8.23)$$

$$5V_2O_5 \cdot H_2O = H_2V_{10}O_{28}^{4-} + 4H^+ + 2H_2O \quad (8.24)$$

$$H_2V_{10}O_{28}^{4-} = HV_{10}O_{28}^{5-} + H^+ \quad (8.25)$$

$$HV_{10}O_{28}^{5-} = V_{10}O_{28}^{6-} + H^+ \quad (8.26)$$

The vanadate precipitation reaction is shown in (8.12), while the calcination reaction of red cake is shown in (8.13).

8.2.2.3 Typical process operation and technical indexes

The V-bearing stone coal is crushed to a granularity <0.1 mm and then mixed with calcium salt ($CaCO_3$ limestone or active lime CaO). The mixture is roasted at high temperature to achieve oxidation and calcification. The vanadium in the stone coal is converted into water-soluble calcium pyrovanadate, calcium orthovanadate, and calcium metavanadate compounds, which are leached with dilute sulfuric acid or alkali solutions. The aqueous V-containing solution is purified and then acid is added to adjust the solution pH. Finally, V_2O_5 is obtained after hydrolysis.

8.2.2.3.1 Typical process of extracting vanadium pentoxide by acid leaching

Depending on the proportion of ingredients, V-bearing stone coal is mixed with lime and SM-1 agent in fractions of 16% and 1.3%, respectively, of the stone coal. The mixture is roasted at 950°C for 3 h, giving a conversion efficiency of 87.6%. The material is then leached with 4% sulfuric acid at room temperature for 12 h with a solid−liquid ratio of (3−4):1, with a leaching efficiency of 85%. The leach liquor is purified by extraction; 15% N263 and 3% dioctanol are used as the extractant, while the diluent is sulfonated kerosene. The phase ratio of extraction (O/A) is 1:(2−3), and four stages of countercurrent extraction are performed. The reverse-extractant is a NH_4Cl and $NaCl$ solution. A sodium polyvanadate solution is obtained with a concentration is 30 g/L. The vanadate is precipitated using NH_4Cl, which is stirred at room temperature for 2−3 h, and then stored for 12 h before drying. The calcination temperature is 550°C, and V_2O_5 is obtained with a purity of 99.5%, and total yield of 65%.

8.2.2.3.2 Typical process of extracting vanadium pentoxide by alkaline leaching

V-containing stone coal is mixed with lime and pelletized, where the stone coal is previously finely ground to 200 mesh, while 70% of the lime is finely grinding to—64.2% 200 mesh. The pellets are then roasted, followed by leaching in soda solution, ion exchange with 717 strong basic anion exchange resin and precipitation of vanadate using ammonium sulfate. The vanadium red cake is then roasted and dried to obtain the V_2O_5 product.

Process operation and technical indexes Roasting: Particle size of stone coal: at least 70% of the particles <200 mesh; addition of lime powder: 2% of the stone coal mass; roasting time: 2−3 h; roasting temperature: 850°C. Leaching: calcined clinker particle size: at least 90% of the particle <120 mesh; leached with agitation; addition of pure alkali: 6% of the calcined clinker mass; leaching time: at least 2 h; leaching temperature: above 75°C; leaching liquid−solid ratio: 1.5−2:1. Direct yield of vanadium: the recovery rate of vanadium from the roasting process was 100%; the direct yield of vanadium was 71.97%; and the direct yield of vanadium was 99.86%. The desorption rate of vanadium was 99.98%, while in the first column it was 90.98%. The direct yield of vanadium was 99.71%. The direct yield of vanadium in red vanadium was 99.32%.

Direct recovery of vanadium (1) The vanadium recovery from the roasting process is 100%; (2) The vanadium recovery from the leaching process is 71.97%; (3) The vanadium recovery from the resin adsorption process is 99.86%; (4) The vanadium recovery from the resin desorption process is 99.86%, with 90.98% recovery from the first column desorption; (5) The vanadium recovery from vanadate precipitation is 99.71%; (6) The vanadium recovery from red cake roasting is 99.32%. The vanadium recovery from the entire process is 64.776%, giving a total recovery of 71.1%.

Consumption of major materials Depending on the proportion of the ingredients: (1) lime powder: 2% of V-bearing stone coal; (2) soda: 5.70% of V-bearing stone coal and 6.0% of calcined clinker; (3) ammonia water: 6.87 kL/t V_2O_5 product; (4) HCl: 2.11 kL/t V_2O_5 product; (5) sulfuric acid: 0.33 kL/t product V_2O_5; (6) ammonium sulfate: 0.15 t/t product V_2O_5.

8.2.3 Direct acid leaching

The composition of V-bearing stone coal is complex, and the occurrence of vanadium varies. It mainly occurs as carbonaceous roscoelite, siliceous vanadium clay, and intermediate types between these two. The V-bearing stone coal can also be divided into weathered shale and carbonaceous shale, where the use of the direct acid leaching depends on the weathering.

8.2.3.1 Process details

In general, the valuable elements in the V-containing shale cannot be separated by leaching in strong acid as many other components are also leached and dissolved in the leaching liquid. Sulfuric acid, fluosilicate acid, or a combination of these can be used for leaching. The acid leaching liquid is cooled, purified, and pH adjusted. The vanadium is extracted by transformation and neutralization, followed by reverse extraction. The reverse extraction liquid is precipitated to produce ammonium vanadate. The direct utilization efficiency of sulfuric acid is about 40%. Fig. 8.4 shows the typical process of extracting vanadium pentoxide from stone coal by direct acid leaching (oxidation, neutralization, and vanadate precipitation). Fig. 8.5 shows a typical process flowchart of extracting vanadium pentoxide from stone coal by direct acid leaching (oxidation, ion exchange, and vanadate precipitation).

8.2.3.2 Process principles

Analysis of the valence of vanadium in stone coal showed that generally only V^{3+} and V^{4+} are present in stone coal from different regions, and it is very rare to find V^{2+} and V^{5+}. Although V^{4+} can be present in higher concentrations than V^{3+} in individual local stone coals, the vanadium in stone coal from most areas mainly contains V^{3+}. The V^{3+} in stone coal is found within the sandwich layer of the clay dioctahedron, which replaces Al^{3+}. This type of silicate aluminate structure is relatively stable, where the V^{3+} in the stone coal is

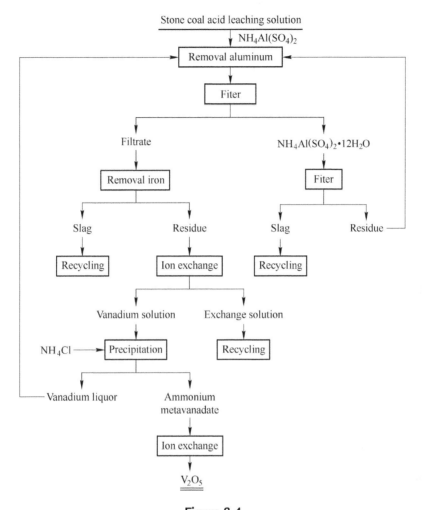

Figure 8.4
Typical process of extracting V_2O_5 from stone coal by direct acid leaching (oxidation, neutralization, and vanadate precipitation).

not easily dissolved by water, acid, or alkali media, although hydrofluoric acid (HF) can be used to break down the clay mineral crystal structure. Hence, it is generally thought that the V^{3+} cannot be leached, and V^{3+} needs to be oxidized to a higher valence before it can be leached. In stone coal, V^{4+} can occur in vanadium oxide (VO_2), vanadium ions (VO^{2+}), or as subvanadate. VO_2 can partially replace Al^{3+} in the illite clay mineral dioctahedral lattice. These V^{4+} ions cannot be leached with water, acid, or alkali solutions. The free VO^{2+} ions in the stone coal are insoluble in water, but soluble in acid, in which they produce vanadoxy salt VO^{2+}, which is stable and blue.

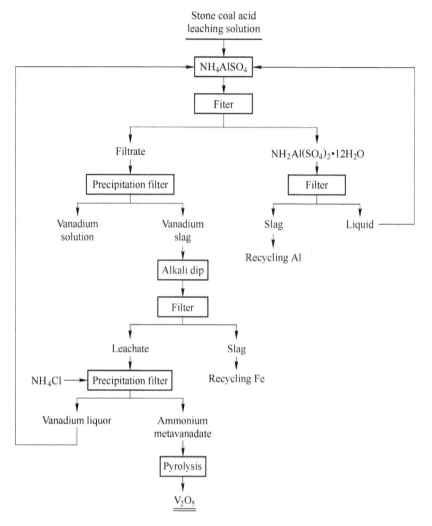

Figure 8.5
Typical process flowchart for extracting V_2O_5 from stone coal by direct acid leaching (oxidation, ion exchange, and vanadate precipitation).

The V^{5+} ionic radius is too small to exist in the clay dioctahedron. V^{5+} in stone coal mainly occurs as free V_2O_5 or vanadate crystals ($xM_2O \cdot yV_2O_5$), which are soluble in acid. The chemical reactions that occur during extraction of vanadium by direct acid leaching are as follows:

$$VO_2 + H_2SO_4 = VOSO_4 + H_2O \tag{8.27}$$

$$Al_2O_3 + 3H_2SO_4 = (Al_2SO_4)_3 + 3H_2O \tag{8.28}$$

$$FeO + H_2SO_4 = FeSO_4 + H_2O \quad (8.29)$$

$$Fe_2O_3 + 3H_2SO_4 = Fe_2(SO_4)_3 + 3H_2O \quad (8.30)$$

Fluosilicic acid in the solution is first decomposed to generate HF in the acid solution. The stable cations in the mineral crystal lattice are immediately acted upon by F^- ions from HF ionization under the acidic conditions; this plays an important role in accelerating dissolution of the cations in the mineral.

$$H_2SiF_6 + 2H_2O \rightarrow 6HF + SiO_2 \quad (8.31)$$

$$HF \rightarrow H^+ + F^- \quad (8.32)$$

Removal of iron by oxidation

$$Fe^{2+} \rightarrow Fe^{3+} \quad (8.33)$$

$$Fe^{3+} + 3OH^- = Fe(OH)_3 \downarrow \quad (8.34)$$

Ammonium sulfate ($3NH_4Al(SO_4)_2 \cdot 12H_2O$) is precipitated with addition of ammonium hydroxide.

Precipitation, dissolution, and enrichment of vanadium

$$VO^{2+} + H_2O = [VOOH]^+ + H^+ \quad (8.35)$$

$$[VOOH]^+ + H_2O = VO(OH)_2 + H^+ \quad (8.36)$$

$$VO(OH)_2 + OH^- = VO(OH)_3^- \quad (8.37)$$

The vanadate precipitation reaction is shown by (8.12), while the calcination reaction is shown by (8.13).

8.2.3.3 Typical process operation and technical indexes

Extracting V_2O_5 from stone coal by direct acid leaching involves acid leaching at atmospheric pressure, acid leaching at room temperature and atmospheric pressure, heap leaching, and pressure acid leaching. The typical composition of stone coal used for extracting vanadium pentoxide by direct acid leaching is shown in Table 8.4. The stone

Table 8.4: Typical composition of stone coal used for extracting vanadium pentoxide by direct acid leaching.

V_2O_5	SiO_2	Al_2O_3	Fe	ZnO	MgO	K_2O	TiO_2	C	Volatile
0.83	65.88	16.32	1.21	1.50	0.88	5.77	0.53	8.1	

coal is leached with 30% sulfuric acid at 100°C and atmospheric pressure, with a solid−liquid ratio of 1:2. The leaching efficiency of vanadium from the stone coal is 81%.

The best parameters for sulfuric acid leaching of stone coal from Guizhou, China are as follows: 30% dosage of sulfuric acid, 95°C leaching temperature, 24 h leaching time, and 1:1 solid-to-liquid ratio during leaching. The conditions give a V_2O_5 leaching efficiency of 68%. The best parameters for fluosilicic acid leaching are as follows: 20% dosage of fluosilicate, leaching temperature of 95°C, leaching time of 8 h, leaching liquid−solid ratio of 1:1, resulting in a V_2O_5 leaching efficiency up to 80%. The best parameters for the sulfuric acid leaching of stone coal from Gansu, China are as follows: 30% sulfuric acid, 95°C leaching temperature, 24 h leaching time, and 1:1 liquid−solid ratio, giving a leaching efficiency for vanadium from the stone coal of 40%. The best parameters for fluosilicic acid leaching are as follows: 30% dosage of fluorosilicate, 95°C leaching temperature, 12 h leaching time, liquid−solid ratio of 1:1, giving a V_2O_5 leaching efficiency of 60%. The best parameters for mixed-acid leaching of the stone coal from Guizhou, China are as follows: 15% sulfuric acid and 8% fluosilicic acid, leaching time of 12 h, leaching temperature of 95°C, and leaching liquid-to-solid ratio of 1:1, resulting in a leaching efficiency of V_2O_5 up to $\sim 80\%$.

The best parameters for mixed-acid leaching of stone coal from Gansu, China are as follows: 15% sulfuric acid and 15% fluorosilicic acid, leaching time of 16 h, leaching temperature of 95°C, leaching liquid-to-solid ratio of 1:1, giving a leaching efficiency of V_2O_5 up to $\sim 55\%$. The experimental results from the separation and enrichment phase of vanadium extraction showed that the optimized extraction parameters are 10% P204 + 5% TBP + 85% sulfonated kerosene, extraction liquid pH of 3.0, 5 min contact time, phase ratio O/A of 1:1, extraction temperature of 30°C. Single-stage extraction of the mixed-acid leaching liquid resulted in an extraction efficiency of over 60%. After seven stages of extraction the extraction efficiency was 99%. The best parameters for reverse extraction were as follows: 10% sulfuric acid in the antiextractant, 10 min contact time, and phase ratio O/A of 10:1, resulting in an antiextraction efficiency of V_2O_5 of over 70%. After five stages of antiextraction the antiextraction efficiency of V_2O_5 was above 99%.

In the stage of vanadate precipitation and preparation of vanadium products, $NaClO_3$ is used for oxidation, the pH is controlled to 2, the precipitation temperature is 95°C, and the precipitation time is 2 h, resulting in a vanadate precipitation efficiency of 95%. H_2O_2 can also be used for oxidation at a pH of 4, vanadate precipitation temperature of 95°C, and precipitation time of 5 h, resulting in a vanadate precipitation efficiency of 95%. The vanadium products are obtained after precipitation and roasting, which achieves the V_2O_5 that meet the standard of Chinas national vanadium products.

8.2.4 Salt-free roasting

8.2.4.1 Process

Salt-free roasting is used dependent on the specific metallurgical properties of the stone coal. Vanadium in stone coal is mainly found in roscoelite with trivalent vanadium. Without the addition of any raw chemical materials during salt-free roasting, thermal decomposition reactions occur in the stone coal to produce soluble vanadium components.

8.2.4.2 Roasting reactions

The chemical reactions are as follows:

$$KAl_2[VSi_3O_{10}OH]_2 \cdot CaCO_3 + O_2 \rightarrow Al_2O_3 + SiO_2 + CaO + CaO + KVO_3 + V_2O_5 \\ + Ca(VO_3)_2 + H_2O + CO_2 \tag{8.38}$$

Leaching reaction: KVO_3 in the roasted clinker is water soluble and V_2O_5 is soluble in dilute acid. The chemical reactions during KVO_3 leaching are as follows:

$$KVO_3 + H_2O \rightarrow K^+ + VO_3^- + H_2O \tag{8.39}$$

Dilute sulfuric acid leaching

$$V_2O_5 + H_2SO_4 \rightarrow (VO_2)_2SO_4 + H_2O \tag{8.40}$$

$$Ca(VO_3)_2 + 2H_2SO_4 \rightarrow (VO_2)_2SO_4 + CaSO_4 + 2H_2O \tag{8.41}$$

$$Ca_2V_2O_7 + 3H_2SO_4 \rightarrow (VO_2)_2SO_4 + 2CaSO_4 + 3H_2O \tag{8.42}$$

$$Ca_3(VO_4)_2 + 4H_2SO_4 \rightarrow (VO_2)_2SO_4 + 3CaSO_4 + 4H2O \tag{8.43}$$

Alkali leaching

$$Ca(VO_3)_2 + Na_2CO_3 + H_2O = 2NaVO_3 + CaCO_3 \downarrow \tag{8.44}$$

$$Ca_2V_2O_7 + Na_2CO_3 + H_2O = NaVO_3 + CaCO_3 \downarrow \tag{8.45}$$

$$Ca_3(VO_4)_2 + Na_2CO_3 + H_2O = NaVO_3 + CaCO_3 \downarrow \tag{8.46}$$

The vanadate precipitation reaction is shown by (8.12), while the calcination reaction is shown by (8.13).

8.2.4.3 Typical process operation and technical indexes

Control of the salt-free roasting process is very important. If the roasting temperature is too low, inadequate oxidation will directly affect the leaching efficiency of vanadium. If the temperature is too high, CaO, SiO_2, and $Na_2O(K_2O)$ in the stone coal easily produce a silicate melt, where low-valence vanadium is surrounded by the silicate, which is not conducive to the transformation of low-valence vanadium. The best roasting temperature is 800°C–850°C. If the roasting time is too short, oxidation is not sufficient, while if it is too long, secondary reactions involving the stone coal and silicon–oxygen enveloping phenomenon occur, which is not conducive to the conversion of low-valence vanadium; the best roasting time is a minimum of 3 h. The best parameters are as follows: roasting temperature of 850°C, roasting time of 3 h, particle size of 74 μm occupied (73%), and leaching liquid–solid ratio of 1:1.5. As the leaching temperature increases, the solubility of sodium vanadate increases, which is favorable for leaching; the best leaching temperature is above 90°C. Longer leaching times result in higher leaching efficiencies. For example, a vanadium leaching efficiency of above 70% is achieved for the optimized leaching time of >3 h. The typical chemical composition of stone coal used in salt-free roasting is shown in Table 8.5.

Table 8.5: Typical chemical composition of stone coal used in salt-free roasting.

V_2O_5	SiO_2	Al_2O_3	Fe	CaO	MgO	K_2O	Na_2O	C	Volatiles
1.13	59.52	5.62	2.09	0.44	0.5	0.91	0.16	10.8	4.59

After the leaching solution settles the pH is adjusted to 9–10 with sulfuric acid. After adding the silica-removal agent the concentration of SiO_2 is reduced to 2 g/L after purification by washing, resulting in a vanadium recovery of 99.5%. After silicon removal, sulfuric acid is added to adjust the pH to 7, and it is adsorbed with strongly alkaline 717 anion exchange resin. The tail liquid after adsorption contains no less than 0.01 g/L, where the adsorption efficiency of V_2O_5 is 99.5%. The vanadium is desorbed with 20 g/L sulfuric acid and 80 g/L HCl. The desorbed V-containing liquid has 45 g/L V_2O_5, corresponding to a desorption efficiency of V_2O_5 of 99.5%.

The roasted clinker from sodium-salt-free roasting is soaked with 3%–7% industrial alkali at 65°C–95°C for 3–5 h. Then, the leaching liquid is fed into a reaction kettle to recover white carbon black. Sulfuric acid is added to adjust the pH to 5–6, and the liquid is heated to 75°C–85°C and stirred for 0.5–1.5 h. It is then filtered, washed with hot water (65°C–75°C), and then conventional pulping of the filter cake is performed. Spray drying is used to produce white carbon black and the filtrate and water are recycled for extracting vanadium.

8.3 Comparison of vanadium extraction processes from stone coal

China has built several factories that extract vanadium from V-bearing stone coal, where each plant has developed a process of vanadium extraction with certain characteristics depending on the properties of the ore. Table 8.6 compares the vanadium extraction technology for stone coal, where the differences between these processes are significant.

Table 8.6: Comparison of vanadium extraction processes from stone coal.

Process of extracting V	Sodium salt roasting and water leaching	Sulfuric acid leaching with intermediate salt	Oxidizing–roasting, sulfuric acid leaching, and extraction	Oxidizing–roasting, acid leaching, and separation
Grade of stone coal V_2O_5 (%)	>1	1.73	0.88	1.16
Roasting conversion efficiency (%)	55			75.1
Leaching efficiency (%)	90	88.86	76.1	80.1
Extraction efficiency (%)		98.1	98.4	
Reverse extraction efficiency (%)		98.16	99.4	
Vanadate precipitating efficiency (%)	96	99	98.8	95.82
Thermal decomposition efficiency (%)	98	98	98	98
Product grade V_2O_5 (%)	>98	>98	99.24	99.5
Total recovery (%)	45	82.25	70.70	51.65

CHAPTER 9

Extracting vanadium pentoxide from minerals and composite materials containing vanadium

Chapter Outline
9.1 Extracting vanadium pentoxide from V-bearing titanoferrous magnetite 228
   9.1.1 Process details 228
   9.1.2 Extraction reactions 229
   9.1.3 Process operation, equipment, and typical process indicators 230
9.2 Extracting vanadium pentoxide from V-containing oil residue 232
   9.2.1 Extraction technology 232
   9.2.2 Extraction reactions 233
9.3 Extracting vanadium pentoxide from carnotite 233
9.4 Extracting vanadium pentoxide from alumina-rich slag 234
9.5 Extracting vanadium pentoxide from vanadinite 234
9.6 Extracting vanadium pentoxide from uranium molybdenum vanadium ore 235
9.7 Extracting vanadium pentoxide from kakoxene 235
9.8 Extracting vanadium pentoxide from V-bearing limonite 236
9.9 Extracting vanadium pentoxide from oil refining slag 236
9.10 Extracting vanadium pentoxide from bauxite 236
9.11 Vanadium extraction from spent catalysts 237
   9.11.1 Extraction process 237
   9.11.2 VS catalyst from petroleum cracking 238
   9.11.3 Catalyst from crude oil desulfurization 239
   9.11.4 Catalyst from sulfuric acid making 239
9.12 Vanadium extraction from fly ash 239

Depending on the vanadium content of mineral resources from different regions, different vanadium-extraction technologies are required to process the raw material, which can have multiple complex and variable compositions and be classified. Vanadium-extraction technology includes three approaches, where vanadium pentoxide is extracted from low-grade raw materials; high-vanadium raw materials; or a combination of vanadium-extraction and precious metal recycling, which results in balanced enrichment, transformation, and recycling. All vanadium-extraction technologies have characteristics of chemical metallurgy and metallurgical chemistry. Suitable parameters are selected to treat the V-containing raw material, which allow the V-containing compounds to be converted into water soluble salts, or substances soluble in alkali or acid. The V-containing raw material is roasted and oxidized to convert low-valence vanadium into high-valence vanadium and produce soluble sodium vanadate. High-temperature roasting results in structural transformation of V-containing raw materials, where soluble vanadate is produced as a stable intermediate compound. The vanadium and other mineral components in the V-containing raw material are separated, and soluble vanadium enters the liquid phase, while the insoluble compounds are retained in the slag. Identifying a suitable process for extracting vanadium is the main target for processing V-containing raw materials.

The vanadium content of minerals and composite materials is generally low. Hence, processing of V-containing raw materials and those rich in precious metals is performed to combine vanadium extraction with valuable metal recovery. In general, the composition of V-containing materials is more miscellaneous, containing both the primary mineral, and the secondary vanadium raw material. It is difficult to balance the different functions of recycling and extraction; however, acid leaching can provide a liquid phase to achieve both. According to the characteristics of the liquid phase, different metal salts are precipitated, where inorganic precipitation and organic extraction are combined. Vanadium products are produced, while valuable metallic elements are enriched and recycled. With the expansion of applications of vanadium, V-containing products have been accumulated by industries, which need to be recalled and updated. These secondary regenerated V-containing raw materials may be considered hazardous chemical waste and need to be treated in a timely and effective way following the current environmental protection protocols. There is a dual motivation for urgent vanadium recovery and waste disposal. Vanadium extraction via chemical metallurgy processes has the advantages of high recovery in a short time; however, it requires the processing and enrichment of raw materials containing vanadium. Before South Africa scaled up development of its vanadium industry in the 1960s, chemical metallurgy processes for extracting vanadium played an important role in the vanadium industry [8,16,29,31,38,43–45,53,54,72,77,78,86,87].

9.1 Extracting vanadium pentoxide from V-bearing titanoferrous magnetite

The extraction of vanadium pentoxide (V_2O_5) from V-containing iron concentrate is mainly based on the lack of vanadium mineral resources in the initial stage. A second factor is selecting processes that optimize the yields of both vanadium pentoxide and products for ironmaking.

9.1.1 Process details

V-bearing titanoferrous magnetite is ground and dressed by magnetic separation to obtain V-containing iron concentrate, which contains 0.50%−2.0% V_2O_5 and 50%−65% Fe. The concentrate is roasted with sodium salt and vanadium is extracted by water leaching. After extraction of V_2O_5 the V-containing iron concentrate is used as a raw material for ironmaking. If the iron concentrate contains a high level of vanadium, it can be extracted first as V_2O_5. Then the iron concentrate with high vanadium content is mixed with finely ground sodium salt, then pelletized or briquetted, followed by high-temperature roasting and water leaching. The vanadium leaching liquid is purified and processed, and the vanadate is precipitated to produce V_2O_5. Then the leaching residue is returned to the ironmaking process after treatment.

The V-containing iron concentrate is mixed with sodium sulfate, briquetted, roasted, and leached with water. Sodium vanadate enters the solution, and sulfuric acid is added to precipitate the vanadate, which is filtered and roasted to produce V_2O_5. The water-leached pellets are used as a raw material for ironmaking. South Africa's Highveld company is a production plant typical of Western countries that uses both processes. Rautaruukki is a Finnish state-owned steel company, and its subordinate companies Otanmaki and Mustavaara recover vanadium from V-containing iron concentrate. The total recovery from raw ore to industrial V_2O_5 is about 50%. Because magnetite and rutile are disseminated very finely in V-bearing titanoferrous magnetite, the latter cannot be separated by ore dressing methods. Mustavaara dressed ore contains 1.60% V_2O_5, and its processing method is similar to that used during Otanmaki ore processing. However, because of its high silicate content, the concentration of water-soluble silicate in the leaching liquid from the sodium-salt-roasted clinker is high. Therefore it must be treated by a desilication process before precipitation. Because of the high sodium content, not all of the material can be used for steel production.

9.1.2 Extraction reactions

The aim of the wet roasting and leaching process is to oxidize the low-valence vanadium and then convert it into water soluble vanadate. Various roasting equipment can be used to achieve sodium treatment and oxidation. Granules or pellets are prepared by finely grinding V-containing iron concentrate and the sodium additive (alkali, mirabilite, or Na_2SO_4). The pellets are then oxidized and roasted in a furnace, where low-valence vanadium in the V-containing iron concentrate is oxidized to V_2O_5:

$$V_2O_3 + O_2 \rightarrow V_2O_5 \quad (9.1)$$

$$Na_2SO_4 \rightarrow Na_2O + SO_2 + \frac{1}{2}O_2 \quad (9.2)$$

$$V_2O_5 + Na_2O \rightarrow 2NaVO_3 \quad (9.3)$$

In the process of water leaching, $NaVO_3$ enters the solution and is separated from most of the insoluble material. Then vanadate is precipitated from the purified V-containing solution. The chemical reaction during this process is the same as that occurring during extraction of V_2O_5 by sodium salt roasting.

9.1.3 Process operation, equipment, and typical process indicators

The process for extracting V_2O_5 from V-bearing titanoferrous magnetite is divided into four methods, depending on the main equipment used: sodium salt roasting is performed either in a shaft furnace, fluidized bed, rotary kiln, or chain rotary kiln.

9.1.3.1 Sodium salt roasting in a shaft furnace

Sodium salt roasting in a shaft furnace for extracting V_2O_5 is used in the Otanmaki and Mustavaara vanadium plants, which are owned by the Finnish steel company Rautaruukki. V-containing iron concentrate is mixed with sodium salt and then pelletized and roasted in the shaft furnace. The raw material composition (%) used by the Otamatti plant is TFe 68.4, TiO_2 3.2, V_2O_5 1.125, SiO_2 0.4, CaO 0.06, MgO 0.24, and Al_2O_3 0.5. The V-containing iron concentrate is ground to 85% of 0.038 mm, and then Glauber's salt (Na_2SO_4) or Na_2CO_3 is added at concentrations of 2.2%−2.3% or 1.6%−1.8%, respectively, of the V-containing iron concentrate. A mixing drum is used for pelletizing, with a diameter of 2.7 m, length of 9 m, and angle of 7 degrees. Pellets of 13−16 mm in diameter are obtained and roasted in a circular shaft furnace, which is 3−3.3 m in diameter and 15 m high. The roasted clinker is leached in 20 leaching tanks. The leaching tanks are made of welded steel plate with external heat insulation. They have a diameter of 2.5 m, height of 12.5 m, and volume of 60 m^3 and can be loaded with 80 t of roasted pellets. After leaching the pellets are processed and sent to a blast furnace for ironmaking.

The leaching liquid contains 20−25 g/L vanadium, where the vanadate is precipitated in six 10 m^3 precipitation tanks in the presence of sulfuric acid and ammonium sulfate at 363K. The tail liquid from vanadium precipitation contains 0.08 g/L vanadium, which is discharged after further treatment. The operating efficiency of the shaft furnace is 90%, and the heat consumption is 18−20 L heavy oil per ton pellets. The steam consumption is 600 L/t V_2O_5 and the power consumption is 3300 kWh/t V_2O_5. The purity of the produced V_2O_5 is 99.5%, and the vanadium yield is 78%.

The vanadium production process at the Mustavara is similar to that of the Otamatti. The composition (%) of the V-containing iron concentrate is TFe 63, V_2O_5 1.64, TiO_2 6.5, SiO_2 2.5, CaO 1.0, MgO 1.0, and Al_2O_3 1.1. The V-containing iron concentrate is ground to 90% of 0.038 mm. The production capacity of the shaft furnace is 40 t/h of pellets, and it has an operating efficiency of 85%. The leaching efficiency of vanadium is 97%, deposition efficiency of vanadium is 99.5%, and V_2O_5 purity is 99.8%.

9.1.3.2 Sodium salt roasting in a fluidized bed

In 1980 a vanadium plant was established by Agnew Clough Ltd., Australia to produce 1620 t/a of V_2O_5. The raw ore used by this factory contains 0.5%–1% V, where the grade of V_2O_5 is increased to 2% and SiO_2 is decreased to below 1% after the removal of silicon and aluminum by dressing. This V-containing iron concentrate is mixed with Na_2SO_4 and then pelletized into 3-mm granules, which are roasted in a fluidized bed furnace. The roasted clinker is leached with water, and the leaching solution contains 40–50 g/L V_2O_5, with a solution pH of 10. The leaching solution is neutralized to pH 7, and after the removal of Al and Si, ammonium sulfate is added to precipitate ammonium polyvanadate. Ammonium polyvanadate is calcined to produce V_2O_5, and ammonia is recovered from the furnace gas to make ammonium sulfate for circulatory system.

9.1.3.3 Sodium salt roasting in a rotary kiln

The rotary kilns used in industry for sodium salt roasting have a diameter of 2.3–2.5 m, length of ≥ 40 m, kiln body tilt of 2%–4%, and rotational speed of 0.4–1.08 rpm. Crude oil, natural gas, or gas is used as the heat source, with a residence time in the charge furnace of 2.5 h and material recovery of ~95%. The rotary kiln can be divided into the preheating, firing, and cooling zones. The preheating zone length has a length of about 10–15 m and a temperature of 627°C–927°C; the firing zone has a length of 15–20 m and temperature of 1073°C–1173°C, and the cooling zone is 5–8 m long, with a temperature of 523°C–773°C. The mixed furnace charge first enters the preheating zone of the kiln, where the water is removed, and the material is preheated. Part of the furnace charge is oxidized and decomposed. The preheated furnace charge moves forward along the rotary kiln into the firing zone, and depending on the process requirements, the furnace material is decomposed, oxidized, and nitrified to produce roasted clinker. The roasted furnace material enters the cooling zone after conversion is complete. After the discharge of the roasted clinker from the kiln the large furnace material is screened and separated from the roasted clinker and transferred through the inclined tube into a wet ball mill to produce the leaching paste.

The V-containing iron concentrate from Chengde ore is used as a raw material for extracting V_2O_5 in a second smelter in Shanghai, China. The raw material composition (%) is TFe 58, $V_2O_5 > 0.72$, $TiO_2 \leq 8$, and $SiO_2 \leq 2.5$. The V-containing iron concentrate is ground, where the -0.075 mm granules are more than 65% of the total. This material is mixed with sodium carbonate at a fraction of 6.5%–10% of the concentrate. Then the furnace material is roasted for 4–4.5 h in the rotary kiln at 1373K–1473K. The roasted clinker is leached with hot water at 363K. The leaching pulp is thickened and filtered. The residue contains 0.1%–0.2% V_2O_5, and the filtrate contains 7.14–10 g/L V_2O_5. The vanadate is precipitated with the addition of acid, resulting in a V_2O_5 recovery of 74.41%.

9.1.3.4 Sodium salt roasting in chain-and-grid conveyer rotary kilns

The German aluminum company (VAW; Vereinigte Aluminium Werke, AG) and South Africa's Transvaal Alloys Ltd. jointly built a vanadium plant in Middelburg. V-containing iron concentrate and sodium sulfate are mixed and pelletized, then heated and precured by kiln tail gas at 1173K. Then the material is placed into the rotary kiln and roasted at 1543K for 60–110 min, where the conversion efficiency of vanadium can reach 92%. Hot water is used in a large leaching tower with a special sealing device for countercurrent leaching. The leaching solution contains 35 g/L V_2O_5, 1 g/L SiO_2, and 3–7 g/L dust solids. The extracted slurry is filtered to remove Si, and ammonium sulfate is added to precipitate vanadate.

1. The composition of V-containing iron concentrate from V-bearing titanoferrous magnetite used in the Ventra vanadium plant of South Africa is TFe 50%–60%, V_2O_5 2.5%, TiO_2 8%–20%, Al_2O_3 1%–9%, and Cr_2O_3 1%. A rotary kiln is used in the roasting process for oxidation and transformation.
2. In the Former Soviet Union and Agnew Clough, Australia, sodium salt roasting in a fluidized bed is used to extract V_2O_5, where 97%–98% of the vanadium in the iron concentrate is converted to soluble vanadium.
3. In Otanmaki, Finland, the original mineral composition is Fe 40%, TiO_2 15.5%, V 0.26% (V_2O_5 0.71%), which is roasted in a shaft furnace with a conversion efficiency of 80%–90%.

9.2 Extracting vanadium pentoxide from V-containing oil residue

There is vanadium in crude oil and tar sand, although they are not listed in some countries as a vanadium resource. Crude oil is a potential vanadium resource, where the vanadium content varies globally. Venezuela, Mexico, Canada, and the United States have V-containing crude oil reserves with 220–400 ppm and are some of the countries with the world's highest vanadium content in oil.

9.2.1 Extraction technology

Countries such as the United States, Japan, Germany, Canada, and Russia extract V_2O_5 from oil residue and oil ash. The final vanadium product is usually V_2O_5, but it can also be directly refined to produce ferrovanadium. There are many methods of vanadium extraction, which mainly depend on the specific ingredients and properties of the raw materials. When extracting V_2O_5 from V-containing oil residue, the main processes include acid leaching, alkali treatment, and pyrogenic processes.

Venezuelan crude oil contains 0.06% V_2O_5 and Mexican crude oil contains 0.02%–0.044% V_2O_5. These V-containing crude oils are burned and the vanadium is concentrated in the ash. For example, Venezuelan crude oil boiler dust contains 35.28% V_2O_5.

In Canada, sulfuric acid leaching is used for recovering vanadium from the ashes of oil. Ash collected by an electrostatic precipitator is leached in a tank with a volume of 11,360 L with sulfuric acid. The leaching slurry is filtered, and then the filtrate is heated to 366K and sodium hypochlorite ($NaClO_4$) is added to oxidize the low-valence vanadium into pentavalent vanadium. When the filtrate changes from blue to yellow, the solution pH changes from 0.3 to 1.7, and the vanadate is precipitated by the ammonium salt. The sediment is calcined at 593K, and then flake V_2O_5 is produced in an oxidizing atmosphere at 1373K.

9.2.2 Extraction reactions

The main chemical reactions are as follows:

Acid leaching process:

$$V_2O_5 + 6HCl \rightarrow 2VOCl_2 + 3H_2O + Cl_2 \tag{9.4}$$

$$V_2O_5 + H_2SO_4 \rightarrow (VO_2)_2SO_4 + H_2O \tag{9.5}$$

Oxidation of $NaClO_4$:

$$VOCl_2 + NaClO_4 \rightarrow NaVO_3 + 2NaCl + Cl_2 \tag{9.6}$$

Precipitation and calcination:

$$NaVO_3 + NH_4Cl \rightarrow NH_4VO_3 + NaCl \tag{9.7}$$

$$2NH_4VO_3 \rightarrow V_2O_5 + 2NH_3 + H_2O \tag{9.8}$$

When extracting V_2O_5 using alkali treatment, the V-containing oil residue is mixed with Na_2CO_3 for pelletizing and roasting, followed by leaching with water. Alternatively, NaOH solution is used in direct pressurized leaching, where the vanadate is precipitated from the solution with an alkaline ammonium salt. The leach solution can be purified and enriched by extraction or ion exchange. During the pyrogenic process of vanadium extraction the V-containing oil residue is mixed with Na_2CO_3 or NaCl, which is smelted in a furnace in the presence of sulfide or sulfate to obtain vanadium slag and nickel matte. This is followed by a wet process for vanadium extraction.

9.3 Extracting vanadium pentoxide from carnotite

Fig. 9.1 shows a flowchart of vanadium extraction from carnotite. During this process the vanadium can be directly leached by sulfuric acid, or the ore can be roasted, and then leached with aqueous solutions of dilute hydrochloric acid or sulfuric acid. In this case, 80% of the vanadium and uranium in the ore are dissolved and then separated with tertiary amine, quaternary amine, or alkyl phosphate solvent.

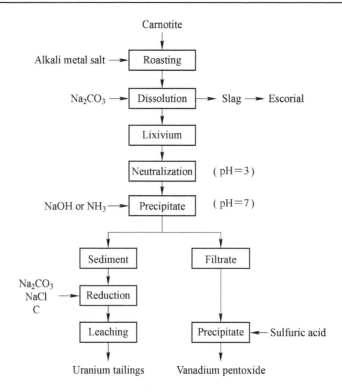

Figure 9.1
Flowchart of vanadium extraction from carnotite.

9.4 Extracting vanadium pentoxide from alumina-rich slag

High alumina slag refers to slag with a high alumina content, which is obtained by smelting of ferrovanadium, especially slag iron interface slag. Na_2CO_3 is used as a roasting additive and $MgSO_4$ as a conversion agent during roasting at high temperature. Then V_2O_5 is extracted by ammonium bicarbonate leaching. The vanadium recovery is as high as 65%–75% and the V_2O_5 purity can reach 98%. The leaching liquid is clarified and filtered easily and quickly. The technological process is simple, and treatment with sodium desnator is not required, in which the precipitating vanadium supernatant is treated and standardized discharge. No wastewater is produced in this process, and the sediment contains up to 80%–84% Al_2O_3, which can be used as a refractory material or raw material for the production of aluminate products.

9.5 Extracting vanadium pentoxide from vanadinite

Vanadium pentoxide is extracted from the concentrate of vanadinite or descloizite by sulfuric acid leaching. The Abinabithi ore in the Ottawa region of south-west Africa consists of vanadinite (3Pb$_3$(VO$_4$)$_2 \cdot$ PbCl$_2$), descloizite (Pb, Zn)$_3$(VO$_4$)$_2 \cdot$ (Pb, Zn)(OH)$_2$, cerussite (PbCO$_3$), and other minerals, along with 1.39% V$_2$O$_5$ and 7.17% Pb. The recovery of vanadium is 89.4%. The concentrate is obtained by flotation, which contains 10.75% V$_2$O$_5$ and 49.98% Pd. The recovery of vanadium in the dressing is 89.4%. V$_2$O$_5$ is extracted from the concentrate of vanadinite by sulfuric acid leaching.

The descloizite deposit in broken hills, Zambia is a red soil weathered ore, which is beneficiated by gravity. The concentrate contains 16.57% V$_2$O$_5$, which is commercialized and used for production of vanadium slag. The middlings containing 8.5% V$_2$O$_5$ and slime is ground to less than 0.07 mm and leached with sulfuric acid via agitation in a leaching tank to obtain a solution containing 30–40 g/L V$_2$O$_5$. After adding MnO$_2$ to the solution, all vanadium is oxidized from V^{4+} to V^{5+}. The solution is transferred to a precipitation tank and heated to 363K. Then the pH is adjusted to 1.5–3 to precipitate vanadium pentahydrate. The resulting V$_2$O$_5$ purity is more than 90%.

9.6 Extracting vanadium pentoxide from uranium molybdenum vanadium ore

The raw materials processed by the manufacturers of vanadium in the United States are mainly vanadite (K$_2$(VO$_2$)$_2$(V$_2$O$_8$) $\cdot$ 3H$_2$O or K$_2$O $\cdot$ 2UO$_2 \cdot$ V$_2$O$_5 \cdot$ 3H$_2$O), uranium molybdenum vanadium ore, and kakoxene. Recently, large vanadite deposits were found in calcareous rock in Yeelirrie, Western Australia, while vanadite deposits were also found in Shaanxi and Hunan provinces of China. The process of producing vanadium from vanadite used by Union Carbide in the United States (the world's largest mining company) involves calcining, leaching, precipitation, reduction, and releaching. The vanadium and uranium leaching efficiencies are 70%–80% and 90%–95%, respectively. The V-containing solution produced by acid and alkaline leaching can be separated via ion exchange, solvent extraction, or selective precipitation methods.

9.7 Extracting vanadium pentoxide from kakoxene

Kakoxene is smelted in an ore-smelting furnace to produce phosphorus and phosphorus fertilizer, in which the byproduct is V-containing ferrophosphorus. The kakoxene used in the Kerr–McGee chemical company contains 3.26%–5.2% V, 24.7%–26.6% P, 59.9%–68.5% Fe, 3.4%–5.7% Cr, and 0.84%–1.0% Ni. The kakoxene is ground to less than 0.42 mm, mixed with soda ash of 1.4 time demand in extracting vanadium and salt of 0.1 times demand in extracting vanadium, and roasted in a rotary kiln at 770°C–800°C. The vanadium is converted into water-soluble sodium vanadate and the calcined clinker is leached in boiling water. Vanadium, chromium, and phosphorus enter the leaching solution. Sodium phosphate is extracted from the filtrate after filtration, where the coarse material can be purified until it reaches the product requirements. The crystallized mother liquor of sodium phosphate contains >0.98 g/L phosphorus, which can be mixed with $CaCl_2$ to precipitate calcium phosphate ($CaPO_4$) and then hydrolyzed to recover vanadium. Then the mother solution from precipitating vanadium is mixed with lead nitrate to precipitate lead chromate. The recovery of vanadium, chromium, and phosphorus using the process can reach 85%, 65%, and 94%, respectively.

9.8 Extracting vanadium pentoxide from V-bearing limonite

The composition of V-bearing monolite is 0.5%–2.5% V_2O_5, 20%–40% Fe, and 3%–65% SiO_2, where the ore is mainly composed of pyrite, hematite, and gangue. The gangue is mainly quartz, followed by mud and a small amount of sericite. Vanadium does not occur as an independent mineral in limonite but appears with the iron and mud because of ion adsorption. The main processing steps include crushing, ball milling, roasting, leaching, and precipitation of Na_4VO_3 or V_2O_5. Studies have shown that the V_2O_5 content in limonite varies, where the conversion efficiency is influenced by the mineral components, particularly the CaO content; increasing the CaO content influences the conversion of vanadium. Increasing the roasting temperature can improve the conversion efficiency of vanadium. The temperature at which the highest vanadium conversion efficiency occurs depends on the vanadium content of the ore.

9.9 Extracting vanadium pentoxide from oil refining slag

The United States is the largest producer of vanadium and has been producing vanadium from oil ash since the late 1980s. Companies such as Amax and CRI Ventures in the United States process oil refining slag to recover vanadium, molybdenum, cobalt, nickel, and aluminum. During processing the oil refining slag is mixed with caustic soda and ground, followed by high-pressure leaching. Components of the oil refining slag are oxidized at high temperature and pressure, where sulfur is converted into sulfide, most hydrocarbons are decomposed, while vanadium and molybdenum dissolve in solution. After filtration and separation, vanadium and molybdenum are recovered from the solution. Then Na_2CO_3 or NaCl is added and the oil refining slag is smelted in the furnace in the presence of sulfides and sulfates to produce vanadium slag and nickel mattes.

9.10 Extracting vanadium pentoxide from bauxite

The term "bauxite" mainly refers to bauxite with high vanadium content, although various vanadium grades occur. In the Bayer aluminum extraction process, some V, Fe, As, and P impurities exist as their sodium salts, which enter the aluminum solution. The vanadium is enriched in the red mud after primary filtration, up to a concentration of 0.3% V_2O_5. Na_2CO_3 is used as a roasting additive, and V_2O_5 is extracted by oxidizing and roasting the red mud, or via alkali leaching. During aluminum precipitation, V_2O_5 in the aluminum leaching solution is enriched in the liquid phase, which is then precipitated to obtain a Bayer salt sludge containing 6%–20% V_2O_5. After leaching with water a $NaVO_3$ solution is obtained, which is used in the common vanadium-extraction process.

9.11 Vanadium extraction from spent catalysts

Catalysts prepared using vanadium and a carrier material can change the rate of certain chemical reactions but do not participate in the reaction itself. Such a vanadium catalyst, V_2O_5. NH_4VO_3, is used to replace platinum for sulfuric acid production, where SO_2 is converted to SO_3. In the petroleum industry, vanadium is mainly used in the form of versus (VS) as a pyrolysis catalyst and desulfurizer. Vanadium is used in the rubber industry as a catalyst (VCl_4) for ethylene and propylene.

9.11.1 Extraction process

During recycling of waste vanadium sulfide catalyst the material is roasted to obtain a soluble material, which is leached at high temperature with ammonia. The V-containing waste catalyst is added to a pressure boiler and heated for 4 h with 1–14 mol/L ammonium hydroxide at 473K to dissolve ammonium vanadate in the aqua ammonia. The temperature of the ammonium vanadate filtrate drops to 323K, where the ammonium vanadate crystallizes and precipitates. The crystal slurry is then filtered, washed, and dried. The ammonium vanadate is roasted at 473K–873K to obtain V_2O_5. The mother liquor is returned to the leaching cycle.

V_2O_5 is extracted from waste V-containing catalyst by alkali leaching. The waste catalyst is leached for 1–6 h with NaOH or Na_2CO_3 solution at 363K–378K and then filtered and separated. The filtrate is injected with ammonia and carbon dioxide, and the temperature is maintained at 298K–308K. Depending on the dosage, 1.5–5 mol ammonia is added to 1 mol of vanadium, and the solution is adjusted to a pH of 6–9. After treatment with ammonia a temperature of 308K is maintained to precipitate ammonium vanadate.

In general, spent V-containing catalyst is roasted and oxidized at 1073K, where the roasted clinker contains 10.88% V, 5.49% Mo, 2.03% Co, 1.94% Ni, and 35.48% Al. Depending on the ingredients, 150 g of roasted clinker is mixed with 300 mL of 15% NaOH solution, which is then leached and agitated for 3 h at 333K. The leaching liquor is then filtered at 323K and cooled from 323K to 278K to precipitate vanadate crystals. The mother solution from precipitating vanadate is returned to the leaching process. The crystals are washed and dried, and V_2O_5 is produced after calcination. The roasted clinker can also be used in the acid leaching process. In addition to vanadium in the catalyst, Mo, Ni, Co, and other valuable elements enter the solution, and the vanadium is recovered by extraction−separation methods.

Direct acid leaching solutions contain a large number of Fe ions in addition to vanadium, which can complicate purification of the solution. Vanadium is oxidized to high-valence vanadium by precalcining, and the obstacles was reduced and eliminated by conversion of vanadium. Because the waste catalyst itself contains 10% potassium sulfate, the process of oxidation, roasting, and water leaching can be classified depending on the use of Na-free salt or sodium salt. The roasting temperature of 900°C in Na-free salt roasting achieves the highest conversion efficiency ($\sim$80%). The conversion efficiency of vanadium is not high, regardless of the roasting temperature. During sodium salt roasting the catalyst waste containing vanadium is mixed with 5% Na_2CO_3 and roasted for 2 h at 800°C. The conversion efficiency of vanadium can reach 92%. The calcined clinker is leached in two stages: first, water leaching, followed by acid leaching or alkali leaching. During this process, potassium salt, sodium salt, and nearly 80% of the vanadate enters the mildly acidic solution during water leaching. This solution has fewer impurities, is easy to handle, and can be recycled with the potassium salt. The purpose of acid or alkali leaching is to dissolve the insoluble vanadium salt in water as much as possible to increase vanadium recovery. The vanadium in the solution is separated by N235 extraction, where an alkali is used for reverse extraction. Ammonium vanadate is precipitated with the addition of NH_4Cl and then calcined to produce V_2O_5.

9.11.2 VS catalyst from petroleum cracking

Waste vulcanized V-containing catalyst is roasted and then treated using high-temperature ammonia leaching. The roasted clinker is added to a pressure boiler, heated for 4 h with 1−14 mol/L ammonium hydroxide at 473K to dissolve the ammonium vanadate in the aqua ammonia. After separation the temperature of the ammonium vanadate filtrate is decreased to 323K to crystalize the ammonium vanadate. The crystal slurry is filtered, washed, and dried. The ammonium vanadate is calcined at 473K−873K to obtain V_2O_5. The crystallization mother liquor is returned to the leaching cycle. In addition to the above methods, V_2O_5 is extracted from waste catalyst containing vanadium by alkali leaching. The waste catalyst is leached for 1−6 h with NaOH or Na_2CO_3 solutions at 363K−378K, and then filtered and separated. The filtrate is injected with ammonia and carbon dioxide and kept at the temperature of 298K−308K. Depending on the dosage, 1.5−5 mol ammonia is added to 1 mol of vanadium, and the solution is adjusted to a pH of 6−9. After treatment with ammonia a temperature of 308K is maintained to precipitate ammonium vanadate. Then the filtrate is returned to the desorber. Steam is used to drive NH_3 and CO_2 from the liquid, which is then returned to the leaching process. The recovery of vanadium and ammonium sulfate is treated similarly.

9.11.3 Catalyst from crude oil desulfurization

The AMR is a big producer of vanadium from waste catalyst from oil companies, where the quantity obtained from the treatment of spent catalyst accounts for 50% of the national average. They annually process 16,000 t of spent catalyst and recover 1500 t of V_2O_5, more than 1000 t of Mo, 400–600 t of Ni, 110–180 t of Co, and some Al_2O_3.

9.11.4 Catalyst from sulfuric acid making

In the catalytic process used by the sulfuric acid industry a complex is created between the As_2O_5 in the SO_2 gas and the V_2O_5 in the catalyst, and the normal operation temperature of the catalyst is 480°C. The complex is volatilized with the gas, where the volatiles account for 40%–50% of the total V_2O_5. In addition, V_2O_5, K_2SO_4, and SiO_2 are present. The compositions of fresh and spent catalysts are compared in Table 9.1.

The treatment of waste catalyst can be used in the industrial application of direct acid leaching and sodium salt roasting processes. Direct acid leaching reduces the consumption of acid and alkali materials, which reduces solution impurities and free acid. Two stages of leaching are applied: the first is weak acid leaching, followed by strong acid leaching. The strong acid leaching solution is added to the used catalyst after weak acid leaching. The leaching efficiency of vanadium after the two stages of leaching can reach 88.5%–91.1%. The V_2O_5 content of the leaching residue can be reduced to 0.59%. When the acid concentration is increased to 80–100 g/t, the V_2O_5 content of the slag can be reduced to 0.3% V_2O_5. The solution is purified by N235 or P204 extraction, where alkali is used for reverse extraction. Ammonium vanadate is precipitated by the addition of NH_4Cl and calcined to obtain V_2O_5.

9.12 Vanadium extraction from fly ash

Fly ash specifically refers to the soot collected after burning fuel and coal, which is enriched with vanadium and can be used as a source of vanadium. Fly ash is generally treated by leaching with water, alkali, or acid solutions, where vanadium is separated from other impurities. The purification process is chosen according to the composition of the impurities. After purification, vanadate is precipitated from the vanadium solution.

Table 9.1: New catalyst components.

Component	V_2O_5	K_2SO_4	SiO_2
New catalyst	9–10	20–22	20
Spent catalyst	5–6	10–12	80

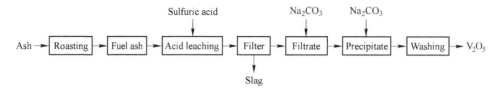

Figure 9.2
Typical process of extracting vanadium from fly ash.

The V-containing fly ash must be dissolved and transformed, where acid leaching, alkaline leaching, decarbonization-roasting water leaching, and high-pressure acid leaching are used to create a homogeneous V-containing system. Vanadium extraction can be combined with metal recovery. The typical process of extracting vanadium from fly ash is shown in Fig. 9.2.

There is trace vanadium in most global oil supplies, where vanadium is enriched during the oil refining process and concentrated in heavy oil. The heavy oil is mainly used for power generation and industrial heating, where the fly ash is collected during combustion processes with a boiler and furnace. The vanadium is enriched further by the high-temperature combustion process and concentrated in the fly ash, which can be used as a raw material for extracting vanadium. Typical fly ash components of heavy oil are shown in Table 9.2.

In power-generation processes, heavy oil is used as fuel, where ~30% of the fly ash is produced from the electric dust collector. During the electric dust collection process the acidic flue gas is neutralized using ammonia; hence, the fly ash from the electric dust collector contains a considerable amount of ammonium sulfate. The main components of fly ash are carbon particles, which also contain Fe, V, and Ni. The fly ash containing nickel usually has proportional V and Ni contents.

Boiler ash contains 4.4%–19.2% V and 0.2%–0.5% Ni. Boiler ash is ground to 0.15 mm and leached with 8 mol/L NaOH solution at 112°C for 4 h. In the three stages of countercurrent leaching the vanadium leaching efficiency is 43%, 16%, and 43%, respectively, where the leaching liquid is used directly for vanadate precipitation. The leaching residue is leached with 88 mol/L HCl solution, where Ni, Fe, and Mg enter the leaching liquid. Iron is extracted with 25% TBP kerosene, the raffinate is adjusted to pH 6, and V and Ni are extracted with

Table 9.2: Typical fly ash components of heavy oil.

Water	C	V	Fe	Na	Si	Ni	Al	S	P
1.9	33.1	2.6	3.5	2.3	2.5	1.3	2.1	12.6	0.1
0.9	39.6	3.3	3.9	2.4	0.7	1.2	0.4	11.3	0.1
1.5	79.1	1.3	4.0	0.2	0.8	0.7	0.6	1.6	0.1

25% LIX64N kerosene. Ni is reverse extracted with 0.3 mol/L HCl and vanadium is reverse extracted with 6 mol/L HCl, giving a vanadium recovery of 80%.

Boiler ash also can be treated by acid leaching, that is, boiling in 1 mol/L H_2SO_4, where the solid–liquid ratio is controlled to 2 mL/g. The leaching residue is washed with water three times, where the washing solid–liquid ratio is 3 mL/g. The washing water is combined with the filtrate and an excess 25% of oxidant is added, then the pH is adjusted to 2.3 with sodium carbonate and the solution is precipitated for 1 h. As there is high sodium content in the sediment, it is washed with a H_2SO_4 solution (around pH 2) and the solid–liquid ratio is controlled to 10 mL/g. After drying, it can be used as a raw material for ferrovanadium production.

Boiler ash contains a high carbon content and is treated with a process of carbon oxidation and sodium salt roasting to extract vanadium pentoxide, which is oxidized by roasting for decarburization, and then the vanadium oxide is combined with sodium salt to produce sodium vanadate, which is leached with water or alkali solution. During the roasting process a temperature above 960°C is maintained. This results in a high degree of vanadium and nickel volatilization and vanadium bronze production. The deformation of heavy oil fly ash begins at 1150°C, softening occurs at 1190°C, while melting occurs at 1260°C. Hence, a roasting temperature of 850°C is considered appropriate. The leaching efficiency of vanadium can reach 99%, where the vanadium precipitation efficiency is 89%, and the total yield of vanadium is 83%. Because of the excessive roasting during the process of fly ash natrification and decarbonization, the fly ash itself has temperature variability, and the fine particles can easily be agglomerated, which complicates the roasting process. The fly ash is acid leached under pressure at 200°C with an oxygen partial pressure of 1.5 MPa, acid concentration of 60 g/L, liquid–solid ratio of 1:1, and leaching time of 15 min. The leaching efficiency of Ni and V is >95%. Fe is removed from the leaching solution at 200°C by hydrolysis precipitation, while nickel is separated by electrolysis. The solution is neutralized and ammonium vanadate is precipitated with ammonium salt, where ammonium vanadate is calcined to produce V_2O_5. Table 9.3 shows the typical chemical composition of fly ash from heavy oil boiler electric dust collection and cyclone dust removal.

For the treatment of fly ash containing 30%–40% ammonium sulfate, 0.41% V, and 1% Ni, ammonia leaching process is used. A solution of 0.25 mol/L NH_3 + 1 mol/L $(NH_4)_2SO_4$ is used to leach the fly ash. First, nickel is leached to produce Ni-ammonia ions, where the leaching efficiency is 60%. Vanadium is leached with NaOH, where the leaching efficiency

Table 9.3: Typical fly ash composition from heavy oil boiler dust collection and cyclone dust removal.

Fly ash	C	V	Fe	Na	Mg	Ni	NH_4	SO_4^{2-}
Electric dust collection	56.7	0.41	0.55	0.41	2.55	1.02	7.72	29.1
Cyclone dust removal	63.2	1.91	1.96	1.50	0.07	0.80		24.8

Table 9.4: Typical components of fly ash from coal–water slurry combustion (%).

Water	C	V	Fe	Na	Ni	Al	S
5.2	7.1	11.7	0.6	1.1	2.5	3.2	13.2

is 80%. After solid–liquid separation, vanadate is precipitated from the leaching liquid by the addition of ammonium salt.

Bituminous rocks also contain trace amounts of vanadium, which is combined with water and coal to produce a coal–water slurry (CWS). Fly ash is produced during burning of coal. The typical components of fly ash from CWS combustion are shown in Table 9.4.

For the treatment of fly ash from CWS, it is leached in boiling 2 mol/L H_2SO_4, where the solid–liquid ratio is controlled to 2 mL/g. The leaching residue is washed three times with water, where the washing solid–liquid ratio is 3 mL/g. The washing water is mixed with the filtrate and an excess 25% of oxidant is added, and then the pH is adjusted to 2.3 with sodium carbonate, and precipitation is performed for 1 h. As there is high sodium content in the sediment, it is washed with a H_2SO_4 solution ($\sim$pH 2) and the solid–liquid ratio is controlled to 10 mL/g. After drying, it can be used as a raw material for ferrovanadium production.

CHAPTER 10

Ferrovanadium

Chapter Outline

10.1 Electro-silicon thermal process 244
- 10.1.1 Process details 244
- 10.1.2 Process reactions 244
- 10.1.3 Raw materials and ingredient requirements 247
- 10.1.4 Typical process operation and technical indexes 250
- 10.1.5 Main process steps 251
- 10.1.6 Factors influencing the production of ferrovanadium 253
- 10.1.7 Common problems and their solutions 254
- 10.1.8 Electric furnace requirements 255

10.2 Aluminum thermal method 255
- 10.2.1 Reduction process 255
- 10.2.2 Process details 256
- 10.2.3 Reduction reactions 257
- 10.2.4 Raw materials and ingredient requirements 257
- 10.2.5 Furnace characteristics 259
- 10.2.6 Typical process operation and technical indexes 260
- 10.2.7 Factors affecting the process 261

10.3 Direct alloying of vanadium slag to produce ferrovanadium 262
- 10.3.1 Direct alloying technology 262
- 10.3.2 Direct alloying reactions 262
- 10.3.3 Raw material and ingredient requirements 263
- 10.3.4 Typical process operation and technical indexes 263
- 10.3.5 Factors affecting the process 265

10.4 Ferrosilicon–vanadium alloy 266

Ferrovanadium is an important alloy product containing vanadium, which is typically used as a furnace ingredient in steel metallurgy processes. It is common to use ferrovanadium with either 40%, 60%, or 80% V, where the uncertainty in the vanadium content needs to be small, and the C, S, and P contents are strictly limited. The producers can select the other ingredients depending on the requirements of the user. Ferrovanadium production methods mainly include the electro-silicon thermal process, aluminum thermal process, and direct alloying of vanadium slag. The silicon thermal method can produce medium-grade ferrovanadium and ferrosilicon–vanadium alloys, while the aluminum thermal process can produce high-quality ferrovanadium alloy. The aluminum thermal process and electric aluminum thermal process were developed to address the needs of safe operation. Direct alloying of vanadium slag can omit the step of extracting vanadium from vanadium slag, which can result in cost savings, although the product quality is low as C, Si, S, P, and Cr impurities remain. In general, the vanadium content in ferrovanadium is above 40%, and the material has a melting point of 1480°C, solid density of 7.0 t/m³, bulk density of 3.3–3.9 t/m³, and used blocksize of <200 [9,13,37,39,52,55,56,68,80,86].

10.1 Electro-silicon thermal process

In 1894 Mwain invented the technology for reducing vanadium oxide. Fig. 10.1 shows the process flowchart of ferrovanadium production by the electro-silicon thermal method.

10.1.1 Process details

In the ferrovanadium production process using the electro-silicon thermal method, flake vanadium pentoxide (V_2O_5) is used as the raw material, while ferrosilicon containing 75% Si and small amounts of Al are used as reducing agents. The ingredients are smelted in a basic electric arc furnace. After two-phase reduction and refining, ferrovanadium is produced. In the reduction phase, all of the reducing agent and 60%–70% of the flake V_2O_5 are fed into the electric furnace, where silicon thermal reduction is carried out in the presence of high-calcium oxide slag. When the V_2O_5 content of the slag is <0.35%, the slag is tapped and enters the refining process. The flake V_2O_5 and lime are fed into the process, where excess Si and Al in the alloy are removed to meet the compositional requirements of the ferroalloy. The slag tapped in the later stage of refining is rich in V_2O_5 (up to 8%–12%). When the next furnace starts to feed, it will be used again. The alloy is generally cast as a cylindrical ingot, which is finished after cooling, molding, crushing, and cleaning.

The oxygen potential diagram of vanadium oxide is given in Fig. 10.2. The $\Delta G^\theta - T$ plot for the reduction reaction of vanadium oxide (V_2O_5, V_2O_3) by Al, Si, and C is shown in Fig. 10.3.

10.1.2 Process reactions

In the electro-silicon thermal process, the main reductants include Si, Al, and the C electrodes. The Fe in ferrosilicon remains as the dissolution medium of vanadium and iron element of vanadium–iron–silicide in the alloy. The electric heating and exothermic reaction are the sources of heat. The chemical reaction of carbon reduction is as follows:

$$1/5 V_2O_5 + C = 2/5 V + CO \tag{10.1}$$

$$\Delta G^\theta = 208,700 - 171.95T$$

$$2/5 V_2O_5 + 9/5 C = 4/5 VC + CO_2 \tag{10.2}$$

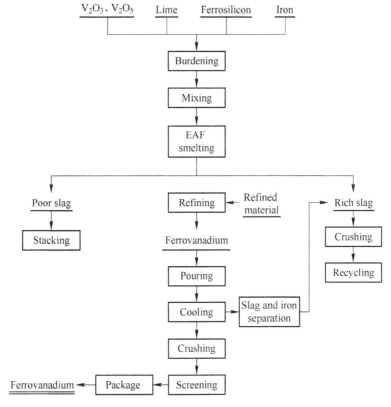

Figure 10.1

Process flowchart of ferrovanadium production by the electro-silicon thermal method.

$$\Delta G^\theta = 150250 - 160.6T$$

Carbon is generally a weak reducing agent, where reaction (10.1) is more difficult than that shown in (10.2). Carbon reduction of V_2O_5 during the production of ferrovanadium can only produce VC, resulting in high-carbon ferrovanadium. When sufficient carbon participates during the production of ferrovanadium, the carbon content in the ferrovanadium can reach 5%, or as high as 7%–8%, which affects the quality of the ferrovanadium and its application. The chemical reaction of silicon reduction is as follows:

$$2/5 V_2O_5 + Si = 4/5 V + SiO_2 \tag{10.3}$$

$$\Delta G^\theta = -326,026 + 75.2T$$

$$2/3 V_2O_3 + Si = 4/3 V + SiO_2 \tag{10.4}$$

$$\Delta G^\theta = -105,038 + 54.8T$$

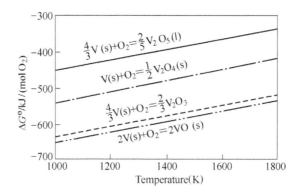

Figure 10.2
Oxygen potential diagram of vanadium oxide.

$$2VO + Si = 2V + SiO_2 \tag{10.5}$$

$$\Delta G^\theta = -25,414 + 50.5T$$

$$V_2O_5 + Si = V_2O_3 + SiO_2 \tag{10.6}$$

$$\Delta G^\theta = -646,023 + 101.53T$$

Low-valence vanadium oxide (VO and V_2O_3) and silicon oxide can easily form silicates and delay the reduction of vanadium oxide. In general, CaO is added to take part in the reaction and produce calcium silicate, allowing the low-valence vanadium to participate in the reaction to a greater extent.

In the presence of calcium oxide, the silicon reduction reaction is as follows:

$$2/5 V_2O_5 + 2CaO + Si = 4/5 V + 2CaO \cdot SiO_2 \tag{10.7}$$

$$\Delta G^\theta = -470,000 + 75.0T$$

$$2/3 V_2O_3 + Si + 2CaO = 4/3 V + 2CaO \cdot SiO_2 \tag{10.8}$$

$$\Delta G^\theta = -250,000 + 54.0T$$

$$2VO + Si + 2CaO = 2V + 2CaO \cdot SiO_2 \tag{10.9}$$

$$\Delta G^\theta = -41,070 + 12.03T$$

The aluminum thermal reduction reaction is as follows:

$$2/5 V_2O_5 + 4/3 Al = 4/5 V + 2/3 Al_2O_3 \tag{10.10}$$

$$\Delta G^\theta = -540,097 + 24.8T$$

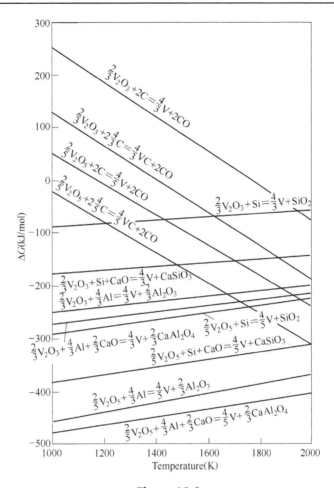

Figure 10.3
$\Delta G^0 - T$ plot of the reduction of vanadium oxide (V_2O_5, V_2O_3) by aluminum, silicon, and carbon.

$$2/3V_2O_3 + 4/3Al = 4/3V + 2/3Al_2O_3 \qquad (10.11)$$

$$\Delta G^\theta = -319,453 + 63.96T$$

$$2VO + 4/3Al = 2V + 2/3Al_2O_3 \qquad (10.12)$$

$$\Delta G^\theta = -316,760 + 65.8T$$

10.1.3 Raw materials and ingredient requirements

10.1.3.1 Raw material requirements

Flake V_2O_5 must meet the national metallurgical standard, where the particle size cannot exceed 200 mm, with a flake thickness <5 mm. The ferrosilicon must meet the national standard of 75SiFe, with a particle size of 20–30 mm. The aluminum particles have a size of 30–40 mm. The steel or carbon scrap steel and iron from magnetic separation of vanadium slag have $\omega(C) < 0.5\%$ and $\omega(P) < 0.35\%$. The lime has an effective $\omega(CaO) > 85\%$, effective $\omega(P) < 0.5\%$, and particle size of 30–50 mm.

10.1.3.2 Ingredients

Flakes of V_2O_5 are used as the V-containing material, which are designed for the production of 1 t of ferrovanadium. V_2O_5 has a molecular weight of 182, while V has a molecular weight of 102. The theoretical amount of $V_2O_5 = 1 \times$ (vanadium content of ferrovanadium)/(102/182). Considering factors such as process losses and a typical recovery of 93%–95%, the actual amount of V_2O_5 used during ferrovanadium smelting is equal to $1 \times$ (vanadium content of ferrovanadium)/(102/182) $\times$ 1.07.

$$W_{V_2O_5} = 1 \times \frac{W_V \times 182}{102}$$

Here W_{V2O5} refers to the amount of vanadium pentoxide (the theoretical value) of ferrovanadium in smelting (the theoretical value) and W_v refers to vanadium content in ferrovanadium.

$$W_{V_2O_5} = 1 \times \frac{W_V \times 182 \times 1.07}{102}$$

Here W_{V2O5} refers to the amount of vanadium pentoxide (the actual value) of ferrovanadium in smelting (the actual value) and W_v refers to vanadium content in ferrovanadium.

Both siliceous and aluminum reductants are commonly used. Silicon accounts for 80% of the reductant reaction in the production of ferrovanadium, where the silicon coefficient of the ingredients is generally 1.10, while aluminum accounts for 20%, and the aluminum coefficient of the ingredients is 1.30.

The chemical reaction shown in (10.3) gives a theoretical demand for silicon of 0.385 t for reduction of 1 t V_2O_5. The theoretical requirement for ferrosilicon is equal to $1 \times$ (vanadium content of ferrovanadium)/(102/182) $\times$ 1.07 $\times$ 0.385/0.75, while the actual demand for ferrosilicon = $1 \times$ (vanadium content in ferrovanadium)/(102/182) $\times$ 1.07 $\times$ 0.385/0.75 $\times$ 1.10.

$$W_{Fe-Si} = 1 \times \frac{W_V \times 182 \times 1.07 \times 0.385}{102 \times 0.75}$$

Here W_{Fe-Si} refers to the theoretical requirement of the ferrosilicon, W_v refers to vanadium content in ferrovanadium.

$$W_{Fe-Si} = 1 \times \frac{W_V \times 182 \times 1.07 \times 0.385 \times 1.10}{102 \times 0.75}$$

Here W_{Fe-Si} refers to the actual demand for ferrosilicon and W_v refers to vanadium content in ferrovanadium.

The chemical reaction shown in (10.8) is used to calculate the theoretical demand for aluminum, which is 0.5 t per 1 t V_2O_5. The theoretical requirement for aluminum is equal to $1 \times$ (vanadium content in ferrovanadium)/(102/182) $\times 1.07 \times 0.5$/(purity of fine aluminum). The actual requirements for fine aluminum $= 1 \times$ (vanadium content in ferrovanadium)/(102/182) $\times 1.07 \times 0.5$/(purity of fine aluminum) $\times 1.3$.

$$W_{Al} = 1 \times \frac{W_V \times 182 \times 1.07 \times 0.5}{102 \times W_{pureAl}}$$

Here W_{Al} refers to the theoretical requirement of aluminum, W_v refers to vanadium content in ferrovanadium, and W_{pureAl} refers to purity of fine aluminum.

$$W_{Al} = 1 \times \frac{W_V \times 182 \times 1.07 \times 0.5 \times 1.3}{102 \times W_{pureAl}}$$

Here W_{Al} refers to the actual requirements of fine aluminum, W_v refers to vanadium content in ferrovanadium, and W_{pureAl} refers to purity of fine aluminum.

Iron in the ferrovanadium smelting process is the main purpose of molten iron medium, which is obtained by melting the Fe-containing raw material and remained from ferrosilicon. The iron in the ferrovanadium is not volatile at the temperature range used for smelting. Except for very small amounts in the slag residue, all iron enters the alloy. In general, a coefficient of ingredients of 1.00 is selected.

Demand for scrap iron $= 1 \times [1 -$ Ferrovanadium containing vanadium (%) $-$ impurity (%)]-iron from ferrosilicon.

$$W_{scrap\ Fe} = 1 \times (1 - W_V - W_{impurity}) - W_{Fe}$$

Here $W_{scrap\ Fe}$ refers to demand for scrap iron, W_v refers to vanadium content in ferrovanadium, $W_{impurity}$ refers to impurity (%), and W_{Fe} refers to iron brought into the system from ferrosilicon.

The amount of iron added to the system by the ferrosilicon is equal to the actual requirement of silicon iron $\times [1 -$ silicon fraction in ferrosilicon (%)].

$$W_{Fe} = W_{Fe-Si} \times (1 - W_{si})$$

Here W_{Fe} refers to amount of iron brought into the system from ferrosilicon, W_{Fe-Si} refers to the actual requirement of ferro silicon, and W_{si} refers to silicon quality fraction in ferrosilicon (%).

The amount of added lime is calculated based on the basicity requirement. The actual amount of lime required is equal to $1 \times$ (vanadium content in ferrovanadium)/(102/182) $\times$ $1.07 \times 0.385/0.75 \times 1.10 \times 075 \times (62/28) \times$ the basicity/(CaO mass fraction of lime) $\times 110\%$.

$$W_{lime} = 1 \times \frac{W_V \times 182 \times 1.07 \times 0.385 \times 1.10 \times 0.75 \times 62}{102 \times 0.75 \times 28} \times \frac{B}{W_{CaO}} \times 110\%$$

Here W_{lime} refers to the actual requirement of lime, W_v refers to vanadium content in ferrovanadium, B refers to the basicity, and W_{CaO} refers to CaO mass fraction of lime.

10.1.4 Typical process operation and technical indexes

The two steps in the process of ferrovanadium smelting using the electro-silicon thermal method are reduction and refining. Depending on the requirements, the furnace charge can contain both raw materials and refined materials. The reduction process can be divided into two stages and three reduction steps. The first step is reduction of the vanadium oxide with ferrosilicon and Al to produce ferrovanadium silicon with high silicon content. The second step is the refining phase, where high-V_2O_5 slag is used to refine ferrovanadium silicon and reduce the silicon content to obtain ferrovanadium. After completion of a smelting cycle, the furnace is repaired, and then the Fe-containing material is fed into the furnace. After electrifying the refined slag from the previous smelting cycle is returned to the furnace and added to one batch of raw material (flake V_2O_5, lime, and most of the ferrosilicon). When the molten bath is formed, power is delivered at full load, and the furnace material quickly melts. After all of the material is melted, the power is appropriately reduced, and the remaining ferrosilicon is added. Then, the V_2O_5 in the slag is reduced with the addition of aluminum. The molten bath is stirred until the V_2O_5 content in the slag is <0.35%, and then the slag is tapped. After deslagging the second batch of raw materials (flake V_2O_5 and lime mixture) is added. After all of the ingredients are melted, they are thoroughly stirred. First, ferrosilicon, then the aluminum block is added to reduce the slag. When the V_2O_5 content in the slag is <0.35%, the slag is tapped. The number of batches used during the reduction period is determined by the required vanadium content (40%–60%) of the ferrovanadium. The production of ferrovanadium with 50%–60% V, and <2.0% Si require three reduction batches. The composition of the waste residue produced during reduction is 0.2%–0.35% V_2O_5, 45%–55% CaO, 25%–28% SiO_2, 5%–8% MgO, and 5%–10% Al_2O_3.

After reduction the refining process is started. A mixture of V_2O_5 and lime is added to the furnace. The main purpose is to reduce the silicon content and increase the vanadium content. After the alloy in the furnace reaches the required composition, the refined V_2O_5-

Table 10.1: Consumption of raw materials in the production of FeV40 (kg/t).

V_2O_5 (100%)	FeSi75	Aluminum ingots	Scrap steel	Graphite electrode	Magnesite brick	Magnesite	Lime	Water
735.6	340	130	250	28	130	130	1540	80

containing slag is tapped and returned to the next furnace. The composition of the slag in the refining stage is 8%–12% V_2O_5, 45%–50% CaO, 23%–25% SiO_2, and ∼10% MgO. The ferrovanadium is cast into an ingot, which is naturally cooled over 15–20 h, and then demolded, refined, and broken into a specified size, before being packaged and placed in storage. The consumption of raw materials in the production of FeV40 is shown in Table 10.1. The compressed air consumption is 500 m³/t, the overall electricity consumption is 1600 kWh/T, and the smelting power consumption is 1520 kWh/T.

The electro-silicon thermal method is used to produce ferrovanadium with 40%–60% vanadium. The recovery of vanadium is about 97%. The production of ferrovanadium (V 40%) consumes ∼750 kg of flake vanadium pentoxide, ∼370 kg of ferrosilicon (75% Si), 100–110 kg of aluminum, 380–400 kg of scrap steel (iron material), and ∼1300 kg of lime. The electricity consumption is 1500–1600 kWh/t.

10.1.5 Main process steps

The repairing mass must be prepared in phase I reduction in the furnace, which is made according to the desired proportion, for example, the brine:magnesia brick powder:magnesia ratio is 1:3:5. The lining repair defects should be found in the operating clearance, where fast repair at high temperature and a good taphole are required to remove the defects. Part of the refining slag is interlayed after repairing. After adding the scrap steel to the furnace, the electrode and the power supply system are checked. The liquid refining slag of the previous furnace is introduced, and then the primary mixture is added near the center, which increases the current rapidly, depending on the arc, and reaches the peak. After melting the furnace charge until it is smooth, the ferrosilicon is added for reduction, and the basicity is adjusted with lime. Then, aluminum block is added for further reduction and the current changes depending on the reduction intensity. The V_2O_5 content in the furnace slag is ≤0.35%. The lean slag is tapped, and the power supply is maintained with a small current and low voltage. Fast process control is required in the early stage, while it can be slow in later periods, which prevents the slag from being mixed out. A sample is taken to analyze the poor slag.

In the second phase of the reduction, the applied power is increased to the highest point with the addition of the second-phase mixture. When the furnace charge is melted, ferrosilicon is added for reduction, and the basicity is adjusted. Then, aluminum and ferrosilicon are continuously added to deplete the slag. The V_2O_5 content in the furnace

Table 10.2: Alloy composition in the reduction phase (%).

Composition	V	Si	C	P	S
	31–37	3–4	≤0.6	<0.08	<0.05

Table 10.3: Compositional requirements of the refined alloy (%).

Composition	V	Si	C	P	S
	>40	<2	<0.75	<0.1	<0.06

Table 10.4: Compositional requirements of the refined slag (%).

Composition	V_2O_5	CaO	SiO_2	MgO	CaO/SiO_2
	8–13	45–50	23–25	8–15	1.8–2.0

slag is ≤0.35%. The lean slag is tapped, and the power supply is maintained with a small current and low voltage. The third phase of reduction repeats the process of the second phase. In the third phase of reduction, samples are taken to analyze the Si, V, C, S, and P contents in the alloy. The basicity is adjusted, and then ferrosilicon and aluminum block are added to further dilute the slag. The alloy composition in the reduction phase is shown in Table 10.2.

The alloy composition is mainly adjusted during the refining period, which is the same as the second phase of reduction. Power is supplied with a high current and voltage to smelt the furnace charge, and the basicity is adjusted. The current is adjusted depending on the smelting intensity. Samples are taken to analyze the Si, V, C, S, and P contents in the alloy. When the composition of the alloy is suitable, it is discharged. The process uses a small voltage and large current to maintain the temperature of the furnace, and the refining slag is removed via the slag taphole. The iron taphole is opened after the power is turned off. The compositional requirements of the refined alloy are shown in Table 10.3, while that of the refined slag is shown in Table 10.4.

When pouring ferrovanadium, a hot metal ladle is preheated and cleaned, and then dry river sand is placed in the bottom. Depending on the ladle temperature, and the exhaust and mold size, the casting speed is controlled. The metal bath surface is maintained around 100 mm from the top of the mold. Demolding occurs around 80 min after casting, where late pouring is required to avoid slag mixing with the alloy. After demodulation, ferrovanadium ingots are immediately sent to a water cooling bath, where the water occupies about two-thirds of the bath volume. The cooling time is usually 30–40 min. After adding the alloy ingot, the water is filled into the bath. The alloy ingot is quickly dried after cooling and weighed in a warehouse.

10.1.6 Factors influencing the production of ferrovanadium

The smelting of ferrovanadium by the electro-silicon thermal method has two main indexes, the ferrovanadium product quality and the technical and economic indicators. The ferrovanadium product quality is related to the choice of product standards and relevant operational parameters, where the control of the product quality depends on the materials, ingredient mixtures, and appropriate operation. The technical and economic indexes of the smelting process are mainly related to the operation control and equipment performance. The C, S, and P contents in the ferrovanadium product are limited. With respect to impurity control, the selected materials (such as the V_2O_5, ferrosilicon, and lime) must meet the requirements of the standards, where the impurities entering the furnace charge are restricted. The impurities in the slag and their volatility should be carefully considered with reference to similar products and technologies.

The electro-silicon thermal process for smelting ferrovanadium has strict requirements for the raw materials and is a favorable process for producing industrial V_2O_5 with high purity and few impurities. The P in the raw materials is nonvolatile at the high temperatures used in the smelting process; considering its equilibrium distribution coefficient in the metal and slag phase, about 85% of the P will enter the alloy phase. Hence, raw material control is very important. S is mainly present as Na_2SO_4. Some of the S in the raw materials can be volatile during high-temperature smelting, where reactions such as (10.13) occur. The remaining S will be distributed between the slag phase and the alloy depending on the equilibrium distribution coefficient:

$$2Na_2SO_4 + Si + V_2O_5 = 2Na_2O \cdot SiO_2 + 2SO_2\uparrow + V_2O_3 \qquad (10.13)$$

Due to the low melting point (570°C) of Na_2O, SiO_2 may be generated during smelting, resulting in thinning of the slag. This reduces its calorific capacity, limits increases in the furnace temperature, increases thermal radiation losses, and reduces the lifetime of the refractory furnace wall and roof. In addition, the difference between the melting points of the alloy and slag increases, making slag dilution difficult. Reaction (10.10) is an endothermic reaction that generates SO_2 gas and produces bubbles during smelting. The Na_2O content in flake V_2O_5 should be as low as possible as it can be reduced by Al, resulting in unnecessary Al consumption. If the lime is not cooked thoroughly, or if the water is poorly, the energy will be consumed during the smelting process, and the basicity of the slag will be reduced.

Silicon thermal reduction is an exothermic reaction. The smelting temperature is influenced by the ingredients, silicon oxidation, applied power, and the heat preservation and operation of the process. In general, the smelting temperature is in the range of 1600°C–1650°C. If the temperature is too high, it can cause vanadium loss via volatilization, and increase energy consumption. Basicity (CaO/SiO_2) can replace and balance low-valence vanadium oxide entering the slag. Lower basicity can impede silicon reduction as excess calcium oxide can react with V_2O_5 to generate calcium vanadate. In contrast, a high basicity increases the slag viscosity, which complicates vanadium ferroalloy production and separation and increases material consumption. Hence, the basicity is controlled in 2.0–2.2 in general.

10.1.7 Common problems and their solutions

10.1.7.1 Adjusting the vanadium content of the ferrovanadium

The vanadium content varies due to the mixing of ingredients or slag loss, where scrap steel is added when the vanadium content is high. The amount of iron scrap = (fraction of vanadium that needs to be reduced) × (amount of alloy in the furnace)/(decrease in alloy vanadium content). Here, the amount of alloy is important. When the vanadium content is low, V_2O_5 should be added. The amount of V_2O_5 = (content of vanadium increased) × (amount of alloy in the furnace) × (182/102)/(V_2O_5 purity). Here, the amount of alloy is important.

10.1.7.2 Adjusting the silicon content of the ferrovanadium

The Si content varies due to the mixing of ingredients or slag loss, where V_2O_5 is added when the Si content is high. Amount of V_2O_5 = (Si content to be reduced) × (amount of alloy in the furnace) × (364/140)/(V_2O_5 purity). Here, the amount of alloy is important. When the silicon content of the alloy is low, the ferrosilicon should be considered. Amount of ferrosilicon = Si content to be increased × amount of alloy in the furnace/mass fraction of Si in ferrosilicon × Si content of alloy after silicon addition.

10.1.7.3 High carbon and phosphorus contents in the alloy

High carbon and phosphorus contents in the alloy are mainly due to the choice of raw materials. A temporary solution is the removal of phosphorus and decarbonization or adding iron scrap. Long-term solutions should update the raw material selection to limit the carbon and phosphorus contents. Scrap steel supplement = carbon content to be reduced × amount of alloy in the furnace/(alloy carbon content reduced)/(carbon content of scrap steel).

10.1.7.4 Boiling and slag loss

Due to reaction cycles, inhomogeneous segregation of the furnace charge, loss of sulfur and carbon due to oxidation, and fast heating and consumption of auxiliary raw materials, boiling and slag loss occurs, mainly near the furnace door, which is the safe hidden trouble. This problem can be limited by better management of raw materials and optimizing the operational parameters and power supply.

10.1.7.5 Lining protection

When the furnace is shut down, it is repaired and slag is removed. To protect the basic furnace hearth and furnace wall, a new furnace hearth and furnace charge must be used after roasting consolidation. Slag is placed at the bottom of the furnace before the next run to protect it. The temperature cannot be too high, and the bottom of the furnace should not be damaged during stirring.

10.1.8 Electric furnace requirements

In general, steelmaking furnaces are used, which contain a graphite electrode with a diameter of 200–250 mm and operate at a reference voltage of 150–250 V and current of 4000–4500 A. The actual production requirements are as follows: transformer of 2500 kVA; voltage of 10,000 V; secondary voltages of 121, 92/210, and 160 V; and a rated current of 6870 A. The specifications of a 3-t arc furnace include electrode Φ 250 mm, center of the heart Φ 760 mm, furnace shell Φ 2900 × 1835 mm, and electrode stroke 1300 mm.

10.2 Aluminum thermal method

Aluminum is one of the strongest known reductants. In 1897 Gordon Schmitt invented the process of smelting ferrovanadium by aluminum thermal reduction. Flake V_2O_5 is reduced to produce ferrovanadium with Al that is a highly exothermic process. The large amount of heat generated (4577 kJ/kg mixture) satisfies the heat demand for completing aluminothermic reduction smelting, while ferrovanadium scraps are added to absorb excess thermal energy and lower the reaction temperature. The addition of lime, magnesia, and fluorite is used to reduce slag viscosity and adjust the basicity. Scrap steel is added as a dissolution medium. The particle size of the alumina and V_2O_5 is adjusted to reduce the reaction speed and splashing loss and increase the yield of vanadium.

Compared with smelting ferrovanadium using the electro-silicon thermal method, the aluminum thermal reduction has the characteristics of low C and high V contents. The equipment used in process of aluminum thermal reduction includes a basic cylindrical reactor structure and a refractory lining with resistance to high-temperature corrosion. In general, a magnesia lining is used or knotted with slag from ferrovanadium smelting. The reaction pot is designed in accordance with the requirements of the target product and furnace capacity, where accurate calculations of the ingredients and heat balance are required. All ingredients are mixed and fed into the reaction pot. After mixing in the reaction tank, ignition initiates the reactions.

10.2.1 Reduction process

Aluminum is used as the reducing agent in the process of aluminum thermal reduction. The reaction occurs in the reaction pot with an alkaline lining, where magnesia lines the furnace base. The reaction pot is lined with magnesium brick. All magnesia used must be consolidated, roasted, and dried. A small amount of ingredients is placed in the reactor, and spark ignition induces reactions in the furnace charge. As the reactions occur, the remaining ingredients are successively added. When the primary reduction reactions ends, a heat settler is added immediately to the slag surface layer to increase the exothermic reaction and maintain the molten state of the slag (composed of ferric oxide and alumina). It also promotes settling of the alloy in the slag while continuing to reduce the V_2O_5, and adsorbing and accumulating suspended alloy droplets in slag.

The alloy is cooled for 16–24 h. After cooling is complete, the furnace is dismantled, and the alloy is separated from the slag. The slag near the alloy is stored apart from the rest of the slag. Depending on the size of the alloy blocks, they are processed and finished. Slag close to the alloy is mainly used as a cooling agent in the smelting processing, and the other slag after processing is used for lining knotting. The aluminothermic method is usually used in the smelting of ferrovanadium with high vanadium content (60%–80%), and the recovery is slightly lower than that of the silicon thermal method that is about 90%–95%. The aluminum thermal reduction process was the first method used to produce ferrovanadium and is still in use. Currently, continuous feeding (lower ignition process) and feeding (upper ignition process) are used. The typical production process of smelting ferrovanadium by aluminum thermal reduction is shown in Fig. 10.4.

10.2.2 Process details

During smelting of ferrovanadium by electro-aluminum thermal reduction, flake V_2O_5, vanadium trioxide, calcium vanadate, vanadate ferrovanadium, or a mixture of these are used as raw materials, while Al is used as the reducing agent. Electricity is supplied to replace aluminum heating to supplement heat and intensify heat addition. In the basic electric arc furnace, the appropriate ferrovanadium products are smelted in two stages of reduction and refining. In the reduction phase the whole reducing agent and V-containing raw material are fed into the electric furnace, where aluminum thermal reduction is carried out in the presence of high-CaO slag. After the reduction is finished, the lean slag is tapped, and the refining agent is added to adjust the alloy composition. When the composition of the ferroalloy meets the requirements, both the slag and alloy are tapped. The refined slag is rich in V_2O_5. When the next cycle starts, the furnace will be fed and used again. The alloy is generally cast after cooling, molding, crushing, and dredging, to give a finished product. The typical process of producing ferrovanadium by electro-aluminum thermal reduction is shown in Fig. 10.5.

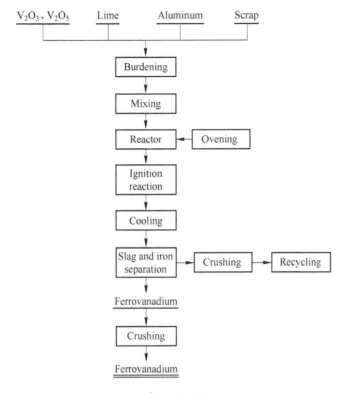

Figure 10.4
Typical process of smelting ferrovanadium by aluminum thermal reduction.

10.2.3 Reduction reactions

In the production of ferrovanadium by aluminum thermal reduction, flake V_2O_5 is used as the V-containing material, while Al is the reductant. The chemical reactions occurring during aluminum thermal reduction are shown by (10.10)–(10.12).

10.2.4 Raw materials and ingredient requirements

The flake vanadium pentoxide should have a V_2O_5 content above 98%, a particle size of <20 mm × 20 mm and thickness of 3–5 mm. The aluminum should have an Al content >98%, granule size <3 mm, and 3–10 mm flake size. The slag is a corundum slag produced in the early stage, with a particle size of 5–10 mm. The scrap steel is a carbon steel with a coil length <15 mm. The compositions of the materials used for smelting ferrovanadium by aluminum thermal reduction are shown in Table 10.5.

The amount of aluminum used in smelting ferrovanadium by aluminum thermal reduction is reasonable. The ingredients are designed to ensure the material balance and product quality.

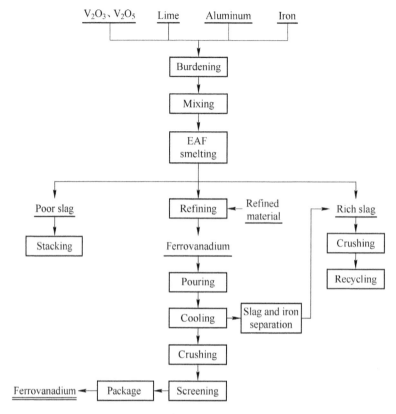

Figure 10.5
Typical process of producing ferrovanadium by electro-aluminum thermal reduction.

Table 10.5: Composition of materials for smelting ferrovanadium by aluminum thermal reduction.

Material	Ω (%)				Particle size (mm)
	P	C	S	Si	
Flake vanadium pentoxide (V_2O_5 <93%)	<0.05	0.05	<0.035		1–3
Aluminum granules (Al > 98%)				<0.2	3
Scrap steel	<0.015	<0.5			10–15
Lime	<0.015				

The best heat of reaction of the furnace charge is 3140–3350 kJ/kg. The theoretical amount of aluminum used in smelting ferrovanadium by aluminum thermal reduction is calculated using the chemical reaction shown in (10.10), where the actual amount of aluminum is 100%–102% of the theoretical value.

Theoretical amount of Al used in smelting ferrovanadium by aluminum thermal reduction = amount of $V_2O_5 \times V_2O_5$ content × aluminum relative atomic mass × 10/the relative atomic mass of $V_2O_5 \times 3$.

Flake vanadium pentoxide containing 98% V_2O_5 is used to produce 80%V ferrovanadium. The 85% recovery of vanadium is calculated assuming 100 kg of flake vanadium pentoxide is added to smelt 80VFe ferrovanadium. Amount of 80VFe obtained = amount of flake $V_2O_5 \times V_2O_5$ grade × 2 × molecular weight of vanadium/molecular weight of V_2O_5 × recovery of vanadium:

$$= 100 \times 0.93 \times 102/182 \times 0.85/0.8 = 55.4 \text{kg}$$

The impurity content in the scrap steel is 5%. The amount of scrap steel added = $55.4 - 55.4 \times 5\% - 100 \times 93\% \times 102/182 \times 0.85 = 8.33$ kg

During smelting of ferrovanadium by aluminum thermal reduction, the heat content is enormous. The heat of reaction of the mixture is usually 4500 kJ/kg. In the case of the production of ferrovanadium containing 75–80%V, this value is controlled to 3100–3400 kJ/kg by cooling and buffering with coolants such as lime and slag. In the process of aluminum thermal reduction, the V_2O_5 purity is ~95%. After melting it is reduced to low-valence vanadium, which reduces the heat content, where the actual reaction heat is 3768 kJ/kg.

The reduction reaction heat of 100 kg V_2O_5 = actual unit reaction heat × (flake V_2O_5 added + aluminum granules added) = $3768 \times (100 + 46) = 550128$ kJ.

The reaction heat of aluminum is 3266 kJ/kg, where the amount of furnace charge = $550{,}128/3266 = 168.44$ kg.

In the case of a furnace charge with 100 kg of V_2O_5, 47.88 kg of Al and 8.33 kg of scrap steel should be added to the cooling material.

$$= \text{furnace charge} - \text{actual addition of } V_2O_5 - \text{actual aluminum addition}$$
$$- \text{actual scrap steel addition}$$
$$= 168.44 - 100 - 47.88 - 47.88 = 12 \text{ kg}$$

The furnace coolant consists of equal parts of lime and returned slag (6 kg of each). The ignition agent uses equal parts of aluminum powder, potassium chlorate (barium peroxide or magnesium powder), and V_2O_5 (i.e., 1:1:1 ratio).

10.2.5 Furnace characteristics

Depending on the furnace type and capacity, cast iron or thick steel plate is used to form the cylindrical shell that is reinforced by external steel fastening clamps or steel hoop reinforcement and divided into upper and lower parts. The smelting furnace contains a furnace lining, knotting, and the oven. The lining is divided into permanent and temporary layers, where the permanent layer is made of magnesium bricks and high-alumina bricks and built in three segments. The temporary layer is knotted with brined returned slag and knotting material. The strength of this material needs to be sufficient to prevent leakage and protect the permanent layer. The overall strength of the knotted layer should be moderate, as it needs to be easily dismantled. The bottom seam of the furnace body and the bottom of the furnace must be sealed and have a high consolidation strength. The bottom layer of the furnace is higher than that of the upper part, which cannot contain low-melting-point materials.

The inner lining of the ferrovanadium smelting furnace is made of magnesium-brick masonry, and the inner wall of the furnace is made from a mixture of finely ground returned corundum slag and brine. The bottom of the furnace can be paved with magnesia that is dried and consolidated. Depending on the requirements of continuous production, the same kind of furnace can be arranged at the same time, which can be placed on a mobile platform car, and one time is a small production cycle. The inner diameter of the furnace is 0.5–1.7 m, and the height is 0.6–1.0 m.

The furnace is selected based on specifications such as the furnace charge batch size, alloy variety, weighing measuring, mixture ingredients, feeding device, inspection, and electric heating and environmental protection facilities. Some factories can be configured at the same time the furnace internal on line detection device.

10.2.6 Typical process operation and technical indexes

10.2.6.1 Smelting ferrovanadium by aluminum thermal reduction

A small part of the mixture is first fed into the reaction pot with an alkaline lining, which is then ignited. After the reaction starts, the remaining charge is added. After the smelting reaction is complete, natural cooling occurs over 16–24 h. The pot reactor is then dismantled, and the ferrovanadium ingot is removed, cleaned, and packaged. It is usually used for smelting ferrovanadium with high vanadium content (60%–80%), where the recovery is slightly lower than that of the silicon thermal process, which is about 90%–95%. The composition of the smelted ferrovanadium is 75%–80% V, 1%–4% Al, 1.0%–1.5% Si, 0.13%–0.2% C, $\leq 0.05\%$ S, and ≤ 0.075 P. The slag composition includes 5%–6% V_2O_5, ~85% Al_2O_3, and ~10% CaO + MgO. During production of 1 t ferro 80% V, 1500–1600 kg of flake vanadium pentoxide (V_2O_5 98%) and 810–860 kg aluminum (Al 98%) are consumed. The recovery of vanadium is 90%–95%.

10.2.6.2 Smelting ferrovanadium by electro-aluminum thermal reduction

To reduce the vanadium content in the slag, after completion of the thermal reaction, the reaction platform car is moved under the electric heater. Then, the heating electrode is inserted to keep the slag in a molten state and enhance settling of the suspended metal particles in the slag. Vanadium oxides in the slag and Al residue continue to react, increasing the vanadium recovery. During smelting of ferrovanadium by electro-aluminum thermal reduction, the V-containing material can be flake V_2O_5, vanadium trioxide, calcium vanadate, iron vanadate, or mixtures of these, while Al is used as the reductant. Electricity is supplied to replace aluminum heating, similar to the silicon thermal reduction process.

10.2.7 Factors affecting the process

In the production of ferrovanadium by aluminum thermal reduction, the main influencing factors are the accuracy of the ingredients, heat distribution, and feeding. These three factors interact and influence each other. The main challenges related to the ingredients include uneven mixing, furnace charge segregation, and inaccurate weighing. The aluminum content should allow the aluminum thermal reduction reaction to be completed; excess aluminum will enter the alloy and increase its Al content. This results in a decrease in the specific gravity of the alloy and the settling rate, resulting in increased inclusion of slag in the alloy, a higher aluminum consumption, and lower vanadium recovery. However, Al deficiency may result in incomplete, inadequate, or slow reaction of the furnace charge, or even termination of the reaction. In addition, the ferrovanadium composition can be affected (abnormally high or low compositions), waste products can be produced, and raw materials can be lost. An imbalance of aluminum, potassium chlorate, and/or coolant in the process of dosing results in insufficient heating, overheating, or severe spatter during the reaction process.

The aluminum thermal reduction reaction starts after ignition. In general, the feeding speed is gradually increased to ensure a smooth reduction process. If feeding is too fast, the temperature increases sharply, and serious splashing of the burden occurs, increasing vanadium and aluminum loss. If the feeding speed is too slow, the low melting temperature results in a slow reaction. As the thermodynamics and kinetics of the reaction are poor under these conditions, premature slag bonding occurs, the vanadium oxide reduction reaction is incomplete, and separation of the slag and alloy is difficult. In addition, condensation of the alloy is poor, the alloy diffuses in the slag, and alloy loss increases, resulting in a lower vanadium recovery. Considering the furnace capacity and shape, the feeding speed is maintained at $160-200$ kg/(m^2 min). The dosage of the coolant (back slag and lime) is usually 20%–40% of the V_2O_5 content.

Vanadium trioxide is a vanadium oxide with a lower valence than that of vanadium pentoxide. Hence, during aluminum thermal reduction, vanadium trioxide reacts during the first reduction with a weak reductant. In the secondary reduction, natural gas or gas is introduced to replace the aluminum reductant, and 40% less Al can be used. Vanadium trioxide is used in aluminum thermal reduction processes to produce ferrovanadium. The low Al content during the reduction process results in insufficient reaction heat, so further heat is added by electrical heating. Vanadium trioxide, or mixtures of vanadium trioxide and vanadium pentoxide, can be used in the production of ferrovanadium by electro-aluminum thermal reduction.

10.3 Direct alloying of vanadium slag to produce ferrovanadium

Direct alloying of vanadium slag is most suitable for the production of ferrovanadium with lower vanadium contents. Iron oxide is reduced with a weak reductant, and vanadium oxide is reduced with a strong reducing agent that can simplify the refining of vanadium and minimize production costs.

10.3.1 Direct alloying technology

In ferrovanadium production by direct alloying of vanadium slag, the iron oxide in the slag is first selectively reduced in an arc furnace with carbon, silicon, or a silicon–calcium alloy. Most of the iron is removed from the vanadium slag to obtain a slag with high V/Fe content. In the second phase the vanadium slag after iron removal is reduced in an electric arc furnace with carbon, silicon, or aluminum to produce ferrovanadium.

10.3.2 Direct alloying reactions

The vanadium slag is directly alloyed, and iron oxide is reduced with carbon, silicon, or silicon–calcium alloys, while vanadium oxide is reduced with silicon or aluminum. The carbon reduction reactions of iron oxide are as follows:

$$Fe_2O_3 + C = FeO + CO \tag{10.14}$$

$$FeO + C = Fe + CO \tag{10.15}$$

$$CO + O_2 = CO_2 \tag{10.16}$$

$$Fe_2O_3 + Si + Ca = Fe + SiO_2 + CaO \tag{10.17}$$

In the case of the vanadium oxide reduction, the carbon thermal reduction reaction is shown in (10.1) and (10.2). The silicon thermal reduction reaction is as shown in (10.3)–(10.6). Low-valence vanadium oxide (VO and V_2O_3) and silicon oxide easily generate silicate that delays vanadium oxide reduction. CaO is often used as it reacts to produce calcium silicate that allows low-valence vanadium oxide to continue participating in the reaction. In the presence of CaO the silicon reduction reaction is as shown in (10.7)–(10.9), and the aluminum thermal reduction reaction is as shown in (10.10)–(10.12).

10.3.3 Raw material and ingredient requirements

Depending on the reduction requirements, a reducing agent is selected. In general, carbon and silicon reductants are used. The carbon can be petroleum coke or another carbon reductant. The silicon reductant is mainly 75SiFe, or a silicon–calcium alloy, where the calcium is also a reducing agent. The primary reduction efficiency of iron oxide is controlled to ∼86%, while the reduction efficiency of vanadium oxide is controlled to ∼5%. The V/Fe ratio increases from 0.20–0.25 to 1.0–1.5. The second reduction is based on the target composition of ferrovanadium, where the corresponding mixtures are calculated.

10.3.4 Typical process operation and technical indexes

In Russia, vanadium converter slag is selectively reduced at 1290°C–1290°C with carbon, where 86% of the iron oxide and 5% of the vanadium oxide are reduced and enter the metallic phase. After separation of the vanadium slag, V/Fe increased from 0.20–0.25 up to 1.0–1.5, which is reduced with 75SiFe and aluminum. The compositions of the initial alloy and refined alloy are shown in Table 10.6.

The US Patent US34202659 describes a procedure where, in the first step, vanadium slag, quartz, and fluxing agent are mixed with carbon, and smelted in a 1200 kVA electric arc furnace to produce ferrovanadium silicon. The vanadium slag consists of $\omega(V_2O_5)$ 17.5%–22.5%, and $\omega(SiO_2)$ 16.74%–17.57%, while the ferrovanadium silicon contains $\omega(V)$ 18.97% and $\omega(Fe)$ 32.16%. The ferrovanadium silicon can be refined by addition of vanadium pentoxide and lime. Ferrovanadium silicon, vanadium pentoxide, and lime are added in the proportion of 120:75:126. The alloy composition after refining in the electric furnace is $\omega(V)$ 44.6%, $\omega(Fe)$ 34.85%, $\omega(Si)$ 16.97%, $\omega(Cr)$ 0.91%, $\omega(Ti)$ 0.92%, $\omega(Mn)$ 0.71%, and $\omega(C)$ 0.23%; the intermediate slag contains $\omega(V)$ 9.1%. A second refining step can be used, where a refining slag and ferrovanadium silicon are refined in an electric furnace to produce an intermediate ferrovanadium silicon alloy, which contains $\omega(V)$ 33.80% and $\omega(Si)$ 23.33%. Vanadium pentoxide and lime are added for secondary refining to obtain a ferrovanadium composition of $\omega(V)$ 55.80% and $\omega(Si)$ 0.78%. The vanadium recovery of this process is 87%.

Table 10.6: Composition of the initial alloy and refined alloy (%).

Item	V	Ti	Cr	Si	Mn
Initial alloy	20–26	3–6	2–4	14–18	10–15
Refined alloy	26–34		4–6		14–18

The US Patent US3579328 describes mixing lime, ferrosilicon, and vanadium slag, which is smelted in a normal steelmaking electric furnace, resulting in a vanadium slag composition of $\omega(V_2O_5)$ 12.2%, $\omega(SiO_2)$ 15.8%, $\omega(Cr_2O_3f)$ 0.5%, $\omega(MnO)$ 1.1%, $\omega(FeO)$ 37.2%, $\omega(P_2O_5)$ 0.05%, $\omega(CaO)$ 0.5%, and $\omega(TiO_2)$ 4%. The selected ferrosilicon is 75SiFe. The calcined lime or dolomite lime, vanadium slag, lime, and ferrosilicon are mixed in a proportion of 1000:700:90. The furnace charge is melted and smelted for 1.5 h at 1600°C–1700°C, producing 1400 kg of intermediate slag and 425 kg of steel. The slag basicity is set to 1.0–2.0. The MgO and Al_2O_3 mass fractions in the slag are limited to 2%–0% and 2%–20%, respectively. The intermediate slag is poured into a slag ladle that can be moved after preheating. In this process, 83 kg of 75SiFe is fed over 18 min, the reduction temperature is 1650°C, and 169 kg of ferrovanadium is produced. The alloy composition is as follows: $\omega(V)$ 32.1%, $\omega(Fe)$ 32.1%, $\omega(Si)$ 6.4%, $\omega(Cr)$ 1.7%, $\omega(Ti)$ 0.2%, and $\omega(Mn)$ 2.3%. The vanadium recovery is 79%.

The US Patent US4165234 and German patent DIN2810458 describe smelting ferrovanadium by direct alloying in a bottom-blown converter fed with 6 t of vanadium slag, where the oxygen and air nozzles are allocated at the bottom of a 10 t converter. The vanadium slag composition is $\omega(V)$ 11.2% and $\omega(Fe)$ 42%. The oxygen is blown at a rate of 30 m^3/min, and the air is blown at a rate of 15 m^3/min into the converter. The vanadium slag is melted, while oxygen is blown into the converter during the final 20 min. Adding 550 kg of lime, the temperature increases to 1500°C, and melting is performed for 45 min. The slag composition is $\omega(V)$ 9.52% and $\omega(Fe)$ 37%. Water vapor is blown at a rate of 12 m^3/min into the furnace, while natural gas is blown at a rate of 4 m^3/min. Then, 550 kg of 75SiFe and 550 kg of lime are added, and the blowing rate of the steam is reduced. The slag [ω V_2O_5] is removed, and oxygen is blown at a rate of 35 m^3/min and natural gas is blown at 3 m^3/min into the melt in the furnace. The molten metal contains $\omega(V)$ 17.6% and $\omega(Si)$. After blowing for 20 min, 2.2 t of slag [$\omega(V)$ 28% and $\omega(Fe)$ 10%] and 1.5 t steel [$\omega(V)$ 0.12% and $\omega(C)$ 0.05%] are obtained. The molten steel is poured out, and then 800 kg of aluminum block and 1000 kg of lime are added to reduce the slag. Water vapor is blown at a rate of 12 m^3/min into the furnace, and natural gas is blown at 2 m^3/min. The highest reduction temperature is 1700°C. Finally, 1 t ferrovanadium [$\omega(V)$ 43.4%] is produced.

In the Canadian patent 860886, ferrovanadium is produced directly with vacuum carbon reduction, and the vanadium slag consists of $\omega(V)$ 14.4%, $\omega(FeO)$ 38.5%, $\omega(SiO_2)$ 20.01%, $\omega(Cr_2O_3)$ 2.3%, $\omega(TiO_2)$ 8.1%, $\omega(MnO)$ 2.2%, $\omega(MgO)$ 1.4%, $\omega(P_2O_5)$ 0.05%, and $\omega(CaO)$ 0.3%. Both the slag and petroleum coke are crushed to <0.043 mm, and then mixed and agglomerated, and loaded in a vacuum resistance furnace at a vacuum pressure of 0.1333 Pa. The mixture is then heated to 1480°C over 3 h. During the heating process the pressure is increased to 27 Pa, and the temperature is held for 10 min. Then the power is turned off, and the pressure is decreased to 8 Pa. The obtained ferrovanadium contains ω(V) 24.84%−26.42%, ω(Fe) 42.15%−43.15%, ω(Si) 8.50%, ω(Cr) 8.5%, ω(Al) 2.9%−3.0%, ω(Mn) 0.02%−0.04%, and ω(Ca) 0.25%− 0.10%. If it is necessary to improve the grade of the alloy, it can be refined with the addition of vanadium pentoxide.

In Austria, Bach's factory (TCW) produces ferrovanadium directly from vanadium slag. First, vanadium slag is selectively reduced in a furnace with carbon or silicon to remove iron. The directly reduced iron is used for steelmaking, while the prereduced vanadium slag is mixed with silicon reductant and ferrovanadium silicon (from the reduction of slag to produce ferrovanadium and reduced slag). The reduced slag is processed using silicon reduction to produce ferrovanadium silicon alloy, which is returned to the production of ferrovanadium, while the lean slag is rejected. The ferrovanadium product contains ω(V) 45%, ω(Si) 4.3%, ω(Mn) 1.1%, and ω(C) 0.7%.

10.3.5 Factors affecting the process

The production of ferrovanadium by direct alloying of vanadium slag simplifies vanadium refining. The ferrovanadium produced is not a standard product, and the quality cannot compare with production of vanadium pentoxide or vanadium trioxide. In addition, the ferrovanadium composition is complex and highly dependent on the raw material composition. The primary use of such products is determined by the particular application but can be adapted to the production of cast iron containing V and Ti, which can meet the product quality requirements of different customers for their specific processes. There are various processes in the production of ferrovanadium by direct alloying of vanadium slag, but most are still in the experimental stage. Industrial production facilities and practical applications are limited to the Trebach plant in Austria and the Kristner Spiegel plant.

In China the Jinzhou ferroalloy factory and Panzhihua steel have tested and studied the production of ferrovanadium using Panzhihua vanadium slag via direct alloying. With progress in vanadium production technology, resource expansion, product quality standards, and environmental protection laws, few similar production facilities exist. The main parameters affecting the ferrovanadium quality are the corresponding qualities of the vanadium slag and reducing agent. Different manufacturers use slightly different processes to produce vanadium slag. During alloying of vanadium slag, any oxide impurities are reduced and dissolved in the alloy, where some residual reductant content is remained. The factors influencing the economics of the process include the temperature, equipment selection, energy consumption, and the reducing agent type and content. Although direct alloying of vanadium slag is an immature vanadium processing technology, it has the function of recycling other metals, such as Cr.

Table 10.7: Ferrosilicon−vanadium alloy composition produced using carbon reductant.

V	Si	Mn	Ti	Cr	C	S	P
8−13	10−30	5−6	0.9−3.0	2−2.5	0.5−1.5	0.003−0.006	0.03−0.10

10.4 Ferrosilicon−vanadium alloy

In this process, vanadium slag and silica are used as raw materials, coke is the reducing agent, ferrosilicon is the reducing agent, and lime is the fluxing agent.
Ferrosilicon−vanadium alloys can be produced by smelting in an electric arc furnace. By controlling the amount of carbonaceous reductant, ferrosilicon−vanadium alloy compositions are produced, as shown in Table 10.7. If silicon iron is used as reducing agent, the content of C in the silicon−vanadium alloy is <0.5%. The reaction principle is the same as that for ferrovanadium.

CHAPTER 11

Applications of vanadium in the steel industry

Chapter outline
11.1 Classification of steel   268
11.2 Main elements in steel and their effects   270
      11.2.1 Main elements   270
      11.2.2 Role of alloying elements   270
11.3 Role of vanadium in steel   275
      11.3.1 Behavior of vanadium in steel   276
      11.3.2 Effect on microstructure and heat treatment of steel   279
      11.3.3 Effect on physical, chemical, and technological properties of steel   280
      11.3.4 Effect on mechanical properties of steel   280
11.4 Applications of vanadium in steel   280
      11.4.1 Processing performance characteristics of V-containing steel   280
      11.4.2 Requirements for continuous casting of V-containing steel   281
      11.4.3 Thermomechanical control   281
      11.4.4 V-bearing quenched and tempered steel   282
      11.4.5 Common quenched and tempered steel   284
      11.4.6 Substitutes for quenched and tempered steel   284
      11.4.7 V-containing quenched and tempered steel   285
      11.4.8 Nonquenched and tempered steel   285
      11.4.9 Vanadium microalloyed structural steel   289
      11.4.10 Heat-resistant steel   314
      11.4.11 Stainless steel   318
11.5 Vanadium in cast iron   319

Iron and steel refer to steel and pig iron, both of which are alloys with iron and carbon as the main elements. Iron and steel are the most widely used metal materials in industry. When vanadium is added to steel, it has a strong affinity with carbon, nitrogen, and oxygen and can generate stable compounds with these elements. Vanadium mainly exists as its carbide in steel, where its main role is to refine the structure and grains of the steel, which reduces its sensitivity to overheating, and increases its strength and toughness. During the heat treatment of steel, the compounds produced by vanadium, carbon, and nitrogen in the steel can strongly prevent the grains from growing, resulting in hot-processed steel with a fine structure at room temperature, high strength, and good plasticity and toughness. When austenite is dissolved at high temperature, the hardenability can be increased. Conversely, in the presence of carbides, the hardenability is reduced. Vanadium increases the tempering stability of quenched steel and produces a secondary hardening effect. Vanadium can refine the grains, increase the strength and yield ratio and low-temperature toughness after normalizing, and improve the welding performance of ordinary low-alloy steel. Vanadium is often used together with manganese, chromium, molybdenum, tungsten, and other elements in structural steels as it reduces the hardenability under general heat-treatment conditions.

Vanadium is mainly used to improve the strength and yield ratio of steel, refine grains, and reduce overheating sensitivity. As vanadium can refine grains in carburized steel, the steel can be directly quenched after carburization without requiring secondary quenching. Vanadium can increase the strength and yield ratio of spring steel and bearing steel; in particular, it can increase the ratio limit and elastic limit, reduce the sensitivity to decarbonization during heat treatment, and improve the surface quality. In bearing steel containing vanadium (without chromium), there is high carbide dispersion, resulting in a good performance. Vanadium refines grains in tool steels to reduce overheating sensitivity and increase tempering stability and wear resistance, thus extending tool life.

Vanadium is an important additive in steel. Worldwide, the consumption of vanadium in the steel industry accounts for about 85% of the total production. The vanadium content in steel is generally no more than 0.5%, except in high-speed tool steel. Practice shows that adding 0.1% vanadium to structural steel can increase its strength by 10%−20%, reduce the structural weight by 15%−25%, and reduce its cost by 8%−10%. If high-strength V-containing steel is used, the weight of the metal structure can be reduced by 40%−50%, while the cost is 15%−30% lower than that of ordinary structural steel. Adding vanadium to steel can also improve its welding performance. Vanadium and titanium are added to steel to provide good corrosion resistance and seismic performance [14−17,25,28].

11.1 Classification of steel

Iron and steel products can be classified into pig iron, cast iron, and steel. Pig iron is an iron−carbon alloy with a mass fraction of carbon more than 2%, which is classified as pig iron for steelmaking (low silicon) and pig iron for casting according to the application. Pig iron is divided into ordinary pig iron and special pig iron (including natural alloy pig iron and ferroalloy) depending on the chemical composition. Cast iron is an Fe−C alloy with a carbon content of >2 wt.% (commonly 2.5−3.5 wt.%). Cast iron is usually remelted using a cupola and other equipment for casting machine parts. According to the color of the fracture, classifications of gray cast iron, white cast iron, and hemp cast iron are done. Cast iron is classified into ordinary cast iron and alloy cast iron according to its chemical composition. In addition, cast iron is classified as ordinary gray cast iron, inoculated cast iron, malleable cast iron, nodular cast iron, vermicular cast iron, and cast iron with special properties according to production process and microstructure.

Steel is an Fe−C alloy with a mass fraction of carbon <2%. There are many kinds of steel, and their classification is quite complicated. According to the use of steel, they can be divided into three categories: structural steel, tool steel, and high-performance steel. Based on the quality of the steel, they are classified as (1) ordinary steel (P 0.045%, S 0.050%); (2) high-quality steel (P and S contents of 0.035%); (3) high-grade fine steel (P 0.035%, S 0.030%). (1) Carbon steel is classified according to its chemical composition: (A) low-carbon steel (C 0.25%); (B) medium-carbon steel (C 0.25%−0.60%); and (C) high-carbon steel (C 0.60%); (2) alloy steel: (A) low-alloy steel (total content of alloying elements 5%); (B) medium-alloy steel (total content of alloying elements >5%−10%); and (C) high-alloy steel (> 10% total alloying elements). Steels can also be classified by the machining or shaping process used: (1) forged steel; (2) cast steel; (3) hot-rolled steel; and (4) cold-drawn steel. Steels classified by their metallographic structure include: (1) annealed steel: (a) hypoeutectic steel (ferrite + pearlite); (b) eutectoid steel (pearlite); (c) hypereutectoid steel (pearlite + cementite); and (d) lewisite steel (pearlite + cementite); (2) normalized steel: (a) pearlite steel; (b) bainitic steel; (c) martensitic steel; and (d) austenitic steel; (3) no phase change or partial phase change occurs. According to their use, steels are classified as (1) construction and engineering steels: (a) ordinary carbon structural steels; (b) low-alloy structural steel; and (c) steel reinforcement; (2) structural steel: (a) machinery steel including tempered structural steel; (b) surface hardening structural steels, including carburized steels, ammoniated steels, and surface hardening steels; (c) easily cut structural steel; (d) steel for cold plastic forming, including cold stamping steel, cold heading steel, spring steel, and bearing steel; (3) tool steel: (a) carbon tool steel; (b) alloy tool steel; (c) high-speed tool steel; (4) high-performance steel: (a) stainless and acid-proof steel; (b) heat-resistant steels, including oxidation-resistant steel, hot-strength steel, and air valve steel; (c) electrothermal alloy steel; (d) wear-resistant steel; (e) low-temperature steel; and (f) electrical steel; and (5) industrial steel, such as bridge steel, ship steel, boiler steel, pressure vessel steel, and agricultural machinery steel.

The overall classification of steel is ordinary steel and structural carbon steel: (a) Q195; (b) Q215 (A, B); (c) Q235 (A, B, and C); (d) Q255 (A, B); and (e) Q275; (b) low-alloy structural steels, (c) general structural steels for specific purposes; (2) high-quality steel (including high-grade fine steel), structural steel, (a) high-quality carbon structural steel; (b) alloy structural steel; (c) spring steel; (d) easily cut steel; (e) bearing steel; (f) high-quality structural steel for specific purposes; (b) tool steel: (a) carbon tool steel; (b) alloy tool steel; and (c) high-speed tool steel. (c) Special-performance steel: (a) stainless and acid-proof steel; (b) heat-resistant steel; (c) electrothermal alloy steel; (d) steel for electrical purposes; (e) high-manganese wear-resistant steel.

11.2 Main elements in steel and their effects

11.2.1 Main elements

The required quality and performance of the steel depend on the application, where different uses require different elemental compositions. Higher carbon contents result in harder steels, but the plasticity and toughness are poor. Sulfur is a harmful impurity in steel, as high sulfur contents can make the material brittle during high-temperature and pressure processing (hot brittleness). Phosphorus can significantly decrease the plasticity and toughness of the steel, especially at low temperature; this phenomenon is called cold brittleness. Sulfur and phosphorus are strictly controlled in high-quality steel. In contrast, low-carbon steel contains high sulfur and phosphorus contents, which can make the steel brittle, but improves its machinability. Manganese can increase the strength of steel, reduce the negative effects of sulfur, and increase its hardenability. High-alloy steel (high-manganese steel) with high-manganese content has good wear resistance and other physical properties. Silicon can increase the hardness of steel, but the plasticity and toughness decrease. Electrical steel contains some Si, which improves the soft magnetic properties. Tungsten can improve the hot hardness and heat strength of steel and increase its wear resistance. Chromium can improve the hardenability and wear resistance of steel, while increasing its corrosion and oxidation resistance. Vanadium can refine the grain structure of steel, which increases the strength, toughness, and wear resistance. When it is melted to form austenite at high temperature, the hardenability of the steel can be increased. On the contrary, when vanadium exists as its carbide, the hardenability of the steel will be reduced. Molybdenum can significantly increase the hardenability and thermal strength of steel, prevent temper brittleness, and enhance the remanence and coercivity. Titanium can refine the grain structure of steel, resulting in an increase in its strength and toughness. In stainless steel, titanium can eliminate or reduce intergranular corrosion. Nickel can increase the strength and toughness and improve the hardenability of the steel. High contents of Ni can significantly change some of the physical properties of steel and alloys and increase the corrosion resistance of steel. Trace amounts of boron in steel (0.001%–0.005%) can improve the hardenability of the steel. Aluminum can refine the grain structure of the steel and prevent aging of low-carbon steel. It can also improve the oxidation resistance, wear resistance, and fatigue strength of steel. Copper shows outstanding effects in improving the atmospheric corrosion resistance of ordinary low-alloy steel, especially when used with phosphorus.

11.2.2 Role of alloying elements

The properties of the steel depend on the phases, phase composition, phase structure, volume composition, and distribution of the various phases in the steel, which are all dependent on the specific alloying elements.

11.2.2.1 Alloying elements and alloy steel

Alloy steel is classified into high-quality and special-quality alloy steels. The production of high-quality alloy steel requires special consideration of the quality and performance, which are strictly controlled during the production process. There are many different kinds of alloy steel, which can be divided into steel for building structures, mechanical steels (alloy structural steel, alloy spring steel, and bearing steel), tool steel [(mold) tool steel and high-speed tool steel] and special-performance steel (acid-proof stainless steel, heat-resistant and not peeling steel, and nonmagnet steel). According to the content of the alloying element, alloy steel can be divided into low-alloy steel (alloy content <5%), medium-alloy steel (alloy content 5%–10%), high-alloy steel (alloy content >10%). Considering the main alloying element, alloy steel can be divided into chrome steel, nickel steel, chromium nickel steel, and nickel chrome molybdenum steel. In addition, steels can be classified as ferrite steel, pearlite steel, bainite steel, martensite steel, austenite steel, hypoeutectoid steel, and eutectoid steel. Alloying elements commonly used in alloy steel include Si, Mn, Cr, Ni, Mo, W, V, Ti, Nb, Zr, Co, Al, Cu, B, and Re. For example, P, S, and N play the role of alloying elements in steel under specific conditions. According to the tendency of each element to produce carbides in the steel, they are divided into three categories: the first category includes strong carbide-forming elements, such as V, Ti, Nb, and Zr. As long as there is enough carbon to form their own carbides, only in the absence of carbon or at high temperature, their atomic state is entered into the solid solution. In the case of the second type of carbide-forming elements, such as Mn, Cr, W, and Mo, some fraction of the elements enter the solid solution in their atomic state, while the remainder produces displacement-type alloyed cementites, such as $(Fe,Mn)_3C$, $(Fe,Cr)_3C$. If the alloy content is above a certain limit, except for Mn, the elements will form carbides, such as $(Fe,Mn)_3C$ and $(Fe,Cr)_3C$. The third category is noncarbide-forming elements, such as Si, Al, Cu, Ni, and Co, which generally exist in the solid solution of austenite and ferrite in the atomic state.

Active alloying elements include Al, Mn, Si, Ti, and Zr, which can easily combine with N and O in steel to form stable oxides and nitrides, all of which generally exist in steel in the form of inclusions. Mn and Zr are also mixed with sulfide. Different types of intermetallic compounds can be formed when the steel contains enough elements, such as Ni, Ti, Al, and Mo. Some alloying elements, such as Cu and Pb, exist in a relatively pure metallic phase if the content exceeds the solubility in steel.

11.2.2.2 Influence on the phase transition point

Some elements such as Mn, Ni, C, N, Cu, and Zn can change the phase transition temperature and phase transformation point, expand the gamma phase (austenite) area, reduce the A_3 temperature, and increase the A_4 temperature. In contrast, elements such as Zr, B, Si, P, Ti, V, Mo, W, and Nb reduce the gamma phase (austenite) area, increase the A_3 temperature, and decrease the A_4 temperature. Only Co can increase both the A_3 and A_4

temperature simultaneously. Cr has a unique effect; when the Cr content in steel is less than 7%, the A_3 temperature decreases and the A_4 temperature increases. When the Cr content in steel is above 7%, the A_3 point temperature increases and the position of the eutectoid point S changes. The elements that reduce the gamma phase (austenite) area can increase the temperature of the eutectoid point S, while the opposite occurs for elements that increase the gamma phase (austenite) area. All alloy elements can make C content of eutectoid point reduced, where the eutectoid point S shifts to the left. However, high contents of carbide-forming elements V, Ti, Nb, W, and Mo move the eutectoid point S to the right. The shape, size, and location of the gamma phase can be modified. In general, higher alloying element contents (e.g., Ni and Mn) result in more significant changes, where the gamma phase zone can be extended to room temperature, or the steel can form a single-phase austenitic structure. In addition, gamma phase with high Si or Cr contents can be reduced in size or even disappear, producing ferrite steel at any temperature.

11.2.2.3 Influence on phase transition during heating and cooling

During heat treatment of the steel, phase transformations occur that convert nonaustenite phases to austenite (i.e., the austenitization process), which is associated with the diffusion of carbon. The noncarbide-forming elements, such as Co and Ni, can reduce the activation energy of carbon in austenite, and accelerate austenite formation. Strongly carbide-forming elements, such as V, Ti, and W, can greatly hinder diffusion of carbon in steel, which significantly slows the austenitization process.

During cooling of the steel, supercooled austenite decomposition occurs, including pearlite phase transformation, bainite transition, and martensite phase transformation. Due to the existence of several kinds of alloying elements in the steel, the influence of the phase changes during cooling is very complex. Most alloying elements (except for Co and Ni) slow isothermal austenitic decomposition, although the role of the different alloy elements varies. Noncarbide-forming elements (Si, Cu, P, and Ni), and a small amount of carbide-forming elements (V, Ti, W, and Mo) participate in the transformation from austenite to perlite, and the transformation from austenite to the bainite. However, these changes are minor, resulting in the transition curve moving to the right.

For high contents of carbide-forming elements (V, Ti, Cr, W, and Mo) the transition from austenite to pearlite is significantly delayed, while the transition from austenite to bainite is not significantly affected. When the content of such elements increases to a certain extent, a substable zone of supercooled austenite occurs between the two transition regions.

The effect of alloying elements on the martensite transition temperature M_s (initial transition temperature) and M_f (final transition temperature) is significant. Most elements reduce M_s and M_f, where carbon has the largest effect, followed by Mn, V, and Cr, while Co and Al increase M_s and M_f.

11.2.2.4 Influence on grain size and hardenability

There are many factors influencing the austenitic grain size; the deoxidizing and alloying of steel are related with the austenitic grain size. Some alloying elements do not form carbides, such as Ni, Si, Cu, and Co, and their inhibition of austenite grain growth is weak, while Mn and P tend to promote grain growth. Carbide-forming elements such as W, Mo, and Cr moderately prevent austenitic grain growth, and strongly carbide-forming elements such as V, Ti, Nb, and Zr greatly inhibit austenite growth and promote grain refinement. Although Al belongs to the group of noncarbide-forming elements, it is one of the most common elements and plays a role in grain refinement and determining the temperature at which coarsening begins. The hardenability of steel depends on the chemical composition and grain size. With the exception of Co and Al, most alloying elements are melted and enter the solid solution to varying degrees.

11.2.2.5 Role of alloying elements in structural steel

Generally, structural alloy steels are divided into tempered structural steels and surface hardening structural steels. Almost all alloying elements (e.g., Mn, Cr, Mo, Ni, Si, C, N, and B), except for Co, can increase the hardenability of steel, where Mn, Mo, Cr, and B have the strongest effect, followed by Ni, Si, and Cu. However, carbide-forming elements such as V, Ti, and Nb can only increase the hardenability of steel when they are dissolved in austenite. Alloy elements in the steel can affect the steel tempering process by hindering the diffusion of atoms in the steel compared with carbon steel at the same temperature, which generally delays martensite decomposition and enhances carbide agglomeration and growth. This improves the tempering stability of steel and the tempering softening resistance. The effects of V, W, Ti, Cr, Mo, and Si are most significant, while the effects of Al, Mn, and Ni are not as obvious. Steel containing high contents of carbide-forming elements, such as V, W, and Mo, is tempered at 500°C–600°C, resulting in small carbide precipitate particles being produced, such as V_4C_3, Mo_2C, and W_2C. These carbides replace part of the relatively bulky alloying cementite, which increases the strength of the steel, rather than decreasing it. Secondary hardening is achieved, where Mo can prevent and slow down the temper brittleness of steel. The strengthening and toughening of steel can also be influenced, where Ni strengthens the ferrite by solid-solution strengthening, while Mo, V, Nb, and other carbide-forming elements increase the yield strength of steel via both dispersion hardening and solid-solution strengthening. The strengthening effect of carbon is the most significant, where addition of alloying elements can refine the austenite grains and increase the strengthening effect of grain boundaries. Ni increases the toughness of steel, while Mn facilitates austenite grain coarsening. In steels sensitive to temper brittleness, reducing S and P contents can improve the purity and toughness of the steel.

11.2.2.6 Influence of nitrogen on steel

The atomic radius of nitrogen (0.075 nm) is smaller than that of iron (0.172 nm). Hence, nitrogen easily enters the steel lattice to produce interstitial solid solutions, resulting in lattice distortion and asymmetric strains in the body-centered cubic α-Fe due to interstitial solute atoms. This has a strong solid-solution strengthening effect in low-carbon ferritic steel and austenitic stainless steel, which is 10–100 times greater than the substitutional solid-solution atoms. Vanadium and nitrogen in steel have a compound strengthening effect, where vanadium can combine with free nitrogen in the steel. When the nitrogen and carbon in vanadium steel are saturated, the carbonitrides precipitate along dislocation lines and hinder their movement. The enhancement of interstitial solid solution atoms is greater than that of substitutional solid-solution atoms.

Vanadium is the most suitable element for producing stable and strong precipitation, because vanadium has high solubility in the carbonitride in steel, with low solid-solution temperature and high solubility at high temperature. Vanadium nitride has superior solubility, where the nitrogen in steel can generate a large number of dispersed fine carbon nitride particles with vanadium. The presence of nitrogen leads to precipitation strengthening. Precipitation strengthening and grain refinement strengthening can significantly improve the strength of steel and improve or maintain good plasticity and toughness of the steel.

Nitrogen is one of the permanent elements in nontempered steel. During smelting of nonquenched and tempered steel, it is necessary to ensure that there is an appropriate amount of stable N in the steel. In general, N is harmful to steel, for example, considering aging, and is usually associated with various brittle failures. However, N also has some beneficial effects on steel, especially for nonquenched and tempered steel. The behavior and function of microalloying elements in nonquenched and tempered steels are largely realized by carbon and nitrogen compounds formed by microalloying elements and nitrogen.

Nitrogen strongly enhances austenite formation and stabilization. It has strong affinity with Ti, Nb, V, and other elements and can form extremely stable interstitial phases. Nitrides and carbides can dissolve each other to form carbonitrides. Nitrides can also dissolve each other to form compound nitrides. These compounds are usually present as fine particles, producing dispersion strengthening effects that enhance the strength of the steel. Nitrogen can combine with Al in steel to form AlN. Nitrides such as AlN, TiN, and NbN can effectively prevent austenite grain coarsening to obtain fine ferrite grains, which increase the toughness of steel.

The role of N in nonquenched and tempered steel is mainly in enhancing precipitation and refining grains. In particular, for nonquenched and tempered microalloyed steels containing V, the yield strength increases by 5 MPa with the addition of 0.001% N. By increasing the

N content in steel the precipitation range of carbonitride is extended, and the effect of the microalloying elements is enhanced. VN in steel is a strengthening phase, but can also inhibit the migration of austenite grain boundaries and refine the austenite, ferrite, and pearlite grains. During the phase transition, it also plays a major role in further refining ferrite grains. Therefore the presence of both V and N results in obvious precipitation strengthening and toughening effects.

11.3 Role of vanadium in steel

Among steel materials, ferrovanadium and vanadium–nitrogen alloys are the most important consumer vanadium products, accounting for 85%–90% of the total consumption of vanadium. The uses of vanadium in steel are shown in Table 11.1. It is common to use ferrovanadium or vanadium–nitrogen alloy as additives in steel products. As the product quality of ferrovanadium and vanadium–nitrogen alloy is different, the utilization efficiency of vanadium varies, and some tendencies must be taken into account. When used as an alloy and microalloy additive in steel, vanadium shows many favorable functions. Adding vanadium to steel can improve its properties remarkably. Vanadium is used in steel because it can form stable compounds with carbon in the steel, for example, V_4C_3. This can refine the structure and grain size of the steel; increase the grain coarsening temperature; significantly improve the performance of the steel; and improve its strength, toughness, corrosion resistance, wear resistance, and impact load capacity. In addition, studies have shown that, when vanadium is added to steel, the solid-solution strengthening effect is not significant, while the following properties are significantly improved: (1) the V(CN) solubility in the austenitic product is high, where vanadium in low-, medium-, and high-carbon steel has a strong precipitation strengthening effect; (2) vanadium addition prevents

Table 11.1: Effects of vanadium in steel.

Steel type	Amount of added vanadium (%)	Main effect	Added form	Application
High-strength steel	0.02–0.06	Grain refinement, increase strength	Ferrovanadium	Oil pipes, pressure vessels, ships, bridges
High-speed tool steel	1.00–5.00	Wear resistance, high strength	Ferrovanadium	Cutting tools
Alloy tool steel	0.10–0.20	Replacing molybdenum	Ferrovanadium	Cutting tools (including bearing steel)
Heat-resistant steel	0.15–0.25	Increase high-temperature strength	Ferrovanadium	Turbine blades
Heat-resistant alloy steel	1.00–5.10	Increase high-temperature strength	vanadium	Turbine nozzles and blades

austenite grain growth during heating, inhibiting the recrystallization of deformed austenite and promoting the $\gamma \rightarrow \alpha$ phase change for refining grains and try other ways, which results in fine-grained steel; (3) vanadium has an obvious influence on the transformation of supercooled austenite. Unlike most alloying elements, vanadium does not delay ferrite transformation but does delay bainite and pearlite transformation. In addition, vanadium can improve the hardenability of steel and has twice the contribution of the same content of molybdenum.

When used as an alloy and microalloy additive in steel, vanadium shows many favorable functions. Vanadium has many special functions in different types of steels, which have various applications. During the heat treatment of steel, vanadium addition can increase its ability to temper and increase the hardness of high-speed steel. In addition, the creep resistance can be improved in hot-strength steel, while it can increase corrosion resistance and limit strain aging in corrosion-resistant steel.

11.3.1 Behavior of vanadium in steel

Vanadium compounds with carbon and nitrogen have a face-centered cubic crystal structure, where the molar ratio of C:N:V is 1:1:1 in an ideal state. As a result of lattice defects, vacancies occur on carbon and nitrogen sites in the vanadium compounds, where the chemical ratio of C and N atoms in vanadium carbides and nitrides is often 0.75–1. In steel production the proportion of C in vanadium carbide is close to the lower limit of 0.75, and the vanadium carbide is V_4C_3. Meanwhile, proportion of N is close to the upper limit of 1, where the vanadium nitride is VN. Vanadium carbide and vanadium nitride have the same crystal structure, and their lattice parameters are similar. Hence, it is difficult to accurately distinguish vanadium carbides from vanadium nitrides in the precipitated V-containing phase in steel. Generally, they are expressed as vanadium carbonitride (VC_xN_y) and V(C,N), where $(x + y)$ is 0.75–1.

The enhancement of microalloying elements may be due to the dispersion of fine carbonitride and inhibition of its grain growth. Considering grain refinement, to maintain fine austenite grains before the phase transition, carbonitride should be partially insoluble in the austenite or be precipitated during hot rolling. Considering precipitation strengthening, the microalloying elements should be dissolved in the solid solution of austenite and precipitated during the transformation of austenite/ferrite or after phase transformation to obtain fine precipitates (3–5 nm) and achieve dispersion strengthening.

11.3.1.1 Austenite precipitation

Austenitic precipitation mainly occurs as inclusions within the grain interior and at austenitic grain boundaries. For standard compositions of V-containing steel, at the temperatures above 1000°C used for finishing, almost all of the vanadium is precipitated in the ferrite phase, with minimal precipitation in austenite. In the case of high vanadium and nitrogen contents, some small amount of vanadium may precipitate in the austenite. During controlled rolling, some solid-solution vanadium can be precipitated in the form of V(C,N) in austenite by deformation induction. For V–Ti microalloyed steel, composite-phase particles are formed in austenitic during continuous casting or reheating. The precipitation of V(C,N) particles in austenite is accelerated by nitrogen addition to V-containing microalloyed steel.

MnS inclusions occur in steel, which are favorable sites for V(C,N) particles to precipitate in austenite. V(C,N) relies on MnS inclusions as nucleation sites, and grows into square VN precipitate phase grains. For vanadium microalloyed steel at temperatures below 850°C, VC is completely solid soluble in austenite. In the case of high vanadium and nitrogen contents and a deformation temperature below 1000°C, small amounts of VN precipitates can be formed in the steel.

The strain-induced precipitation of V(C,N) in austenite grains depends on the content of vanadium and nitrogen in the steel, the deformation temperature, and the deformation strain. Increasing the deformation can effectively promote the precipitation of V(C,N) particles in austenite. Relatively large V(C,N) particles precipitate from the austenite, which cannot play a role in precipitation strengthening. On the contrary, the precipitation of vanadium in the austenite reduces the content of solid-solution vanadium in the matrix, which reduces the amount of V(C,N) precipitated from the ferrite and weakens the precipitation strengthening effect. However, V(C,N) particles precipitated from austenite provide effective sites for ferrite nucleation and enhance ferrite nucleation within the grain, resulting in ferrite grain refinement.

11.3.1.2 Ferrite precipitation

V(C,N) is precipitated in ferrite during or after the phase transition. The precipitated phases include fibrous, interphase, and random precipitations. Fibrous precipitations move with the γ/α interface or are precipitated randomly in the ferrite to form dispersion distribution, which strengthens the ferrite. For special typical structural steel, precipitation in the interphase region occurs at high temperature, where random precipitation is observed at low temperatures, typically below 700°C.

Fibrous precipitation parallel to the γ/α interface produces lamellar phase precipitation at a certain distance from the interface layer, where the fiber bundles are perpendicular to the γ/α interface. V(C,N) precipitates with a fibrous morphology are formed via a variation of eutectoid transformation, which follows the $\gamma \rightarrow \alpha + V(C,N)$ transition. This is because of changes in the vanadium concentration gradient at the $\alpha/V(C,N)$ interface. The balance of $\gamma/V(C,N)$ and γ/α determines the vanadium concentration gradient parallel to the interface, which leads to a transverse vanadium distribution from γ to $\alpha + V(C,N)$. Fibrous precipitation is not the main precipitation type in vanadium microalloyed steel. The addition of Mn and Cr can significantly increase fibrous precipitation of VC.

In vanadium microalloyed steel, interphase precipitation of carbon nitrides in ferrite is the main precipitation type, where the precipitation phase forms parallel to the γ/α interface and grows as a single habitual surface. In all kinds of V-containing steel with different carbon contents, V(C,N) can first be precipitated in the interphase of eutectoid ferrite, pearlite, and ferrite. Carbide precipitation is dominant, where heterogeneous nucleation of VC or V(C,N) is associated with the structural characteristics of the phase interface. The phase transition temperature, cooling rate, and steel composition have remarkable effects on the precipitation of V(C,N), particularly the morphology, spacing, and size of the grains. One of the characteristics of interphase precipitation is that smaller precipitated phases occur at lower temperatures. When the V and C contents in steel are both high, larger volume fractions of carbides are precipitated, but with smaller average precipitate particle size, and smaller layer spacing between phases. The particle size and layer spacing of carbide precipitated between phases increase with increasing phase transition temperature. In steel with low vanadium and carbon contents the influence of temperature is more obvious. Mn, Ni, and Cr in steel can delay interphase precipitation and reduce the interphase precipitation temperature. With increasing nitrogen content in the steel the V(C,N) precipitated phase has a smaller average particle size and smaller layer spacing. With decreasing phase transition temperature the layer spacing between phases decreases. V(C,N) in V–N microalloyed steel is precipitated in the same sample and grain. The precipitation patterns are diverse.

Carbonitride in V-containing steel can precipitate in the eutectoid ferrite, or in the pearlite and ferrite. The random precipitation of V(C,N) particles in ferrite mainly results in a lamelliform morphology. The ferrite matrix in line with the B–N relations, carbon nitrides in the V-containing steel mainly nucleate along lines of dislocations, and can also produce homogeneous precipitation in ferrite grains. A study of V-containing steel treated at 600°C and 700°C showed isothermal transformation; with increasing nitrogen content in the steel, the carbon fraction x in the VC_xN_y precipitated phase decreases rapidly. For a N content in the steel above 0.010%, x in the VC_xN_y phase is less than 10%, resulting in a nitrogen-rich V(C,N) precipitated phase. When the nitrogen content in steel is less than 0.005%, the carbon fraction in V(C,N) increases significantly. The nitrogen content in the steel has a significant influence on the precipitation of V(C,N) in ferrite. As the nitrogen content increases from 0.005% to 0.05%, the number of the precipitated particles increases, while the particle size decreases greatly.

The carbon content in steel plays an important role in precipitation strengthening. An increase in the content of solid-solution carbon in the ferrite significantly increases the chemical driving force for precipitation of V(C,N) and promotes nucleation of V(C,N). As the carbon content in steel increases, the γ/α phase transformation kinetics are weakened, while carbon diffusion in the austenite is delayed, which increases the activity of carbon in the ferrite. Subsequently, the supersaturation time of carbon in the ferrite phase is delayed, resulting in a large number of V(C,N) grains being nucleated in the ferrite, and a dense precipitate phase is obtained. The addition of an extra 0.01% C in V-containing steel can enhance precipitation, giving an increase in strength of 5.5 MPa.

11.3.1.3 Bainite precipitation

Bainite in steel is one of the most complex phases. Compared with the polygonal ferrite phase change, the phase transformation temperature of bainite is lower, while the transformation rate is higher. Bainite and ferrite are metastable, and there may be a large amount of supersaturated carbon, which greatly increases the chemical driving force of carbonitride precipitation in bainite and ferrite. In V−N microalloyed low-carbon bainite steel, fine V(C,N) particles are precipitated in bainite and ferrite. The morphology of the precipitated particles is related to the morphology of the bainite; the precipitated phase in granular bainite is dispersed and randomly distributed, while in lamellar bainite it is distributed in layers, similar to interphase precipitation. V(C,N) in the bainite and ferrite is precipitated as fine, uniform, and stable particles that are formed during the lath formation process of bainite and ferrite, and simultaneously during phase transition. The thin lamelliform precipitated V(C,N) is dominant in bainite and ferrite, which maintains a coherent or semicoherent relationship with bainite and ferrite. In V-containing low-carbon (0.1%) steel, the precipitation of V(C,N) in bainite and ferrite shows three types of precipitated phases: along the dislocation lines and precipitated phase; similar interphase precipitation with a laminar distribution; and spherical precipitated phases. Most precipitates occur as a precipitated phase along dislocation lines.

11.3.1.4 Precipitation during tempering

When the vanadium content in steel reaches a critical value, significant secondary hardening effects in the range of 550°C−650°C are observed during tempering, where large quantities of vanadium carbides are precipitated. The vanadium carbide precipitated during tempering is V_4C_3, which is mainly precipitated by nonhomogeneous nucleation along dislocation lines.

11.3.2 Effect on microstructure and heat treatment of steel

Vanadium and iron can produce a continuous solid solution, which strongly reduces the austenitic phase area, such as in 304L stainless steel for seamless pipes. Vanadium has a strong affinity with carbon, nitrogen, and oxygen and mainly exists in carbides, nitrides, and/or oxides in steel. Controlling the austenitizing temperature can change the content of vanadium in the austenite, the amount of undissolved carbides, and the grain size of steel, which allows the hardenability of the steel to be adjusted. Due to the formation of stable dissolved vanadium carbides, the microstructure of the steel remains fine at high temperatures, greatly reducing the sensitivity of the steel to overheating.

11.3.3 Effect on physical, chemical, and technological properties of steel

Vanadium can be added to high-iron ferronickel alloy to improve the permeability after appropriate heat treatment. Vanadium addition to permanent magnetic steel can improve magnetic coercivity. Vanadium can be added up to 5.7 times the carbon content. When carbon is fixed in vanadium carbides, the stability of steel in hydrogen-containing environments at high temperature and pressure can be greatly increased; this strong effect is similar to that of Nb, Ti, and Zr. In stainless and acid-proof steel, V can improve the anti-intergranular corrosion performance, but not as significantly as Ti and Nb. The formation of vanadium oxide reduces the oxidation resistance of steel at high temperature. V-containing steel significantly increases the deformation resistance at low processing temperatures. In addition, vanadium improves the welding performance of steel.

11.3.4 Effect on mechanical properties of steel

Addition of a small amount of vanadium refines the grain size and increases the toughness of steel, which is particularly beneficial for low-temperature steel. When high vanadium contents result in the formation of aggregated carbides, the strength is reduced. The precipitation of carbides reduces the toughness at room temperature. When the carbide is dispersed and precipitated by appropriate heat treatment, vanadium increases the high-temperature endurance strength and creep resistance of steel. Vanadium carbide is the hardest and most wear-resistant metal carbide. Dispersion of vanadium carbide increases the hardness and wear resistance of tool steels.

11.4 Applications of vanadium in steel

Vanadium is widely used in low-alloy steel, alloy structural steel, spring steel, bearing steel, alloy tool steel, high-speed tool steel, heat-resistant steel, hydrogen-resistant steel, low-temperature steel, among others.

11.4.1 Processing performance characteristics of V-containing steel

The processing performance characteristics of V-containing steel are discussed here. First, the reheating temperature is low. Vanadium carbonitride in the austenite has high solubility at relatively low temperature. Hence, heat treatment ensures nearly complete dissolution of the alloying element during cooling precipitation, resulting in precipitation strengthening, which controls overgrowth of the original austenitic grains, resulting in refined grains and enhanced toughness. Second, the resistance to thermal deformation is small. The recrystallization resistance is the weakest during high-temperature rolling. As the recrystallization termination temperature is relatively low, the resistance to hot rolling is small. The final rolling temperature has little effect on the properties. Vanadium is a precipitated strengthening element in steel. During reciprocating rolling of V-containing steel in the austenitic recrystallization zone, fine austenite grains can be obtained via repeated rolling and recrystallization deformation. Finally, such alloys can adapt to the high nitrogen content in the process of electric steelmaking, where effective nitrogen fixation is achieved by the alloying elements, controlling the free nitrogen content.

11.4.2 Requirements for continuous casting of V-containing steel

V-containing steel is mainly processed using continuous casting processes. The liquid steel is poured into a mold in contact with a cooled copper wall. Fast cooling occurs from the center of the slab to the edge to form layers of small equiaxed isometric crystals, columnar isometric crystals, and central equiaxed isometric crystals. The small isometric crystal layer is located at the slab surface, while the columnar isometric layer is located between the small isometric layer and the equiaxed isometric layer at the center of the casting billet. The width and temperature of the high-temperature plastic trough used in the continuous casting of the V-containing steel control the cooling speed of the secondary cooling area. Constant weak cooling occurs, resulting in the slab surface temperature exceeding that of the carbide precipitation temperature, alpha and beta phase transition temperatures, or the upper critical temperature of embrittlement, avoid the plastic trough area. Selection of an appropriate mold, high frequency, small amplitude, and the use of high-quality slag with limited S and P contents can limit carbide precipitation at grain boundaries. Meanwhile, the austenite grains at the slab surface are refined by secondary cooling. Controlling the mold level reduces crack sensitivity and prevents transverse slab cracks.

11.4.3 Thermomechanical control

The thermomechanical control process for V-containing steel includes recrystallization controlled rolling and controlled rolling. During heating the steel is deformed and recrystallized in the austenite recrystallization zone, where the austenite grains become tapered, and finally, fine equiaxed austenite grains are obtained. This increases the austenitic grain boundary area, which provides sites for phase change and nucleation of austenite to ferrite. In the subsequent phase-change processes, fine ferrite grains are obtained by accelerated cooling, where it is necessary to control the higher austenite grain coarsening temperature, recrystallization end temperature, and lower grain coarsening rate, large content of V(C,N) in austenite and ferrite and ensure sufficient supercooling capacity. In recrystallization controlled rolling the cooling rate is limited to no more than 10–12°C/s. The accelerated cooling end temperature is above 500°C, which avoids bainite and martensite transformation.

11.4.4 V-bearing quenched and tempered steel

The chemical composition of quenched and tempered steel is characterized by a carbon content of 0.3%–0.5%, and one or several alloying elements with low or medium degree of alloying. The role of the alloying elements in steel is mainly to improve its hardenability and ensure that the expected comprehensive properties of the parts are achieved after high-temperature tempering. The heat-treatment process is critical, as above a certain temperature after heating quenching into martensite, and in 500°C–650°C tempering. The metallographic structure after heat treatment is tempered sorbite. The structure has a good combination of strength, plasticity, and toughness.

Commonly used such as quenched and tempered steel, are categorized into four groups according to their hardenability and strength: (1) low hardenability; (2) medium hardenability; (3) higher hardenability; and (4) high hardenability quenched and tempered steel. In addition to the general metallurgical aspects of low and high tissue requirements the quality of tempered steel is mainly determined by the mechanical properties, working reliability, and lifetime, which are closely related to the cold brittle transition temperature, fracture toughness, and fatigue resistance. For specific applications the steel needs wear resistance, corrosion resistance, and some heat resistance. Because tempered steel is subjected to high-temperature tempering as the final process, the stress in the steel can be completely eliminated. Hence, it has low rates of hydrogen embrittlement failure, low notch sensitivity, and high brittle failure resistance. However, it has a unique high-temperature tempering brittleness. Most of the quenched and tempered steels have a medium-carbon alloy structure with a yield strength ($\sigma_{0.2}$) of 490–1200 MPa. In the case of tempered steels that require a good welding performance, low-carbon alloy steels are usually used with $\sigma_{0.2} = 490-800$ MPa, and high plasticity and toughness. The $\sigma_{0.2}$ of a small number of precipitation-hardened tempered steels can exceed 1400 MPa, which are categorized as high-strength and super-high-strength tempered steels.

11.4.4.1 Performance requirements

Quenched and tempered steels need to have (1) good comprehensive mechanical properties (i.e., high strength, good plasticity and toughness, and high fatigue strength) and (2) good process performance, specifically hardenability, to ensure uniform mechanical properties over the cross section of the part.

11.4.4.2 Composition

First, an intermediate carbon content is required, where the carbon mass fraction $w_c = 0.25\% - 0.5\%$ (commonly 0.4%). If w_c is too low, the strength and hardness after quenching and tempering cannot meet the performance requirements. If w_c is too high, the plasticity and toughness of the steel are too low. For some quenched and tempered steels where the strength is the most important requirement, a higher w_c is used. In the case of quenched and tempered steels where plasticity is the main requirement, a lower w_c is used. Second, the alloying elements need to be selected and controlled, where their main function is to increase the hardenability. Contents of $w_{Mn} < 2\%$, $w_{Si} < 2\%$, $w_{Cr} < 2\%$, $w_{Ni} < 4.5\%$, and $w_B < 0.004\%$ are common. The critical diameter for water quenching grade 40 steel is 10–15 mm, while that for oil quenching 40CrNiMo steel is >70 mm. The secondary effect is that in addition to B, it can dissolve into the solid solution (ferrite) and strengthen it. The addition of Ni can also improve the toughness of steel. Auxiliary elements include Mo, W, V, Ti, and other strong carbide-forming elements. The main role of W and Mo is to inhibit the Cr, Ni, Mn, and Si content in quenched and tempered steel to control the high-temperature temper brittleness, while their secondary role is to further improve hardenability. The main role of V and Ti is to form carbides to hinder the growth of austenite grains and enhance the dispersion of fine grains to improve toughening. Almost all alloying elements improve the tempering stability of tempered steel.

11.4.4.3 Heat-treatment characteristics

The main purpose of the preliminary heat treatment of quenched and tempered steels is to ensure good cutting performance of the parts. The preliminary heat-treatment conditions are selected according to the carbon mass fraction and type and quantity of alloying elements. For quenched and tempered steels with low mass fraction of alloying elements, normalizing is performed for low-carbon/alloy steels (e.g., grade 40 steel), while annealing is generally used for high-carbon/alloy steels (e.g., 42CrMo) during preliminary heat treatment to refine the forging structure and improve cutting performance. Quenched and tempered steel with high contents of alloying elements (e.g., 40CrNiMo) produce martensite after air cooling. Their high hardness is not conducive to cutting; hence, high-temperature tempering is performed after air cooling to 650°C–700°C to obtained tempered sorbite, resulting in the hardness dropping to around 200 HBW.

The final heat-treatment process involves quenching and high-temperature tempering. The specific process specifications vary according to the composition and application requirements. The quenching medium and method are determined according to the hardenability of steel and the shape and size of the parts. The tempering temperature depends on the hardness requirements of the quenched and tempered parts. Since the hardness of the parts can indirectly reflect the strength and toughness, only the hardness value is generally stipulated in technical documents. Only very important parts are required to specify other mechanical properties. The working conditions, manufacturing process requirements, mass production characteristics, and shape and size of parts should all be considered when determining the quenching and tempering hardness. When the quenched and tempered parts require high wear resistance and good fatigue performance, the quenching and tempering can be performed after a nitriding treatment, surface hardening, or surface deformation strengthening (such as crankshaft journal rolling strengthening).

Tempered sorbite has better ductility and toughness than lamellar pearlite with the same hardness. If free ferrite appears in the microstructure, the strength and fatigue life is greatly reduced. For example, the eccentric shaft of a diesel engine air pump used in a factory was made from grade 45 steel. Due to the lack of hardenability and high level of free ferrite in the structure, fractures often occurred. Hence, 40Cr steel was used to avoid fractures.

11.4.5 Common quenched and tempered steel

According to their different degrees of hardenability, quenched and tempered alloy steels are divided into three categories: (1) low hardenability quenched and tempered steels with low mass fractions of alloying elements, such as steel grades 40Cr, 40MnB, and 35SiMn. They have an oil quenching critical hardening diameter of 20–40 mm with the following performance indicators: R_m = 800–1000 MPa, R_{eL} = 600–800 MPa, and α_k = 60–90 J/cm^2. Such steels are mainly used for manufacturing parts for medium-load and medium-speed applications, such as automobile steering knuckles, and the gears, shafts, and worms of machine tools. (2) Medium hardenability quenched and tempered steels have higher mass fractions of alloying elements than low hardenability steels. The commonly used steel grades are 40CrNi, 40CrMn, 40CrMnTi, and 35CrMo. The critical diameter of oil quenching is 40–60 mm, while the performance indicators are as follows: R_m = 900–1000 MPa, R_{eL} = 700–900 MPa, and α_k = 50–80 J/cm^2 after quenching. Such steels are mainly used for manufacturing parts with large sections that need to bear heavy loads, such as large motor shafts, automobile engine spindles, and large-section gears. (3) High hardenability quenched and tempered steels have the highest content of alloying elements and include steel grades 40CrMnMo, 30CrNi3, and 45CrNiMoV. The critical diameter of oil quenching is 60–100 mm. After quenching and tempering, their performance indicators are R_m = 1000–1200 MPa, R_{eL} = 800–1000 MPa, and α_k = 60–120 J/cm^2. These steels are mainly used for making high-strength large-section parts that are subjected to impact loads, such as eccentric drive shafts and crankshafts of horizontal forging presses, and high-strength connection bolts.

11.4.6 Substitutes for quenched and tempered steel

Low-carbon martensite steels and low-carbon (alloy) steels (such as carburized steels and low-alloy high-strength steels) can be treated using appropriate intermediate quenching and low-temperature tempering to obtain low-carbon martensite. This material has better comprehensive mechanical properties than common medium-carbon alloy steels after quenching and tempering and has fully developed strengthening and toughening effects.

This results in steels with high strength, along with good plasticity and toughness. For example, using 15MnVB steel instead of 40Cr steel to manufacture automobile connecting rod bolts increases their strength, plasticity, and toughness. This increases the bearing capacity of the bolts by 45%–70%, extends their service life, and meets the design requirements of new high-power models of cars. In another example, 20SiMnMoV steel has been used to replace 35CrMo steel to make lifting rings for oil drilling, which reduced the weight of the rings from 97 to 29 kg, greatly reducing the labor of the drilling workers. The chemical composition of medium-carbon microalloy nonquenched and tempered steel is characterized by the addition of trace amounts ($w_{Me} < 0.2\%$) V, Ti, Nb, and other elements based on the medium-carbon steel composition. Hence, it is called microalloy nonquenched and tempered steel. The outstanding advantage of this steel is the control of rolling or forging processes without quenching and tempering treatment. For example, air-cooled parts can achieve appropriate comprehensive mechanical properties with a ferrite + pearlite microstructure.

11.4.7 V-containing quenched and tempered steel

V-containing steel includes a wide range of quenched and tempered steels. The effects of vanadium in quenched and tempered steel include increased hardenability and tempering stability, precipitation strengthening and secondary hardening, and a lower critical cooling rate. Most V-bearing quenched and tempered steels are medium-carbon (0.2%–0.5%) alloy structural steels with tensile strength of 490–1200 MPa and good plasticity and toughness. To improve the comprehensive properties of the steel (strength, plasticity, and toughness), quenching, high-temperature tempering, and tempering heat treatment are used to refine the microstructure of the steel and evenly distribute the vanadium and molybdenum carbides in the steel. This ensures that the steel has a high strength, while maximizing the toughness as much as possible. The chemical composition of medium-carbon vanadium tempered steel is given in Table 11.2.

Cr−Ni−Mo−V is a typical V-containing carbon steel that is often quenched and tempered at low temperature. It belongs to the classes of high-strength and ultra-high-strength steels, where its specific composition is shown in Table 11.3.

Compared to medium-carbon steel, the carbon content in low-carbon steel is greatly reduced to ensure high strength and toughness, while the welding performance is significantly improved. Cr−Ni−Mo−V is a typical low-carbon steel, where the compositions of common grades are shown in Table 11.4.

11.4.8 Nonquenched and tempered steel

Microalloy nonquenched and tempered steel experiences three stages of development: (1) ferrite−pearlite structure; (2) low-carbon bainite structure; and (3) low-carbon martensite

Table 11.2: Chemical composition of medium-carbon vanadium tempered steel (%).

Steel	C	Si	Mn	P (≤)	S (≤)	Cr	Ni	Mo	V
27MnCrV	0.24–0.30	0.15–0.35	1.00–1.30	0.035	0.035	0.60–0.90	–	–	0.07–0.12
42CrV	0.38–0.46	0.15–0.35	0.50–0.80	0.035	0.035	1.4–1.7	–	–	0.07–0.12
50CrV	0.47–0.55	0.15–0.35	0.80–1.10	0.035	0.035	0.90–1.2	–	–	0.07–0.12
30CrNiMoV	0.26–0.34	0.15c–0.35	0.40–0.70	0.035	0.035	2.3–2.7	–	0.15–0.25	0.10–0.20
34CrNiMoVA	0.32–0.40	0.15–0.35	0.50–0.80	0.035	0.035	1.3–1.7	1.3–1.7	0.40–0.50	0.10–0.20
30CrNi3MoVA	0.27–0.32	0.15–0.35	0.20–0.70	0.035	0.035	1.2–1.7	3.00–3.50	0.40–0.65	0.10–0.20

Table 11.3: Composition of high-strength and ultra-high-strength medium-carbon steel after quenching and low-temperature tempering (%).

Type of steel	C	Si	Mn	Ni	Cr	Mo	V
30M	0.40–0.46	1.45–1.80	0.65–0.90	1.65–2.00	0.70–0.95	0.30–0.50	0.05–0.10
4340V	0.37–0.44	0.20–0.35	0.60–0.95	1.55–2.00	0.60–0.95	0.40–0.60	0.01–0.10
D6AC	0.42–0.48	0.15–0.30	0.60–0.90	0.40–0.70	0.90–1.20	0.90–1.10	0.07–0.15
4330V	0.28–0.33	0.15–0.35	0.65–1.00	1.65–2.00	0.75–1.00	0.35–0.50	0.05–0.10

Table 11.4: Composition of V-containing low-carbon tempered steel with high strength, high toughness, and good weldability (%).

Steel	C	Mn	Si	Ni	Cr	Mo	V	Cu
HY-130	≤0.12	0.60–0.90	0.15–0.35	4.75–5.25	0.40–0.70	0.30–0.65	0.05–0.10	
NS80	≤0.12	0.35–0.90	0.15–0.40	3.50–4.50	0.30–1.00	0.20–0.60	≤0.10	
NS90	≤0.12	0.35–1.00	0.35–1.00	4.75–5.50	0.40–0.80	0.30–0.65	≤0.10	
NS110	≤0.08	0.10–0.75	0.10–0.75	9.0–10.20	0.35–1.00	0.70–1.50	≤0.20	
AK-44	0.08–0.10	0.30–0.60	0.30–0.60	4.3–4.7	0.60–0.90	0.55–0.65	≤0.10	1.20–1.45

structure. Compared with quenched and tempered steel, traditional nonquenched and tempered steel for hot forging has higher strength but lower toughness, which limits its use for strong impact applications. Nonquenched and tempered steel is prepared by adding V and Ti as trace elements. The forging (rolling) process is controlled, where the alloying elements are dispersed and precipitated with C and N to achieve a strength equivalent to quenched and tempered steel. This eliminates the need for quenched and tempering treatment and simplifies the production process. Nonquenched and tempered steel is mainly1 used in automobile engines, tractor engines, air compressor connecting rods, machine tool parts, and shaft parts. Microalloying is the core technology for producing nonquenched and tempered steel. Microalloying elements include V, Nb, Ti, and B, where V is the major element. Among the various nonquenched and tempered steels produced globally, 186 brands with different chemical compositions exist, among which 158 (i.e., 85%) are nonquenched and tempered steels containing vanadium. There are nine Nb-containing steels (i.e., 5%), while there are fourteen V–Nb composite alloys (i.e., 7.5%).

Nonquenched and tempered steel is strengthened by dispersed carbide or nitride particles in the ferrite, which results in diffusion reinforcement when the microalloyed steel is cooled after thermal deformation processing. In addition, the combination of high strength and toughness is ensured by controlling the proportion of pearlite and ferrite, and the pearlite interlamellar spacing, while refining the grain size. Currently, the main disadvantages of such steels are the plasticity and low impact toughness, which limit its application under strong impact conditions. In recent years the automobile industry has developed bainite and martensite type nonquenched and tempered steel to meet the increasing demand for high-toughness quenched and tempered steel. Two different cooling regimes after forging the steel can obtain a bainite + martensite or predominantly martensite structure. The composition is characterized by carbon reduction and proper addition of Mn, Cr, Mo, V, and B to make the steel to be higher than 900 MPa, which has good tensile strength in addition to maintaining sufficient plasticity and toughness.

11.4.8.1 Ferrite−pearlitic microalloyed nonquenched and tempered steel

Ferrite−pearlitic microalloyed nonquenched and tempered steel has the largest market, accounting for more than 60% of the total products. To use carbide precipitation strengthening to achieve the required strength, the carbon content is increased to increase the fraction of pearlite in the microstructure. However, it is difficult to meet the requirements for toughness. Therefore a series of new technologies have been demonstrated for increasing the toughness of ferrite−pearlite microalloyed nonquenched and tempered steel, especially grain refinement techniques. Refining the grains can effectively improve the toughness of steel while maintaining high strength. Al, Ti, and other elements are often added to nontempered steel, where fine aluminum nitride and titanium nitride are precipitated, which pins the austenite grain boundaries and prevents grain growth, inhibits austenite recrystallization during deformation, and refines austenite grains. Nonquenched and tempered steel with a composition of 0.32C−1.0Mn−0.12V−0.024Ti was heated to 1250°C, where the austenitic grains remained above magnitude 5, with a uniform particle size distribution of 0.1 μm. The titanium nitride particles pinned the austenitic grain boundaries and prevented grain coarsening.

During cooling of nonquenched and tempered steel forgings, ferrite tends to nucleate first along austenite grain boundaries, while the remainder of the austenite transforms into pearlite. If a network of ferrite is formed along pearlite grains, the toughness of the steel significantly deteriorates. Researchers from the Japan Iron and Steel Company showed that proper control of the production process to provide many ferrite nucleation sites in the austenitic grain ensured that ferrite nucleates both at the grain boundaries and within the austenitic grains during the phase transition. This results in evenly distributed fine ferrite particles which significantly increase the toughness of the steel.

The precipitation of IGF is related to MnS, along with VN or TiN particles precipitated on MnS, while the precipitation and distribution of MnS are related to the microoxide core in steel. Hence, the characteristics, type, quantity, and size of oxides in the steel determine the quantity and size of MnS particles. The type, quantity, and distribution of oxides formed by different deoxidizing elements are different. The sulfur content in steel should be around 0.06%, which is conducive to the precipitation of IGF. In this case the purpose of sulfur is not to improve cutting performance, but to form composite inclusions with oxides to promote IGF formation.

11.4.8.2 Bainite microalloyed nonquenched and tempered steel

It is advantageous to obtain low-carbon bainite microstructures as they provide high strength and good toughness for nonquenched and tempered steel. In addition, a certain amount of carbon must be present to ensure high strength. To obtain a bainite structure by air cooling, it is necessary to add Mo, Mn, B, and other alloying elements to the steel. Mo has the effect of significantly lowering the transition point to a medium temperature rather than a high temperature. When the Mn content reaches a critical level, the austenitic isothermal curve is changed to the shape of the ε and the C curve is separated. Boron can significantly delay ferrite transformation. Therefore the bainite structure can be obtained over a wide cooling range by adding mixtures of Mo/B or Mn/B. In addition, Mn can reduce the phase transition temperature, and increase the toughness and strength. To overcome the decrease in strength caused by the decrease in carbon content, V, Cr, and other elements are often added to low-carbon bainite steel to increase its strength.

11.4.8.3 Martensitic microalloy nonquenched and tempered steel

The third generation of microalloy nonquenched and tempered steel has a low-carbon tempered martensite structure. Similar to bainite microalloy nonquenched and tempered steel, the low-carbon martensite steel can also achieve high strength and toughness. Such steel has been used in automobile walking parts and construction machinery. In addition, after the development and application of ferrite−pearlite (F−P), bainite (B), and martensite (M) microalloy nonquenched and tempered steels, F−B and F−M microalloy nonquenched and tempered steels were slowly developed and used due to their low cost and excellent performance. With the expanding application of nonquenched and tempered steel, regional and industrial standards for such steels have been developed globally. The grades of nonquenched and tempered steel developed in various countries are shown in Table 11.5, while their corresponding mechanical properties are shown in Table 11.6.

The steel grades and chemical compositions specified in European Standard EN10267−1998 for nonquenched and tempered steels are shown in Table 11.7. The steel grades and chemical compositions specified in the Chinese Standard GB/T15712−2008 for nonquenched and tempered steels are shown in Table 11.8.

11.4.9 Vanadium microalloyed structural steel

Vanadium nitride is the most effective form for adding vanadium and nitrogen to high-strength low-alloy steels. It can strengthen and refine grains more effectively than ferrovanadium, which results in more efficient use of vanadium and nitrogen, reducing the required V content by 20%−40%; thus, reducing the production cost of steel. Several advantages are shown in Table 11.9.

Table 11.5: Nonquenched and tempered steel grades developed globally.

Country	Steel	Chemical composition (mass fraction) (%)						
		C	Si	Mn	S	V	Ti	N
Germany	49MnVS3	0.44–0.50	≤0.6	0.7–1.0	0.04–0.07	0.08–0.13		
	44MnSiVS6	0.42–0.47	0.5–0.8	1.3–1.6	0.02–0.035	0.10–0.15	0.02	
	38MnSiVS6	0.35–0.40	0.5–0.8	1.2–1.5	0.03–0.07	0.08–0.13	0.02	
	27MnSiVS6	0.25–0.30	0.5–0.8	1.3–1.6	0.03–0.05	0.08–0.13	0.02	
Sweden	V-2906	0.43–0.47	0.15–0.40	0.6–0.8	0.04–0.06	0.07–0.10		$(90-140) \times 10^{-4}$
	V-2903	0.30–0.35	1.4–1.6	0.03–0.05	0.03–0.05	0.07–0.12	0.015–0.030	$(150-200) \times 10^{-4}$
	V-2904	0.36–0.40	1.2–1.4	0.04–0.06	0.04–0.06	0.07–0.10	0.015–0.030	$(150-200) \times 10^{-4}$
United Kingdom	BS970-280M01	0.3–0.5	0.15–0.35	0.6–1.5	0.045–0.06	0.08–0.20		
	VANARD	0.3–0.5	0.15–0.35	1.0–1.5	≤0.1	0.05–0.20		
	VANARD850	0.36	0.17	1.25	0.04	0.09		
	VANARD1000	0.43	0.35	1.25	0.06	0.09		
China	YF35MnV	0.32–0.39	0.30–0.60	1.0–1.5	0.035–0.075	0.06–0.13		
	YF40MnV	0.37–0.44	0.30–0.60	1.0–1.5	0.035–0.075	0.06–0.13		
	YF45MnV	0.42–0.49	0.30–0.60	1.0–1.5	0.035–0.075	0.06–0.13		
	F35MnVN	0.32–0.49	0.20–0.40	1.0–1.5	≤0.035	0.06–0.13		$\geq 90 \times 10^{-4}$
	F40MnV	0.37–0.44	0.20–0.40	1.0–1.5	≤0.035	0.06–0.13		

Table 11.6: Mechanical properties of nonquenched and tempered steels developed globally.

Country		Steel	Mechanical properties			
			R_m (MPa)	$A_{k\,room\,temperature}$ (J)	Z (%)	A (%)
Germany		38MnSiVS6	820–925	–	≥25	≥12
Sweden		Vanard	770–1000	–	–	10–18
United Kingdom		850–1000	750–950	–	–	≥12
France		V2905	800–1000	–	–	–
Finland		METASAFE800–1000	770–1000	–	–	–
Japan	Aichi Steel	SVdT30	≥600	70–90	35–40	≥20
	Kawasaki steel	NH30MV	840–1000	42–72	–	19–23
	Kobe steel	KNF23M	810	100	58	24
		KN33M	910	50	50	22
	Sanyo Special Steel	SMnV30TL	735	–	–	–
		TMAX3	980	–	–	–
	Nippon Steel	NQF250–300XM	810–910	98	–	–
		NQ22TiN	760	137	–	–
	Sumitomo Metal	LMIC90F	880	–	–	–
		THF50B	840	104	56	23
	NKK	NCHFC	985	37	–	–
	Mitsubishi Steel	VMC30	960	50	–	–

The chemical composition of precipitated vanadium–nitrogen microalloyed steel is shown in Table 11.10, while its phase structure is shown in Table 11.11.

11.4.9.1 Hot-rolled bar

Developed countries widely use high-strength steel with a strength above 400 MPa, while China generally still uses low-strength 20MnSi II-grade steel. Rebar containing vanadium has a low cost, high strength, good resistance to strain timeliness, good welding performance, high strain, and low cycle fatigue performance. Considering the requirements of high-strength hot-rolled steel, 500–550 MPa reinforcement can replace the existing 350–460 MPa bars. Priority is given to continuous and semicontinuous rolling production processes, where the rolling process in most cases is hot rolling with a high finishing temperature. The steel grade needs to be designed for V/VN alloying and microalloying processes. Vanadium is used as a microalloying element in steel, which produces highly stable carbon nitride that is finely distributed and strengthens the steel matrix. Both ferrovanadium and VN microalloying

Table 11.7: Steel grades and chemical compositions specified by European Standard EN10267–1998 for nonquenched and tempered steels.

Steel		Chemical composition (mass fraction) (%)							
Code	Name	C	Si	Mn	P (max)	S	N	Cr (max)	Mo (max)	V
1.1301	19MnVS6	0.15–0.22	0.15–0.8	1.20–1.60	0.025	0.020–0.060	0.010–0.020	0.3	0.08	0.08–0.20
1.1302	30MnVS6	0.26–0.33	0.15–0.8	1.20–1.60	0.025	0.020–0.060	0.010–0.020	0.3	0.08	0.08–0.20
1.1303	38MnVS6	0.34–0.41	0.15–0.8	1.20–1.60	0.025	0.020–0.060	0.010–0.020	0.3	0.08	0.08–0.20
1.1304	46MnVS6	0.42–0.49	0.15–0.8	1.20–1.60	0.025	0.020–0.060	0.010–0.020	0.3	0.08	0.08–0.20
1.1305	46MnVS3	0.42–0.49	0.15–0.8	1.20–1.60	0.025	0.020–0.060	0.010–0.020	0.3	0.08	0.08–0.20

Table 11.8: Steel grades and chemical compositions specified in Chinese Standard GB/T15712−2008 for nonquenched and tempered steels.

No.	Code	Grade	Chemical composition (mass fraction) (%)									
			C	Si	Mn	S	P	V	Cr	Ni	Cu	Other
1	122358	F35VS	0.32−0.39	0.20−0.40	0.60−1.00	0.035−0.075	≤0.035	0.06−0.13	≤0.30	≤0.30	≤0.30	
2	122408	F40VS	0.37−0.44	0.20−0.40	0.60−1.00	0.035−0.075	≤0.035	0.06−0.13	≤0.30	≤0.30	≤0.30	
3	122468	F45VS	0.42−0.49	0.20−0.40	0.60−1.00	0.035−0.075	≤0.035	0.06−0.13	≤0.30	≤0.30	≤0.30	
4	122308	F30MnVS	0.20−0.33	≤0.80	1.20−1.60	0.035−0.075	≤0.035	0.08−0.15	≤0.30	≤0.30	≤0.30	
5	122378	F35MnVS	0.32−0.39	0.30−0.60	1.00−1.50	0.035−0.075	≤0.035	0.06−0.13	≤0.30	≤0.30	≤0.30	
6	122388	F38MnVS	0.34−0.41	≤0.60	1.20−1.60	0.035−0.075	≤0.035	0.08−0.15	≤0.30	≤0.30	≤0.30	
7	122428	F40MnVS	0.37−0.44	0.30−0.60	1.00−1.50	0.035−0.075	≤0.035	0.06−0.13	≤0.30	≤0.30	≤0.30	
8	122478	F45MnVS	0.42−0.49	0.30−0.60	1.00−1.50	0.035−0.075	≤0.035	0.06−0.13	≤0.30	≤0.30	≤0.30	
9	122498	F49MnVS	0.44−0.52	0.15−0.60	0.70−1.00	0.035−0.075	≤0.035	0.08−0.15	≤0.30	≤0.30	≤0.30	
10	127128	F12Mn2VBS	0.09−0.16	0.32−0.39	2.20−2.65	0.035−0.075	≤0.035	0.06−0.13	≤0.30	≤0.30	≤0.30	B 0.001−0.004

Table 11.9: Properties of steel improved by vanadium nitride addition.

Advantage	Description
Effective strengthening and grain refinement	Compared with vanadium carbide, the nitrogen in vanadium nitride can promote precipitation of vanadium carbonitride, which is rich in nitrogen.
Reduce the amount of vanadium (and, hence, cost)	Less vanadium is used to precipitate vanadium nitride than vanadium carbide.
Improved weldability, notch toughness, and malleability	The required strength grade can be achieved with low-carbon content and few alloy additives.
Effective strengthening of all types of carbon steel	The solubility of carbon vanadium nitride in austenite is very high at 1050°C and is not dependent on the carbon content. Hence, vanadium nitride is just as effective in high-, medium-, and low-carbon steels.
Strain aging and plastic loss	Nitrovan 7, 12, or 16 alloys are used to adjust the V:N ratio to avoid free nitrogen and produce steel with minimal aging.
High and consistent absorption of vanadium and nitrogen	Vanadium nitride is homogeneous in size and compact in mass and is rapidly dissolved in steel. When used in Nitrovan12, an increase of 0.001% vanadium can increase the total nitrogen content by 0.001%.
Good control of deoxidization and reduction of impurity levels to avoid transport loss	High-purity products have low contents of residual elements, such as Si, Al, Cr, and Ni (Table 11.4). Vanadium nitride is packaged in a highly moisture-resistant material and is added directly to the molten steel.

treatments are used to produce high-strength steel bars. The chemical compositions of such steel bars produced using different processes are shown in Table 11.12.

Vanadium is a microalloying element that can produce stable finely dispersed carbonitrides in steel that strengthens the matrix. The chemical composition of 500-MPa high-strength reinforcement is given in Table 11.13. In V-containing steel with ferrovanadium addition

Table 11.10: Chemical composition of precipitated vanadium–nitrogen microalloyed steel.

	M(C,N) mass fraction in steel (%)						
	V	Ti	Mo	Cr	C	N	Σ
Before soaking the billet	0.1151	0.0040	0.0021	0.0024	0.0140	0.0174	0.1550
After soaking the billet	0.0573	0.0043	0.0024	0.0022	0.0049	0.0122	0.0833
1.8-mm strip steel	0.0943	0.0041	0.0023	0.0022	0.0097	0.0168	0.1294
3.2-mm strip steel	0.0772	0.0039	0.0024	0.0022	0.0070	0.0152	0.1079
6.3-mm strip steel	0.0580	0.0038	0.0023	0.0023	0.0041	0.0132	0.0837

Table 11.11: Chemical phase structure of precipitated vanadium–nitrogen microalloyed steel.

	M(C,N) phase structure
Before soaking the billet	$V_{0.973}Ti_{0.035}Cr_{0.019}Mo_{0.009}C_{0.485}N_{0.515}$
After soaking the billet	$V_{0.877}Ti_{0.070}Cr_{0.033}Mo_{0.020}C_{0.321}N_{0.679}$
1.8-mm strip steel	$V_{0.924}Ti_{0.043}Cr_{0.021}Mo_{0.012}C_{0.401}N_{0.599}$
3.2-mm strip steel	$V_{0.911}Ti_{0.049}Cr_{0.025}Mo_{0.015}C_{0.384}N_{0.652}$
6.3-mm strip steel	$V_{0.885}Ti_{0.062}Cr_{0.032}Mo_{0.019}C_{0.267}N_{0.733}$

Table 11.12: Chemical composition of steel bars produced by different processes.

Steel bars	Chemical composition (%)											
	C	Si	S	P	Mn	Ni	Cr	Mo	V	Cu	Sn	Co
Hot-rolled carbon	0.375	0.287	0.029	0.022	1.304	0.064	0.085	0.009	0.003	0.197	0.016	0.00
Hot-rolled vanadium microalloy	0.245	0.154	0.043	0.014	1.029	0.143	0.127	0.023	0.050	0.502	0.020	0.011
Residual heat self-tempering	0.219	0.193	0.047	0.015	0.870	0.106	0.083	0.014	0.001	0.261	0.016	0.010
Hardening process	0.271	0.160	0.046	0.027	0.786	0.099	0.168	0.013	0.001	0.532	0.023	0.001

the V(C,N) precipitated phase contains 35.5% vanadium, accounting for 56.3% of the total vanadium content. In VN-containing steel, 70% of the vanadium occurs as the V(C,N) precipitated phase, while 20% is in a solid solution in the matrix. Hence, the presence of N greatly promotes the precipitation of the V(C,N) phase which enhances precipitation strengthening and fine crystal strengthening.

V–N microalloyed steel is used to improve the precipitation strengthening and fine-grain strengthening ability of V(C,N). V–N steel bar has good anti-strain aging performance; the C and N are solidified, which reduces the probability of a Kirchstein air mass forming, and

Table 11.13: Chemical composition of 500-MPa high-strength reinforcement (%).

Alloying elements	Specifications Φ (mm)	Chemical composition (%)				
		C	Si	Mn	V	N
V–Fe	16–32	0.20–0.25	0.50–0.80	1.35–1.60	0.07–0.12	Remnant
V–N	16–32	0.20–0.25	0.50–0.80	1.35–1.60	0.05–0.07	0.010–0.015
	40	0.20–0.25	0.50–0.80	1.35–1.60	0.07–0.09	0.012–0.018

Table 11.14: Chemical composition of low-cost V−N microalloyed high-strength reinforcement.

Steel grade (MPa)	C	Si	Mn	P	S	V	N
400	0.20−0.24	0.4−0.5	1.4−1.5	约0.030	约0.020	0.020−0.030	0.0080−0.012
500	0.20−0.23	0.45−0.55	1.40−1.55	约0.030	约0.020	0.040−0.060	0.010−0.014

effectively reduces the strain aging sensitivity of the steel bar. V−N microalloyed steel partially replaces vanadium by nitrogen, reducing the amount of required vanadium, which subsequently reduces production costs. Table 11.14 shows the chemical composition of low-cost V−N microalloyed high-strength reinforcement, while Table 11.15 shows the corresponding process conditions and performance of this steel.

11.4.9.2 Wire and rod

The addition of vanadium to hard wire steel reduces the interstitial atomic activity of C and N, resulting in reduced aging sensitivity and significantly higher strength of the wire. The effects of vanadium on the strength and hardness of hard wire steel include (1) refining austenite grains by adjusting the austenite; (2) formation of a solid solution in austenite improves the hardenability and microstructure of the hard wire steel by refining the size of the pearlite group and lamellar spacing; (3) V is precipitated in pearlite and ferrite, which plays a role in strengthening. A typical composition of V-containing hard wire steel is shown in Table 11.16.

11.4.9.3 Heat-treated PC rod

Prestressed Concrete (PC) bar belongs to the high-strength grade of steel rods. PC bar has high toughness, low relaxation, resulting in a high compressive strength of the concrete. In addition, it has good weldability and upsetting property. Medium-carbon steel, high-carbon

Table 11.15: Process and performance of low-cost V−N microalloyed high-strength reinforcement.

Steel grade (MPa)	Specifications Φ (mm)	Final cooling temperature (°C)	R_{eL} (MPa)	R_m (MPa)	A (%)
400	16	810−856	435	605	30.5
	20	800−850	450	615	27
	32	840	455	615	25.5
500	20	700	555	715	25
	25	730−755	525	660	27
	28	710−725	545	695	24.5
	32	690	565	705	21

Table 11.16: Typical composition of V-containing hard wire steel.

Steel	C	Si	Mn	P	S	Cr	Ni	Cu	Al	Ti	Mo	V	O_2	N_2
Standard steel	0.801	0.239	0.72	0.008	0.008	0.27	0.02	0.04	0.001	0.001	0.005	0.058	0.003	0.003
Standard steel	0.804	0.25	0.77	0.007	0.009	0.25	0.03	0.04	0.001	0.001	0.005	0.054	0.002	0.003
Standard steel	0.805	0.245	0.78	0.009	0.007	0.24	0.02	0.04	0.001	0.001	0.006	0.053	0.003	0.003

Table 11.17: Typical chemical compositions of Chinese wire rods (%).

No.	C	Si	Mn	P	S	V	B	Grade
1	0.27	1.65	0.83	0.028	0.028	0.080	–	27Si2MnV
2	0.35	1.26	0.81	0.016	0.030	0.081	–	35Si2MnV
3	0.31	0.78	1.42	0.025	0.021	–	–	30SiMn2
4	0.27	0.46	1.29	0.012	0.008	–	0.092	25SiMnB

steel, low-alloy steel, and spring steel are subjected to vanadium alloying to produce PC bar. Typical chemical compositions of Chinese wire rods are shown in Table 11.17, while their basic mechanical properties are shown in Table 11.18.

11.4.9.4 Shaped steel

There are many types of shaped steel, which are classified according to the processing method as hot-rolled, cold-rolled, extruded, and cold-formed section steel. According to the size and specification, they can be divided into large steel, medium steel, small steel, and wire. According to the shape of the section, the steel is classified as simple or complex section steel. Simple section steel is divided into round steel, square steel, flat steel, and deformed steel bar, while complex section steel can be divided into angle steel, channel steel, heavy rail, light rail, and H-shaped steel.

Rail steel needs both high strength and toughness, which can be provided by vanadium addition. Adding 0.16% V to this steel can increase the strength by 100 MPa after air cooling, or by 200 MPa after controlled cooling. The addition of 0.09% V to the steel can effectively increase the strength while greatly improving the toughness.

The required chemical composition of hot-rolled rail steel is shown in Table 11.19. In hot-rolled rail, vanadium is mainly used in the PD3 grade rail. The chemical composition of PD3 rail is shown in Table 11.20, while its mechanical properties are shown in Table 11.21. Vanadium microalloying is used in high-carbon steel, mainly for the production of rail, bearing steel, tool steel, and die steel. Heavy rail steel is a type of pearlitic steel. The strength, toughness, and plasticity of the rail are controlled by the interlamellar spacing and

Table 11.18: Basic mechanical properties of typical Chinese wire rods.

No.	R_m (MPa)	$R_{p0.2}$ (MPa)	A_8 (%)	Grade
1	1520	1440	8.0	27Si2MnV
2	1600	1530	7.5	35Si2MnV
3	1550	1470	7.5	30SiMn2
4	1500	1420	8.0	25SiMnB

Table 11.19: Required chemical composition of hot-rolled rail steel (%).

Steel grade	Chemical composition (mass fraction) (%)						
	C	Si	Mn	S	P	V	Nd
U74	0.68–0.79	0.13–0.28	0.70–1.00	≤0.030	≤0.030	≤0.030	≤0.010
U71Mn	0.65–0.74	0.15–0.35	1.10–1.40	≤0.030	≤0.030		
U71MnSi	0.66–0.74	0.85–1.15	0.85–1.15	≤0.030	≤0.030		
U71MnSiCu	0.64–0.74	0.70–1.10	0.80–1.20	≤0.030	≤0.030		
U75V	0.71–0.80	0.50–0.80	0.70–1.05	≤0.030	≤0.030	0.04–0.12	
U76NbRE	0.72–0.80	0.60–0.90	1.00–1.30	≤0.030	≤0.030	≤0.030	0.02–0.05
76Mn	0.61–0.79	0.10–0.50	0.85–1.25	≤0.030	≤0.030		≤0.010

size of the pearlite grains. The tensile strength and yield strength increase with decreasing interlamellar spacing. The toughness is related to the size of the pearlite and cementite thickness. The role of trace vanadium in heavy rail steel has been studied, where the results showed that vanadium can refine both austenite grains and the pearlitic structure and change its tissue. This achieves the precipitation strengthening effect and increase the strength and service lifetime of the heavy rail. V-containing rails are produced by Panzhihua Iron and Steel Group Co., including vanadium microalloyed hot-rolled rail, off-line V-containing heat-treated rail, and on-line V-containing heat-treated rail.

V-containing high-strength low-alloy steels for boilers and pressure vessels in various countries are listed in Table 11.22.

Heat- and hydrogen-resistant steels for pressure vessels are listed in Table 11.23.

Table 11.24 lists steels for nuclear applications.

Table 11.25 shows the chemical composition of high-strength construction machinery steels from various countries.

Table 11.20: Chemical composition of PD3 rail (%).

Steel	C	Si	Mn	V	P, S
PD3	0.75–0.81	0.60–0.90	0.75–1.05	0.05–0.12	≤0.035

Table 11.21: Mechanical properties of PD3 rail.

Category	$R_{p0.2}$ (MPa)	R_m (MPa)	A_5 (%)	α_{ku} (J/cm^2)	K_{JC} (MPa m$^{1/2}$)
Required value	≥880	≥1275	≥10	–	–
Actual value	890–1010	1290–1370	10–13	20–32	44.0–50.5

Table 11.22: V-containing high-strength low-alloy steels for boilers and pressure vessels in various countries.

Country	Steel	Chemical composition (%)							Specification (mm)	Yield strength (MPa)	Heat treatment
		C	Si	Mn	Ni	Cr	Mo	V			
China	09Mn2V	≤0.12	0.15–0.50	1.40–1.80	—	—	—	0.04–0.10	6–16 17–16	290 270	Normalizing and tempering
	15MnV	0.10–0.18	0.20–0.60	1.20–1.60	—	—	—	0.04–0.12	6–16 17–25 26–36 37–60	390 375 355 335	Hot-rolling normalization
	15MnVN	≤0.20	0.20–0.60	1.30–1.70	—	—	N: 0.010–0.020	0.10–0.20	6–16 17–36 37–60	440 420 400	Hot-rolling normalization
	14MnMoV	0.10–0.18	0.20–0.50	1.20–1.60	—	—	0.40–0.65	0.05–0.15	30–115	490	Normalized tempering
	07MnCrMoV	≤0.09	0.15–0.40	1.20–1.60	≤0.30	0.10–0.30	0.10–0.30	0.02–0.06	16–50	490	Quenching and tempering
	07MnNiCrMoV	≤0.09	0.15–0.40	1.20–1.60	0.20–0.50	0.10–0.30	0.10–0.30	0.02–0.06	16–50	490	Quenching and tempering
	12MnNiV	≤0.09	0.15–0.35	1.10–1.50	≤0.30	≤0.50	≤0.30	0.02–0.06	14–50	490	Quenching and tempering
	17MnNiVNb								6–25		
United States	SA225C	≤0.25	0.15–0.45	≤1.72	0.37–0.73			0.11–0.20	≤75 >75	485	Hot-rolling normalization
	SA225D	≤0.20	0.08–0.56	≤1.84	0.37–0.73			0.08–0.20	≤75 >75	415 380	Normalization
	ASTMA225G	≤0.25	0.15–0.30	≤1.60			0.40–0.70	0.09–0.14	15	485	Hot-rolling normalization
	ASTMA737A	≤0.20	0.15–0.50	1.00–1.35				≤0.10		345	Normalization
	ASTMA737C	≤0.22	0.15–0.50	1.15–1.50		Nb: 0.05		0.04–0.11		415	Normalization

Japan	SFV245	≤0.20	0.15–0.60	0.80–1.60			≤0.35	≤0.10	≤50 50–100 100–125 125–150	370 355 345 335	
	SFV295	≤0.10	0.15–0.60	0.80–1.60		Nb : 0.05		≤0.10	≤50 50–100 100–125 125–150	420 400 390 380	
	SFV345	≤0.10	0.15–0.60	0.80–1.60		Nb : 0.05	0.1–0.4	≤0.10	≤50 50–100 100–125 125–150	430 430 420 410	
	HITEN-590U	≤0.09	0.15–0.60	0.90–1.40	B : 0.003	≤0.30	≤0.20	≤0.08		450	
	HITEN-610U	≤0.09	0.15–0.40	0.90–1.40	B : 0.003	≤0.30	≤0.20	≤0.08		490	Quenching and tempering
	K-TEN62	≤0.18	≤0.55	≤1.60	0.20–0.60	0.10–0.30	≤0.60	≤0.10		686	Quenching and tempering
	K-TEN80	≤0.18	≤0.55	≤1.50	≤1.60	≤0.80	≤0.30	≤0.10			Quenching and tempering
	Wel-Ten63CF	≤0.09	0.15–0.35	1.00–1.60	≤0.60	≤0.30	≤0.30	≤0.10	≤0.10	490	Quenching and tempering
	Wel-Ten60H	≤0.18	0.15–0.70	0.90–1.50	0.30–1.60		V + Nb : ≤0.10	≤0.10	6–38 38–50	440 410	Normalization

ASTM, American Association for Materials and Testing Standards.

Table 11.23: Heat- and hydrogen-resistant steels for pressure vessels.

Country	Steel	C	Si	Mn	Cr	Mo	V	W	Ti	B
United States	SA542B	0.09–0.18	≤0.50	0.25–0.66	1.88–2.62	0.85–1.15	≤0.03			
	SQ542C	0.08–0.18	≤0.50	0.25–0.66	2.63–3.37	0.85–1.15	0.18–0.33			
	SA542D	0.09–0.18	≤0.50	0.25–0.66	1.88–2.62	0.85–1.15	0.23–0.37			
Germany	47CrMoV	0.15–0.20	0.15–0.35	0.30–0.50	2.70–3.00	0.20–0.30	0.10–0.20			
	20CrMoV	0.17–0.23	0.15–0.35	0.30–0.50	3.00–3.30	0.50–0.60	0.45–0.55			
	20CrMoVW	0.18–0.25	0.15–0.35	0.30–0.50	2.70–3.00	0.35–0.45	0.75–0.85			
China	12Cr2MoV	0.08–0.15	0.17–0.37	0.40–0.70	0.90–1.20	0.25–0.35	0.15–0.30			
	12Cr2MoWVTiB	0.08–0.15	0.45–0.75	0.40–0.65	1.60–2.10	0.50–0.65	0.28–0.42	0.30–0.55	0.08–0.18	0.002–0.008
	12Cr3MoVSiTiB	0.09–0.15	0.60–0.90	0.50–0.80	2.50–3.00	1.00–1.20	0.25–0.35		0.22–0.38	0.005–0.011
	10Cr5MoWVTiB	0.07–0.12	0.40–0.70	0.40–0.70	4.50–6.00	0.48–0.65	0.20–0.33	0.20–0.40	0.16–0.24	0.008–0.014

Table 11.24: Steel for nuclear applications.

Country	Steel	C	Si	Mn	Ni	Cr	Mo	V	Cu
United States	A508-П	≤0.27	0.15–0.35	0.50–0.90	0.50–0.90	0.25–0.45	0.55–0.70	0.01–0.05	<0.10
United States	A508-Ш	≤0.26	0.15–0.40	1.20–1.50	0.40–1.00	<0.25	0.45–0.55	0.01–0.05	<0.10
Russia	15Kh2MFA	0.13–0.18	0.17–0.37	0.50–0.70	≤0.4	2.5–3.0	0.50–0.70	≤0.30	<0.15
Russia	15Kh2NMFA-A	0.13–0.18	0.17–0.37	0.50–0.70	≤0.4	2.5–3.0	0.50–0.70	0.10–0.12	<0.05

Table 11.25: Chemical composition of high-strength construction machinery steels from various countries.

Country	Steel	Chemical composition (%)								Specification (mm)	Yield strength (MPa)
		C	Si	Mn	Ni	Cr	Mo	V	Ti		
China	HQ60	0.09–0.16	0.15–0.50	1.10–1.60	0.30–0.60	≤0.30	0.08–0.20	0.03–0.08		4–40 40–50	450 440
	HQ70	0.09–0.16	0.15–0.40	0.60–1.20	0.30–1.0	0.30–0.60	0.20–0.40	V+Nb≤0.15 B:0.0005–0.0003		18–50	590
	HQ80	0.10–0.16	0.15–0.35	0.60–1.20		0.60–1.20	0.03–0.08	0.03–0.08	B:0.0006–0.005	20–50	585
	HQ90	0.10–0.18	0.15–0.35	0.80–1.40	0.70–1.50	0.40–0.80	0.30–0.60	0.03–0.08	Cu0.15–0.50	8–50	880
United States	A514Q	0.14–0.21	0.15–0.35	0.95–1.30	1.20–1.50	1.00–1.50	0.40–0.60	0.03–0.08		≤63.5 >63.5–152	690 620
	A514B	0.12–0.21	0.20–0.35	0.70–1.00		0.40–0.65	0.15–0.25	0.03–0.08	B: ≤0.005	≤32	690
Germany		≤0.20	0.10–0.15	1.20–1.70	0.40–0.70			0.10–0.13	N: ≤0.020	≤16 16–50 35–55 50–100 100–150	460 450 440 400 390
	STE885	≤0.18	≤0.45	≤1.00	≤1.40	≤0.80	0.20–0.60	≤0.10		≤35	885
Japan	Wel-ten60	≤0.16	0.15–0.55	0.90–1.5	≤0.60	≤0.30	≤0.30	≤0.10		6–50	450
	Wel-ten70	≤0.16	0.15–0.35	0.60–1.20	0.30–1.00	≤0.60	≤0.40	V+Nb≤0.15	B≤0.006	6–50 50–75	685 685
	Wel-ten80	≤0.16	0.15–0.55	0.60–1.20	0.40–0.80	0.40–0.80	0.30–0.60	≤0.10	B≤0.006	6–50 50–100	685 685
	Wel-ten100	≤0.18	0.15–0.55	0.60–1.20	0.70–1.50	0.40–1.50	0.40–0.80	0.30–0.60	B≤0.006	6–32	880
Sweden	WELDOX1300	≤0.25	≤0.5	≤1.4	≤2.0	≤0.8	≤0.7	≤0.08	≤0.005		
	WELDOX1100	≤0.21	≤0.5	≤1.4	≤3.0	≤0.8	≤0.7	≤0.08	≤0.005		
	WELDOX960	≤0.20	≤0.5	≤1.6	≤1.5	≤0.7	≤0.7	≤0.06	≤0.005		
	WELDOX900	≤0.20	≤0.5	≤1.6	≤0.1	≤0.7	≤0.7	≤0.06	≤0.005		
	WELDOX700	≤0.20	≤0.6	≤1.6	≤2.0	≤0.7	≤0.7	≤0.09	≤0.005		

The chemical composition and specifications of various V-containing wear-resistant steels are given in Table 11.26.

Table 11.27 lists V-containing high-strength low-alloy structural steels.

Table 11.28 lists vanadium microalloyed strip steel for automotive applications.

Table 11.29 lists V-containing weathering steels.

The chemical compositions of high-strength structural steel with yield strengths of 350–410 MPa are given in Table 11.30.

The mechanical properties of high-strength structural steels with yield strengths of 350–410 MPa are given in Table 11.31.

The chemical composition of high-strength structural steel with a yield strength of 550 MPa is given in Table 11.32.

Table 11.33 shows the main V-containing alloy structural steels.

The chemical composition of 60CrV7 spring steel is given in Table 11.34.

11.4.9.5 Tool steels

There are three main types of tool steels, namely, hot working, cold working, and plastic molding steels. Vanadium is mainly used in hot working tool steels, which contain 1% vanadium and are resistant to mechanical and thermal shock at high temperatures, while maintaining a certain hardness and strength during high-speed cutting and other thermal applications. The composition of cold working steel alloys varies greatly, but only a small amount of vanadium is added. Vanadium is almost never added to plastic molding steels. The composition of tool steels varies greatly, and the relationship between performance and price must be considered when selecting alloying elements for a specific steel. Currently, there are many kinds of tool steels with high alloying, but not all of them need vanadium addition. Table 11.35 shows the major V-containing tool steels.

Table 11.36 shows the effect of vanadium on the overall properties of tool steels.

Note: (1) 840°C oil quenching, 427°C tempering; (2) steel composition: 0.49% C, 0.76% Mn, 0.21% Si, and 1.07% Cr; (3) steel composition: 0.50% C, 0.79% Mn, 0.31% Si, and 0.20% V. The grades and chemical composition of the alloy tool steels are given in Table 11.37.

The grades and chemical compositions of alloy tool steels are given in Table 11.38.

The chemical compositions of low-alloy tool steels containing vanadium are given in Table 11.39.

Table 11.26: Chemical composition and specifications of various V-containing wear-resistant steels.

Country	Steels	Chemical composition (%)									Specification (mm)	HB
		C	Si	Mn	Ni	Cr	Mo	V	B		
China	20MnVK	0.17–0.24	0.17–0.37	1.20–1.60				0.07–0.20			
	25MnVK	0.22–0.30	0.50–0.90	1.30–1.60				0.06–0.13			
	NM 360	≤0.26	0.20–0.24	≤1.60	0.30–0.60	0.80–1.20	0.15–0.50	≤0.10	≤0.005	12–15	360
Japan	WeltenAR320	≤0.22	≤0.35	0.60–1.20	0.40–1.50	0.40–0.80	0.15–0.60	≤0.10	≤0.005	6–100	321
	WeltenAR360	≤0.22	≤0.35	0.60–1.20	0.40–1.50	0.40–0.80	0.15–0.60	≤0.10	≤0.005	6–75	361
	WeltenAR400	≤0.24	≤0.35	0.60–1.20	0.40–1.50	0.40–0.80	0.15–0.60	≤0.10	≤0.005	6–32	401
Sweden	HardHiTuf	≤0.20	≤0.50	≤1.60	≤2.0	≤0.70	≤0.70	≤0.06	≤0.005	40–70	310–370
		≤0.20	≤0.60	≤1.60	≤2.0	≤0.70	≤0.70	≤0.09	≤0.005	70–130	310–370

Table 11.27: V-containing high-strength low-alloy structural steels.

Grade	Steel	C	Si	Mn	P (≤)	S (≤)	V	Ti	N	Re
Q315	09MnV	≤0.12	0.20–0.55	0.8–1.20	0.045	0.045	0.04–0.12			
Q345	12MnV	≤0.15	0.20–0.55	1.0–1.20	0.045	0.045	0.04–0.12			
Q390	15MnV	0.12–0.18	0.20–0.55	1.2–1.60	0.045	0.045	0.04–0.12			
Q420	15MnVN	0.12–0.20	0.20–0.55	1.30–1.70	0.045	0.045	0.10–0.20		0.10–0.20	
Q440	14MnVTiRe	≤0.18	0.20–0.55	1.30–1.60	0.045	0.045	0.04–0.10	0.09–0.16		0.02–0.20

Table 11.28: Vanadium microalloyed strip steel for automotive applications.

Steel	C	Si	Mn	P (≤)	S (≤)	V	Nb	Al	N
Domex590	≤0.12	≤0.4	≤1.5	0.03	0.01	0.09–0.12	0.03–0.04	0.02–0.05	0.012–0.016
Domex590	≤0.12	≤0.4	≤1.65	0.03	0.01	0.10–0.15	0.03–0.04	0.02–0.05	0.012–0.016
VAN-60	≤0.15	≤0.30	≤1.20	0.03	0.01	≤0.010	≤0.04	0.02–0.05	
09SiVL	0.08–0.15	0.70–1.0	0.45–0.75	0.03	0.01	0.04–0.10		0.02–0.05	Remainder
W510L	0.08–0.11	0.10–0.30	1.30–1.50	0.03	0.01	0.05–0.06	0.01–0.03	0.02–0.05	Remainder
P510	0.08–0.12	0.40–0.70	0.90–1.30	0.03	0.01	0.06–0.09		0.02–0.05	Remainder

Table 11.29: V-containing weathering steels.

Country	Steel	C	Si	Mn	P	S	Cr	Cu	V	Re
China	08CuPV	0.12	0.20–0.40	0.20–0.50	0.07–0.12	0.04		0.25–0.45	0.02–0.08	0.02–0.20
United States	CORTEN-B	0.10–0.19	0.16–0.30	0.90–1.25	0.04	0.05	0.40–0.65	0.25–0.40	0.02–0.10	
Germany	KT52-3	0.08–0.12	0.25–0.50	0.90–1.20	0.05–0.09	0.04	0.50–0.80	0.30–0.50	0.04–0.10	

Table 11.30: Chemical compositions of high-strength structural steels with a yield strength of 350−410 MPa.

Minimum yield strength (MPa)	C	Si	Mn	V	N	Al
275	0.04−0.07	≤0.03	0.30−0.35	0.015−0.030	0.009−0.013	0.02−0.05
310	0.04−0.07	≤0.03	0.50−0.60	0.025−0.035	0.01−0.014	0.02−0.05
340	0.04−0.07	≤0.03	0.70−0.80	0.045−0.055	0.012−0.016	0.02−0.05
380	0.04−0.07	≤0.03	1.00−1.15	0.055−0.065	0.013−0.017	0.02−0.05
410	0.04−0.07	≤0.03	1.20−1.30	0.075−0.085	0.015−0.019	0.02−0.05

The toughening components of V-containing impact-resistant cutting tools are given in Table 11.40.

The influence of vanadium on the mechanical properties of die steel include: (1) vanadium dissolved in austenite hinders growth of austenite grains; (2) precipitation strengthening and secondary hardening of the steel. The chemical compositions of low-vanadium die steels treated using oil quenching or common air quenching are given in Table 11.41.

The chemical compositions of hot die steels with medium and high vanadium contents are given in Table 11.42.

The chemical compositions of hot die steels with medium and high vanadium contents are shown in Table 11.43.

Table 11.44 shows the chemical composition of special wear-resistant die steel with ultrahigh vanadium content.

Vanadium is an indispensable alloying element for high-speed tool steel, including W18Cr4V and W6Mo5Cr4V2. Both cold-worked and hot-worked steels contain vanadium, such as Cr6WV and Cr4W2MoV, with an average vanadium content of 0.47%−0.68%. Adding vanadium to steel refines the grains and increases toughness,

Table 11.31: Mechanical properties of high-strength structural steels with yield strengths of 350−410 MPa.

Specification (mm)	Chemical composition (%)				Yield strength (MPa)	Tensile strength (MPa)	Elongation (%)	Ferrite grain size (μm)
	C	Mn	V	N				
6.0	0.04	0.9	0.08	0.0136	460	522	27	6.5
9.6	0.04	0.7	0.05	0.0120	420	500	25	11.5
9.6	0.05	0.6	0.03	0.0100	364	462	36	11.5

Table 11.32: Chemical composition of high-strength structural steel with a yield strength of 550 MPa.

Producer	C	Si	Mn	V	N	Al	Nb	Mo
Crawfordsville	0.03–0.06	0.30–0.40	1.45–1.55	0.11–0.13	0.018–0.022	0.02–0.05	0.015–0.025	
Gallatin	0.056–0.075	0.02–0.15	1.25–1.45	0.12–0.13	0.015–0.0205	0.013–0.035		0.010–0.055

Table 11.33: Main V-containing alloy structural steels.

Steels	Composition system	Grade	w (V) (%)
Alloy structural steel	MnV	20MnV	0.07–0.12
	SiMnMoV	20SiMn2MoV	0.05–0.12
		25SiMn2MoV	0.05–0.12
		37SiMn2MoV	0.05–0.12
	MnVB	15MnVB	0.07–0.12
		20MnVB	0.07–0.12
		40MnVB	0.05–0.10
	CrMoV	12CrMoV	0.05–0.30
		35CrMoV	0.10–0.20
		12Cr1MoV	0.15–0.30
		25Cr2MoVA	0.15–0.30
		25Cr2Mo1VA	0.30–0.50
	CrV	40CrV	0.10–0.20
		50CrVA	0.10–0.20
	CrNiMoV	45CrNiMoVA	0.10–0.20
Spring steel		55SiMnVB	0.08–0.16
		60Si2CrVA	0.10–0.20
		50CrVA	0.10–0.20
		30W4Cr2VA	0.50–0.80

Table 11.34: Chemical composition of 60CrV7 spring steel.

C	Mn	Si	S	P	Cr	Ni	Mo	Cu	Al	V	N
0.62	0.55	1.56	0.006	0.011	0.660	0.05	0.01	0.11	0.230	0.190	0.012

hardness, thermal hardness, and wear resistance, while also reducing the tendency for cracking. The grades and chemical compositions of Chinese high-speed tool steel are given in Table 11.45.

The basic properties of high speed steels (HSS) carbides are given in Table 11.46.

The chemical composition of general high-speed steel is given in Table 11.47, while Table 11.48 shows the composition of semihigh-speed steel.

Table 11.35: Major V-containing tool steels.

Steels	C	Mn	Si	Cr	W	Mo	Co	V
W–Cr–V high-speed tool steel	0.70–1.55			3.5–5.0	9–19	0–1.0		0.7–5.0
W–Cr–V–Co high-speed tool steel	0.70–1.55			3.5–5.5	12–21	0–1.0	0.50–16.0	0.7–5.0
Mo–W–V H high-speed tool steel	0.70–1.55			3.5–4.0	1.0–7.5	2.0–10.0		0.8–5.0
Mo–W–V–Co high-speed tool steel	0.75–1.60			3.5–5.0	1.0–11.0	3.0–9.0	4.0–13.0	1.5–5.5
W–Cr–V tool steel	0.30–1.50	<0.04	<0.35	0.30–7.00	0.80–9.00			0.20–0.70
W–Cr–Si–V tool steel	0.60–0.60	0.20–0.40	0.80	0.80–1.10	1.80–4.20			0.15–0.30
Cr–V tool steel	0.80–1.25	0.30–0.60	<0.35	0.45–0.75				0.15–0.30
Cr–W–Mo–V tool steel	0.40–0.50	0.30–0.50	0.50–0.80	1.20–1.50	0.40–0.60	0.30–0.50		0.75–0.85
Cr–Mo–V tool steel	1.45–1.70	<0.35	<0.40	11–12.5		0.40–0.60		0.15–0.30
Cr–Mn–Si–W–V tool steel	0.40–0.50	1.35–1.65	0.80–1.20	2.5–3.0	0.80–1.20			0.75–0.85
Si–Cr–V tool steel	0.40–0.50	<0.40	1.20–1.60	1.3–1.3				0.15–0.30
Si–Mn–Mo–V tool steel	0.45–0.55	0.50–0.70	1.50–1.80	0.2–0.4		0.30–0.50		0.20–0.30
Si–Mn–V tool steel	0.55–1.50	0.70–1.250	0.70–1.10					0.15–0.30
Mn–Cr–W–V tool steel	0.90–1.05	1.00–1.30	<0.35		0.40–0.60			0.16–0.30
Mn–V tool steel	0.85–0.95	1.20–2.00	<0.35	0.4–0.6				0.10–0.25

Table 11.36: Effect of vanadium on the overall properties of tool steels.

Property (1)	Steel A (2)	Steel B (3)	Property (1)	Steel A (2)	Steel B (3)
Tensile strength (MPa)	1553	1634	Area reduction (%)	35.9	43.1
Yield point (MPa)	1411	1576	Vickers hardness HV	457	481
Yield strength/tensile strength ratio	0.911	0.963	Ehrlich shock value (J)	16.6	16.6
Elongation (%)	9.6	10.4			

The chemical composition of American Institute of Steel Standards (AISI) (United States) high-vanadium high-speed steel is given in Table 11.49.

The chemical composition of V3N nitrogenous super-high-speed steel is given in Table 11.50.

The chemical composition of powder metallurgy die steels is given in Table 11.51.

11.4.10 Heat-resistant steel

Heat-resistant steel is special steel that has high strength and corrosion resistance at high temperature. Depending on the application of the heat-resistant steel, it is classified as refractory steel or oxidation-resistant steel. The working temperature range of refractory steel is 450°C–900°C, over which the steel needs to have good resistance to creep, fracture, and oxidation, while withstanding cyclic fatigue forces. The working temperature range of oxidation-resistant steel is 500°C–1200°C, over which the steel needs to have excellent oxidation resistance and high-temperature corrosion resistance, while withstanding low loads and creep fracture. Heat-resistant steel is mainly strengthened by addition of alloying elements, which increases the stability of carbides precipitated during tempering. During heat treatment, relatively stable strengthening phases can be produced, where the emphasis is on adding alloying elements to form carbides.

Vanadium is used as the alloying element for narrowing the austenite phase area and expanding the alpha phase area as it is a strong carbide-forming element. Adding vanadium in steel after heat treatment results in the VC and VN phases being precipitated in the range of 500°C–700°C and improvement in the heat resistance of the steel. The chemical composition (%) (GB/T1221–2007) of heat-resistant V-containing steel rods is given in Table 11.52.

Table 11.53 shows the chemical composition of V-containing heat-resistant steel plates and steel strips (%) (GB/t4238–2007).

Table 11.37: Grades and chemical composition of alloy tool steels.

Code	No.	Steel group	Grade	C	Si	Mn	P	S	Cr	W	Mo	V	Other
T30100	1-1	Measuring and cutting tools	9SiCr	0.85–0.95	1.20–1.60	0.30–0.60	0.030	0.030	0.95–1.25				Co ≤ 1.00
T30000	1-2		8MnSi	0.75–0.85	0.30–0.60	0.80–1.10	0.030	0.030					
T30060	1-3		Cr06	1.30–1.45	≤ 0.40	≤ 0.40	0.030	0.030	0.50–0.70				
T30201	1-4		Cr2	0.95–1.10	≤ 0.40	≤ 0.40	0.030	0.030	1.30–1.65				
T30200	1-5		9Cr2	0.80–0.95	≤ 0.40	≤ 0.40	0.030	0.030	1.30–1.70				
T30001	1-6		W	1.05–1.25	≤ 0.40	≤ 0.40	0.030	0.030	0.10–0.30	0.80–1.20			
T4024	2-1	Impact-resistant tool steel	4CrW2Si	0.35–0.45	0.80–1.10	≤ 0.40	0.030	0.030	1.10–1.30	2.00–2.50			
T40125	2-2		5CrW2Si	0.45–0.55	0.50–0.80	≤ 0.40	0.030	0.030	1.10–1.30	2.00–2.50			
T40126	2-3		6CrW2Si	0.55–0.65	0.50–0.80	≤ 0.40	0.030	0.030	1.10–1.30	2.20–2.70			
T40100	2-4		6CrMnSi2Mol	0.50–0.45	1.75–2.25	0.60–1.00	0.030	0.030	0.10–0.50		0.20–1.35	0.15–0.35	
T40300	2-5		5CR3Mn1SiMolV	0.45–0.55	0.20–1.00	0.20–0.90	0.030	0.030	3.00–3.50		1.30–1.80	≤ 0.35	
T21200	3-1	Cold-working die steel	Cr12	2.00–2.30	≤ 0.40	≤ 0.40	0.030	0.030	11.5–13.0				
T21202	3-2		Cr12Mo1V1	1.40–1.60	≤ 0.60	≤ 0.60	0.030	0.030	11.00–13.00		0.70–1.20	0.50–1.10	
T21201	3-3		Cr12MoV	1.45–1.70	≤ 0.40	0.40	0.030	0.030	11.00–12.50		0.40–0.60	0.15–0.30	
T20503	3-4		Cr5Mo1V	0.95–1.05	≤ 0.4	≤ 1.00	0.030	0.030	4.75–5.50		0.90–1.40	0.15–0.50	
T2000	3-5		9Mn2V	0.85–0.95	≤ 0.4	1.72–2.00	0.030	0.030				0.10–0.25	
T20111	3-6		CrWMn	0.90–1.05	≤ 0.4	0.80–1.10	0.030	0.030	0.90–1.20	1.20–1.60			Nb: 0.20–0.35
T20110	3-7		9CrWMn	0.85–0.95	≤ 0.4	0.80–1.20	0.030	0.030	0.50–0.80	0.50–0.80			
T20431	3-8		Cr4W2MoV	1.12–1.25	≤ 0.4	≤ 0.4	0.030	0.030	3.50–4.00	1.90–2.00	0.80–1.20	0.80–1.10	
T20432	3-9		6Cr4W3Mo2VNb	0.60–0.70	≤ 0.4	≤ 0.4	0.030	0.030	3.80–4.40	2.50–3.50	1.80–2.50	0.80–1.20	
T20465	3-10		6W6Mo5Cr4V	0.55–0.65	≤ 0.4	≤ 0.6	0.030	0.030	3.70–4.30	6.00–7.00	4.50–5.50	0.70–1.10	
T20104	3-11		7CrSiMnMoV	0.65–0.75	0.85–1.15	0.65–1.05	0.030	0.030	0.90–1.20		0.20–0.50	0.15–0.30	

Table 11.38: Grades and chemical compositions of alloy tool steels.

Code	No.	Steel group	Grade	Chemical composition (%)									
				C	Si	Mn	P	S	Cr	W	Mo	V	Al	其它
T20102	4-1	Hot-working die steel	5CrMnMo	0.50–0.60	0.25–0.60	1.20–1.60	0.030	0.030	0.60–0.90		0.15–0.30			
T20103	4-2		5CrNiMo	0.50–0.60	≤0.40	0.50–0.80	0.030	0.030	0.50–0.80		0.15–0.30			Ni: 1.40–1.80
T20280	4-3		3Cr2W8V		≤0.40	≤0.40	0.030	0.030	2.20–2.70	7.50–9.00		0.20–0.25		
T20403	4-4		5Cr4Mo3SiMnVAl	0.47–0.57	0.80–1.10	0.80–1.10	0.030	0.030	3.80–4.30		2.80–3.40	0.80–1.20	0.30–0.70	
T20323	4-5		3Cr3Mo3W2V	0.32–0.42	0.60–0.90	≤0.65	0.030	0.030	2.80–3.30	1.20–1.80	2.50–3.00	0.80–1.20		
T20452	4-6		5Cr4W5Mo2V	0.40–0.50	≤0.40	≤0.40	0.030	0.030	3.40–4.40	4.50–5.30	1.50–2.10	0.70–1.10		
T20300	4-7		8Cr3	0.75–0.85	≤0.40	≤0.40	0.030	0.030	3.20–3.80					
T20101	4-8		4CrMnSiMoV	0.35–0.45	0.80–1.10	0.80–1.10	0.030	0.030	1.30–1.50		0.40–0.60	0.20–0.40		
T20303	4-9		4Cr3Mo3SiV	0.35–0.45	0.80–1.20	0.25–0.70	0.030	0.030				0.25–0.75		
T20501	4-10		4Cr5MoSiV	0.33–0.43	0.80–1.20	0.20–0.50	0.030	0.030				0.30–0.60		
T20502	4-11		4Cr5MoSiV1	0.32–0.45	0.80–1.20	0.20–0.50	0.030	0.030	4.75–5.50		1.10–1.75	0.80–1.20		
T20520	4-12		4Cr5W2VSi	0.32–0.42	0.80–1.20	≤0.40	0.030	0.030	4.50–5.50	1.60–2.40		0.60–1.00		

Table 11.39: Chemical composition of low-alloy tool steels containing vanadium.

Steel	Code	AISI	Chemical composition (%)							
			C	Mn	Si	Cr	V	Mo	Ni	Co
Cr-system	211	L2,L3	0.65–1.10	0.10–0.90	0.25	0.70–1.70	0.20	–	–	–
	220	L2	0.45–0.65	0.30	0.25	0.70–1.20	0.20	–	–	–
	221	–	0.45–0.66	0.70	0.25	0.70–1.20	0.20	0.25	–	–
	224	–	0.55	0.90	0.25	1.10	0.10	0.45	–	–
	225	–	0.45	0.85	0.30	1.15	0.10	0.55	–	–
	226	–	0.45	0.30	0.25	1.60	0.25	1.10	–	–
Ni-system	242	–	0.55	0.55	0.80	1.00	0.15	0.75		–
	243	–	0.55	0.90	1.00	0.40	0.15	0.45	2.70	–
Si-system	310	S2	0.65	0.50	1.00	–	0.20	0.50	–	–
	311	–	0.55	0.50	1.00	–	0.20	0.50	–	–
	312	S4	0.55	0.80	2.00	0.25	0.20	–	–	–
	313	S5	0.55	0.80	2.00	0.25	0.20	0.40	–	–
	314	S6	0.45	1.40	2.25	1.50	0.30	0.40	–	–
	315	–	0.55	0.90	2.00	0.25	0.25	1.20	–	–
W-system	353	F1	1.25	0.30	0.30	0.35	0.15	–	–	–
	354	F2	0.90	0.25	0.25	0.35	0.15	–	–	–

AISI, American Institute of Steel Standards.

Table 11.40: Toughening components of V-containing impact-resistant cutting tools.

Steel	AISI	Chemical composition (%)							
		C	Mn	Si	Cr	Ni	V	W	Mo
320	S1	0.45	0.25	0.25	1.40		0.25	2.25	
321	S1	0.55	0.25	0.25	1.40		0.25	2.50	0.30
399	S1	0.55	0.25	0.90	1.40		0.25	2.25	0.50

AISI, American Institute of Steel Standards.

Table 11.54 shows the chemical composition (%) (GB/T5310−2008) of heat-resistant V-containing steel tubes for boilers.

Thermostable tougheners (%) (GB/T8732−2004) for gas turbine blades are given in Table 11.55.

Table 11.56 shows the thermostable tempering composition (%) (GB/t12773−2008) of V-containing steels for valves of internal combustion engines.

Table 11.41: Chemical compositions of low-vanadium die steels treated using oil quenching or common air quenching.

Code	AISI	Chemical composition (%)							
		C	Mn	Si	Cr	V	W	Mo	Other
410	O1	0.95	1.20	0.25	0.50	0.20	0.50		
411	O2	0.95	1.60	0.25	0.20	0.15		0.30	
413	O7	1.20	0.25	0.25	0.60	0.20	1.60	0.25	
420	A2	1.00	0.60	0.25	5.00	0.25		1.00	
429		1.00	0.60	0.25	3.00	0.25	1.05	2.20	1.00 Ti

AISI, American Institute of Steel Standards.

Table 11.42: Chemical composition of hot die steels with medium and high vanadium content.

Code	AISI	Chemical composition (%)								
		C	Mn	Si	Cr	Ni	V	W	Mo	Other
422	A3	1.25	0.60	0.26	6.00	—	1.00	—	1.00	—
427	A8	0.55	0.30	1.00	5.00	—	0.40	1.25	1.25	—
428	A9	0.50	0.40	1.00	5.00	1.50	1.00	—	1.40	—
430	D2	1.50	0.30	0.25	12.00	—	0.60	—	0.80	—
431	D4	2.20	0.30	0.25	12.00	—	0.50	—	0.80	—
432	D3	2.20	0.30	0.25	12.00	0.50	0.60	—	—	—
434	D5	1.50	0.30	0.50	12.50	0.35	0.50	—	1.00	2.00Co
435	D1	1.00	0.30	0.25	12.00	—	0.60	—	0.80	—

AISI, American Institute of Steel Standards.

Table 11.57 gives the chemical composition (%) of heat-resistant structural steel containing vanadium (GB/T3077–1999).

Table 11.58 shows the thermostable tempering compositions of V-containing steel for rotors (JB/T1265–85, JB/T1265–93).

11.4.11 Stainless steel

Vanadium is both a carbide-forming and ferrite-forming element. In martensite stainless steel containing 12% Cr, vanadium promotes the formation of the precipitated carbide phase and enhances the secondary hardening effect. Table 11.59 shows the chemical composition (%) of 00Cr22Ni13Mn5Mo2N stainless steel.

Table 11.60 shows the chemical composition of 90Cr18MoV steel.

Table 11.43: Chemical composition of hot die steels with medium and high vanadium contents.

Steel	AISI	Chemical composition (%)							
		C	Mn	Si	Cr	V	W	Mo	Co
511	–	0.95	0.30	0.30	4.00	0.50	–	0.50	–
512	–	0.60	0.30	0.30	4.00	0.75	–	0.50	–
520	H11	0.35	0.30	1.00	5.00	0.40	–	1.50	–
521	H13	0.35	0.30	1.00	5.00	1.00	–	1.50	–
522	H12	0.35	0.30	1.00	5.00	0.40	1.50	1.50	–
523	–	0.40	0.60	1.00	3.50	1.00	1.25	1.00	–
524	H10	0.40	0.55	1.00	3.25	0.40	–	2.50	4.25
531	H19	0.40	0.30	0.30	4.25	2.00	4.25	0.40	0.50
532	–	0.45	0.30	1.00	5.00	0.50	3.75	1.00	–
536	H23	0.30	0.30	0.50	12.00	1.00	12.00	–	–
540	H21	0.35	0.30	0.30	3.50	0.50	–	9.00	–
541	H20	0.35	0.30	0.30	2.00	0.50	–	9.00	–
543	H22	0.35	0.30	0.30	2.00	0.40	–	11.00	–
544	–	0.30	0.30	0.30	3.50	0.40	3.60	12.00	–
545	525	0.25	0.30	0.30	4.00	1.00	–	15.00	–
546	–	0.40	0.30	0.30	3.50	0.40	–	14.00	–
547	H24	0.45	0.30	0.30	3.00	0.50	–	15.00	–
549	H26	0.50	0.30	0.30	4.00	1.00	–	18.00	–
550	H15	0.35	0.30	0.40	3.75	0.75	1.00	6.00	–
551	H15	0.40	0.30	0.50	5.00	0.75	1.00	5.00	–
552	H43	0.55	0.30	0.30	4.00	2.00	–	8.00	–
553	H42	0.65	0.30	0.30	3.50	2.00	6.40	5.00	–
554	H41	0.65	0.30	0.30	4.00	1.00	1.50	8.00	–
556	–	0.10	0.30	0.30	2.50	0.50	4.00	5.00	25.00

AISI, American Institute of Steel Standards.

Table 11.61 shows the chemical composition of V-containing Fe-based and Fe−Ni-based alloys (%) (GB/T14992−2005).

11.5 Vanadium in cast iron

Cast iron for industrial use generally contains 2%−4% carbon, mainly in the form of graphite and sometimes as cementite. In addition to carbon, cast iron also contains 1%−3%

Table 11.44: Chemical composition of special wear-resistant die steel with ultra-high vanadium content.

Code	AISI	Chemical composition (%)						
		C	Mn	Si	Cr	V	W	Mo
440	A7	2.3	0.50	0.50	5.25	4.75	1.10	1.10
441	–	2.20	0.40	0.30	4.00	4.00	–	–
442	D7	2.40	0.40	0.40	12.5	4.00	–	1.10
443	–	1.50	0.30	0.30	17.25	4.00	–	–
445	–	1.40	0.40	0.30	0.50	3.75	–	–
446	–	3.25	0.30	0.30	1.00	12.00	–	–
447	–	2.70	0.70	0.40	8.25	4.50	–	–
448	–	1.10	–	1.00	5.25	4.00	–	–
449	–	2.45	0.50	0.90	5.25	9.75	–	–

AISI, American Institute of Steel Standards.

Si, along with Mn, P, S, and other elements. Alloyed cast iron also contains Ni, Cr, Mo, Al, Cu, B, V, and other elements. Carbon and silicon are the main elements affecting the microstructure and properties of cast iron. Ordinary cast iron is obtained by adding appropriate alloying elements (such as Si, Mn, P, Ni, Cr, Mo, Cu, Al, B, V, and Sn). The matrix structure of cast iron is modified by the addition of alloying elements, resulting in enhanced heat resistance, wear resistance, corrosion resistance, low-temperature resistance, and elimination of magnetic behavior. These are used for manufacturing parts and components for mining, chemical equipment, instruments, and meters. Vanadium mainly exists in cast iron in the following three states: (1) solid solution in α-Fe; (2) precipitated phase; (3) block compound. Vanadium is distributed in α-Fe, cementite, and alloy carbides and nitrides, and rarely as oxides. Vanadium in α-Fe, pearlite, and cementite occurs in a solid solution, while the vanadium in carbides and nitrides is in a combined state. Vanadium has a strong affinity for C and N. The carbon content of cast iron is high and carbides can be formed at all temperature ranges. In addition, the nitrogen in cast iron easily produces nitrides and carbonitrides, along with vanadium carbides.

During the solidification process a considerable part of the vanadium in the cast iron is precipitated in the form of massive carbides, nitrides, and carbonitrides. After solidification the solubility of vanadium decreased gradually with decreasing temperature, which promotes continuous precipitation of vanadium carbides during cooling, which are dispersed in the cast iron matrix. Table 11.62 shows the grades, mechanical properties, and uses of wear-resistant cast iron used for guide rails of machine tools, while Table 11.63 shows the corresponding chemical compositions.

Table 11.45: Grades and chemical compositions of Chinese high-speed tool steels.

No.	Code	Grade	C	Mn	Si	S	P	Cr	V	W	Mo	Co
1	T63342	W2Mo8Cr4V	0.95–1.03	≤0.40	≤0.45	≤0.030	≤0.030	3.8–4.5	2.20–2.50	2.70–3.00	2.50–2.90	
2	T64340	W3Mo3Cr4VSi	0.83–0.93	0.20–0.40	0.70–1.00	≤0.030	≤0.030	3.8–4.5	1.20–1.80	3.50–4.50	2.50–3.50	
3	T51841	W18Cr4V	0.73–0.83	0.10–0.40	0.20–0.40	≤0.030	≤0.030	3.8–4.5	1.00–1.20	17.2–18.7		
4	T62841	W2Mo8Cr4V	0.77–0.87	≤0.40	≤0.70	≤0.030	≤0.030	3.8–4.5	1.00–1.40	1.40–2.00	8.00–9.00	
5	T62942	W2Mo9Cr4V2	0.95–1.05	0.15–0.40	≤0.70	≤0.030	≤0.030	3.8–4.5	1.75–2.20	1.50–2.10	8.20–9.20	
6	T66541	W6Mo5Cr4V2	0.80–0.90	0.15–0.40	0.20–0.45	≤0.030	≤0.030	3.8–4.40	1.75–2.20	5.50–6.75	4.50–5.50	
7	T66542	CW6Mo5Cr4V2	0.86–0.94	0.15–0.40	0.20–0.45	≤0.030	≤0.030	3.8–4.5	1.75–2.10	5.90–6.70	4.70–5.20	
8	T66642	W6Mo6Cr4V2	1.00–1.10	≤0.40	≤0.45	≤0.030	≤0.030	3.50–4.5	2.30–2.60	5.90–6.70	5.50–6.50	
9	T69341	W9Mo3Cr4V	0.77–0.87	0.20–0.40	0.20–0.40	≤0.030	≤0.030	3.5–4.5	1.30–1.70	8.50–9.50	2.70–3.30	
10	T66543	W6Mo5Cr4V3	1.15–1.25	0.15–0.40	0.20–0.45	≤0.030	≤0.030	3.8–4.4	2.70–3.20	5.90–6.70	4.70–5.20	
11	T66545	CW6Mo5Cr4V3	1.25–1.32	0.15–0.40	≤0.70	≤0.030	≤0.030	3.8–4.50	2.70–3.20	5.90–6.70	4.70–5.20	
12	T66544	W6Mo5Cr4V4	1.25–1.40	≤0.40	≤0.45	≤0.030	≤0.030	3.75–4.5	3.70–4.20	5.20–6.00	4.20–5.00	
13	T66546	W6Mo5Cr4V2Al	1.05–1.15	0.15–0.40	0.20–0.60	≤0.030	≤0.030	3.8–4.40	1.75–2.20	4.50–5.50	4.50–5.50	Al: 0.80–1.20
14	T71245	W12Cr4V5Co5	1.50–1.60	0.15–0.40	0.15–0.40	≤0.030	≤0.030	3.75–5.00	4.50–5.25	11.75–13.00	—	4.75–5.35
15	T76545	W6Mo5Cr4V2Co5	0.87–0.95	0.15–0.40	0.20–0.45	≤0.030	≤0.030	3.8–4.5	1.70–2.10	5.90–6.70	4.70–5.20	4.50–5.00
16	T76438	W6Mo5Cr4V4Co8	1.23–1.33	≤0.40	≤0.70	≤0.030	≤0.030	3.8–4.5	2.70–3.20	5.90–6.70	4.70–5.30	8.00–8.80
17	T77445	W7Mo4Cr4V2Co8	1.05–1.15	0.20–0.60	0.15–0.50	≤0.030	≤0.030	3.75–4.5	1.75–2.25	6.25–7.00	3.25–4.25	4.75–5.75
18	T72948	W2Mo9Cr4V2Co8	1.05–1.15	0.15–0.40	0.15–0.65	≤0.030	≤0.030	3.5–4.25	0.95–1.35	1.15–1.85	9.00–10.00	7.75–8.75
19	T71010	W10Mo4Cr4V3Co10	1.20–1.35	≤0.40	≤0.45	≤0.030	≤0.030	3.8–4.5	3.00–3.50	9.00–10.00	3.20–3.90	9.50–10.50

Table 11.46: Basic properties of HSS carbides.

Metal elements	Atomic radius ratio r_e/r_m	Carbide		Lattice structure	Melting point (°C)	Hardness HRC[a]
		Category	Chemical formula			
Zr	0.48	MC	ZrC	Face-centered cubic	3500	2840
Ti	0.554	MC	TiC	Face-centered cubic	3200	2850
Nb	0.53	MC	NbC	Face-centered cubic	3500	2050
V	0.57	MC	VC	Face-centered cubic	~2750	2010
W	0.55	M_2C	W_2C	Close-packed hexagonal Complex cubic	2750 —	—
Mo	0.56	M_6C	Mo_2C Fe_3Mo_3C	Close-packed hexagonal Complex cubic	2700 —	1480 —
Cr	0.6	M_7C_3 $M_{23}C_6$	Cr_7C_3 $Cr_{23}C_6$	Complex cubic Complex cubic	~1670 ~1550	2100 1650
Fe	0.61	M_3C	Fe_3C	Complex cubic	~1600	~1300

[a] HRC (Rockwell hardness), an academic concept developed by S. P. Rockwell in 1921, is the hardness value of metallic materials measured using the Rockwell hardness tester. This value has no unit and is represented only by the code "HR". HRC is the hardness obtained by 150 Kg load and 120° Diamond cone impactor. It is used for extremely hard materials. For example: hardened steel and so on.

The morphology and distribution of vanadium in cast iron are affected by the chemical composition and cooling rate. The research results show that obvious blocky compounds are formed when the vanadium content is above 0.1%. The size and shape of the block gradually change with increasing V content, from bone stick, triangle, and square forms into Y-shaped, irregular polygon, and pattern-shaped particles, where the number and size of the particles also increase. The cooling speed mainly affects the particle size, which increases with decreasing cooling rate.

Table 11.47: Chemical composition of general high-speed steel.

Steel	C	Mn	S, P	Si	Cr	W	V	Mn
W18Cr4V (Ti)	0.70–0.80	0.10–0.40	≤0.030	0.20–0.40	3.80–4.40	17.50–19.00	1.00–1.10	≤0.30
W6Mo5cR4V2 (M2)	0.80–0.90	0.15–0.40	≤0.030	0.20–0.40	3.80–4.40	5.50–6.75	1.75–2.20	4.50–5.50
W9Mo3Cr4V	0.77–0.87	0.20–0.40	≤0.030	0.20–0.40	3.80–4.40	8.50–9.50	1.30–1.70	2.70–3.30
W2Mo9Cr4V2 (M7)	0.97–1.05	0.15–0.40	≤0.030	0.20–0.55	3.50–4.40	1.40–2.10	1.75–2.25	8.20–9.20
M1	0.78–0.88	0.15–0.40	≤0.030	0.20–0.50	3.50–4.00	1.40–2.10	1.00–1.35	8.20–9.20
M10	0.84–1.05	0.15–0.40	≤0.030	0.20–0.45	3.75–4.50		1.80–2.20	7.75–8.50

Table 11.48: Semihigh-speed steel compositions.

Symbol	Code	Chemical composition (%)						
		C	Mn	Si	Cr	V	W	Mo
360	0-4-4-1	0.80	0.25	0.25	4.00	1.10	–	4.25
361	0-4-4-2	0.90	0.25	0.25	4.00	2.00	1.00	4.25
362	0-4-4-3	1.20	0.25	0.25	4.00	3.15	–	4.25
363	0-4-4-4	1.40	0.25	0.25	4.00	4.15	–	4.25
364	3-2 (1/2)-4-2	0.95	0.25	0.25	4.00	2.30	2.80	2.50
365	1-2-4-2	0.90	0.25	0.25	4.00	2.25	1.00	2.00
366	1 (1/2)-1 (1/2)-4-3	1.20	0.25	0.25	4.00	2.90	1.40	1.60
367	2-1-4-2	0.95	0.25	0.25	4.00	2.20	1.90	1.10
368	2 (1/2)-2 (1/2)-4-4	1.10	0.25	0.25	4.00	4.00	2.50	2.60
369	2-5-4-1	0.95	0.25	0.25	4.00	1.20	1.70	5.00

Table 11.49: Chemical composition of AISI (United States) high-vanadium high-speed steel.

Steel	C	W	Mo	Cr	V	Co
M3 (1)	1.00–1.10	5.00–6.75	4.75–6.50	3.75–4.50	2.25–2.75	
M3 (2)	1.15–1.25	5.00–6.75	4.75–6.50	3.75–4.50	2.25–2.75	
M4	1.25–1.40	5.25–6.50	4.25–5.50	2.75–4.75	2.25–2.75	
T15	1.50–1.60	11.75–13.00	≤ 1.00	4.50–5.25	2.25–5.00	4.75–5.25

Table 11.50: Chemical composition of V3N nitrogenous super-high-speed steel.

C	W	Mo	Cr	V	Co	N
1.15–1.25	11.00–12.50	2.70–3.20	3.50–4.10	2.50–3.10	–	0.04–0.10

Table 11.51: Chemical composition of powder metallurgy die steels.

Steel grade		C	Cr	W	Mo	V	Co	Other	HRC[a]
Cold working die steel	CPM9V	1.78	5.25		1.30	9.00		S : 0.03	53–55
	CPM10V	2.45	5.25		1.30	9.75		S : 0.07	60–62
	CPM440V	2.15	17.50		0.50	5.75			57–59
	Vanadis4	1.50	8.00		1.50	4.00			59–63
Hot working die steel	CPMH13	0.40	5.00		1.30	1.05			42–48
	CPMH19	0.40	4.25	4.25	0.40	2.10	4.25		44–52
	CPMH19V	0.80	4.25	4.25	0.40	4.00	4.25		44–56
High-speed tool steel	ASP23	1.28	4.20	6.40	5.00	3.10			65–67
	ASP30	1.28	4.20	6.40	5.00	3.10	8.50		66–68
	ASP60	2.30	4.00	6.50	7.00	6.50	10.50		67–69
	CPMRexM3HCHS	1.30	4.00	6.25	5.00	3.00		S : 0.27	65–67
	CPMRexT15HS	1.55	4.00	12.25		5.00	5.00	S : 0.06	65–67

[a]HRC (Rockwell hardness), an academic concept developed by S. P. Rockwell in 1921, is the hardness value of metallic materials measured using the Rockwell hardness tester. This value has no unit and is represented only by the code "HR". HRC is the hardness obtained by 150 Kg load and 120° Diamond cone impactor. It is used for extremely hard materials. For example: hardened steel and so on.

Table 11.52: Chemical composition (%) (GB/T1221−2007) of heat-resistant V-containing steel rods.

Element	14Cr11MoV (1Cr11MoV)	18Cr12MoVNbN (2Cr12MoVNbN)	15Cr12WMoV (1Cr12WMoV)	22Cr12NiWMoV (2Cr12NiWMoV)	13Cr11Ni2W2MoV (1Cr11Ni2W2MoV)	18Cr11NiMoNbVN (2Cr11NiMoNbVN)	06Cr15Ni25Ti2MoAlVB (0Cr15Ni25Ti2MoAlVB)
C	0.11−0.18	0.15−0.20	0.12−0.18	0.20−0.25	0.10−0.16	0.15−0.20	≤0.08
Si	≤0.50	≤0.50	≤0.50	≤0.50	≤0.60	≤0.50	≤1.00
Mn	≤0.60	0.50−1.00	0.50−0.90	0.50−1.00	≤0.60	0.50−0.80	≤2.00
Cr	10.00−11.50	10.00−13.00	10.00−13.00	10.00−13.00	10.50−12.00	10.00−12.00	13.50−16.00
Mo	0.50−0.70	0.30−0.90	0.50−0.70	0.75−1.26	0.35−0.50	0.60−0.90	1.00−1.50
V	0.25−0.40	0.10−0.40	0.15−0.30	0.20−0.40	0.18−0.30	0.20−0.30	0.10−0.50
Ti							1.90−2.35
B							0.001−0.010
Ni	≤0.60	≤0.60	0.40−0.80	0.50−1.00	1.40−1.80	0.30−0.60	24.00−27.00
Al						≤0.30	≤0.35
Nb		0.20−0.60				0.20−0.60	
N		0.05−0.10				0.04−0.09	
W			0.70−1.10	0.75−1.25	1.50−2.00		
P	≤0.035	≤0.035	≤0.035	≤0.040	≤0.035	≤0.030	≤0.040
S	≤0.030	≤0.030	≤0.030	≤0.030	≤0.030	≤0.025	≤0.030

Table 11.53: Chemical composition of V-containing heat-resistant steel plates and steel strips (%) (GB/t4238−2007).

No.	Grade	C	Si	Mn	Cr	Mo	V
1	22Cr12NiMoWV	0.20−0.25	≤0.50	0.50−1.00	11.00−12.50	0.90−1.25	0.20−0.30
2	06Cr15Ni25Ti2MoAlVB	≤0.08	≤1.00	≤2.00	13.50−16.00	1.00−1.50	0.100.50

No.	Grade	Ti	B	Ni	Al	W	P	S
1	22Cr12NiMoWV			0.50−1.00		0.90−1.25	≤0.025	≤0.025
2	06Cr15Ni25Ti2MoAlVB	1.90−2.35	0.001−0.010	24.00−27.00	≤0.35		≤0.040	≤0.040

Table 11.54: Chemical composition (%) (GB/T5310−2008) of heat-resistant V-containing steel tubes for boilers.

No.	Grade	C	Si	Mn	Cr	Mo	V	Ti	B
1	12Cr1MoVG	0.08−0.15	0.17−0.37	0.40−0.70	0.9−1.20	0.25−0.35	0.15−0.30		
2	12Cr2MoWVTiB (G102)	0.08−0.15	0.45−0.75	0.45−0.65	1.60−2.10	0.50−0.65	0.28−0.42	0.08−0.18	0.0020−0.0080
3	07CrMoW2VNbB (T/P23)	0.04−0.10	≤0.50	0.10−0.60	1.90−2.60	0.05−0.30	0.20−0.30		0.0005−0.0060
4	08Cr2Mo1VTiB (T/P24)	0.05−0.10	0.15−0.45	0.30−0.70	2.20−2.60	0.90−1.10	0.20−0.30	0.06−0.10	0.0015−0.0070
5	12Cr3MoVSiTiB	0.09−0.15	0.60−0.90	0.50−0.80	2.50−3.00	1.00−1.20	0.25−0.35	0.22−0.38	0.0050−0.0110
6	10Cr9Mo1VNbN (T/P91)	0.08−0.12	0.20−0.50	0.30−0.60	8.00−9.00	0.85−1.05	0.18−0.25		
7	10Cr9MoW2VNbN (T/P92)	0.07−0.13	≤0.50	0.30−0.60	8.50−9.50	0.30−0.60	0.15−0.25		0.0010−0.0060
8	10Cr11MoW2VNbBN (T/P122)	0.07−0.14	≤0.50	≤0.70	10.00−12.50	0.25−0.60	0.15−0.30		0.0005−0.0050
9	11Cr9Mo1W1VNbBN (T/P911)	0.09−0.13	0.10−0.50	0.30−0.60	8.50−9.50	0.90−1.10	0.18−0.25		0.003−0.0060

No.	Grade	Ni	Al	Cu	Nb	N	W	P	S
1	12Cr1MoVG							≤0.025	≤0.010
2	12Cr2MoWVTiB (G102)						0.30−0.55	≤0.025	≤0.015
3	07CrMoW2VNbB (T/P23)		≤0.030		0.02−0.08	≤0.030	1.45−1.75	≤0.025	≤0.010
4	08Cr2Mo1VTiB (T/P24)		≤0.02			≤0.012		≤0.020	≤0.010
5	12Cr3MoVSiTiB							≤0.025	≤0.015
6	10Cr9Mo1VNbN (T/P91)	≤0.040	≤0.040		0.060−0.10	0.03−0.070		≤0.020	≤0.010
7	10Cr9MoW2VNbN (T/P92)	≤0.040	≤0.040		0.04−0.09	0.030−0.070	1.50−2.00	≤0.020	≤0.010
8	10Cr11MoW2VNbBN (T/P122)	≤0.050	≤0.040	0.30−1.70	0.04−0.10	0.04−0.100	1.50−2.50	≤0.020	≤0.010
9	11Cr9Mo1W1VNbBN (T/P911)	≤0.040	≤0.040		0.06−0.10	0.040−0.090	0.90−1.10	≤0.020	≤0.010

Table 11.55: Thermostable tougheners (%) (GB/T8732−2004) for gas turbine blades.

No.	Grade	C	Si	Mn	Cr	Mo	V	Ni
1	1Cr11MoV	0.11−0.18	≤0.50	≤0.60	11.00−11.50	0.50−0.70	0.25−0.40	≤0.60
2	1Cr12W1MoV	0.12−0.18	≤0.50	0.50−0.90	11.00−13.00	0.50−0.70	0.15−0.30	0.40−0.80
3	2Cr12MoV	0.18−0.24	0.10−0.50	0.30−0.80	11.00−12.50	0.80−1.20	0.25−0.35	0.30−0.60
4	2Cr11NiMoNbVN	0.15−0.20	≤0.50	0.50−0.80	10.0−12.0	0.60−0.90	0.20−0.30	0.30−0.60
5	2Cr12NiMo1W1V	0.20−0.25	≤0.50	0.50−1.00	11.00−12.50	0.90−1.25	0.20−0.30	0.50−1.00

No.	Grade	Al	Cu	Nb	N	W	P	S
1	1Cr11MoV		≤0.30				≤0.030	≤0.025
2	1Cr12W1MoV		≤0.30			0.70−1.10	≤0.030	≤0.025
3	2Cr12MoV		≤0.30				≤0.030	≤0.025
4	2Cr11NiMoNbVN	≤0.03	≤0.10	0.20−0.60	0.04−0.09		≤0.020	≤0.015
5	2Cr12NiMo1W1V				0.90−1.25		≤0.030	≤0.025

Table 11.56: Thermostable tempering composition (%) (GB/t12773−2008) of V-containing steels for valves of internal combustion engines.

No.	Grade	C	Si	Mn	Cr	Mo	V	Ti	Cu
1	85Cr18Mo2V	0.80−0.90	≤1.00	≤1.50	16.5−18.5	2.00−2.50	0.30−0.60		≤0.30
2	86Cr18W2VRe	0.82−0.92	≤1.00	≤1.50	16.5−18.5	—	0.30−0.60		≤0.30
3	61Cr21Mn10Mo1V1Nb1N	0.57−0.65	≤0.25	'9.50−11.50	20.00−22.00	0.75−1.25	0.75−1.00		≤0.30

No.	Grade	Ni	Al	Re	Nb	N	W	P	S
1	85Cr18Mo2V							≤0.040	≤0.30
2	86Cr18W2VRe			≤0.20			2.00−2.50	≤0.035	≤0.30
3	61Cr21Mn10Mo1V1Nb1N	≤1.50			1.00−1.20	0.40−0.60		≤0.050	≤0.30

Table 11.57: Chemical composition (%) of heat-resistant structural steel containing vanadium (GB/t3077–1999).

No.	Grade	C	Si	Mn	Cr	Mo
1	50CrVA	0.47–0.54	0.17–0.37	0.50–0.80	0.80–1.10	
2	12CrMoV	0.08–0.15	0.17–0.37	0.40–0.70	0.30–0.60	0.25–0.35
3	12Cr1MoV	0.08–0.15	0.17–0.37	0.40–0.70	0.90–1.20	0.25–0.35
4	35CrMoV	0.30–0.38	0.17–0.37	0.40–0.70	1.00–1.30	0.20–0.30
5	25Cr2MoVA	0.22–0.29	0.17–0.37	0.40–0.70	1.50–1.80	0.25–0.35
6	25Cr2MoVA	0.22–0.29	0.17–0.37	0.50–0.80	2.10–2.50	0.90–1.10
7	20Cr3MoWV (GB3077–1988)	0.17–0.27	0.20–0.40	0.25–0.60	2.40–3.30	0.35–0.55

No.	Grade	V	Ni	W	P	S
1	50CrVA	0.10–0.20				
2	12CrMoV	0.15–0.30				
3	12Cr1MoV	0.15–0.30				
4	35CrMoV	0.10–0.20				
5	25Cr2MoVA	0.15–0.30				
6	25Cr2MoVA	0.30–0.50				
7	20Cr3MoWV (GB3077–1988)	0.60–0.85	≤ 0.50	0.30–0.50	≤ 0.035	≤ 0.030

Table 11.58: Thermostable tempering compositions of V-containing steel for rotors (JB/T1265−85, JB/T1265−93).

No.	Grade	Chemical composition (%)					
		C	Si	Mn	Cr	Mo	V
1	30Cr1Mo1V	0.27−0.34	0.17−0.37	0.70−1.00	1.05−1.35	1.00−1.30	0.21−0.29
2	30Cr2MoV	0.22−0.32	0.30−0.35	0.50−0.80	1.50−1.80	0.60−0.80	0.20−0.30
3	28CrMoNiVE	0.25−0.30	≤0.30	0.30−0.80	1.10−1.40	0.80−1.00	0.25−0.35

No.	Grade	Chemical composition (%)				
		Ni	Al	Cu	S	P
1	30Cr1Mo1V	≤0.50	≤0.010		<0.012	<0.012
2	30Cr2MoV	≤0.30		≤0.20	<0.015	<0.018
3	28CrMoNiVE	0.50−0.75	≤0.010	≤0.20	<0.012	<0.012

Table 11.59: Chemical composition (%) of 00Cr22Ni13Mn5Mo2N stainless steel.

C	Si	Mn	S	P	Cr	Ni	Mo	N	Co	V
≤0.06	≤1.0	4−6	≤0.03	≤0.04	20.5−23.5	11.5−13.5	1.5−3.0	0.2−0.3	0.1−0.3	0.1−0.3

Table 11.60: Chemical composition of 90Cr18MoV steel.

C	Si	Mn	S	P	Cr	Mo	V
0.85−0.95	≤0.80	≤0.80	≤0.030	≤0.040	17−19	1.00−1.30	0.07−0.12

Table 11.61: Chemical composition of V-containing Fe-based and Fe−Ni-based alloys (%) (GB/T14992−2005).

No.	Grade	C	Si	Mn	Cr	Mo	V	Ti
1	GH1016	≤0.08	≤0.60	≤1.80	19.00−22.00			
2	GH2036	0.34−0.40	0.30−0.80	7.50−9.50	11.50−13.5	1.10−1.40	1.25−1.55	≤0.12
3	GH2132	≤0.08	≤1.00	≤2.00	13.5−16.0	1.00−1.50	0.10−0.50	1.75−2.30
4	GH2136	≤0.06	≤0.75	≤0.35	13.0−16.0	1.00−1.75	0.01−0.10	2.40−3.20

No.	Grade	B	Ni	Al	Nb	N	W	P	S
1	GH1016		32.00−36.00			0.13−0.25	5.00−6.00	≤0.020	≤0.015
2	GH2036		7.0−9.0		0.25−0.50			≤0.035	≤0.030
3	GH2132	0.001−0.010	24.0−27.0	≤0.40				≤0.030	≤0.020
4	GH2136	0.005−0.025	24.5−28.5	≤0.35				≤0.025	≤0.025

Table 11.62: Grades, mechanical properties, and uses of wear-resistant cast iron used for guide rails of machine tools.

Cast iron	Grade	mechanical property				Use
		Tensile strength σ_b (MPa) $\geq$	Bending strength σ_{bb} (MPa)	Deflection f (mm)	Hardness HBS	
(V, Ti)- containing wear-resistant cast iron	MTVTi20	200	400	3.0	160–240	Casting of small and medium machine tool guides
	MTVTi25	250	470		160–240	
	MTVTi30	300	540		170–240	
(P, Cu, Ti)-containing wear-resistant cast iron	MTPCuTi15	150	330	2.5	170–229	Precision machine beds, columns, and tables
	MTPCuTi20	200	400		187–235	
	MTPCuTi25	250	470		187–241	
	MTPCuTi30	300	540	2.8	187–255	
High-P wear-resistant cast iron	MTP15	150	330	2.5	170–229	Common machine beds, slide plates, and tables
	MTP20	200	400		179–235	
	MTP25	250	470		187–241	
	MTP30	300	540	2.8	187–255	
(Cr,Mo,Cu)-containing wear-resistant cast iron	MTCrMoCu25	250	470		185–230	Small/ medium precision instrument machine tool beds and other guide-rail castings
	MTCrMoCu30	300	540	3.0	200–250	
	MTCrMoCu35	350	610	3.5	220–260	
(Cr,Cu)-containing wear-resistant cast iron	MTCrCu25	250	470	3.0	185–230	
	MTCrCu30	300	540	3.2	200–240	
	MTCrCu35	350	610		210–250	

Table 11.63: Chemical composition of wear-resistant cast iron used for guide rails of machine tools.

Cast iron	Grade	C	Si	Mn	P ≤	S ≤	Cr	Mo	Cu	Other
(V, Ti)-containing wear-resistant cast iron	MTVTi20	3.3–3.7	1.4–2.2	0.5–1.0	0.3	0.12	—	—	—	V ≥ 0.15 Ti ≥ 0.05
	MTVTi25	3.1–3.5	1.3–2.0	0.5–1.1	0.3	0.12	—	—	—	V ≥ 0.15 Ti ≥ 0.05
	MTVTi30	2.9–3.3	1.2–1.8	0.5–1.1	0.3	0.12	—	—	—	V ≥ 0.15 Ti ≥ 0.05
(P, Cu, Ti)-containing wear-resistant cast iron	MTPCuTi15	3.2–3.5	1.8–2.5	0.5–0.9	0.35–0.6	0.12	—	—	0.6–1.0	Ti0.09–0.15
	MTPCuTi20	3.0–3.4	1.5–2.0	0.5–0.9	0.35–0.6	0.12	—	—	0.6–1.0	Ti0.09–0.15
	MTPCuTi25	3.0–3.3	1.4–1.8	0.5–0.9	0.35–0.6	0.12	—	—	0.6–1.0	Ti0.09–0.15
	MTPCuTi30	2.9–3.2	1.2–1.7	0.5–0.9	0.35–0.6	0.12	—	—	0.6–1.0	Ti0.09–0.15
High-P wear-resistant cast iron	MTP15	3.2–3.5	1.6–2.2	0.5–0.9	0.6–0.65	0.12	—	—	—	—
	MTP20	3.1–3.4	1.5–2.0	0.5–0.9	0.6–0.65	0.12	—	—	—	—
	MTP25	3.0–3.2	1.4–1.8	0.5–0.9	0.6–0.65	0.12	—	—	—	—
	MTP30	2.9–3.2	1.2–1.7	0.5–0.9	0.6–0.65	0.12	—	—	—	—
(Cr,Mo,Cu)-containing wear-resistant cast iron	MTCrMoCu25	3.3–3.6	1.8–2.5	0.7–0.9	0.15	0.12	0.10–0.20	0.20–0.35	0.7–0.9	—
	MTCrMoCu30	3.0–3.2	1.6–2.1	0.8–1.0	0.15	0.12	0.10–0.25	0.25–0.45	0.8–1.1	—
	MTCrMoCu35	2.9–3.1	1.5–2.0	0.8–1.0	0.15	0.12	0.15–0.25	0.35–0.50	1.0–1.2	—
(Cr,Cu)-containing wear-resistant cast iron	MTCrCu25	3.2–3.5	1.7–2.0	0.7–0.9	0.30	0.12	0.15–0.25	—	0.6–0.8	—
	MTCrCu30	3.0–3.2	1.5–1.8	0.8–1.0	0.25	0.12	0.20–0.35	—	0.7–1.0	—
	MTCrCu35	2.9–3.1	1.4–1.7	0.8–1.0	0.25	0.12	0.25–0.35	—	0.9–1.1	—

CHAPTER 12

Vanadium metal

Chapter Outline

12.1 Metallic vanadium   333
    12.1.1 Thermal reduction of vanadium oxide   334
    12.1.2 Carbon thermal reduction method   339
    12.1.3 Hydrogen reduction   344
    12.1.4 Reduction of vanadium chloride   346

12.2 Vanadium refining   350
    12.2.1 Vacuum refining methods   351
    12.2.2 Molten salt electrolytic refining   353
    12.2.3 Thermal dissociation of iodides   355
    12.2.4 Regional smelting   356

12.3 Application of vanadium metal   357

Vanadium metal and alloy products include vanadium metal, vanadium—nitrogen alloy, vanadium—aluminum alloy, and vanadium carbide. Vanadium and its alloys are closely related to iron, steel, and nonferrous metal products with respect to their application scale and characteristics and are, hence, important vanadium products. Vanadium metal and alloy products are produced mainly from vanadium pentoxide, vanadium trioxide, and vanadium halide as raw materials, which are processed via reduction using reductants such as carbon, hydrogen, and alkali metals. The alkali metals K and Na are particularly active, but the reactions are not easy to control. Hence, Ca and Al are generally chosen for vanadium oxide reduction. Mg can be used for the vanadium halide reduction, while ferrovanadium chlorination produces vanadium chloride. Vanadium metal and alloy products are thereby produced by metal reduction. According to the product quality and process requirements, metal or hydrocarbon reductants are selected [3,6,7,12,14,25,27,68].

12.1 Metallic vanadium

Thirty years after the discovery of vanadium, British chemist Roscoe reduced vanadium chloride using hydrogen to obtain metallic vanadium with a purity of 96%. Dissolved elements (C, N, H, and O) affect the properties of vanadium. In particular, they increase the hardness and brittleness, which is far from the expectation of good ductility of vanadium.

The process of reducing vanadium oxide to metal vanadium using metal or carbon reduction is an important part of the vanadium metallurgical process. There are four main methods: calcium thermal reduction, vacuum carbon thermal reduction, magnesium thermal reduction of chloride, and aluminothermic reduction. In industry, calcium thermal reduction and the combination of aluminothermic reduction and vacuum electron beam remelting (developed in the 1970s) are used to produce pure vanadium for nuclear reactors. In addition, vacuum carbon thermal reduction is used.

12.1.1 Thermal reduction of vanadium oxide

Fig. 12.1 shows the relationship between the free energy of metal oxides and temperature. It can be seen that the order of stability of the vanadium oxides is

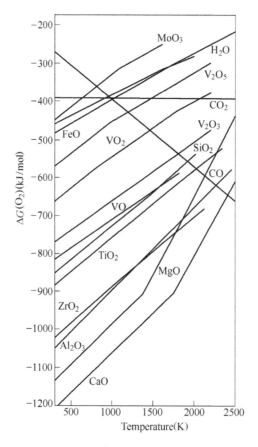

Figure 12.1
Relationship between the free energy of metal oxides and temperature.

VO > V_2O_3 > V_2O_4 > V_2O_5. When carbon reduction is applied to V_2O_5 as the raw material, successive reduction occurs. VO is the most difficult oxide to be reduced. The main oxides in the V–O system are V_2O_5, V_2O_4, V_2O_3, and VO; their standard free energies of formation are:

$$2V(s) + O_2(g) = 2VO(s) \ (1500K - 2000K) \tag{12.1}$$

$$\Delta G_1^\ominus = -803,328 + 148.78T \ (J/mol)$$

$$\frac{4}{3}V(s) + O_2(g) = \frac{2}{3}V_2O_3(s) \ (1500K - 2000K) \tag{12.2}$$

$$\Delta G_2^\ominus = -800,538 + 150.624T \ (J/mol)$$

$$V(s) + O_2(g) = \frac{1}{2}V_2O_4(s) \ (1500K - 1818K) \tag{12.3}$$

$$\Delta G_3^\ominus = -692,452 + 148.114T \ (J/mol)$$

$$\frac{4}{3}V(s) + O_2(g) = \frac{2}{5}V_2O_5(l) \ (1500K - 2000K) \tag{12.4}$$

$$\Delta G_4^\ominus = -579,902 + 126.91T \ (J/mol)$$

12.1.1.1 Metal thermal reduction

Metal thermal reduction is the process of producing metal or its alloys by reducing a compound using another metal with stronger activity. Compounds containing the metal to be reduced include metal oxides, sulfides, chlorides, fluorides, or molten salts, or aggregates or concentrates of these compounds. Excess reductant is separated from the metal–reductant mixture produced by the reaction using slag-making stratification, distillation, or pickling methods. For metal reductants the reduction intensity of the metal elements is generally proportional to the order of metal activity. Metals with lower activity lose electrons less easily, resulting in weaker reduction. The further left the element is in the same period, the more metallic it is. The lower the element is in the same family, the more metallic it is.

Metal reducing order:

$$K > Ca > Na > Al > Mg > Mn > Zn > Cr > Fe > Ni > Sn > Pb > (H)$$
$$> Cu > Hg > Ag > Au > Pt$$

Order of metal cation oxidation:

$$K^+ < Ca^{2+} < Na^+ < Mg^{2+} < Al^{3+} < Mn^{2+} < Zn^{2+} < Cr^{3+} < Fe^{2+} < Ni^{2+} < Sn^{2+}$$
$$< Pb^{2+} < H^+ < Cu^{2+} < Hg^{2+} < Fe^{3+} < Ag^+ < Pt^{2+} < Au^{2+}$$

The metal thermal reduction reaction is as follows:

$$MeX + Me' = Me + Me'X + Q \tag{12.5}$$

Here, MeX is the compound of the metal to be reduced, Me' is the reductant, Me is the obtained metal, and Q is the thermal effect of the reaction. When the heat released by the metal thermal reduction reaction is sufficient to maintain the temperature required for the reaction, the reaction occurs via self-thermal reduction. If excessive reaction heat is released, inert substances can be added to slow down the reaction. If the released heat is not sufficient to maintain the reaction, special heating additives or electric heating are required.

To control the reduction reaction, product quality, and safety, metal thermal reduction is generally carried out under the protection of inert gas, where the molten salt or slag is processed in special containers and electric furnaces. In this case the contamination of the products by the reaction vessels should be avoided. The metal thermal reductant needs to have strong reduction ability. The chemical affinity of the reductant for the nonmetallic components of the compound should be greater than that of the metal. The standard Gibbs free energy of reduction for the reaction is negative. Metal thermal reductants can easily be processed and purified. The reduction products and resulting metals have the characteristics of easy separation, low cost, safety, and reliability. The thermodynamic data for vanadium oxide metal reduction are shown in Table 12.1.

Commonly used metal reductants for vanadium oxide include Ca, Mg, and Al. The reduction product of vanadium oxide is a mixture of vanadium metal and the oxidation product of the

Table 12.1: Thermodynamic data for vanadium oxide metal reduction.

	Equation	ΔH/(kJ/g)	ΔH^θ/(kJ/g)
Main reaction	$V_2O_5 + 5Ca = 2V + 5CaO$	−1621	−4.240
	$V_2O_5 + 3Ca = 2V + 3CaO$	−684	−2.532
	$V_2O_5 + 5Mg = 2V + 5MgO$	−1456	−4.800
	$3V_2O_5 + 10Al = 6V + 5Al_2O_3$	−3735	−4.579
	$V_2O_3 + 2Al = 2V + Al_2O_3$	−459	−2.249
Reaction between promoter and diluent	$Ca + I_2 = CaI_2$	−533	−1.814
	$Ca + S = CaS$	−476	−6.598
	$3BaO + 2Al = 3Ba + Al_2O_3$	−1410	−2.510
	$KClO_3 + Al = KCl + Al_2O_3$	−1251	−7.068
	$NaClO_3 + Al = NaCl + Al_2O_3$	−1285	−8.028

Table 12.2: Physical properties of metal oxides during metal thermal reduction.

Metals and oxides	Al	Ca	Mg	V	Al_2O_3	CaO	MgO	V_2O_3	V_2O_5
Melting point (°C)	660	842	650	1910	2050	2615	2825	1957	678
Boiling point (°C)	2520	1494	1090	3409					

reductant, along with residual reductant. After slag stratification and acid washing and separation, crude vanadium metal products are obtained.

All reactions occurring during the reduction (or thermal reduction) of vanadium oxides by Ca, Mg, and Al have negative free energy. If the thermodynamic conditions are favorable, reactions occur depending on the preset conditions. Vanadium is not reductive or reversible to CaO, MgO, and Al_2O_3. The physical properties of the metal oxides used in the metal thermal reduction process are shown in Table 12.2. During thermal reduction of vanadium oxide with Ca, Mg, and Al, the reaction products include metal vanadium, excess reductant, CaO, MgO, and Al_2O_3. The boiling point of the reductant is above 2000°C, where the difference between the melting points of vanadium metal and its oxide is 140°C–915°C, with separation conditions. The boiling point of metallic vanadium is 3409°C. During the process of metal thermal reduction, toxic or dangerous substances should not be produced.

12.1.1.1.1 Calcium thermal reduction method

Calcium thermal reduction is used for producing vanadium metal on an industrial scale. V_2O_5 or V_2O_3 is used as the raw material, while calcium scrap metal is used as the reducing agent. The amount of added calcium is 60% of the theoretical amount. Calcium scraps are mixed with V_2O_5 or V_2O_3 and added to a magnesium oxide crucible in a steel reaction tank purged with inert gas. Iodine (or sulfur) is added as the heating agent. The amount of added iodine is generally 0.2 mol/1 mol of produced vanadium.

$$V_2O_5 + 5Ca \rightarrow 2V + 5CaO + 1620.07 \text{ kJ} \tag{12.6}$$

$$V_2O_3 + 3Ca \rightarrow 2V + 3CaO + 683.24 \text{ kJ} \tag{12.7}$$

$$V_2O_5 + Ca = V_2O_4 + CaO \tag{12.8}$$

$$V_2O_4 + Ca = V_2O_3 + CaO \tag{12.9}$$

$$V_2O_3 + Ca = VO + CaO \tag{12.10}$$

$$VO + CaO = V + CaO \tag{12.11}$$

$$V_2O_5 + 3Ca = 3V + 3CaO \tag{12.12}$$

Since the reaction is exothermic, external heating stops after the reaction begins and the temperature automatically increases to 2173K. Plastic metal vanadium blocks or vanadium particles are washed with water, and the yield of vanadium is ~74%. When adding Al to the charge, the yield of vanadium can be increased to 82%–97.5%. High-purity V_2O_5 was used as the raw material, with 50%–60% of the theoretical amount of Ca metal, and iodine as a flux and heating agent, which was placed in a sealed reactor or "reaction bomb" for reaction. Dense metal ingots or frits were obtained, which contained ~0.2% C, 0.02%–0.08% O, 0.01%–0.05% N, and 0.002%–0.01% H.

12.1.1.1.2 Aluminothermic reduction method

V_2O_5 is reacted with high-purity Al in the "response" to produce dense V–Al alloy, followed by removal of Al at 1790°C under high vacuum. Using vacuum electron beam remelting, residual Al and impurities such as dissolved oxygen are removed, resulting in a vanadium purity >99.9%. Higher purity vanadium ingots can also be obtained using two stages of electron beam smelting.

Germany uses aluminothermic reduction to produce crude metal vanadium. In this method, vanadium pentoxide and pure Al are placed in the reaction bomb to produce V–Al alloy. The coarse metal vanadium containing 94%–97% vanadium is prepared by removing Al from the vanadium alloy at 2063K. When aluminum is used as the reductant, the reduction reaction of vanadium oxide is as follows:

$$3V_2O_5 + 2Al = 3V_2O_4 + Al_2O_3 \tag{12.13}$$

$$3V_2O_4 + 2Al = 3V_2O_3 + Al_2O_3 \tag{12.14}$$

$$3V_2O_3 + 2Al = 6VO + Al_2O_3 \tag{12.15}$$

$$3VO + 2Al = 3V + Al_2O_3 \tag{12.16}$$

$$3V_2O_5 + 10Al = 6V + 5Al_2O_3 \tag{12.17}$$

Vanadium trioxide is also a stable vanadium oxide and can be used as raw material for the production of vanadium metal. The reduction reaction of vanadium oxide is as follows:

$$3V_2O_3 + 2Al = 2V + Al_2O_3 \tag{12.18}$$

When potassium chlorate is used as a catalytic heat accelerator, the reduction reaction of vanadium oxide is as follows:

$$KClO_3 + 2Al = KCl + Al_2O_3 \tag{12.19}$$

12.1.1.1.3 Magnesium thermal reduction

The purity of Mg metal is high and the price is lower than that of Ca. Magnesium chloride generated by the reaction is more volatile than calcium chloride, so the reduction with Mg is more efficient than that with Ca. The reduction process is as follows: (1) preparation of crude vanadium tetrachloride by chlorination of ferrovanadium containing 80% vanadium; (2) ferric chloride in crude vanadium tetroxide is removed by distillation; (3) vanadium tetrachloride is converted into VCl_3 using a cylindrical magnesium reflux; (4) vanadium oxytrichloride ($VOCl_3$) in VCl_3 is removed by distillation; (5) after cooling, vanadium trichloride is crushed and placed in a reduction reaction tank, where Mg is added to reduce VCl_3 into vanadium under the protection of argon; (6) vacuum distillation is used to remove magnesium and magnesium chloride from the vanadium; (7) residual magnesium chloride in the vanadium metal is removed by washing with water to obtain vanadium powder after drying.

The reduction operation is carried out in a mild steel crucible, which is placed in a mild steel pot and heated using gas. The pickling magnesium ingot is first added to the crucible, and then three times the amount of vanadium trichloride is added. The reduction temperature is controlled to 1023K–1073K. The reaction speed is assessed by monitoring the temperature. If the reaction is slow, Mg is added and the mixture is kept at room temperature for about 7 h. Each batch can produce 18–20 kg of vanadium metal. Then, the crucible is removed, placed in a distillation furnace, slowly heated to 573K, and then kept at this temperature. When the pressure reaches 0.1333–0.6666 Pa, the temperature is increased to 1173K–1223K for 8 h. Then, the sponge-like vanadium is rapidly cooled to room temperature. The purity of the obtained vanadium sponge is 99.5%–99.6%, and the vanadium yield is 96%. The Mg thermal reduction reaction with VCl_2 is as follows:

$$VCl_2 + Mg = MgCl_2 + V \tag{12.20}$$

12.1.2 Carbon thermal reduction method

Carbon or carbide is used as the reducing agent to reduce oxides in metallurgical raw materials to produce metals, alloys, or intermediate products. Nonmetallic reductants for vanadium oxide include carbon, hydrogen, gas, silicon, and natural gas. Carbon thermal reduction can be divided into carbon thermal vanadium oxide reduction, carbon thermal vanadium chloride reduction, and vacuum carbon thermal reduction methods. Generally solid carbon reduction is used, with graphite or carbon powder as reductant. In addition, gas-based reduction is used with gas or natural gas as reductant. Coke oven gas mainly contains HCH and CO but also more complex components compared to natural gas. Hence, the coke oven gas composition depends on the coal type and coking conditions. The chemical reactions involved in the carbothermic reduction of vanadium oxide are as follows:

$$V_2O_5 + C = 2VO_2 + CO\uparrow \tag{12.21}$$

$$\Delta G_T^\ominus(C) = 49,070 - 213.42T \, (J/mol)$$

$$2VO_2 + C = 2V_2O_3 + CO\uparrow \tag{12.22}$$

$$\Delta G_T^\ominus(C) = 95,300 - 158.68T \, (J/mol)$$

$$V_2O_3 + C = 2VO + CO\uparrow \tag{12.23}$$

$$\Delta G_T^\ominus(C) = 239,100 - 163.22T \, (J/mol)$$

$$VO + C = 2V + CO\uparrow \tag{12.24}$$

$$\Delta G_T^\ominus(C) = 310,300 - 166.21T \, (J/mol)$$

$$V_2O_5 + 7C = 2VC + 5CO\uparrow \tag{12.25}$$

$$\Delta G_T^\ominus(C) = 79,824 - 145.64T \, (J/mol)$$

12.1.2.1 Carbon thermal reduction reactions

Production of vanadium metal from vanadium oxide by carbon reduction is achieved above 1700°C, where some stable vanadium carbide (VC or VC_2) can be produced. The stability of CO is higher than that of vanadium oxide, which is reduced in the following sequence during carbon thermal reduction: V_2O_5, V_2O_4, V_2O_3, VO, V(O)s, and finally V. The formation free energy of vanadium oxide and carbon oxide is shown in Fig. 12.2.

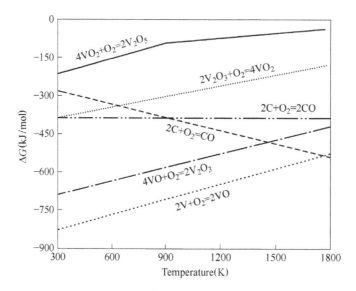

Figure 12.2
Formation free energy of vanadium oxide and carbon oxide.

The carbon thermal reduction of VC follows the order: VC → VC$_2$ → V(C)s → V. The basic chemical reaction of carbon thermal reduction can be expressed as follows:

$$\frac{1}{y}V_xO_y + C = \frac{x}{y}V + CO \tag{12.26}$$

When the temperature is below 1000°C, the chemical reaction is as follows:

$$V_2O_5 + CO = 2VO_2 + CO_2 \tag{12.27}$$

$$2VO_2 + CO = V_2O_3 + CO_2 \tag{12.28}$$

When the temperature is above 1000°C, the chemical reaction is as follows:

$$V_2O_3 + 5C = 2VC + 3CO \tag{12.29}$$

$$2V_2O_3 + VC = 5VO + CO \tag{12.30}$$

$$VO + 3VC = 2V_2C + CO \tag{12.31}$$

$$VO + V_2C = 3V + CO \tag{12.32}$$

12.1.2.2 Vacuum carbon thermal reduction

Vacuum carbon thermal reduction is one of the most important methods for preparing forging vanadium. First, V_2O_5 is reduced to V_2O_3 with hydrogen and then mixed with carbon black. After repeated high-temperature reduction steps in a vacuum furnace, vanadium blocks are produced, which contain about 0.02% carbon and 0.04% oxygen. This material is malleable at room temperature. Vanadium metal can also be purified by iodide thermal decomposition, to give a vanadium purity of 99.95%. Vanadium chloride can also be reduced in a hydrogen atmosphere at 1000°C to obtain forging vanadium. The carbon reduction reaction of VO is as follows:

$$VO(s) + C(s) = V(s) + CO(g) \tag{12.33}$$

$$2V(s) + O_2(g) = 2VO(s)(1500K - 2000K) \tag{12.34}$$

$$\Delta G_{22}^{\theta} = -803,328 + 148.78T \; (J/mol)$$

$$2C(s) + O_2(g) = 2CO(g) \tag{12.35}$$

$$\Delta G_{23}^{\theta} = -225,754 - 173.028T \; (J/mol)$$

Carbonaceous reductants are used to reduce V_2O_5 or V_2O_3. Under standard conditions the highest reduction temperature $T_{initial} = 1794.77K$ (1521.77°C). To reduce the initial reduction temperature, the pressure of the gas in the system is decreased to reduce the partial pressure of CO (P_{CO}). The initial reduction temperature depends on P_{CO}.

In industrial vacuum carbon thermal reduction processes, V_2O_5 powder is evenly mixed with high-purity carbon powder, then 10% camphor ethyl ether solution or alcohol is added. This mixture is pressed cake and then placed in a vacuum carbon resistance furnace or induction furnace. After the vacuum pressure in the furnace reaches 6.66×10^{-1} Pa, the temperature is increased to 1573K and reacted at this temperature for 2 h. After cooling the reaction products are crushed. An appropriate amount of VC or vanadium oxide is added according to the composition of the first reduction product for the second reduction cycle. The vacuum pressure in the second reduction furnace is 2.66×10^{-2} Pa, and the temperature is controlled to 1973K–2023K. The furnace is kept at the desired temperature for a period of time. The composition of the vanadium metal obtained by vacuum carbon reduction (mass fraction/%) is generally: V 99.5, O 0.05, N 0.01, and C 0.1. The vanadium yield can reach 98%–99%.

12.1.2.3 Preparation of vanadium by multistep carbon thermal reduction

In general, carbon thermal reduction is carried out at high temperature. Considering the high-temperature volatility of metallic vanadium and the vanadium oxide raw material, a high carbon ratio is used in a closed reactor. The V-containing raw material can be V_2O_5, V_2O_3, or VC. The reduction steps are as follows: the intermediate product is removed, crushed, finely ground, and then dehydrogenated, and then this sequence is repeated. The C/O ratio is adjusted depending on the ingredients. The raw materials are mixed for pelletizing. The pellets are used as the charge for the next process, until metallic vanadium is obtained. Fig. 12.3 shows the process of carbon reduction of V_2O_5 to produce vanadium metal.

Acetylene black is mixed with V_2O_5 with a mole ratio of $x(O)/x(C) = 1.25$. The raw material composition is (V_2O_5 + 4C), which is reduced at 450°C–540°C to produce V_2O_4. Then, the raw material composition is adjusted to (V_2O_4 + 3.5C), and this mixture is heated to 1350°C, evacuated to 10 Pa, and reacted to form VC, which contains 86%–87% V, 5%–6% C, and 7%–8% O. Carbon black or V_2O_3 is added to adjust the O/C ratio to 1, and then the mixture is heated to 1500°C and evacuated to 0.1 Pa. After 3 h, coarse vanadium is obtained, which contains 96%–97% V, 1%–1.5% C, and 2%–3% O. The O/C ratio is adjusted to 1, and then the mixture is heated to 1700°C and evacuated to 0.001 Pa. After 12 h, ductile vanadium is obtained, which contains 99.6% V, 0.12% C, and 0.06% O.

V_2O_3 and VC are used as raw materials, depending on the product requirements and process. The ingredients are placed in an induction furnace crucible, evacuated to 0.05 Pa, and heated at 1450°C for 8 h. Then, the furnace is evacuated to 0.01 Pa and held at a temperature of 1500°C for 9 h to sinter the material and produce C–OV block. In a resistance furnace, it is heated to 1650°C, evacuated to 0.002 Pa, and reacted for 2 h. Then, the VC control components are added, and then the mixture is heated to 1675°C, evacuated

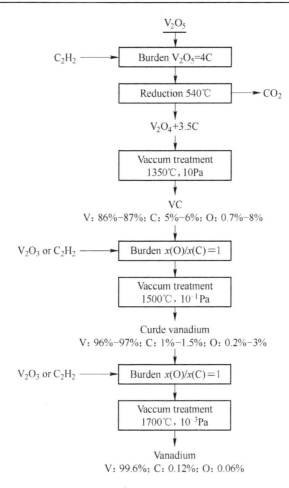

Figure 12.3
Vanadium production using carbon reduction of V_2O_5.

to 0.005 Pa, and reacted for 3 h to obtain ductile vanadium with impurity levels are 0.01% N, 0.12% C, and 0.014% O. The process flowchart for multistep carbon thermal reduction of V_2O_3 and VC to prepare vanadium metal is shown in Fig. 12.4.

The preparation of metallic vanadium by carbon thermal reduction can use plasma arc technology. A mixture of V_2O_5 and graphite powder is prepared and then pelletized to produce wafers with diameter of 15 mm and thickness of 8 mm. The mole ratio of the burden is $[x(O)/(C)x] = 0.8-17$, which is placed in a water-cooled copper crucible, which is loaded into a transmitting electric arc furnace and then heated to 2100°C−2800°C. The furnace is evacuated and argon gas is added to dilute the CO concentration in the charge and maintain a low CO partial pressure. V_2O_5 is reduced and melted rapidly; after 45 s the content of melted vanadium is above 90%, while after 10 min, it is above 96%. An alloy

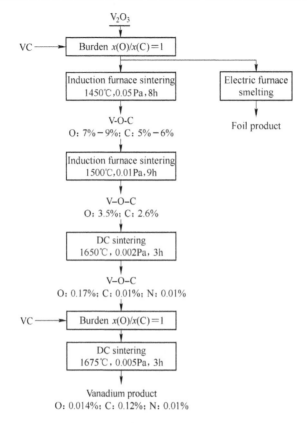

Figure 12.4
Process flowchart for multistep carbon thermal reduction of V_2O_3 and VC to prepare vanadium metal.

containing 2.3% C and 1.8% O is produced. The conversion efficiency of vanadium is 87%, which can reach 90% after optimization.

First, crude vanadium is produced by smelting (V_2O_5 + C) pressing block $[x(O)/x(C)]$ = 1.20–1.25. An Ar/H_2 mixture with a hydrogen ratio of 25% is injected to remove C and O in the alloy. Smelting deoxidization using a plasma arc is favorable, as the decarbonization ability is general. Crude vanadium is generally high in O and low in C.

12.1.3 Hydrogen reduction

Vanadium oxide can be reduced by hydrogen to obtain metallic vanadium. The reactions are carried out from high valence to low valence, and finally vanadium metal is obtained. In pure dry hydrogen the vanadium pentoxide is heated to 600°C and kept at this temperature for 3 h and then heated to 900°C–1000°C and kept at this temperature for 5 h. Finally, metallic vanadium is produced after cooling the furnace. Table 12.3 shows the free energy change of the hydrogen reduction reaction of vanadium oxide in the standard state ($=A + BT$).

Table 12.3: Free energy change of the hydrogen reduction reaction of vanadium oxide in the standard state.

Reaction	A (J)	B (J/K)
$V_2O_5 + H_2 = 2VO_2 + H_2O$	− 91630	− 68.20
$V_2O_5 + 2H_2 = V_2O_3 + 2H_2O$	− 166,732	− 76.15
$V_2O_5 + 3H_2 = 2VO + 3H_2O$	− 14,244	− 111

The hydrogen reduction reactions involving vanadium oxides are as follows:

$$V_2O_5 + H_2 = V_2O_4 + H_2O \qquad (12.36)$$

$$V_2O_4 + H_2 = V_2O_3 + H_2O \qquad (12.37)$$

$$V_2O_3 + H_2 = 2VO + H_2O \qquad (12.38)$$

$$VO + H_2 = V + H_2O \qquad (12.39)$$

$$H_2 + O_2 = 2H_2O \qquad (12.40)$$

$$V_2O_5 + H_2 = 2VO_2 + H_2O \qquad (12.41)$$

$$V_2O_5 + 2H_2 = V_2O_3 + 2H_2O \qquad (12.42)$$

$$V_2O_5 + 3H_2 = 2VO + 3H_2O \qquad (12.43)$$

Fig. 12.5 shows the relationship between the free energy change of vanadium oxide during hydrogen reduction and temperature.

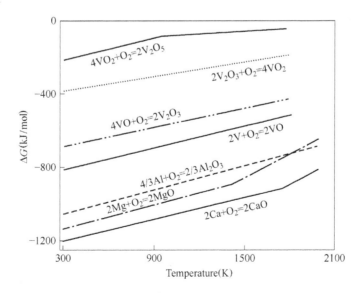

Figure 12.5
Relationship between the free energy change of vanadium oxide during hydrogen reduction and temperature.

12.1.4 Reduction of vanadium chloride

Fig. 12.6 shows the relationship between the free energy of metallic chloride and temperature. Vanadium chloride can be reduced by a metal or hydrocarbon reductant to produce vanadium metal. Metal reducing agents include Li, Mg, Na, Ca, and K and are selected based on comprehensive analysis of the reaction process control, product processing, cost of the material, and overall product quality. The activity of K and Na is high, resulting in difficulties with respect to sourcing the material, and its safe use and storage. The low boiling point and high volatility of Na metal are problematic as the reactor pressure can be increased by the Na vapor, which can react with chlorides of other metals and generate spontaneous combustion compounds. Hence, metal reductants for practical applications are limited to Ca and Mg, and calcium metal and reduction products $CaCl_2$ with high boiling point. Conventional distillation and separation methods are not suitable for treating the reaction products. If the product is cleaned, it can be contaminated. Hydrocarbon reductants are limited to hydrogen, where the reactor is generally vacuum sealed and filled with inert gas in stages. Commonly used vanadium chlorides include high-purity VCl_4, VCl_3, and VCl_2. All raw materials are purified, degassed, and dried. VCl_4 is generally mixed with $VOCl_3$. The crucible material must have good structural integrity at

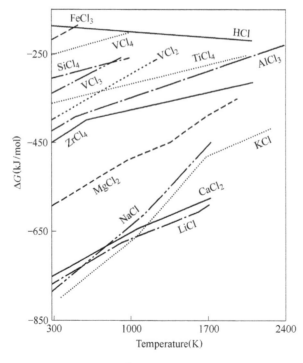

Figure 12.6
Relationship between free energy and temperature of metallic chlorides.

Table 12.4: Thermodynamic and physical properties of vanadium chloride when reduced by Ca and Mg.

Reaction	$\Delta H^\ominus_{298}$ (kJ)	$\Delta H^\ominus_{298}$ (kJ/g)
$VCl_4 + 2Ca = V + 2CaCl_2$	−1022	−3.745
$VCl_4 + 2Mg = V + 2MgCl_2$	−713	−2.954
$VCl_3 + 3Mg = 2V + 3MgCl_2$	−803	−2.072
$VCl_2 + Mg = V + MgCl_2$	−198	−1.365
$VCl_3 + 3Na = V + 3NaCl$	−678	−2.760

Table 12.5: Physical properties of metals and their chlorides used for metal thermal reduction.

Metals and their chlorides	Al	Ca	Mg	V	Na	NaCl	CaCl$_2$	MgCl$_2$	VCl$_4$
Melting point (°C)	660	842	650	1910	98	801	772	714	−26
Boiling point (°C)	2520	1494	1090	3409	882	1465	2000	1418	148.5

high temperature and under the strongly reducing atmosphere to prevent melting, etching, and peeling at high temperature.

The thermodynamic and physical properties of vanadium chloride reduced by Ca and Mg are shown in Table 12.4.

The physical properties of metals and their chlorides used for metal thermal reduction are shown in Table 12.5. Generally, the recovery of vanadium by reduction processes is above 80%, where the recovery can exceed 90% when the control technology is optimized. High-purity vanadium chloride is required to ensure high-quality vanadium metal products. Iron chloride and silicon chloride can affect the product quality and should be separated and removed. In addition, the VOCl$_3$ content should be reduced to minimize the oxygen content in the alloy. The thermodynamics of the metal thermal reduction need to be satisfied; vanadium can be prepared by reducing vanadium chloride under appropriate kinetic reaction conditions.

12.1.4.1 Hydrogen reduction of vanadium chloride

Hydrogen reduction is carried out in a stepwise manner, where the vanadium chloride is reduced from high valence to low valence. The temperature of the reduction reaction is above 300°C. The use of low-valence vanadium chloride requires a temperature higher than 1500°C to obtain vanadium metal by hydrogen reduction. The relationship between the free energy and temperature during the reduction of vanadium chloride by hydrogen is shown in Fig. 12.7.

The specific chemical reactions are as follows:

$$2VCl_4 + H_2 = 2VCl_3 + 2HCl \quad (12.44)$$

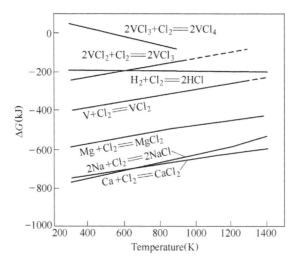

Figure 12.7
Relationship between the free energy and temperature during hydrogen reduction of vanadium chloride.

$$2VCl_3 + H_2 = 2VCl_2 + 2HCl \quad (12.45)$$

$$2VCl_2 + H_2 = 2V + 2HCl \quad (12.46)$$

$$Cl_2 + H_2 = 2HCl \quad (12.47)$$

Hydrogen reduction is the process of making metals by reducing metal oxides with hydrogen at high temperature. The reduction of vanadium chloride by hydrogen is a slow process. Roscoe, a British scientist, used hydrogen to reduce vanadium chloride to obtain crude vanadium with a purity of 95%. Vanadium chloride (VCl_3 or VCl_2) is placed in a Pt reaction vessel. The crude vanadium metal is obtained by heating with hydrogen for 40 h to white-heat temperature. Tyzack used VCl_3 as a raw material, which was heated in a muffle furnace. The VCl_3 in an Mo reaction boat reacted with hydrogen after chip uranium purification. First, VCl_3 is reduced to produce VCl_2 at 400°C–500°C. Then, after a long reduction period at 1000°C, a mild sintering vanadium fuse is obtained, where the V purity was up to 99.99%. Vanadium metal was cut into 0.049 mm below then pressed into a block and vacuum sintered at 1750°C. Vanadium was then obtained by calendaring. Fig. 12.8 shows the process of preparing vanadium from vanadium chloride by hydrogen reduction.

12.1.4.2 Calcium thermal reduction of vanadium chloride

The calcium thermal reduction reaction with VCl_2 is as follows:

$$VCl_2 + Ca = MgCl_2 + V \quad (12.48)$$

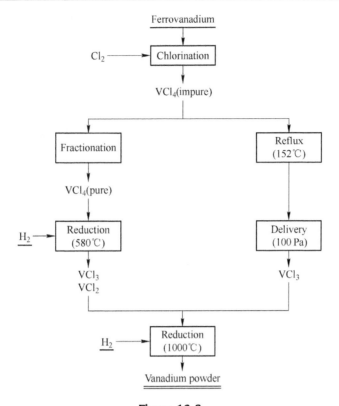

Figure 12.8
Process of preparing vanadium from vanadium chloride by hydrogen reduction.

Calcium metal and its reducing product $CaCl_2$ have high boiling points, which are not suitable for conventional distillation and separation of reaction products. If the product is cleaned, it can be contaminated, so these methods are generally not used to prepare vanadium metal.

12.1.4.3 Magnesium thermal reduction of vanadium chloride

Vanadium chloride can be reduced using Mg, where the reactions are conducted in three stages, according to sequence $VCl_4 \rightarrow VCl_3 \rightarrow VCl_2 \rightarrow V$. In the first stage, VCl_4 is reduced to VCl_3. In the second stage, VCl_3 is reduced to VCl_2. In the third stage, VCl_2 reduced to V. The specific chemical reactions are as follows:

$$VCl_4 + 2Mg \rightarrow V + 2MgCl_2 \quad (12.49)$$

$$2VCl_3 + 3Mg \rightarrow 2V + 3MgCl_2 \quad (12.50)$$

$$VCl_2 + Mg \rightarrow V + MgCl_2 \quad (12.51)$$

Two crucibles are placed one on top of the other in a special reactor equipped with two heating units. The lower crucible is filled with the Mg metal reducing agent, while the upper crucible is loaded with VCl_4. Then, the furnace is filled with inert gas and sealed before heating. The metal Mg and VCl_4 are melted and vaporized, where the vapors react to form $MgCl_2$-containing slag and sponge vanadium; this is an exothermic reaction. According to reaction (12.12–12.34), generally the reaction rate is controlled based on the evaporation rate of the two materials, to avoid overheating and high internal pressure. After completion of the reaction, the pressure drops, and the reactor is cooled and then opened. The reaction material is taken out and placed in a vacuum distillation reactor. The pressure, temperature, and time are kept at 0.01 Pa, 825°C, and 15–17 h, respectively, where excess Mg and magnesium chloride are removed by distillation to obtain metallic vanadium with a yield of 50%–70%.

British scientists used VCl_3 as a raw material, which was processed using a sealed steel reactor. The lower crucible was filled with metal Mg, while an external tank was filled with VCl_3, where the feeding and flow were controlled by a valve connected to the reactor. Depending on the ingredients, Mg is added and the furnace is heated to 700°C for drying. The furnace is evacuated for degassing, and an inert gas is injected to maintain positive pressure. During heating at 750°C–780°C, Mg is vaporized, and the VCl_3 input is adjusted with a screw to allow Mg vapor to react with VCl_3. This reaction shown in Eqs. (12.12–12.35) occurs. After 7 h the reaction is complete and the reactants are cooled and fed into a vacuum distillation apparatus, heated to 920°C–950°C, and vacuum distillated for 8 h. The excess Mg is volatilized, and $MgCl_2$ slag and vanadium are separated, where the slag sinks to the bottom of the furnace. The sponge vanadium is distilled under protective dry air cooling, which prevents oxidation of the vanadium particles. The V yield is 96%–98%.

With VCl_2 as a raw material, an integrated reduction and distillation furnace is used. The crucible for VCl_2 is placed in the integrated vacuum furnace, and amorphous Mg is added with an excess of 40%–50%. The furnace is evacuated, filled with inert gas, and heated to 520°C–570°C to induce the reaction. The reaction shown in Eqs. (12.12–12.36) occurs, which releases heat and increases the temperature to 100°C, continuously heated to 900°C. The reaction is complete after 2 h, and the furnace is cooled to room temperature. Cleaned up the inversion after charging, it is placed upside down in the furnace, evacuated, sealed, heated to 950°C, and distilled for 16 h. The gas is condensed in the interlining, where some of the Mg and $MgCl_2$ is distilled and enters the condensate collection tank. After distillation the vanadium sponge is cleaned in dry air to prevent vanadium oxidation formation. The yield of vanadium is 95%–98%.

12.2 Vanadium refining

Vanadium metal has impurity defects because of contamination from the processing environment, which affect the quality of the metallic vanadium products, further processing, and applications. These impurities include C, N, O, C, Al, Ca, Cr, Cu, Fe, Mo, Ni, Pb, Ti, and Zn. Some of these are included during the production of the raw material and appear as residual inclusions, while others are from contamination from the external environment. The vanadium metal adsorbs contaminants from the gas phase and the surface of processing equipment. If the content of O, N, C, and other nonmetallic impurities in the crude vanadium metal is high, its plasticity is poor. After removing impurities by refining, the plasticity of vanadium metal can be improved. The purity of refined vanadium can reach 99.9%, where high-purity vanadium can reach 99.99% after secondary electrolytic refining. Currently, vanadium refining methods used in industry include vacuum refining, molten salt electrolytic refining, iodide thermal dissociation, and regional smelting. It may be possible to use electron beam regional melting and electromigration refining in the future.

12.2.1 Vacuum refining methods

The metal impurities in crude vanadium mainly exist in the form of solutes. Vacuum refining methods include thermal vacuum processing and high-temperature vacuum refining, and vacuum heat treatment using distillation, dehydrogenation, denitrification, and deoxidization. Distillation purification depends on the vapor pressure of the metal impurity. The vapor pressures of different metals at 2200K are presented in Table 12.6. The evaporation rate of impurities from crude vanadium metal is related to the molecular mass, concentration, activity coefficient, vapor pressure, and absolute temperature. It is generally considered that vacuum distillation is complex; even though the vapor pressure is high, the low concentrations can limit the speed of vacuum distillation.

The main nonmetallic impurities are N_2, H_2, and O_2, which exist as crystal gap compounds in vanadium. Methods for removing these impurities mainly include high-vacuum and heat treatments. Degassing is related to the partial pressure, diffusion coefficient, diffusion surface area, and particle size, where the partial pressure is affected by the concentration and temperature. A gas diffusion coefficient of 10^{-9} m^2/s can be maintained at room temperature, while at 500°C the diffusion coefficient of H_2 in reduced vanadium is more than 10^{-8} m^2/s. Reduced vanadium contains 100×10^{-4}% H_2, which can be removed at 500°C–1000°C. At higher temperatures the degassing speed can increase rapidly after the reduced vanadium melts. Deoxygenation of reduced vanadium is achieved using thermal vacuum treatment and carbon deoxygenation. Fig. 12.9 shows the partial pressure of dissolved oxygen in vanadium metal. The partial pressure of oxygen is proportional to the oxygen concentration and increases with increasing temperature.

Table 12.6: Metal vapor pressures at 2200K.

Metal	V	Al	Ca	Cr	Cu	Fe	Mo	Ni	Pb	Ti	Zn
Vapor pressure (Pa)	3	3×10^3	130×10^3	800	3×10^3	300	0.003	160	120	10	130×10^3

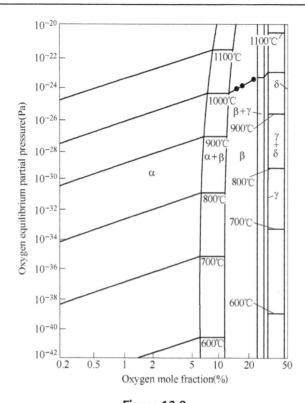

Figure 12.9
Partial pressure of dissolved oxygen in vanadium metal.

The reason is that the whole deoxidization process is the volatilization of vanadium in the form of a suboxide, characterized by vanadium loss during deoxidizing. The chemical reaction is as follows:

$$[O]_{(l)} + V = [VO]_{(g)} \tag{12.52}$$

Here, $[O]_{(l)}$ is the oxygen dissolved in the vanadium melt and $[VO]_{(g)}$ is vanadium oxide in the gas phase.

The Si content in vanadium metal is low. However, the volatility of Si is higher than that of vanadium, so it is generally difficult to remove alone and needs to be removed via an oxidation reaction. Carbon removal is mainly performed by reaction with oxygen (i.e., decarbonization by excess oxygen), where excess oxygen is consumed by the vanadium.

It is very difficult to remove nitrogen from vanadium metal as the V–N solid solution is relatively stable. The equilibrium partial pressure of nitrogen is relatively low. The mole fraction of nitrogen is less than 1% in the solid solution, which is close to the melting point of vanadium, where the partial pressure of nitrogen is less than that of vanadium. Under thermal vacuum conditions (2000°C−2100°C, 2.7 × 10^{-3} Pa), the mass fraction of N in reduced vanadium can be reduced to below 0.3%. Nitrogen contents greater than 0.3% can be removed, while in the case of those below 0.3%, evaporation of vanadium is higher than that of nitrogen, resulting in vanadium loss and an increase in the nitrogen content in the vanadium melt.

Typical methods of comprehensive purification of vanadium include reduction of V$_2$O$_5$ with high-purity Al to obtain V−Al alloy. Then, the V−Al alloy is crushed and heated in a vacuum furnace to 1973K to remove the Al and obtain sponge vanadium. After the sponge vanadium is pressed into an ingot, the residual Al, O, Fe, and other volatile impurities are further removed using electron beam furnace smelting, where 99.9%-pure vanadium can be obtained.

12.2.2 Molten salt electrolytic refining

Coarse vanadium can be classified according to its mode of production and divided into coarse vanadium produced by: carbon thermal reduction, aluminothermic reduction, and calcium thermal reduction. As different reductants are used, the impurities are different. Separation during molten salt electrolysis refining is mainly based on the potential of the molten salt ion in the current. Generally, coarse vanadium is cast into an electrode, where the electrolyte is a low-melting-point chlorine salt system, mainly based on K, Na, Ga, Ba, and Li chlorides. The electrolytic refining reactions are as follows:

$$\text{Anodic reaction}: V(\text{crude}) + 2Cl^- = VCl_2 \quad (12.53)$$

$$\text{Cathode reaction}: VCl_2 + 2e = V(\text{refined}) + 2Cl^- \quad (12.54)$$

$$\text{Overall reaction}: V(\text{crude}) = V(\text{refined}) \quad (12.55)$$

Vanadium metal produced by aluminothermic reduction is used as a soluble anode in LiCl−KCl−NaCl−VC. Electrolytic refining is carried out in a molten salt system. The operating conditions to produce 99.2%-pure vanadium metal include a working temperature of 893K, voltage of 0.3 V, total current of 20−25 A, and cathode current density of 3200−3700 A/dm^2. Before electrolytic refining a small amount of chlorine gas is usually injected into the tank to ensure that the electrolyte contains VCl$_2$. The electrolyzer is sealed with Ar gas. The crucible in the electrolyzer is made of metal, such as Mo or Ni, and acts as a positive electrode. The cathode is generally made of Mo. During refining, with the dissolution of anodic vanadium, the surface of the crude vanadium is gradually oxidized and passivated, resulting in a drop in the current efficiency and poor product quality. Generally, the furnace should be shut down after dissolution of the anodic crude vanadium reaches 30%. The electrolytically refined product is then washed in water. The purity of vanadium can reach 99.5%−99.9%, which can be processed into metal products. The cathode current efficiency of electrolytic refining is 88%−94%. Electrolytic refining is best for deoxidization and desiliconization reactions, followed by removal of iron and aluminum, while chromium removal is the most difficult. Currently, low-Cr high-purity vanadium is prepared from low-Cr crude vanadium. Vanadium with purity of 99.99% and cathode current efficiency of 89%–92% can be produced by secondary electrolytic refining with a cell voltage of 0.3–0.54 V and cathode current density of 33.4–37.7 A/dm^2 with an electrolyte containing 51% KCl, 41% LiCl, and 8% VCl$_2$.

During refining of metal vanadium by the calcium thermal reduction, 51% KCl, 41% LiCl, and 8% VCl$_2$ electrolyte is used, where the crude vanadium metal is used as the electrolytic anode. The process conditions include a temperature of 620°C, voltage of 0.54 V, cathodic current density of 3300 A/cm^2, electric product number of 17, electrolytic of 9690 A h, and cathode current efficiency of 92%. After refining electrolysis, 9.2 kg of the vanadium anode is consumed, while 8.5 kg is deposited on the cathode. Refined vanadium metal is produced at the secondary refining anode using a temperature of 620°C, voltage of 0.3 V, cathodic current density of 3800 A cm^2, electric product number of 11, electrolytic 6600 A h, and cathode current efficiency of 89%. During refining electrolysis, 6 kg of the vanadium anode is consumed, resulting in 5.6 kg of cathodic deposition.

The vanadium metal obtained by carbon thermal reduction generally contains VC and V$_2$C. For decarbonization, chloride is used as the electrolyte, and VC is decomposed via molten salt electrolysis process. The chemical reaction of V$_2$C decomposition is as follows:

$$V_2C = 2V + C \quad \Delta G^\ominus = 143 \text{ kJ} \tag{12.56}$$

The decomposition reaction of VC is as follows:

$$VC = VC_{0.88} + 0.12C \quad \Delta G^\ominus = 47 \text{ kJ} \tag{12.57}$$

VC$_{0.88}$ is a stable phase, where its decomposition reactions are as follows:

$$VC_{0.88} = V + 0.88C \quad \Delta G^\ominus = 96 \text{ kJ} \tag{12.58}$$

$$\text{Anodic reaction:} V + 2Cl^- = VCl_2 \tag{12.59}$$

$$VCl_2 + Cl^- = VCl_3 + e \tag{12.60}$$

$$2VCl_3 + V_2C = 3VCl_2 + VC_{0.88} + 0.12C \tag{12.61}$$

$$V_2C + 2Cl^- = VCl_2 + VC_{0.88} + 0.12C + 2e \tag{12.62}$$

$$\text{Cathodic reaction:} VCl_2 + 2e = V(\text{refined}) + 2Cl^- \tag{12.63}$$

Total electrolytic reaction:

$$V_2C = V(\text{refined}) + VC_{0.88} + 0.12C \tag{12.64}$$

Typical commercial reduced vanadium products contain 85% V, 10% C, and 5% of other impurities, including O, Fe, and Cr. For electrolytic refining of vanadium metal by carbon thermal reduction, a system containing 48% $BaCl_2$, 31% NaCl, and 21% KCl, with the addition of 5%–12% VCl_2, is used. A temperature of 670°C is used, with a cell voltage of 0.4–1.3 V (or 0.2–0.7 V), cathode current density of 2150–9700 A/cm^2, and cathode current efficiency of 70% (87%) to achieve a vanadium yield of 84% (77%).

After a period of electrolysis the V_2C electrode is transferred to the VC electrode after about 50% of the vanadium has been electrolyzed. The electrolytic efficiency decreases, mainly because of the increased content of carbides, O, Fe, and Cr after vanadium electrolysis, and the contamination of the cathodic vanadium. Electrolytic refining of VC-type vanadium metal is used to obtain vanadium metal containing 99% vanadium. An electrode is made, and the NaCl–LiCl–VCl_2 system is used as electrolyte, with a temperature of 620°C. Electrolytic refining is carried out using a cathode current density of 130×10^3 A/cm^2 to obtain 99.80%-pure vanadium metal. For electrolytic refining of VC-type vanadium metal, an Mo bucket electrode can be used. The vanadium metal is placed in a porous graphite tube, which is suspended in the center of the bucket electrode. The 45% NaCl—45% LiCl—10% VCl_2 system is used as electrolyte, the voltage is adjusted from 0.2 to 1.2 V, with a current of 23–90 A and electrolytic charge of 1500–2500 A h. This results in vanadium being deposited at the Mo cathode barrel wall. The electrode is transformed during refining. The porous graphite tube is removed and replaced with an Mo cathode, and refining is performed with a voltage of 0.08 V, current of 10 A, and electrolytic charge of 70 A h. The vanadium metal deposited on the Mo rod has a purity of 99.86%, which satisfies the processing requirements.

12.2.3 Thermal dissociation of iodides

Iodide refining uses iodide gasification, precipitation, and regasification to achieve thermal dissociation. Iodine has a melting point of 113°C and boiling point of 684°C, while VI_2 sublimates at 750°C. The O, C, and N impurities do not react with iodine at 800°C–900°C, and possible impurities in the iodine products do not decompose at 1000°C–1400°C. Decomposable impurities can be volatilized as follows:

$$V(crude) + I_2 = VI_2 \quad (12.65)$$

$$VI_2 = V(refined) + I_2 \quad (12.66)$$

During typical thermal dissociation of iodine, crude vanadium and iodine are first placed in an iodinating reactor made from Mo at a temperature of 1073K. The thermal dissociation temperature of vanadium wire is 1573K. A typical composition of the product [mass fraction cc, (%)] is V99.95, Cr < 0.007, Fe < 0.015, Si < 0.005, Ca < 0.002, Cu < 0.003, Ni < 0.002, Mg < 0.002, Ti < 0.002, C0.015, H < 0.001, N < 0.0005, and O < 0.004. This method is used in small batch production.

12.2.4 Regional smelting

Vanadium metal can be refined using vacuum and different melting methods, such as vacuum sintering, induction melting, arc melting, and electron beam melting. Vacuum sintering requires high temperature, high vacuum, and high-strength crucibles that do not contaminate the product. For example, during vacuum sintering of vanadium metal by aluminum thermal reduction, the raw materials are first chopped into small pieces and then loaded in a tantalum crucible, which is placed in an induction furnace with a temperature of 1700°C and vacuum pressure of 6×10^{-3} Pa, and then sintered for 8 h. After processing the oxygen content decreased significantly, which is attributed to its removal by aluminum; aluminum oxide vapor is produced and separated from the vanadium metal, such as C, Ca, Fe, and Zn removal effect is obvious. The temperature cannot exceed 1820°C, to protect the crucible. Situation is improved, vanadium pellets are placed in a hanging Mo basket, which is loaded in the tantalum crucible and insulated with zirconium felt. A liquid nitrogen ring is used to produce artificial oil vapor between the oil diffusion pump and electric furnace, preventing contamination from the furnace atmosphere. The temperature is set to 1700°C, with a vacuum pressure of 6×10^{-4} Pa, and reaction time of 8 h. After aluminum thermal reduction the Al content in the pure vanadium drops from 10.1% to 0.5%. Meanwhile, the C content decreased from 130×10^{-4} to 25×10^{-4}, Fe content drops from 810×10^{-4} to 170×10^{-4}, N content decreases from 60×10^{-4} to 25×10^{-4}, the O content drops from 2900×10^{-4} to 130×10^{-4}, and the Si content decreased from 500×10^{-4} to 180×10^{-4}.

High-strength noncontaminating crucibles are required for induction smelting, which need to withstand corrosion by liquid vanadium at high temperature, without introducing impurities. To date, only cerium sulfide can satisfy these conditions. However, considering the possibility of sulfur contamination, it is considered that vacuum induction purification methods are not suitable for vanadium. Arc furnace melting has limited refining efficiency and uses an inert atmosphere, while consumable or nonconsumable electrodes are used for melting metal vanadium and casting. During smelting, H, Al, and Mo are partially removed, while the removal of O and N is not significant. Nonconsumable electrodes, such as thorium-plated tungsten electrodes, are used for casting small vanadium ingots.

Vacuum suspension zone smelting and refining is also used. A vanadium rod with a diameter of 4.4 mm can be obtained after six cycles at a vacuum pressure of 1.333 MPa, with a smelting zone length of 6–10 mm, and melting zone speed of 57.16 mm/h.

12.3 Application of vanadium metal

Vanadium metal generally has high purity, and different specifications, such as high-purity vanadium metal, silver luster, particles, and flakes. Vanadium is used for electronic material components, analysis standard test materials, and superalloys and new alloys suitable for the development of products for the aerospace and atomic industries. Ultrafine high-purity vanadium metal powder is listed, which has various specifications as a catalyst reagent or for powder metallurgy. High-purity vanadium slug (5 μm−1 mm) is the best shielding material.

Vanadium is widely used in aerospace, electronics, informatics, ocean, materials science, and other fields. Vanadium is used as a protective material in atomic reactors. In the aerospace and aviation industries, it is used for the manufacture of rockets, missiles, spacecraft shells, large spacecraft, space ferry structural materials, aircraft brakes, navigation parts, and as a high-energy fuel additive for aircraft, spacecraft, missile, rocket, and jet aircraft applications. In the metallurgical industry, vanadium is added to steel alloys, and used to make refractory materials and special glasses. It is also used in integrated circuits and antennas.

The main applications include (1) aerospace aircraft, aviation aircraft, atomic energy equipment, high-tension alloy parts, jet engines, special motor parts, aircraft, and other landing wheel frames; (2) raw materials for processing various special steels; (3) raw material for various test alloys; (4) amorphous metal processing; (5) high-speed bearings and submarine cables in the metallurgical industry because of its good corrosion resistance and high electrical conductivity; (6) high-strength cutting tools, diamond cutting knives, diamond glass knives, diamond cutting machines for road construction; (7) various catalysts; (8) electronic parts; (9) various reagents, analytical standard reagents, and reducing agents; and (10) as a new material, vanadium metal is increasingly valued. Vanadium metal is an indispensable precious material in the atomic energy, aviation and aerospace, and metallurgical industries. Vanadium is highly valued in the atomic energy industry. As vanadium has a high melting point and is particularly resistant to high temperatures, it is the best material for neutron reflectors in nuclear reactors. Vanadium is also an excellent aerospace material. For every kilogram increase in weight of a satellite, the total weight of the carrier rocket increases by about 500 kg. The structural materials required to build rockets and satellites are light and strong. Vanadium metal has strong heat absorption and stable mechanical properties.

Vanadium alloys are produced by mixing vanadium with other alloying elements. Vanadium alloy has a small fast neutron absorption cross section, good corrosion resistance to liquid metal Li, Na, and K, as well as good strength and plasticity, good processing performance, and low embrittlement and good dimensional stability under irradiation. It is an important structural material for nuclear reactors. Typical vanadium alloys include V−15Ti−7.5Ct, V−15Cr−5Ti, V−10Ti, V−20Ti, and V−9Cr−3Fe−1.5Zr−0.05C. These vanadium alloys are used as fuel jackets and structural elements for liquid-metal-cooled fast breeder reactors. Other vanadium alloys include Vanstar7 (V−9Cr−3Fe−ZrC), Vanstar8 (V−9Cr−3Ta−ZrC), Vanstar9 (V−6Fe−3Nb−ZrC), V−40Nb−1.3Zr, V−9Cr−10W−1.5Zr, and V−9Cr−10Ta−1.25Zr.

According to USBM, more than 90% of vanadium in nonsteel alloys is used to produce nonferrous alloys and magnetic alloys, among which Ti alloys account for the majority. Vanadium in Ti alloy can be used as a fortifier and stabilizer. When 4% V is added to Ti alloy, the alloy has good ductility and formability. The two most important Ti alloys are Ti6Al4V (containing 4% V) and Ti8Al1Mo1V, which account for 50% of the Ti alloy market. These alloys are used to make jet engines, high-speed aircraft skeletons, and rocket engine casings. Vanadium is usually added to Ti alloys in the form of V−Al-based alloys.

The absorption cross section of fission neutrons in vanadium is very small, so it is used in the nuclear industry. The main nonaerospace use of high-performance V-containing alloys is in covers and shield walls of nuclear fusion reactors. Current research has focused on the field of vanadium alloys. Compared with the other alloys, at 700°C, vanadium alloys retain good ductility and strength and have minimal neutron radiation decay, which can prevent radiation. Vanadium has good corrosion resistance to liquid Li and Na (used as coolants). The LiVCrTiSi series (containing 0.15% Cr, 0.20% Ti, and <1% Si) is the main vanadium alloy developed to date, among which the most important V-containing alloy is V−5Ti−5Cr. Vanadium can be added to many other alloys to increase strength and ductility. For example, V is added to Cu-based alloys to control the gas composition and microstructure, while it is added to Al alloys to produce internal combustion engine pistons, and to some Ni-based superalloys to produce steam turbines and blades.

CHAPTER 13

Characteristics and technical quality standards of main vanadium products

Chapter Outline
13.1 **Vanadium slag**   360
    13.1.1 Properties   360
    13.1.2 Applications   360
    13.1.3 Quality standards   360
    13.1.4 Technical standards   363
13.2 **Vanadium trioxide**   363
    13.2.1 Properties   364
    13.2.2 Risks   364
    13.2.3 First aid measures   364
    13.2.4 Safety data   365
13.3 **Vanadium pentoxide**   365
    13.3.1 Physical properties   365
    13.3.2 Chemical properties   367
13.4 **Vanadium pentoxide standard**   367
    13.4.1 Chinese national standard for vanadium pentoxide (GB3283-87)   367
    13.4.2 High-purity vanadium pentoxide Chinese standard (GB-3283-87)   371
    13.4.3 Chinese national standard for vanadium pentoxide (super-pure grade)   372
13.5 **Ammonium vanadate**   373
    13.5.1 Ammonium metavanadate   373
    13.5.2 Ammonium polyvanadate   373
13.6 **Ferrovanadium**   375
    13.6.1 Properties   375
    13.6.2 Physical state   376
    13.6.3 Standard   376
13.7 **Vanadium nitride**   389
    13.7.1 Properties   389
    13.7.2 Chinese national standard for vanadium nitride and vanadium carbide   389
    13.7.3 Panzhihua Steel standard for vanadium nitrogen alloy   390
    13.7.4 Vanadium nitride from Vametco mining company, South Africa   390
    13.7.5 Vanadium carbide from the United States   390
13.8 **Vanadium metal**   391
    13.8.1 Properties   391
    13.8.2 Applications   391
    13.8.3 Chinese national standard   391

13.9 Vanadium aluminum alloy   391
  13.9.1 Properties   391
  13.9.2 Applications   392
  13.9.3 Chinese national standard   392
  13.9.4 German industrial standard   394

13.1 Vanadium slag

13.1.1 Properties

Vanadium slag is a product of devanadium and enrichment of V-containing molten iron. Ordinary vanadium slag is generally gray-black, with a density of 3.7–3.8 g/cm^3. V-containing converter slag has a loose atypical structure, which is composed of fine-grained cementing products with metallic iron particles. The oxide phases are olivine, pyroxene, and spinel minerals. Around 30% of the vanadium exists as vanadium spinel minerals (Fe, Mn, Mg)O · (V, Fe, Al, Cr)O$_3$, and enriched in 30% of Mn and 30% of Fe. MgO is uniformly distributed in the slag between silicate and vanadium spinel phases. Vanadium is extracted from different sources of molten iron using various processing equipment, resulting in different grades of vanadium slag, where the vanadium pentoxide content varies between 5% and 25% [1–4,14–17,23,25,41–45,67,68,86,87].

13.1.2 Applications

Vanadium slag is mainly used for extracting vanadium pentoxide in downstream plants, followed by direct alloying.

13.1.3 Quality standards

1. Chinese national standard
   Table 13.1 shows the Chinese national standard for vanadium slag (GB5062-85). The metallic Fe content in bulk vanadium slag should not exceed 22%.
2. Physical state
   Vanadium slag should be delivered in bulk or powder form, where the particle size of lumpy vanadium slag should not exceed 200 mm × 200 mm. The particle size and metallic Fe content of powdery vanadium slag should be agreed upon by both parties.
3. Delivery requirements
   The delivered vanadium slag should not contain obvious impurities.
4. Test method
   Sampling: Bulk vanadium slag sampling is performed according to the method specified in Appendix A (supplementary parts).

Table 13.1: Chinese national standard for vanadium slag (GB5062-85).

Brand			Vanadium slag 11	Vanadium slag 13	Vanadium slag 15	Vanadium slag 17	Vanadium slag 19	Vanadium slag 21
Code name			FZ11	FZ13	FZ15	FZ17	FZ19	FZ21
	V_2O_5		10.0–12.0	>12.0–14.0	>14.0–16.0	>16.0–18.0	>18.0–20.0	>20.0
Chemical composition	P	Group-1	Less than 0.08					
	CaO	Group-2	0.35					
		Group-3	0.70					
		Group-1	1.0					
		Group-2	1.5					
		Group-3	2.5					
	SiO_2	Group-1	22.0					
		Group-2	24.0					
		Group-3	34.0					
		Group-4	40.0					

Sample preparation: The preparation of bulk vanadium slag samples is performed according to the method specified in Appendix 3 (supplementary parts).

Determination of Fe content: The determination of metallic Fe content in the bulk vanadium slag is performed according to the current test methods of the specific factory.

5. Chemical analysis

    Chemical analysis is performed according to the vanadium slag chemical analysis method YB 547-67. Other test methods for powdery vanadium slag can be agreed upon by both parties.

6. Inspection rules

    The delivered vanadium slag should be inspected according to a vehicle inspection, where the vanadium slag order should be delivered in a single batch. The inspection and acceptance of vanadium slag quality should be conducted by the technical supervision department of the supplier. The buyer has the right to reinspect the goods. If there is any objection, they should return the goods to the supplier within 1 month of the delivery date.

7. Packing, transportation, and quality certificate

    Bulk vanadium slag should be transported in bulk or in open trucks. If required by the buyer, it can be shipped in covered wagons or simple covered wagons. The packaging and transportation of powdery vanadium slag should be determined by both parties through consultation. The delivered vanadium slag should be accompanied by a sample for retesting and a quality certificate. The quality certificate should indicate the following: (1) vanadium slag brand number, class, batch, chemical composition, and metallic Fe content; (2) weight and reference quantity; (3) vehicle number and delivery date; and (4) name of supplier and inspector code.

8. Sampling method of bulk vanadium slag (Appendix A, supplementary parts)

    Samples shall be taken with a spade in the carriage of goods. The sample shall be divided into two layers, where the heights of the upper and lower layers are three-fourths and one-fourth, respectively, of the height of the vanadium slag material. The sampling amount at each sampling point should be equal and not less than 10 kg. The sampling amount for each batch of vanadium slag should not be less than 1% of the total weight of the batch. The particle size distribution of vanadium slag samples should represent the actual particle size distribution of the batch. The parties agree that it is permissible to store vanadium slag quantitatively and take samples in advance before loading. After loading the samples from this batch of vanadium slag will be merged with the samples of this batch.

9. Appendix B: Preparation of bulk vanadium slag samples (supplementary parts)

    Vanadium slag samples for testing shall be prepared using all samples from the same delivery batch after multistage crushing and shrinking. The samples are crushed using a special high-manganese steel plate crusher or by hand. The samples should be laid flat on the steel plate and the four-point method (taking the opposite angle) should be used to reduce, according to the following table.

13.1.4 Technical standards

1. Technical requirements

Vanadium slag is classified into seven grades according to the vanadium pentoxide content, where the grade and chemical composition should meet the standards shown in Table 13.2. Vanadium slag shall be delivered in bulk with a particle size not exceeding 200 mm. When the buyer has special requirements for vanadium slag size, it shall be agreed upon by both parties. The content of iron and other metals in the delivered vanadium slag shall not exceed 20%.

13.2 Vanadium trioxide

Appearance: gray-black crystal or powder
Melting temperature (°C): 1970
Boiling point (°C): 3000
Relative density (water = 1): 4.87 (18°C)
Heat of combustion (kJ/mol): N/A
Critical temperature (°C): N/A
Critical pressure (MPa): N/A
Flash point (°C): N/A
Ignition temperature (°C): N/A

Table 13.2: Vanadium slag grades and chemical compositions (%).

Brand	V_2O_5	SiO_2			P			CaO/V_2O_3		
		Class-1	Class-2	Class-3	Class-1	Class-2	Class-3	Class-1	Class-2	Class-3
		Less than								
FZ9	8.0–10.0	16	20	24	0.13	0.3	0.5	0.11	0.16	0.22
FZ11	>10.0–12.0									
FZ13	>12.0–14.0									
FZ15	>14.0–16.1									
FZ17	>16.0–18.2									
FZ19	>18.0–20.0									
FZ21	>20.0									

Note: Moisture content is not a delivery condition, but the supplier should provide the measurement results for the batch to the buyer.

13.2.1 Properties

CAS No.: 1314-34-7
Formula: V_2O_3
Molecular weight: 149.88

13.2.2 Risks

Health hazards: After inhalation, it causes cough, chest pain, hemoptysis, and a metallic taste in the mouth. It can irritate eyes, induce tears, and irritate skin. Oral ingestion causes stomach upset, abdominal pain, vomiting, and weakness. The victim's tongue coating is dark green.

Environmental hazards: It is harmful to the environment and can pollute water.
Combustion and explosion risks: This product is noncombustible, poisonous, and irritating.

13.2.3 First aid measures

Skin contact: Remove contaminated clothing and rinse with plenty of running water.
Eye contact: Lift the eyelid and rinse with running water or saline. Go to a doctor.
Food intake: Drink enough warm water to induce vomiting. Go to a doctor.
Disposal of waste: Waste shall be disposed of using a safe burial method. Reuse containers where available or bury them in designated areas.
Packaging method: Ordinary wooden box around a screw-end glass bottle, iron-covered glass bottle, plastic bottle, or metal drum (can).
Dangerous goods no.: 61028
UN no.: 2860
Precautions for transportation: Before transportation the packaging container should be checked to see if it is complete and sealed. During transportation, it should be ensured that the container does not leak, collapse, fall, or be damaged. Mixed transportation with acid, oxidizing materials, food, or food additives is strictly prohibited. Transport vehicles should be equipped with emergency handling equipment in the case of a leakage. The goods should be protected from exposure, rain, and high temperature during transport. Road transport must follow the prescribed route, avoiding residential areas and densely populated areas.
Operating instructions: Closed operation and local exhaust. Operators must be specially trained and strictly follow the operating procedures. It is recommended that the operator wears a dust mask (full cover), a rubber cloth gas jacket, and rubber gloves. Keep away from inflammable and combustible materials. Avoid producing dust. Avoid contact with acids. Light loading and unloading should be carried to prevent damage to the

packaging and container. Equipment for emergency treatment of leaks should be available. Empty containers can leave harmful residue.

Storage instructions: Store in a cool, ventilated warehouse. Keep away from fire and heat. It should be stored separately from inflammable, acidic, and edible materials.

Solubility: Insoluble in water. Soluble in nitric acid, hydrofluoric acid, and hot water.

Prohibited content: Hot nitric acid.

Contact control/personal protection:

Chinese MAC (mg/m^3): 0.1(dust); 0.02 (smoke)

Former Soviet Union MAC (mg/m^3): 0.5

This product is noncombustible, poisonous, and irritating.

13.2.4 Safety data

Table 13.3 shows the material safety data sheet for vanadium trioxide.

13.3 Vanadium pentoxide

Chemical name: vanadium pentoxide;
Alias: vanadium anhydride;
Chemical formula: V_2O_5;
Relative molecular weight: 182.00;
Chemical category: inorganic substance—metal oxide;
Control type: vanadium pentoxide (highly toxic);
Storage: sealed for storage.

13.3.1 Physical properties

Appearance: orange-yellow, red-brown crystalline powder, or gray-black flake;
CAS no.: 1314-62-1;
Melting point (°C): 690;
Relative density (water = 1): 3.35;
Boiling point (°C): 1750 (decomposition);
Formula: V_2O_5;
Molecular weight: 182.00;
Solubility: slightly soluble in water, insoluble in ethanol, soluble in concentrated acids and bases. Fig. 13.1 gives the structure of vanadium pentoxide.

Table 13.3: Material safety data for vanadium trioxide.

Packaging, storage, and transportation	Hazard category	Class 6.1 toxic substances
	Dangerous goods packaging label	14
	Packing category	II
	Precautions for storage and transportation	Store in a cool, ventilated warehouse away from fire and heat. Special custody. Keep container sealed and prevent exposure to moisture. The material should be stored separately from acidic and edible materials. Avoid mixed shipment with food, seeds, feed, and all kinds of daily necessities. Do not smoke, drink, or eat at the operation site. Handle with care to prevent damage to packaging and containers. Pay attention to personal protection during packing and handling.
Toxicity hazard	Exposure limits	China MAC: 0.1 mg/m^3 (dust); 0.02 mg/m^3 (smoke) Former Soviet Union MAC: 0.5 mg/m^3 US TWA: 0.05 mg (V_2O_5)/m^3 US STEL: Not set.
	Ingestion pathway	Inhalation of eat.
	Toxicity	LD50: 130 mg/kg (mouse oral ingestion).
	Health hazard	Harmful when inhaled, ingested, or absorbed through the skin. Has irritating effect on the eyes, skin, mucous membrane, and the upper respiratory tract.
First aid and protective measures	Skin contact	Rinse thoroughly with soap and water. Go to a doctor.
	Eye contact	Open eyelids and rinse with running water for 15 min. Go to a doctor.
	Inhalation	Leave the area quickly and find fresh air. Go to a doctor.
	Ingestion:	Drink adequate amounts of water and induce vomiting. Go to a doctor.
	Engineering control	Closed operation, partial exhaust.
	Respiratory protection	Wear a respirator when exposed to dust.
	Eye protection	Wear chemical safety glasses.
	Protective clothing	Wear an appropriate uniform.
	Hand protection	Wear protective gloves when necessary.
	Other	Smoking, eating, and drinking are prohibited in the work environment. Shower and change after work. Pay attention to personal hygiene.
Leak management		Isolate the area contaminated by the leakage and set warning signs around it. Emergency workers are recommended to wear self-contained breathing apparatus and chemical protective suits. Do not touch the spill directly. Cover the spill with wet sand and pour it deep into the open. In the case of a large leakage, sweep carefully to avoid dust and place the material into a bag. The contaminated ground should be washed with soap or haihe agent, and the cleaning liquid should be placed in the waste water system.

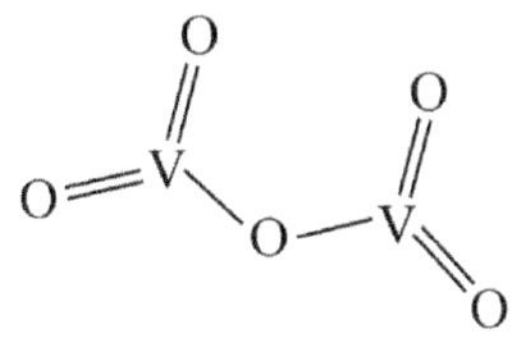

Figure 13.1
Structure of vanadium pentoxide.

13.3.2 Chemical properties

Vanadium pentoxide is an amphoteric oxide, with mainly acidic characteristics. Above 700°C, vanadium pentoxide is highly volatile, while in the range of 700°C–1125°C, it is decomposed into oxygen and vanadium tetroxide. This property makes it a suitable catalyst for organic and inorganic reactions. Vanadium pentoxide is a strong oxidizer and is easily reduced to a variety of low-valence oxides. It is slightly soluble in water and easily forms stable colloidal solutions, while it dissolves in basic solutions to form vanadate and dissolves in acids to form VO_2^+ ions instead of pentavalent vanadium ions. The maximum allowable amount in air is less than 0.5 mg/m^3.

Table 13.4 shows the material safety data table for vanadium pentoxide.

13.4 Vanadium pentoxide standard

13.4.1 Chinese national standard for vanadium pentoxide (GB3283-87)

This standard is applicable to flake or powdery vanadium pentoxide extracted from vanadium slag or other V-containing minerals through roasting, leaching, precipitation, decomposition, and melting for metallurgical and chemical applications.

13.4.1.1 Technical requirements

Grades and chemical compositions.

1. The product is divided into three grades according to its use and vanadium pentoxide content, where the chemical compositions should comply with the provisions shown in Table 13.5.
2. The buyer may, upon special request, negotiate the supply of products with lower impurity levels.
3. At the request of the purchaser, quantification of elements other than those listed in the table may be performed after consultation.

Table 13.4: Material safety data for vanadium pentoxide.

Characteristics	Chinese name: Vanadium pentoxide		Common name: Vanadium pentoxide	
	Molecular formula: V_2O_5		Molecular weight: 182	UN Serial number: 2862
	Dangerous article number: 61028		RTECS serial number	CAS serial number: 1314-62-1
Physical and chemical properties	Appearance: orange-yellow or reddish-brown crystalline powder			
	Melting point: 690°C		Solubility: slightly soluble in water, insoluble in ethanol, soluble in acid and alkali.	
Combustion and explosion hazards	Boiling point (°C) Saturation vapor pressure (kPa) Critical temperature (°C): 1750°C Critical pressure Combustibility: this product does not burn. Flash point (°C) Explosion limit volume fraction (%) Autoignition temperature (°C)		Gas density Relative density: 3.35 Heat of combustion (kJ/mol) Minimum ignition energy (mJ) Combustion products Aggregating harm: none. Stability: stable at room temperature and pressure. Contraband: strong acid, flammable, or combustible substance.	
Toxicity Harm to human body	Dangerous characteristics: there are no special combustion and explosion characteristics. Firefighting method: this product does not burn, fire extinguishing agents can be used around the fire field. High toxicity. LD50: 10 mg/kg (rats through the mouth) LC50. Harmful to the respiratory system and skin. Acute poisoning: can cause nasal, pharynx, and lung irritation. Most workers experience pharynx itching, dry cough, chest tightness, general discomfort, burnout, and other manifestations, while some patients can develop nephritis and pneumonia. Chronic poisoning: long-term exposure can cause chronic bronchitis, kidney damage, and visual impairment.			
First aid	Skin contact: remove contaminated clothing and immediately rinse thoroughly with flowing water. Eye contact: lift eyelid immediately and rinse with flowing water. Inhalation: remove from site to fresh air. Keep warm and give artificial respiration if necessary. Go to a doctor. Ingestion: drink plenty of warm water for accidental ingestion, induce vomiting, and seek medical advice.			
Protective equipment	Maximum permissible concentration: Chinese MAC: 0.1 mg/m^3 (smoke); Soviet MAC: 0.1 mg/m^3 (smoke); US TWA: OSHA0. Engineering control: closed operation, partial exhaust. Respiratory protection: wear a gas mask when the concentration in the air exceeds the limit. Wear self-contained breathing apparatus if necessary. Eye protection: wear chemical safety glasses. Body protection: wear appropriate protective clothing.			

(Continued)

Table 13.4: (Continued)

Characteristics	Chinese name: Vanadium pentoxide		Common name: Vanadium pentoxide	
	Molecular formula: V_2O_5		Molecular weight: 182	UN Serial number: 2862
	Dangerous article number: 61028		RTECS serial number	CAS serial number: 1314-62-1
Leak management	Hand protection: wear protective gloves. Other protection: smoking, eating, and drinking are not allowed on site. After work, shower and change clothes. Store contaminated clothes separately and reuse them after washing. Conduct preemployment and regular employment. Emergency treatment: isolate the leakage area and set warning signs around it. It is recommended that emergency personnel wear positive-pressure self-contained breathing apparatus and chemical protective suits. Avoid direct contact with spills and dust. Collect with a clean shovel in a dry, clean, covered container and transfer to a safe place. Cement, asphalt, or suitable thermoplastic materials can also be used to treat the waste. For large leakages the collected material can be recycled or harmless treatment after waste.			
Storage and transport	Storage: store in a cool ventilated warehouse away from fire, heat, and direct sunlight. The package must be sealed and kept away from moisture. It should be stored separately from alkali, acid, and oxidant materials. Transport: do not use mixed storage and transportation. Handle with care to prevent damage to packaging and containers. Pay attention to personal protection during packing and handling.			
Waste disposal	National and local laws and regulations should be consulted before disposal.			

Table 13.5: Vanadium pentoxide and its chemical constituents (%).

Scope of application	Brand	Chemical constituents								Physical state
		V_2O_5	Si	Fe	P	S	As	$Na_2O + K_2O$	V_2O_4	
		>	<							
Metallurgical uses	$V_2O_5$99	99.0	0.15	0.20	0.03	0.01	0.01	1.0	—	Flake
	$V_2O_5$98	98.0	0.25	0.30	0.05	0.03	0.02	1.5	—	
Chemical uses	$V_2O_5$97	97.0	0.25	0.30	0.05	0.10	0.02	1.0	2.5	Powder Body

Note: The vanadium pentoxide content is converted from the total vanadium content.

13.4.1.2 Physical state

Vanadium pentoxide for metallurgy should be delivered in flake form, with a diameter of no more than 55 mm and a thickness of no more than 5 mm. Vanadium pentoxide used in the chemical industry is delivered as a powder after decomposition.

13.4.1.3 Test method

1. Sampling, chemical analysis shall be performed in accordance with the methods specified in Appendix A.
2. Preparation of samples for chemical analysis follows the method specified in Appendix B.
3. Chemical analysis. The analysis of vanadium pentoxide is performed in accordance with the existing analytical methods of each production plant.

13.4.1.4 Inspection rules

1. Inspection and acceptance of the product quality shall be conducted by the technical supervision department of the supplier, where the buyer has the right to reinspect the product quality in accordance with relevant regulations. Any objection should be made within 30 days after arrival of the goods.
2. Products of the same brand may be classified as a lot of goods delivered in batches, generally between 4 and 10 t, or as agreed by both parties.

13.4.1.5 Packaging, labeling, storage, transport, and quality certificate

1. Packaging

    The product is packaged in iron drums with a protective paint coating on the inside. The net weight of a barrel is generally not more than 250 kg or as agreed by both parties.
2. Labeling, storage, transport, and quality certificate

    Product labels, storage, transport, and quality certificates shall comply with the general provisions of GB 3650-83 for acceptance, packaging, storage, transportation, labeling, and quality certificates of ferroalloys.

13.4.1.6 Sampling of vanadium pentoxide (supplement)

1. Flake vanadium pentoxide

    In 25% of the packaged volumes of each batch of products, samples with approximately equal weights are taken at depths of 100−200 mm below the material surface, with a total amount of not less than 2 kg. Then, the sample is crushed to less than 10 mm, mixed well, and then reduced to 1 kg by quadrature.
2. Vanadium pentoxide powder

    In 25% of the packages of each batch of products, more than half of the thickness of the material layer is inserted in the center of the barrel. The total number of samples with approximately equal number of skewers is at least 1 kg.

13.4.1.7 Sampling method B for vanadium pentoxide (supplementary parts)

1. Flake vanadium pentoxide

   The 1-kg experimental sample is crushed to less than 5 mm, placed in a stainless steel plate, and reduced to 500 g by four points. The sample is then placed in a tamping cylinder and ground to 1.5 mm, then reduced to 250 g using the quadrature method, and then placed in the tamping cylinder again for grinding to less than 1 mm. After reducing the sample size to 60 g using the quadrature method, the material is crushed until it passes through an 80 μ sieve. Two separate bags, one for analysis, and the other sealed for inspection, are prepared.

2. Vanadium pentoxide powder

   After thoroughly mixing 1 kg of powder, it is reduced into 60 g sample using the quartering method, and then finely ground (80 μ). Two separate bags, one for analysis, and the other sealed for inspection, are prepared.

13.4.2 High-purity vanadium pentoxide Chinese standard (GB-3283-87)

The standard for high-purity vanadium pentoxide (GB-3283-87), with a molecular formula V_2O_5, is shown in Table 13.6.

1. High-purity vanadium pentoxide

   Vanadium pentoxide is an orange powder with a relative density of 3.357. It is soluble in acid and alkali solutions, slightly soluble in water, and insoluble in anhydrous ethanol. Vanadium pentoxide vapor is toxic. It has a relative molecular weight of 181.88.

2. High-purity vanadium pentoxide

   High-purity vanadium pentoxide can be used in the metallurgy industry for the production of ferrovanadium, vanadium–aluminum alloy, and other vanadium-bearing alloys. It can also be used in the chemical industry as a catalyst, for printing and dyeing, and for coloring ceramic materials. It is also the main raw material for preparing sulfuric acid catalysts and vanadium compounds.

Table 13.6: High-purity vanadium pentoxide Chinese standard (GB-3283-87).

Grade	Chemical composition								State
	V_2O_5	V_2O_4	Si	Fe	S	P	As	$Na_2O + K_2O$	
	≥	≤							
	98.0	2.5	0.25	0.3	0.03	0.05	0.02	1.0	Powder
	99.0	1.5	0.1	0.1	0.01	0.03	0.01	0.7	
	99.5	1.0	0.08	0.01	0.01	0.01	0.01	0.25	
	99.7		0.02	0.005	0.008	0.008	0.005	0.1	
	99.9		0.01	0.002	0.005	0.005	0.001	0.05	

Table 13.7: Chinese standard for chemical purity of vanadium pentoxide (99.0% vanadium powder) GB 3283-87.

Vanadium pentoxide (powder) 99.0% chemical purity National standard GB3283-87	
Cl ≤ 0.005% Na < 0.10% SO_4 ≤ 0.04%	Fe < 0.03% NH_4 ≤ 0.10% Calcination of weightlessness: 0.25%
Hydrochloric acid insoluble substances and silicate ≤ 0.3% Physical state: powder	

The Chinese national standard GB 3283-87 for chemical purity of vanadium pentoxide (99.0% vanadium powder) is shown in Table 13.7.

13.4.3 Chinese national standard for vanadium pentoxide (super-pure grade)

The molecular formula of vanadium pentoxide powder (99.0%) is V_2O_5, where its standard is shown in Table 13.8.

The molecular formula of vanadium pentoxide powder (99.0%) is V_2O_5, where its standard (GB-3283-87) is given in Table 13.9.

Table 13.8: Standard for vanadium pentoxide (super-pure grade).

Vanadium pentoxide (super pure) Grade: V_2O_5 ≥ 99.5% Impurities: Hydrochloric acid insolubles and silicate ≤ 0.1% Calcination of weightlessness: 0.01% Chloride (Cl): 0.005%, sulfate (SO_4): 0.01% Ammonium salt (NH_4): 0.02%, sodium(Na): 0.02% Iron (Fe): 0.01% Heavy metals (Pb): 0.002%

Table 13.9: Standard GB-3283-87 for vanadium pentoxide powder (99.0%).

Vanadium pentoxide powder 98% Chinese national standard: GB3283-87	
Si < 0.15% P < 0.04% As < 0.01%	Fe < 0.20% S < 0.01% $Na_2O + K_2O$ < 1.0%
Physical state: powder Scope of application	

Table 13.10: Chinese national standard for ammonium metavanadate.

NH_4VO_3 (dry basis) More than	Si	Fe	S	Al	Cl	As	$Na_2O + K_2O$
	Less than						
98.0	0.2	0.3	0.04	0.1	0.2	0.01	0.3
99.0	0.1	0.1	0.03	0.05	0.15	0.01	0.2
99.5	0.08	0.08	0.02	0.03	0.10	0.01	0.01

13.5 Ammonium vanadate

13.5.1 Ammonium metavanadate

13.5.1.1 Chinese standard

Table 13.10 shows China's national standard for ammonium metavanadate.

13.5.1.2 Safety data

The material safety data for ammonium metavanadate are shown in Table 13.11.

13.5.1.3 Properties

Ammonium metavanadate (NH_4VO_3) has a molecular weight of 116.98 and consists of an infinite chain structure of conformal VO_4 tetrahedrons. It is a white to yellowish crystalline powder, which is slightly soluble in cold water, warm ethanol, and ether, soluble in hot water and dilute aqueous ammonia solutions, and not soluble in ammonium chloride. Ammonium metavanadate is converted into toxic vanadium pentoxide when combusted in air. Ammonium metavanadate has a melting point of 210°C (decomposition) and relative density (water = 1) of 2.326.

13.5.2 Ammonium polyvanadate

13.5.2.1 Physical properties

The physical properties of ammonium polyvanadate are shown in Table 13.12.

13.5.2.2 Chemical properties

Ammonium polyvanadate [$(NH_4)_2V_6O_{16}$] has a molecular weight of 597.72 and is slightly soluble in cold water, hot ethanol, and ether, and soluble in hot water and dilute ammonium hydroxide. When combusted in air, it is converted to toxic vanadium pentoxide.

13.5.2.3 Identification information

Table 13.13 shows the identification information for ammonium polyvanadate.

Table 13.11: Material safety data for ammonium metavanadate.

Ammonium vanadate		
Characteristics	Chinese name English name Molecular formula Molecular weight UN number IMDG code page	Ammonium vanadate Ammonium vanadate $NH_4VO_3 \cdot 2H_2O$ 117 2859 6066
Physical and chemical properties Combustion and explosion hazards Packaging and storage	Appearance Main uses Melting point (°C) Combustibility Autoignition temperature (°C) Lower explosive limit (V%) Upper explosive limit (V%) Dangerous characteristics Hazard category Precautions for storage and transport Ingestion path Toxicity Health hazard	White or yellow crystalline powder. Catalysts, dyes, and fast-drying paints. Heating to 210°C results in decomposition to produce vanadium pentoxide. Class 6.1 (highly toxic). Package in a glass box, and store in a cool, dry, and ventilated warehouse. Keep away from fire and store away from combustible and edible materials, and reducing agents. Accidental inhalation. Highly toxic. Rat dosage LD50 160 mg/kg. Accidental ingestion can cause vomiting and diarrhea. Dust can irritate eyes and mucous membranes.
First aid		Patients who inhale the gas should be immediately removed from the contaminated area, kept warm, and sent to the hospital for treatment if necessary. Irritated eyes should be rinsed in water; seek medical treatment for serious cases. If swallowed, immediately gargle water, and then send to hospital for treatment.
Protective measures		Wear protective gear when exposed to steam. In the case of emergency rescue or escape, it is recommended to wear self-contained breathing apparatus.
Leak management		Wear gas mask and gloves. Apply wet sand to the leaked material and place in an open area.

Table 13.12: Physical properties of ammonium polyvanadate.

Appearance	Pale yellow crystalline powder (g/cm³)	Melting point	Decomposition at 350°C
Density Bulk density	3.03 0.9	Particle size	≤ 3 mm

Table 13.13: Identification information for ammonium polyvanadate.

IMDG class	6.1
CAS No.	11115-67-6
UN No.	2861
Tariff No.	2841 9030

13.5.2.4 Applications

Ammonium polyvanadate is mainly used as chemical reagent, catalyst, desiccant, and mordant. In the ceramics industry, it is widely used as a glaze. It can also be used to produce vanadium pentoxide and vanadium trioxide.

13.5.2.5 Chinese national standard for ammonium polyvanadate (Q/JTHJ017-2002)

Table 13.14 shows China's national standard for ammonium polyvanadate.

13.6 Ferrovanadium

Molecular formula: VFe; CAS number: /.

13.6.1 Properties

Vanadium and iron are infinitely soluble in the liquid phase. At low temperature, there is an intermediate VFe phase with a tetrahedral crystal lattice structure. It has a density of 6.7 g/cm^3 and melting temperature of 1450°C. Vanadium in ferrovanadium can produce stable carbide, nitride, and carbonitride compounds with carbon and nitrogen in the steel, which increase the strength, hardness, wear resistance, and toughness of the steel. Ferrovanadium grades can have vanadium contents of 30%, 60%, and 80%.

Table 13.14: Chinese national standard for ammonium polyvanadate (Q/JTHJ017-2002).

Brand	(Decomposed) Chemical composition %						
	V_2O_5	Si	Fe	P	S	As	$Na_2O + K_2O$
	More than	Less than					
APV-01	99.0	0.10	0.10	0.03	0.10	0.01	1.0
APV-02	98.0	0.10	0.10	0.03	0.10	0.01	1.5
APV-03	98.0	0.25	0.30	0.05	0.20	0.02	1.5

13.6.2 Physical state

In general, ferrovanadium is massive, with a maximum bulk weight of less than 8 kg. The debris in the 10 mm × 10 mm sieve hole shall not exceed 3% of the total weight. If the user has special requirements regarding the particle size, this can be agreed upon.

13.6.3 Standard

13.6.3.1 Chinese national standard (GB4139-87)

The Chinese national standard (vanadium iron GB4139-87) includes the requirements shown in Table 13.15.

1. Physical state
   a. Ferrovanadium is supplied in the form of particles, with a maximum weight not exceeding 8 kg. The fragments passing through a 10 mm×10 mm sieve shall not exceed 3% of the total weight of the batch.
   b. If the buyer has special particle-size requirements, this can be separately negotiated by both parties.
2. Test method
   a. Samples for chemical analysis shall be taken in accordance with the GB 4010-83 methods for sample selection for ferroalloy chemical analysis.
   b. Preparation of samples for chemical analysis shall be conducted in accordance with the GB 4332-84 method for sample preparation for alloy chemical analysis.
   c. The chemical analysis of ferrovanadium shall be performed according to the YB 585-65 chemical analysis method.

Table 13.15: Chinese national standard (ferro vanadium GB4139-87).

Brand	Chemical composition (%)						
	V	C	Si	P	S	Al	Mn
	More than	Less than					
FeV 40-A	40	0.75	2	0.1	0.06	1	
FeV 40-B	40	1	3	0.2	0.1	1.5	
FeV 50-A	50	0.4	2	0.07	0.04	0.5	0.5
FeV 50-B	50	0.75	2.5	0.1	0.05	0.8	0.5
FeV 75-A	75	0.2	1	0.05	0.04	2	0.5
FeV 75-B	75	0.3	2	0.1	0.05	3	0.5

3. Inspection rules

   a. Quality inspection and acceptance, and product quality inspection and acceptance shall comply with the general provisions of GB 3658-83 regarding acceptance, packaging, storage, transportation, labeling, and quality certification of ferroalloys.

   b. Products from each furnace shall be delivered as a separate batch. If the desired quantity is less than one furnace volume, it can be combined with other furnace batches with a difference in vanadium content <2%.

4. Packaging, storage, transportation, labeling, and quality certificate

   a. The product is packed in iron bucket with a net weight of 50 or 100 kg/bbl, as specified in the contract with the customer.
   b. Certificates of storage, transportation, labeling, and quality of the products shall meet the requirements stated in standard GB 3650-85.

5. Additional instructions

   a. This standard was proposed by the Ministry of the Metallurgical Industry, China.
   b. This standard was drafted by Jinzhou Ferroalloy Factory.
   c. The main authors of the standard were Huang Shujie and Bai Fengren.
   d. This standard level is labeled as GB 4139-87y.

6. Standard for ferrovanadium specifications and delivery conditions
   Ferrovanadium specifications and delivery conditions: ISO 5451-1980
   1. Scope and application area
      This international standard describes the delivery requirements and conditions for ferrovanadium, usually used in steelmaking and casting.
   2. Relevant standards
      ISO565 test screen-wire mesh and orifice plate-nominal aperture.
      ISO3713 ferroalloy-take harmony sample-general.
   3. Definition
      Ferrovanadium refers to an intermediate Fe−V alloy obtained by reduction, with a vanadium content not less than 35.0 wt.% and not more than 85.0 wt.%.
   4. Order details

The ferrovanadium order shall include the following:

   a. quantity;
   b. group batch methods;
   c. chemical composition, as indicated in Table 13.16;
   d. range of granularity, as shown in Table 13.17; and
   e. appropriate and necessary requirements for analysis of reports, packaging, etc.

Table 13.16: Chemical composition of ferrovanadium.

Brand	Chemical composition (%)									
	V	Si	Al	C	P	S	As	Cu	Mn	Ni
						≤				
FeV40	35.0–50.0	2.0	4.0	0.10	0.10	0.10				
FeV60	50.0–65.0	2.0	2.5	0.06	0.05	0.05	0.06	0.10		
FeV80	75.0–85.0	2.0	1.5	0.06	0.05	0.05	0.06	0.10	0.50	0.15
FeV80A2	75.0–85.0	1.5	2.0	0.06	0.05	0.05	0.06	0.10	0.50	0.15
FeV80Al4	70.0–80.0	2.0	4.0	0.10	0.10	0.10	0.10	0.10	0.50	0.15

Table 13.17: Ferrovanadium particle granularity.

Grade	Particle size range	Maximum content of superfine particles (wt.%)	Maximum content of large particles (wt.%)
1	2–10	3	10
2	0	3	The grain size in two or three dimensions shall not exceed the maximum limit of the specified grain size range × 1.15.
3	2–50	5	
4	2–25	5	
5	2–10 <2	—	

5. Requirements
a. Batches

Ferrovanadium shall be delivered in batches according to one of the following methods:

i. Furnace batch method

According to the furnace batch method, the general delivery product consists of one furnace product (or part of a continuous furnace product) of ferrovanadium.

ii. Batch by grade

In the grading method a batch of delivered products consists of several furnace loads (or parts thereof) of a certain grade. The variation in vanadium contents between the furnace product (or parts of a continuous furnace product) that make up a batch of delivered product shall not exceed 3% (absolute value).

iii. Mixed batch method

According to the mixed batch method, a batch of delivered products is composed of ferrovanadium from several furnaces (or parts thereof) of one grade. The products should be broken into a size less than 50 mm and mixed. The content of the main components of the furnaces (or parts of the furnaces that make up a batch of delivered products) may fluctuate between the minimum and maximum limits specified by the corresponding ferrovanadium grade.

b. Chemical composition

   The chemical composition of ferrovanadium shall comply with the provisions of Table 13.16, and its particle size shall conform to the ranges of grade 1–4 shown in Table 13.17. The chemical compositions listed in Table 13.16 are only those of the main elements and common impurities. If the buyer requires different or narrower compositional ranges for the main elements and/or special limits for another element, this should be negotiated by the supplier and buyer. The chemical composition listed in Table 13.15 is affected by the accuracy of ferrovanadium sampling and analysis methods (see "6").

c. Granularity range

   Ferrovanadium is supplied in bulk sizes or crushed and screened to give a certain granularity. The size range and allowable deviation shall comply with the provisions of Table 13.17. The finest particle size shall be based on the delivery point to the buyer. Material with the specified size is obtained via sieving with a steel mesh, as specified in ISO565. If the buyer requires a range and/or deviation of granularity different to those shown in Table 13.17, this shall be agreed upon by both parties.

d. External contamination

   External contamination of the ferrovanadium should be avoided as much as possible.

   6. Testing

      a. Sampling for chemical analysis and screening analysis

         The delivery point refers to the location where responsibility for the delivered products is transferred from the supplier to the buyer. If neither the supplier nor the buyer is responsible for transportation, then the delivery point with a high finer grained value shall be agreed upon.

         Sampling for chemical and screening analyses is best performed in accordance with the methods specified in ISO3713, although other sampling methods with similar accuracy may also be used. Unless otherwise agreed, sampling is normally conducted at the supplier's depot. Representatives from both parties may be present wherever the sample is taken. If arbitration sampling is required, it shall be conducted by an arbitrator agreed upon by both parties. Sampling shall be conducted in accordance with ISO3713 or using another method with similar accuracy, if agreed upon by the supplier, buyer, and arbitrator. The arbitration sample shall be accepted by both parties.

      b. Analysis

         Chemical analysis of ferrovanadium is best performed by the method specified in ISO6467, although other chemical analysis methods with similar precision can also be used. Delivery of ferrovanadium products shall be accompanied by the supplier, to provide certificates of analysis that describe the content vanadium, and other elements (if agreed upon, following the details specified in Table 13.15). In addition, representative samples of the products should be delivered with the cargo. When a dispute arises, it may be settled by one of the following two methods.

   c. Chemical analysis

Chemical analyses shall be conducted using the same sample, preferably by the method specified in ISO6467, or by other methods of chemical analysis with similar precision, which shall be agreed upon by both parties. If the difference between the two analysis results is within X, the average value should be adopted. If the difference between the two analysis results is more than $X\%$, and if no other agreement is reached, an arbitration analysis should be conducted. Table 13.16 shows the required chemical composition of ferrovanadium, while Table 13.17 shows the corresponding particle size requirements.

   d. Arbitration analysis

Arbitration analysis is best performed using the method specified in ISO6467 or by other chemical analysis methods with similar precision. The method shall be agreed upon by the supplier, buyer, and arbitrator. If the result of the arbitral analysis is between the two disputed values, or deviates by $y\%$ from one of them, this result is considered the final decision.

7. Delivery and storage

The packing, storage, and transportation of ferrovanadium shall comply with international regulations. *Note*: (1) Sieving analysis of ferroalloy shall follow ISO4551; (2) the X value will be specified before it is specified. This value will be agreed upon by both parties.

13.6.3.2 Ferrovanadium (DIN 17563-65)

13.6.3.2.1 Concept

Ferrovanadium as described in this standard refers to an intermediate alloy with a minimum vanadium content of 50 wt.%, which is produced by reduction of the corresponding raw material or its concentrate.

13.6.3.2.2 Request

1. Chemical composition

Ferrovanadium is supplied in two varieties. The chemical composition of its standard granularity shall comply with the provisions shown in Table 13.18.

2. State

Ferrovanadium is usually supplied in fist-sized "standard granularity." It can also be crushed into other particle size fractions, where the specific size should be determined through consultation.

Table 13.18: Ferrovanadium (DIN 17563-65).

Brand	Material code	Chemical composition (wt.%)							
		V	Al	Si	C	S	P	As	Cu
					Less than				
FeV60	0.4706	50–65	2.0	1.5	0.15	0.05	0.06	0.06	0.10
FeV80	0.4708	78–82	1.5	1.5	0.15	0.05	0.06	0.06	0.10

13.6.3.2.3 Testing

1. Sampling

    Sampling shall be conducted according to the methods prescribed by the Chemical Workers Committee of the West German Iron and Steel Association or the Chemical Workers Committee of the West German Mining and Metallurgy Association. In general, sampling is performed in the shipping unit by a sampler recognized by both parties. If it was agreed that samples shall be taken at the receiving location, the supplier has the right to participate, or invite others to participate, in the sampling process. The sampler shall prepare and seal four samples according to the analysis accuracy requirements. The supplier and buyer shall each take one sample for analysis, while the third sample is for arbitration analysis, and the fourth is reserved as a backup. Arbitration samples shall be stored in a location agreed upon by the parties.

2. Analysis

    It is recommended to conduct ferrovanadium analysis in accordance with the methods prescribed by the Chemical Workers Committee of the West German Iron and Steel Association or the Chemical Workers Committee of the West German Mining and Metallurgy Association. At the time of delivery the supplier shall complete a separate analytical certificate, without reference to the analysis results of the buyer, indicating the vanadium content. This document may also give the content of other elements specified in the table according to the agreement.

3. Exchange of analysis results and arbitration analysis

    For the exchange of analysis results, the content of vanadium and other components to be analyzed shall be agreed upon, and the test samples from the cargo shall be provided for analysis. The analysis results of both parties shall be exchanged within an agreed period. If the difference between the two analysis results does not exceed 0.50%V, the average value of the two analysis results is used. However, if this difference exceeds 0.50%, if there is no other way to obtain consensus, arbitration analysis of the samples shall be conducted by an arbitrator according to the methods prescribed by the Chemical Workers Committee of the West German Iron and Steel Association or the Chemical Workers Committee of the West German Mining and Metallurgy Association.

If the result of the arbitration analysis is between the two values provided by the parties, the arbitration result is considered the final judgment. If the result of the arbitration analysis is outside the range of the two exchanged values, the arithmetic mean of the arbitration value and the closest analysis value is used as the standard. Arbitration analysis shall be conducted to determine the liability of the party whose analysis results deviate most from the arbitral value. If the value determined by arbitration analysis is in the middle of the other two values, the liability shall be borne by both parties.

4. Delivery and storage

    Ferrovanadium shall be shipped and stored in labeled containers.

5. Objections

    Objections are allowed only when external and internal defects have a significant impact on the processing and use of the ferroalloy material. The user shall provide the supplier with the opportunity to confirm that the user's objections are valid through consideration of the samples and supplies in question.

13.6.3.3 Ferrovanadium (JIS G 2308-1986)

1. Scope of application

    This standard is mainly applicable to ferrovanadium used in iron and steel smelting as an additive for modifying the alloy composition.

2. Type and brand number

Type and brand numbers are shown in Table 13.19.

3. Batch methods
    a. Batch methods are classified as the batch method, mixed batch method, or furnace batch method.
    b. When the grading method is adopted, the vanadium component is selected for grading, where the grading interval is no more than 2%.
    c. The batch size is as shown in Table 13.20.
4. Quality
    a. The chemical constituents are shown in Table 13.21, while the chemical composition is specified in Table 13.22.
    b. Granularity

The granularity specifications are shown in Table 13.23.

Table 13.19: Type and brand numbers.

Species class		Brand
Ferrovanadium	No. 1	FV1
	No. 2	FV2

Table 13.20: Batch size.

Batch method	Batch method	Mixed batch method	Furnace batch method
Batch size	≤ 20 t	≤ 5 t	1 furnace volume

Table 13.21: Chemical composition.

Species	Brand		Chemical composition (%)					
			V	C	Si	P	S	Al
Ferrovanadium	No. 1	FV1	75.0–85.0	≤ 0.2	≤ 2.0	≤ 0.10	≤ 0.10	≤ 4.0
Ferrovanadium	No. 2	FV2	45.0–55.0	≤ 0.2	≤ 2.0	≤ 0.10	≤ 0.10	≤ 4.0

Table 13.22: Designated chemical composition.

Species		Chemical composition (%)		
		P	S	Al
Ferrovanadium	Whole class	≤ 0.03	≤ 0.05	≤ 1.0 ≤ 0.5

Table 13.23: Particle size definitions.

Type	Code name	Particle size (mm)
General particle size	G	1–100
Small particle size	S	1–50

5. Testing

   a. The sampling method and sample preparation method for determining the average grade of a batch may adopt the following criteria:
      JIS G 1501: General rules for ferroalloy sampling methods.
      JIS G 1602: Sampling method for sample analysis of ferroalloy composition 2 (ferrotungsten, ferromolybdenum, ferrovanadium, ferrotitanium, and ferroniobium).
   b. The following criteria are used for analytical test analysis:
      JIS G 1301: general rules for ferroalloy analysis methods
      JIS G 1318: ferrovanadium analysis method
   c. Particle size testing shall be conducted in accordance with JIS G 1641 (sample sampling method and particle size determination method for ferroalloy particle size).

6. Confirmation

The results of analytical and granularity tests must comply with the provisions of article 4. In the case of nonconformity, the whole batch of materials represented by the sample shall be deemed unqualified.

7. Documentation

For bulk products the total quantity is divided into batches. The following items must be labeled on the invoice. When loading containers, the label on each container must include the following:
   a. type or brand name;
   b. the chemical composition, percentage of each component, and when the chemical composition was specified;
   c. granularity or its code name;
   d. batch number; and
   e. name of the manufacturer or its abbreviation

13.6.3.4 Standard ASTM A 102-92 for ferrovanadium

1. Range
   a. This standard includes the grade of ferrovanadium.
   b. Values expressed in the inch−pound system are standard. Values in brackets are for reference only.
2. Relevant documents
   a. ASTM standard

   E29 is the recommended method for determining the number of significant digits of test data that meet the standard requirements. The method of chemical analysis of iron alloy is given in E31. In addition, E32 describes the determination of the chemical composition by the sampling method for ferroalloys and steel additives.
3. Orders
   a. Orders for this standard material shall include the following: number, material name, ASTM standard number and year of publication, particle size, and special requirements for packaging, analysis, and reporting, if necessary.
   b. Although ferrovanadium is ordered with a total net weight, the usual basis for payment is a price per pound of vanadium.
4. Chemical composition requirements
   a. The chemical composition of the various grades shall comply with the requirements listed in Tables 13.24 and 13.25.
   b. The production plant shall provide the analysis results of each batch of ferrovanadium elements, as specified in Table 13.24.
   c. The values shown in Table 13.25 are the maximum requirements. At the buyer's request the manufacturer shall provide the analytical values of any of these elements accumulated over a period of time, as agreed upon by both parties.

Table 13.24: Chemical requirements A.

Element	Composition (%)
Vanadium B	75–85
Carbon, max	0.75
Silicon, max	1.5
Aluminum	2.0 (max)
Sulfur, max	0.08
Phosphorus, max	0.08

Table 13.25: Supplementary chemical composition A and B.

Element	Maximum allowable limit (%)
Chromium	0.50
Copper	0.15
Nickel	0.10
Lead	0.020
Tin	0.050
Zinc	0.020
Molybdenum	0.75
Titanium	0.15
Nitrogen	0.20

Table 13.26: Typical particle size requirements.

Particle size requirement
Below 2 in. (50 mm) 2 in.
Below 1 in. (25 mm) × 1 in.
Below 1/2 in. (12.5 mm) × 1/2 in.
Below N0.8 (2.36 mm) × N0.8

5. Particle size
   a. The various grades are supplied in accordance with the particle sizes listed in Table 13.23.
   b. Standard particle sizes of the alloy are listed in Table 13.26. They are delivered with varying degrees of brittleness, and during the process of transportation, storage, and loading and unloading will endure some wear.
6. Sampling
   a. Materials shall be sampled using the E32 method.
   b. Other sampling methods agreed upon by the supplier and buyer may also be used. In the case of a dispute, the E32 method shall be used for arbitration.

7. Chemical analysis
   a. The chemical analysis of materials shall be performed using the ferrovanadium chemical analysis method described in the E31 method or other analytical methods which can yield the same results.
   b. In the case of disputes arising from the use of other analytical methods, the E31 method shall be used for arbitration.
   c. If an analytical method for a particular element is not provided in the E31 standard, the analysis of that element shall be performed in accordance with a method agreed upon by the supplier and buyer.

8. Inspection
   a. The production plant shall provide, free of charge, all reasonable conveniences to the inspectors representing the buyer to ensure that the materials are provided in accordance with the standard.
   b. To determine whether this standard is met, the reported analysis results shall be modified to obtain the same number as the last digit of the limit value in accordance with the modification method described in E29.
   c. To determine the vanadium content of any given batch of ferrovanadium, the modified method specified in footnote A shall be used to modify the reported vanadium value to the nearest 0.1%.
   d. The composition of ferrovanadium B should be within these maximum limits, although it does not need to be analyzed for every batch. The manufacturer shall provide the analysis results for these elements within a period mutually agreed upon by both parties, as required by the buyer.

9. Objections
   a. Any demand for or refusal of bribes shall be made by the buyer to the manufacturer within 45 days from the date of arrival.

10. Packaging and labeling
    a. Ferrovanadium shall be packed in strong containers or shipped in bulk with the guarantee that the material will not be damaged or contaminated during shipment.
    b. When shipping in bulk, it shall be accompanied by appropriate labels indicating the material, grade symbol, ASTM standard number, particle size, batch number and name, and trademark or label of the manufacturer.
    c. When transported in containers, each container shall be labeled with the material number, grade symbols, ASTM standard, particle size, batch number, gross weight, tare weight, net weight, and the manufacturer's name, trademark, or label.

13.6.3.5 US ferrovanadium standard

The US ferrovanadium standard is shown in Table 13.27. The required chemical composition is shown in Table 13.28. The chemical composition listed in the ferrovanadium supplement standard for ferrovanadium in the United States is shown in Table 13.29.

Table 13.27: US ferrovanadium standard.

Grade	Standard granularity (mm)	Limits	
A, B, C, cast iron grade	<50	>50 mm ≤10%	<0.84 mm ≤10%
	<25	>25 mm ≤10%	<0.84 mm ≤10%
	<12.5	>12.5 mm ≤10%	<0.60 mm ≤10%
	<2.36	>2.36 mm ≤10%	<0.074 mm ≤10%

Table 13.28: Chemical composition.

Element	Chemical composition (%)			
	Grade A	Grade B	Grade C	Malleable iron
V	50.0–60.0 or 70.0–80.0	50.0–60.0 or 70.0–80.0	50.0–60.0 or 70.0–80.0	35.0–45.0 or 50.0–60.0
C, max	0.20	1.5	3.0	3.0
P, max	0.050	0.06	0.050	0.10
S, max	0.050	0.050	0.10	0.10
Si, max	1.0	2.5	8.0	8.0 or 7.0–11.0
Al, max	0.75	1.5	1.5	1.5
Mn, max	0.50	0.50	–	

Table 13.29: US standard for ferrovanadium–ferrovanadium: supplementary chemical composition.

Element	Grade A, grade B, grade C, and malleable iron grade. Maximum values (%)
Cr	0.50
Cu	0.15
Ni	0.10
Pb	0.020
Sn	0.050
Zn	0.020
Mo	0.75
Ti	0.15
N	0.20

13.6.3.6 Ferrovanadium from Japan (JIS g2308-1986)

1. Type and code name

Ferrovanadium is used as an alloy additive in steel production. The types and codes are shown in Table 13.30.

2. Chemical composition

The chemical composition is shown in Table 13.31. However, the chemical composition can be specified in Table 13.32.

3. Particle size

The typical particle sizes are given in Table 13.33.

Table 13.30: Ferrovanadium types and codes.

Species		Code name
Ferrovanadium	No. 1	FV1
	No. 2	FV2

Table 13.31: Chemical composition.

Species		Label	Chemical composition (%)					
			V	C	Si	P	S	Al
Ferrovanadium	No. 1	FV1	75.0–85.0	≤0.2	≤2.0	≤0.10	≤0.10	≤4.0
	No. 2	FV2	45.0–55.0	≤0.2	≤0.2	≤0.10	≤0.10	≤4.0

Table 13.32: Designated chemical composition.

Species		Chemical composition (%)		
		P	S	Al
Ferrovanadium	Full species	≤0.03	≤0.05	≤1.0
				≤0.5

Table 13.33: Particle sizes.

Species	Code name	Particle size (mm)
General particle size	g	1–100
Small particle size	s	1–50

13.7 Vanadium nitride

13.7.1 Properties

GB/t20567-2006 vanadium nitrogen alloy

English standard name: vanadium-nitrogen

Chinese Standard name: vanadium nitrogen alloy

Standard no: GB/t20567-2006

Density: The apparent density of vanadium nitrogen alloy should be not less than 3.0 g/cm^3.

Particle size: The particle size of vanadium nitrogen alloy shall be 10–40 mm, and the content of products with grain size <10 mm should not exceed 5% of the total amount.

Packaging: High-strength moisture-proof packaging.

Application: Vanadium nitrogen alloys can be used in structural steel, tool steel, pipe steel, steel reinforcement, and cast iron. The application of vanadium nitrogen alloy in high-strength low-alloy steel can effectively achieve simultaneous vanadium and nitrogen microalloying, promote the precipitation of carbon, vanadium, and nitrogen compounds in steel, and enhance sedimentation strengthening and grain refinement.

Vanadium nitrogen alloy is mainly composed of vanadium nitride, which has two crystal structures. The V3N structure is a hexagonal crystal structure with extremely high microhardness ($\sim$1900 HV), where the melting point is unknown. The second structure is VN, with a face-centered cubic crystal structure, microhardness of $\sim$1520 HV, and a melting point of 2360°C. They both have high wear resistance.

13.7.2 Chinese national standard for vanadium nitride and vanadium carbide

The Chinese national standards for vanadium nitride and vanadium carbide are shown in Table 13.34.

- Density

    The apparent density of vanadium nitrogen alloy should be not less than 3.0 g/cm^3.

- Particle size

The grain size of vanadium nitrogen alloy should be 10–40 mm, and the content of products with grain size <10 mm should not exceed 5% of the total amount.

Table 13.34: Chinese national standards for vanadium nitride and vanadium carbide.

Vanadium nitride, vanadium carbide Vanadium nitride V ≥ 77% N: 10–14			
Si ≤ 0.25	P ≤ 0.03%	S ≤ 0.01%	
C ≤ 7%		Al ≤ 0.2	Mn = 0.05%
Vanadium carbide: VC ≥ 99%			
Total C ≤ 17–19, the free volume C ≤ 0.5			
Fe ≤ 0.5%	Si ≤ 0.5%	Other ≤ 0.5%	

13.7.3 Panzhihua Steel standard for vanadium nitrogen alloy

Table 13.35 shows the internal control standard of vanadium nitrogen alloy in Panzhihua Steel. The chemical composition of vanadium nitrogen alloy (%) is shown in Table 13.34.

13.7.4 Vanadium nitride from Vametco mining company, South Africa

The chemical composition of vanadium nitride produced by Vametco Mining Company in South Africa is shown in Table 13.36.

The physical properties of Nitrovan12 are shown in Table 13.37.

13.7.5 Vanadium carbide from the United States

The chemical composition of vanadium carbide produced by the US Strategic Minerals Company is shown in Table 13.38.

Table 13.35: Internal control standard of vanadium nitrogen alloy from Panzhihua Steel.

Brand	V%	N%	C	Al	S	P
			\multicolumn{4}{c}{Less than}			
VN12	76–82	10–14	10	0.20	0.08	0.06
VN16	76–82	14–18	6	0.02	0.08	0.06

Table 13.36: Chemical composition of vanadium nitride from Vametco Mining Company, South Africa.

Alloy	V	N	C	Si	Al	Mn	Cr	Ni	P	S
Nitrovan7	80	7	12.0	0.15	0.15	0.01	0.03	0.01	0.01	0.10
Nitrovan12	79	12	7.0	0.07	0.10	0.01	0.03	0.01	0.02	0.20
Nitrovan16	79	16	3.5	0.07	0.10	0.01	0.03	0.01	0.02	0.20

Table 13.37: Physical properties of Nitrovan12.

Appearance	Single ball weight (g)	Standard size (mm)			Apparent density (g/cm^3)	Bulk density (g/cm^3)	Density (g/cm^3)
		Length	Width	Height			
Coal ball dark gray metallic	37	33	28	23	3.71	2.00	~4.0

Table 13.38: Chemical composition of vanadium carbide produced by the US Strategic Minerals Company.

Composition	V	C	Al	Si	P	S	Mn
	82–86	10.5–14.5	<0.1	<0.1	<0.05	<0.1	<0.05

13.8 Vanadium metal

13.8.1 Properties

Vanadium metal is a granular crystal with a silvery white metallic luster. It has a density of 6.11 g/cm^3, melting point of 1900°C, and is difficult to melt. Pure vanadium is easy to oxidize at high temperature. Vanadium compounds are toxic to the human body. It has excellent corrosion resistance, with better resistance to hydrochloric acid and sulfuric acid than stainless steel and titanium. Its nuclear performance is good and deformation is easy.

13.8.2 Applications

Vanadium is mainly used to manufacture alloy steel and nonferrous metal alloys, as well as in the production of cathodes, grids, ray targets and getters, and electrode fluorescent tubes.

13.8.3 Chinese national standard

Table 13.39 shows the Chinese standard for vanadium metal (GB/t4310-1984).

13.9 Vanadium aluminum alloy

13.9.1 Properties

This alloy is usually in the form of particles with silver-gray metallic luster. As the content of vanadium in the alloy increases, its metallic luster, hardness, and oxygen content increase. When the content of V is ≥85%, the product is not easily broken and an oxide film is easily formed on the surface during long-term storage. The grain size range for AlV55 and AlV65 grades is 0.25–50.0 mm, while that of the AlV75 and AlV85 products is 1.00–100.0 mm.

Table 13.39: Chinese national standard for vanadium metal (GB/t4310-1984).

Brand	Composition (%)							
	V content, more than	Impurities, less than						
		Fe	Cr	Al	Si	O	N	C
V-1	Margin	0.005	0.006	0.005	0.004	0.025	0.006	0.01
V-2	Margin	0.02	0.02	0.01	0.004	0.035	0.01	0.02
V-3	99.5	0.10	0.10	0.05	0.05	0.08	—	—
V-4	99.0	0.15	0.15	0.08	0.08	0.10	—	—

13.9.2 Applications

Vanadium aluminum alloy is an intermediate alloy, mainly used as an additive for making Ti-based alloys, high-temperature alloys, and some special alloys. V−Al alloys are advanced alloy materials that are widely used in the aerospace field. They have high hardness, flexibility, seawater resistance, and are light. Only a few countries, such as the United States and Germany, have achieved industrial production of such alloys.

13.9.3 Chinese national standard

The GB5063-85 standard is applicable to V−Al, which is made by metal thermal reduction of vanadium oxide and used as an additive in Ti-based alloy, high-temperature alloy, and some special alloys.

13.9.3.1 Technical requirements

1. Brands and chemical constituents.
2. V−Al alloy is divided into four grades according to the vanadium content and impurities. The chemical composition should meet the requirements shown in Table 13.40.
3. The aluminum content of each grade and the oxygen content of AlV75 and AlV85 shall be provided by the supplier with the measured data but is not a basis for acceptance.
4. If the buyer has special requirements for the listed, or other, elements, they may be separately negotiated by both parties.
5. Physical state
   a. The products shall be delivered in bulk and granular form, where the particle size range shall comply with the provisions shown in Table 13.41.
   b. The amount of particles larger than the particle-size range shall not exceed 5.0% of the total weight of the alloy.
   c. If the buyer has special granularity requirements, they shall be separately negotiated by both parties.

Table 13.40: Chinese national standard for vanadium−aluminum alloy GB5063-85.

Brand	Composition (%)					
	V	Fe	Si	C	O	Al
				≤		
AlV55	50.0−60.0	0.35	0.30	0.15	0.20	Margin
AlV65	>60.0−70.0	0.30	0.30	0.20	0.20	Margin
AlV75	>70.0−80.0	0.30	0.30	0.20	−	Margin
AlV85	>80.0−90.0	0.30	0.30	0.30	−	Margin

Table 13.41: Chinese national standard for vanadium—aluminum alloy GB5063-85.

Brand	Particle size range
AlV55	3.0—50.0
AlV65	1.0—100.0
AlV75	1.0—100.0
AlV85	1.0—100.0

13.9.3.1.1 Appearance

The alloy should be strictly refined and its surface must not contain a visible oxide film nor nonmetallic inclusions.

13.9.3.2 Test methods

1. Sampling
   Samples for chemical analysis shall be taken in accordance with Appendix A (supplementary part) of this standard.
2. Sample preparation
   The preparation of samples for chemical analysis shall be performed in accordance with Appendix B (supplementary part) of this standard.
3. Chemical analysis
   Chemical analysis of V—Al alloy is temporarily performed according to the current analysis methods of the specific production plants. If there is any objection, it shall be settled through negotiation.

13.9.3.3 Inspection rules

1. The quality inspection and acceptance of the product shall comply with the general provisions of GB 3650-83 for acceptance, packaging, storage, transportation, labeling, and quality certificates of ferroalloys.
2. Each batch of products shall be delivered in a single delivery, with the maximum quantity of the batch not exceeding 500 kg.

13.9.3.4 Packaging, labeling, storage, transport, and quality certificate

1. Packaging
   The product is packed in iron drums with a net weight of 40 kg/bbl, where the actual weight is indicated when less than 40 kg is needed.
2. Storage, transportation, labeling, and quality certificates
   Storage, transportation, labeling, and quality certificates shall comply with the relevant provisions of GB 3650-83.

13.9.3.5 Sampling method A for vanadium aluminum alloy (supplementary parts)

A.1 Production inspection sampling

The radius of the alloy ingots from each furnace are divided into three equal parts, where three metal columns (including the center line of the ingots) are taken from the cross section. Then, three points are taken from the upper surface to the top of each metal column, where the particle size is no more than 20 mm and the total amount is 400 g.

A.2 Sampling for verification

Three samples are taken from each batch of products. Three samples of no more than 20 mm × 20 mm are evenly taken from different areas, where the samples at each point are approximately equal. The total sample weight is 1.6 kg, where the samples are crushed to a size less than 2 mm. The samples are divided into 400 g lots using the quartile method to form comprehensive samples.

13.9.3.6 Sample preparation method of B−V aluminum alloy (supplementary parts)

All 400 g samples are crushed to a size below 2 mm and ground to <1 mm in a stainless steel bowl. Then, 100 g samples are taken using the quartering method, which are then screened using a 100-mesh screen. These samples are then divided into two equal parts, one for analysis and one for retention.

13.9.3.7 Additional instructions

This standard was proposed by the Ministry of the Metallurgical Industry, China. This standard was drafted by the Jinzhou Ferroalloy Factory, where the main authors were You Baoren and Huang Yadong.

13.9.4 German industrial standard

The chemical composition of V−Al intermediate alloy produced by GFE in accordance with German industrial standard DIN17563 is shown in Table 13.42.

Table 13.42: Chemical composition of vanadium aluminum intermediate alloy.

Brand	V	C	Si	Al
V80Al	85	0.10	1.00	15
V40Al	40	0.10	1.00	60
V40Al60	40−45	0.10	0.30	55−60
V80Al20	75−85	0.05	0.40	15−20

CHAPTER 14

Vanadium series products and functional materials

Chapter Outline
14.1 Vanadium halides 396
    14.1.1 Vanadium oxytrichloride 396
    14.1.2 Vanadium tetrachloride 398
    14.1.3 Vanadium pentafluoride 398

14.2 Vanadates 399
    14.2.1 Ammonium metavanadate 399
    14.2.2 Ammonium polyvanadate 400
    14.2.3 Sodium vanadate 401
    14.2.4 Sodium metavanadate 401
    14.2.5 Potassium vanadate 402
    14.2.6 Bismuth vanadate 402
    14.2.7 Yttrium vanadate 403

14.3 Vanadium oxides 404
    14.3.1 Vanadium dioxide 404
    14.3.2 Vanadium trioxide 405
    14.3.3 Vanadium pentoxide 406

14.4 Vanadium pigments 407
    14.4.1 Zirconium vanadium blue 407
    14.4.2 Bismuth vanadate 408
    14.4.3 Compound pigments 408
    14.4.4 Vanadium carbide powder 408

14.5 Functional materials 408
    14.5.1 Superconducting and photosensitive materials 408
    14.5.2 Film materials 409
    14.5.3 Hydrogen storage materials 411
    14.5.4 Battery materials 412

Vanadium compounds are classified as organic or inorganic compounds of vanadium. Vanadium can form organic compounds via coordination polymerization. The formation mechanism is complicated, but a few compounds with stable fixed molecular structures have been isolated, most of which are still in the research stage. Inorganic vanadium compounds include chlorides, sulfates, oxides, vanadates, carbides, and nitrides. There are various applications for inorganic compounds, as the final product, intermediate product, or a component of a product with a specific function. Functional materials are a new material class. Vanadium coordination can be used to prepare functional materials for applications such as catalysts, pigments, membrane materials, photosensitive materials, hydrogen storage materials, battery materials, and superconducting materials, which has realized the function at different application levels [5,11,25,26,35].

14.1 Vanadium halides

Vanadium can be combined with various halogens to form bivalent, trivalent, and tetravalent halides. The only known pure halide of V^{5+} is VF_5. For the same vanadium halide, the chemical stability decreases with the increasing atomic valence of vanadium. For vanadium with the same valence state, the chemical stability of its halide decreases successively from fluorine to iodine. This indicates that vanadium easily reacts with fluorine and chlorine but has difficulty in reacting with bromine and iodine. Bivalent vanadium halide has good thermal stability, is a strong reducing agent, easily absorbs moisture, and can form $V(H_2O)_2^{6+}$ ions in water. The stability of trivalent and tetravalent vanadium halides is poor. Vanadium tetrachloride is relatively stable. VF_5 is the only confirmed pentavalent vanadium halide. VF_5 is a white solid, which melts at 19.5°C to form a pale yellow liquid, and is a strong oxidizer and fluorider. Pentavalent vanadium has more halogen oxides. The highest vanadium valence is pentavalent. During reaction with chlorine, due to the influence of chlorine's own oxidation, tetravalent vanadium is the highest valence state for vanadium chloride.

14.1.1 Vanadium oxytrichloride

Vanadium oxytrichloride ($VOCl_3$) is an inorganic substance that occurs as a yellow liquid. $VOCl_3$ can be obtained by distillation and is easily hydrolyzed in air. It is a strong oxidant and is used as a reagent in organic syntheses. Vanadium in $VOCl_3$ is pentavalent and has diamagnetism. The molecular configuration is a regular tetrahedron, with an O−V−Cl bond angle of 111 degrees, Cl−V−Cl bond angle of 108 degrees, and bond lengths of V−O and V−Cl of 157 pm and 214 pm, respectively. Vanadium oxytrichloride reacts violently in water and produces chlorine gas. It is soluble in nonpolar solvents, such as benzene, dichloromethane, and hexane. Vanadium oxytrichloride and phosphorus oxytrichloride have some similar chemical properties. However, vanadium oxytrichloride is a strong oxidizer, whereas phosphatides are not. It is completely soluble in many hydrocarbons and halides, such as tetrachlorides of carbon, titanium, and tin. It also dissolves in phosphorus trichloride, phosphorus oxychloride, arsenic trichloride, and antimony trichloride. Vanadium trichloride has good solubility in water, methane, and ether. It is a catalyst for the copolymerization of ethylene propylene rubber and ethylene cyclopentadiene.

$VOCl_3$ has a molecular weight of 173.2999 g/mol and occurs as a yellow liquid with a density of 1.826 g/cm^3 (20°C) (liquid), melting temperature of 76.5°C, and boiling point of 126.7°C. It has a refractive index of 1.63, dielectric constant of 2.898 ± 0.007 (25°C), and conductivity of $9 \times 10^{-12}/\Omega$ cm (20°C). $VOCl_3$ can be prepared by reacting V_2O_5 with chlorine at 600°C, as follows:

$$3Cl_2 + V_2O_5 \rightarrow 2VOCl_3 + 1.5O_2 \qquad (14.1)$$

When V_2O_5 is mixed with carbon and the temperature is maintained at 200°C–400°C, the carbon will act as a deoxidizer. Vanadium trioxide can also be used as a reactant:

$$3 Cl_2 + V_2O_3 \rightarrow 2VOCl_3 + 0.5O_2 \qquad (14.2)$$

A more typical laboratory method is to chlorinate V_2O_5 with $SOCl_2$:

$$V_2O_5 + 3SOCl_2 \rightarrow 2VOCl_3 + 3SO_2 \qquad (14.3)$$

Vanadium oxytrichloride is rapidly hydrolyzed into vanadium pentoxide and hydrochloric acid in water:

$$2VOCl_3 + 3H_2O \rightarrow V_2O_5 + 6HCl \qquad (14.4)$$

$VOCl_3$ can react with alcohols to form alkoxides, especially in the presence of proton acceptors (such as Et3N):

$$VOCl_3 + 3ROH \rightarrow VO(OR)_3 + 3HCl (R = Me, Phelse) \qquad (14.5)$$

They are changed to other V–O–Cl compounds. $VOCl_3$ can also be used to synthesize $VOCl_2$:

$$V_2O_5 + 3VCl_3 + VOCl_3 \rightarrow 6VOCl_2 \qquad (14.6)$$

In addition, vanadium monochlorodioxy can be prepared by a rare reaction with Cl_2O:

$$VOCl_3 + Cl_2O \rightarrow VO_2Cl + 2Cl_2 \qquad (14.7)$$

Above 180°C, VO_2Cl is decomposed into V_2O_5 and $VOCl_3$. Likewise, $VOCl_2$ can be decomposed into $VOCl_3$ and $VOCl$. $VOCl_3$ is a strong Lewis acid that tends to form an adduct with various bases such as acetonitrile and amine. In the resulting adduct, vanadium will change from the original V4 geometric tetrahedron to the V6 octahedron:

$$VOCl_3 + 2H_2(NEt) \rightarrow VOCl_3[H_2(NEt)]_2 \qquad (14.8)$$

$VOCl_3$ is used for olefin polymerization and as a catalyst or precatalyst in the preparation of EPDM.

14.1.2 Vanadium tetrachloride

Vanadium tetrachloride (VCl_4) is the chloride of vanadium (IV). It is a bright red liquid that can be used to prepare many other vanadium compounds, including dicyclopentadienyl chloride vanadium. It forms adducts with many ligands, such as the reaction with tetrahydrofuran to form $VCl_4(THF)_2$. VCl_4 has a molar mass of 192.75 g/mol, occurs as a thick reddish-brown liquid that is sensitive to moisture, releases vapor in humid air, and has a density of 1.816 g/cm^3 (liquid), melting point of $-28°C$, and boiling point of 154°C.

Vanadium tetrachloride is paramagnetic and is one of the few compounds that are liquid and paramagnetic at room temperature. VCl_4 can be prepared by chlorination of vanadium metal. The oxidation of chlorine gas is insufficient to oxidize vanadium to VCl_5. Vanadium tetrachloride is decomposed at its boiling point to form vanadium trichloride and chlorine gas:

$$2VCl_4 \rightarrow 2VCl_3 + Cl_2 \tag{14.9}$$

During organic synthesis, phenols can be coupled in the presence of VCl_4, such as 4,4′-phenol, by reaction with phenol:

$$2C_6H_5OH + 2VCl_4 \rightarrow HOC_6H_4 - C_6H_4OH + 2VCl_3 + 2HCl \tag{14.10}$$

In the rubber industry, VCl_4 can catalyze the polymerization of olefins. VCl_4 reacts with HBr to form VBr_3. In the reaction the reactant is decomposed and Br_2 is released at room temperature through the intermediate product VBr_4.

$$2VCl_4 + 8HBr \rightarrow 2VBr_3 + 8HCl + Br_2 \tag{14.11}$$

14.1.3 Vanadium pentafluoride

Vanadium pentafluoride (VF_5) is the only pentahalogenide of vanadium. In its solid state, it has a structure consisting of an infinite chain of V–F–V cis bridges and VF_6 octahedrons. Electron diffraction and spectroscopy experiments indicated that vanadium pentafluoride is a monomer in the gas phase, with a triangular–dipyramidal spatial configuration and a V–F bond length of 171 pm. Metallic vanadium is heated to 300°C and reacted with fluoride or bromine trifluoride to obtain vanadium pentafluoride. In addition, vanadium metal can be heated to 600°C in a nitrogen atmosphere to achieve tetrafluoride vanadium disproportionation:

$$2V + 5F_2 \rightarrow 2VF_5 \tag{14.12}$$

$$2VF_4 \rightarrow VF_3 + VF_5 \tag{14.13}$$

Vanadium pentafluoride is a strong fluorider that can corrode glass at room temperature:

$$4VF_5 + 5SiO_2 \rightarrow 2V_2O_5 + 5SiF_4 \tag{14.14}$$

14.2 Vanadates

Vanadate usually refers to the combination of vanadate units and cations to form a salt. Vanadate salts include MVO_3, M_3VO_4, and $M_4V_2O_7$, where M is the monovalent metal. Vanadates containing $(V_3O_9)^{3-}$ or $(V_4O_{12})^{4-}$ ions are called metavanadates, while those containing $(V_{10}O_{28})^{6-}$ ions are called decavanadates. Vanadates can be produced with combinations of Bi, Ca, Cd, Cr, Co, Cu, Fe, Pb, Mg, Mn, Mo, Ni, K, Ag, Na, Sn, and Zn. The metavanadates of alkali metals and magnesium are soluble in water, resulting in a pale yellow solution. The metavanadates of other metals are not soluble in water.

The aggregation state of vanadium in solution is related to both the acidity and concentration of the solution. When the concentration of vanadium is very low, vanadium exists as a single nucleus at all pH values. When the concentration of vanadium is high, polymerization reactions can occur, and homopolyacid ions with a high degree of polymerization are formed. The polymerization state is related to the pH value of the solution. At a certain vanadium concentrations, orthovanadate is precipitated from alkaline solutions, pyrovanadate is crystallized from weak alkaline solutions, and polyvanadate is crystallized from near-neutral solutions. When the pH of the vanadium solution is below 1, vanadium mainly exists as VO^{2+} ions, where polyvanadate ions are destroyed. Vanadate is stable only at high potential and high pH. When the pH value is less than 4, vanadium can decrease with the potential, successively forming various cations: VO^{2+}, V^{3+}, $V(OH)^{2+}$, and V^{2+}.

14.2.1 Ammonium metavanadate

Ammonium metavanadate plays an important role in the hydrometallurgical process of vanadium. Ammonium metavanadate is a white or yellowish crystalline powder, which is slightly soluble in water and ammonia but difficult to dissolve in cold water. When ammonium salts are present in an aqueous solution, the solubility of the vanadate decreases due to the common ion effect. This phenomenon is widely used in the hydrometallurgy of vanadium. Ammonium metavanadate is stable at room temperature and easily decomposes when heated. Its decomposition in the air is as follows:

$$6NH_4VO_3 = (NH_4)_2 0.3V_2O_5 + 4NH_3 + 2H_2O \quad (14.15)$$

$$2NH_4VO_3 = V_2O_5 + 2NH_3 + H_2O \quad (14.16)$$

At a lower temperature, some ammonia remains in the decomposed solid products. At high temperatures the solid product decomposes to form V_2O_5.

14.2.1.1 Appearance

Ammonium metavanadate (NH_4VO_3) has a molecular weight of 116.98 and an infinite chain structure consisting of a conformal VO_4 tetrahedron. It occurs as a crystalline white to yellowish powder that is slightly soluble in cold water, hot ethanol, and ether; soluble in hot water and dilute ammonia water; and insoluble in ethanol, ether, and ammonium chloride. Ammonium metavanadate is converted to toxic vanadium pentoxide when burned in air. Ammonium metavanadate has a melting point of 210°C (decomposition) and relative density (water = 1) of 2.326.

14.2.1.2 Preparation method

Sodium vanadate solution is adjusted with hydrochloric acid to pH 7.5–8, heated to 70°C–80°C, and then reacted with ammonium chloride solution. Precipitates of ammonium metavanadate are obtained after centrifugation, washing, and drying.

14.2.1.3 Applications

Ammonium metavanadate is mainly used as a chemical reagent, catalyst, desiccant, and mordant, while it is also used as a pigment for glass and ceramics. It is widely used as a glaze in the ceramic industry. It can also be used to manufacture vanadium pentoxide and can be converted to other vanadium products.

14.2.2 Ammonium polyvanadate

14.2.2.1 Chemical properties

Ammonium polyvanadate ($(NH_4)_2V_6O_{16}$) has a molecular weight of 597.72 and is slightly soluble in cold water, hot ethanol, and ether and soluble in hot water and dilute ammonium hydroxide. When burned in the air, it produces toxic vanadium pentoxide.

14.2.2.2 Physical properties

The physical properties of ammonium polyvanadate are shown in Table 14.1.

14.2.2.3 Applications

Ammonium polyvanadate is used to produce vanadium pentoxide and vanadium trioxide, which can be used as pigments and catalysts.

Table 14.1: Physical properties of ammonium polyvanadate.

Appearance	Pale yellow crystalline powder	Melting point	350°C (decomposition)
Density	3.03 g/cm^3	particle size	≤3 mm
Bulk density	0.9 g/cm^3		

14.2.3 Sodium vanadate

Sodium vanadate is the most important vanadate in vanadium metallurgy. Sodium metavanadate ($NaVO_3$), sodium pyrovanadate ($Na_4V_2O_7$), and sodium orthovanadate (Na_3VO_4) are relatively common. They are soluble in water and form hydrates. For example, sodium metavanadate can crystallize out of solution to form anhydrous crystals above 350°C, while below 35°C, $NaVO_3 \cdot 2H_2O$ is precipitated. The solubility of $NaVO_3$ increases with temperature.

14.2.3.1 Appearance

Na_3VO_4 has a molecular weight of 183.91, melting point of 850°C–866°C and can contain 12 crystal water molecules. It is colorless, with a hexagonal prism structure. Na_3VO_4 is soluble in water, insoluble in ethanol and is toxic. In a boiling water solution, sodium metavanadate can be formed. A neutral aqueous solution of Na_3VO_4 is clear and transparent, while acidic solutions are transparent yellow and strongly oxidizing.

14.2.3.2 Application

Na_3VO_4 can be used as a catalyst, paint drier, mordant, corrosion inhibitor, and in the production of other vanadium salts.

14.2.3.3 Preparation method

Na_3VO_4 can be prepared by melting vanadium pentoxide with appropriate amounts of alkali metal carbonate or roasting with ferrovanadium powder and charcoal powder, followed by reaction with sodium hydroxide.

14.2.3.4 Occurrence in nature

Sodium vanadate is a colorless monoclinic prism that is soluble in water. In nature, it is found in rare metamunirite and munirite.

14.2.4 Sodium metavanadate

Sodium metavanadate ($NaVO_3$) is a colorless crystal with a monoclinic prism structure. It is soluble in water and found in nature in rare metamunirite and munirite minerals. Sodium metavanadate is prepared by dissolving vanadium pentoxide in sodium hydroxide solution, followed by concentration and crystallization.

$$V_2O_5 + 2NaOH \rightarrow 2NaVO_3 + H_2O \tag{14.17}$$

Sodium metavanadate is used as a chemical fertilizer and mordant, and in photography and gas desulfurization.

14.2.5 Potassium vanadate

Potassium vanadate (KVO_3) has a molecular weight of 138.04. At purities of $\geq 98\%$, it has a density of 2.84 g/mL (25°C) (lit) and melting point of 520°C and occurs as a white to slightly yellow crystal. It can be used as chemical reagent, catalyst, drier, and mordant.

14.2.6 Bismuth vanadate

Bismuth vanadate ($BiVO_4$) is a bright yellow inorganic chemical with a molecular weight of 323.92, CAS number of 14059-33-7, and purity specification of 99.9% and appears as a yellow powder. It is also known as 184 yellow dye. It was one of the first in the new generation of environment-friendly materials when it was developed BASF, Germany. Bismuth vanadate does not contain harmful heavy metal elements and is considered an environment-friendly and low-carbon metal oxidant. Bismuth vanadate has many kinds of lattice structures, resulting in different properties and applications.

14.2.6.1 Applications

Bismuth vanadate is also known as bismuth yellow, 184 yellow, molybdenum bismuth yellow, titanium bismuth yellow, bismuth yellow 184, and bismuth vanadate yellow. The chemical compositions are Bi–V–Mo–O and Bi–V–O. Bismuth vanadate has a temperature tolerance of <600°C, and oil absorption of 15–30, light resistance grade 8, and heat resistance grade 4–5. It has good acid and alkali resistance, good dispersibility, and fineness up to 600 mesh. It is insoluble in organic materials, such as ethyl acetate, xylene, and ethanol.

14.2.6.2 Structure

The various known lattice structures of bismuth vanadate are as follows, where their applications differ according to the structure: (1) the tetrahedral crystal is a zirconium silicate crystal with a very light yellow color; (2) the orthogonal crystal is a vanadate-type crystal, while bismuth vanadate is brown; (3) the monoclinic crystal can be used as a yellow pigment.

14.2.6.3 Preparation method

1. Precipitation from an aqueous solution: high-purity Bi^{3+} and V^{5+} salt solutions are used to produce Bi–V oxide–hydroxide colloids under certain conditions. The temperature is increased to crystallize and precipitate crude pigment. The performance can be improved by coating with phosphates or oxides.
2. Calcination: a certain proportion of oxide and a small amount of accelerant are mixed. This can be achieved using a dry gel instead of a matrix, where the pigment crystal is

produced when calcined above 600°C. The material is washed in an alkaline solution to remove the soluble ingredients.

14.2.6.4 Applications

Applications of bismuth vanadate include (1) various yellow traffic signs; (2) additive for paint and ink; (3) coloring for automobile finishes, rubber and plastic products, and other high-performance products; (4) food, toys, and other fields as it is highly environment-friendly and nontoxic; and (5) photocatalysis and other fields.

14.2.6.5 Functionality

Bismuth vanadate is one of the most outstanding functional materials in the coating industry. The latest application of bismuth vanadate is as an alternative to chrome and cadmium pigments. However, it has a wide range of applications. Bismuth vanadate can be dispersed easily in water-based coatings, powder, or oil-based coatings. Bismuth vanadate can withstand a temperature of 200°C, which is considered a high-temperature pigment in the coating industry.

Bismuth molybdate or bismuth molybdate yellow is also derived from bismuth vanadate. Bismuth vanadate is a yellow phase, while bismuth molybdate is greenish. Bismuth vanadate can be used to replace cadmium yellow or chrome yellow for traffic signs, indoor decoration, and other applications. Bismuth vanadate is environment-friendly and nontoxic, while titanium and bismuth are known to be nontoxic. Bismuth vanadate is also a photocatalyst material. Photocatalysts form conditions where light can react, without changing the material itself. The chemical reaction produced by photocatalysis is similar to photosynthesis in plants. Under solar irradiation the photocatalyst absorbs some of the ultraviolet rays while simultaneously decomposing harmful inorganic substances and organisms in the air. Photocatalysts can play a role in air purification and also prevent the harmful effects of ultraviolet radiation on the human body.

14.2.7 Yttrium vanadate

Yttrium vanadate (YVO_4) has a relative molecular mass of 203.8. YVO_4 laser crystals (Nd:YVO_4 laser crystals) belong to the tetrahedral crystal system, with a space group of $D2d$-$2m$. Monocrystalline growth is usually carried out by melt pulling. Along a shaft cut the Nd^{3+}:YVO_4 crystal has an absorption coefficient 3.5 times that of Nd^{3+}:YAG crystal at 809 nm, while the laser emission cross section is 2.7 times that of Nd^{3+}:YAG crystal. It is an important material for diode (LD) and small pumped lasers and is usually formed by pulling to grow crystals with good mechanical and physical properties. Because of its wide transmission band and large birefringence, it is an ideal material for optical polarizers. In many practical applications, optical fiber isolators, annular mirrors, optical displacers, separators, Glan polarizers, and other polarizers can be made from YVO_4 instead of calcite ($CaCO_3$) and TiO_2.

14.2.7.1 Main properties

The main properties are transparent band: 0.4–5 μm; symmetry: $D4h$; crystal cell parameters: $a = b = 0.712$ nm, $c = 0.629$ nm; density: 4.22 g/cm³; hydrolysis: nonhydrolysis; coefficient of thermal expansion: $\alpha a = 4.43 \times 10^{-6}$/K; $\alpha c = 11.37 \times 10^{-6}$/K; thermal conductivity: //C: 5.23 W/m K and ⊥C 5.10 W/m K; crystal structure: tetrahedral system; crystallinity: normal optical uniaxial crystal; and refractive index: double refractive index. There are 24 atoms in the lattice unit cell.

14.2.7.2 Applications

Yttrium vanadate is widely used in the optical fiber communication field and is a key material in passive optical communication devices, such as optical isolators, ring devices, optical rotators, retarders, and polarizers.

14.3 Vanadium oxides

Vanadium can combine with oxygen to form many oxides, although the main recognized ones are V_2O_5, V_2O_4, V_2O_3, and VO. VO_2 is an amphoteric oxide that forms tetravalent vanadate with a base. Vanadium pentoxide is an amphoteric oxide with strong acid behavior, which has a greater tendency to form vanadate with a base.

14.3.1 Vanadium dioxide

Vanadium dioxide (VO_2) is a metal oxide with a characteristic phase change at a phase transition temperature of 68°C. Structural changes before and after the phase transition temperature of infrared light that generates the from the transmission to the reflection of the reversible. This behavior has been used for intelligent temperature control and applied to the preparation of thin films.

14.3.1.1 Appearance

VO_2 is a dark blue crystalline powder with a monoclinic crystal structure. It has a density of 4.260 g/cm³ and melting point of 1545°C. VO_2 is insoluble in water and soluble in acids and bases. When dissolved in an acid, it cannot produce tetravalent ions but produces positive bivalent vanadium oxide ions. When VO_2 is heated in a dry hydrogen flow, vanadium trioxide can be formed, while vanadium pentoxide can be produced by air oxidation or nitric acid dissolution. Vanadium pentoxide can be dissolved in alkali to form vanadite. It can be made by reducing vanadium pentoxide using carbon, carbon monoxide, or oxalic acid. It is used as a pigment for glass and ceramics.

14.3.1.2 Applications

Vanadium dioxide is well known in the field of materials for its rapid and sudden phase transition at 68°C. It is widely used in optical, electronic, and photoelectric devices due to its conductive properties. When it is used as an insulator at relatively low temperature, it shows the phenomenon of multiphase competition.

Research assistant, Alexander Terseleuf, from the University of Tennessee cooperated with French scientists at the Center for Nanophase Materials Sciences, Oak Ridge National Laboratory, and used condensed matter physics theory to successfully explain the phase behavior of VO_2. They showed that multiphase competition in VO_2 was purely due to lattice symmetry; during cooling the VO_2 lattice can "fold" in different ways, and various folded forms were observed. Vanadium dioxide is a heat-sensitive functional material, the outstanding characteristics of which are mainly reflected in its metal–insulator phase change characteristics near room temperature that is expected to facilitate its application in many fields.

14.3.2 Vanadium trioxide

Vanadium trioxide (V_2O_3) is a gray-black crystalline powder with a metallic luster. It has an orthorhombic structure, a melting point of 2070°C, and density of 4.843 g/cm^3. Vanadium trioxide is an alkaline refractory oxide. It can conduct electricity with a resistance at 1100°C of 551×10^4 Ω. It slowly absorbs oxygen in the air and transforms into vanadium tetroxide; when heated in air, it burns fiercely. V_2O_3 is insoluble in water and alkali, but soluble in acid, and is a strong reducing agent. It is produced by reduction of vanadium pentoxide by hydrogen, carbon, or carbon monoxide, or via thermal decomposition of vanadium pentoxide at 1750°C in air, or by calcining ammonium vanadate. V_2O_3 can be used as a dye for glass and ceramics.

Using high-purity V_2O_5 powder as a raw material, vanadium dioxide and vanadium trioxide can be prepared via reduction reactions. V_2O_3 is generated by direct hydrogen reduction of vanadium pentoxide. Hydrogen is used to directly reduce vanadium pentoxide, or carbon (carbon black or graphite) is added with a certain molar ratio to reduce vanadium pentoxide under the protection of an inert gas. At the reaction temperature, hydrogen reduction of vanadium pentoxide can generate V_6O_{13}, VO_2, V_3O_5, V_2O_3, and various low-valence oxides, such as V_2O_3. In addition to ensuring the appropriate stoichiometric ratio, the ideal conditions for producing V_2O_3 include a dry pure hydrogen atmosphere, where the vanadium pentoxide is treated at 600°C for 3 h, and then heated to 900°C–1000°C, kept at this temperature for 5 h, and finally cooled in the furnace. The effects of carbon black and graphite reduction are similar, but vanadium pentoxide is, respectively, reduced. The products of the reduction include vanadium dioxide and vanadium pentoxide. For preparation of vanadium dioxide, the ideal parameters include nitrogen/argon gas protection, where the mixture of vanadium pentoxide and carbon black powder (C:V_2O_5 = 1:2) is heated to 600°C and treated for 3 h, then heated to 800°C–850°C, treated for 5 h, followed by cooling in the furnace. In addition, vanadium pentoxide can be reduced using copper as the reductant. Copper reduction can produce impurities that cannot be separated, so it is not an ideal process for preparing low-valence vanadium oxide.

14.3.3 Vanadium pentoxide

Vanadium pentoxide (V_2O_5) is also known as alum anhydride, vanadic acid anhydride, vanadium (acid) anhydride, alum pentoxide, vanadium oxide, anhydrous vanadic acid, and vanadium oxide (V). It can occur as an orange-yellow or reddish-brown crystalline powder, or gray-black flakes. It has a melting point of 690°C, relative density (water = 1) of 3.35, boiling point of 1750°C (decomposition), and molecular weight of 182.00. V_2O_5 is slightly soluble in water, insoluble in ethanol, and soluble in concentrated acids and bases. It is an amphoteric oxide but shows mainly acidic behavior. Above 700°C, it is highly volatile. In the range of 700°C–1125°C, vanadium pentoxide is decomposed into oxygen and vanadium tetroxide, which makes it suitable for use as an organic and inorganic reaction catalyst. V_2O_5 is a strong oxidant and is easily reduced to a variety of low-valence oxides. It is slightly soluble in water and easily forms a stable colloidal solution. It is highly soluble in alkaline solutions, where vanadate (VO_3^-) ions can be produced in weakly alkaline solutions. It is soluble in strong acids (generally soluble at pH = 2), where vanadate ions are not produced, but oxyvanadium ions in the same valence state (VO^{2+}) are generated. The maximum allowable content in air is <0.5 mg/m³. Vanadium pentoxide is amphoteric, soluble in alkali to form polyvanadate, and soluble in nonreducing acid to form a pale yellow solution containing VO^{2+}:

$$V_2O_5 + 2HNO_3 \rightarrow 2VO_2(NO_3) + H_2O \quad (14.18)$$

When it reacts with an alkali, such as sodium hydroxide, the product is colorless sodium vanadate (Na_3VO_4) when excess alkali is present. When the acidity increases, the color gradually changes from colorless to orange to red. When the pH is 9–13, the ions are mainly HVO_4^{2-} and $V_2O_7^{4-}$, while $V_4O_{12}^{4-}$ and $HV_{10}O_{28}^{5-}$ are formed when the pH is below 9. Brown vanadium pentoxide hydrate is precipitated at pH 2.

Vanadium pentoxide reacts with thionyl chloride to obtain vanadium oxychloride:

$$V_2O_5(s) + 3SOCl_2(l) \rightarrow 2VOCl_3(l) + 3SO_2(g) \quad (14.19)$$

The V(V) in V_2O_5 can be reduced to V (IV), where blue $VO(H_2O)_5^{2+}$ vanadium oxy ions are obtained in acidic medium. Using hydrochloric acid and hydrobromic acid as reducing agents, the oxidation products are the corresponding halogen monomers:

$$V_2O_5(s) + 6HCl + 7H_2O \rightarrow 2[VO(H_2O)_5]_2 + + 4Cl^- + Cl_2 \quad (14.20)$$

Solid V_2O_5 can be reduced by oxalic acid, carbon monoxide, or carbon dioxide to produce dark blue solid vanadium dioxide in the presence of excess reducing agent. Intermediate products include mixtures of V_4O_7 and V_5O_9 complex oxides, where the final vanadium trioxide product is black. In solution, vanadate or VO_2^+ ions are reduced by zinc amalgam, and a series of color changes occur. Finally, VII ions are obtained:

$$\text{Colorless } VO^{3-} \rightarrow \text{yellow } VO_2^+ \rightarrow \text{blue } VO_2^+ \rightarrow \text{green } V^{3+} \rightarrow \text{purple } V^{2+}$$

Vanadium pentoxide reduction produces different final products when different reductants are used. The results of temperature resistance tests showed that the produced vanadium trioxide and vanadium dioxide both had obvious temperature-dependent phase changes, accompanied by a change in resistance.

14.4 Vanadium pigments

Bismuth yellow vanadate pigment and bismuth yellow vanadomolybdate pigment were developed for paint applications. The pigment has a pure color, high hiding power, weather resistance, and heat resistance. It is used to prepare nontoxic yellow car paint or for outdoor architectural coatings. In the ceramic industry, V_2O_5 is widely used as a glaze and has good application prospects.

14.4.1 Zirconium vanadium blue

Vanadium zirconia is a type of pigment used in Zr-based ceramic glaze, which is widely used in sanitary ceramics, household ceramics, and the enamel industry. Vanadium zirconia has good color stability, high temperature stability, a bright color, chemical corrosion resistance, and a wide range of applications. In addition, it can be mixed with many other pigments to produce blue, green, and yellow coloring agents.

Sky blue high-temperature pigment for ceramic materials has a zirconium quartzite structure. Common coloring materials include V_2O_5 and ammonium bivanadate. The common parent materials are zirconia and quartz powder, where natural zirconium quartz can also be used as the raw material. Vanadium zirconium blue is a synthetic colored mineral with vanadium as the chromogenic element and $ZrSiO_4$ zirconium quartzite as the carrier, which has a sky blue color. It is mainly used for building sanitary ceramic glaze. It can also be used for underglaze decoration or as a coloring material for blank. It has good adaptability as a glaze but is not suitable for use in reducing atmospheres. After calcining, natural zirconia is crushed and sifted to a particle size of 250–400 mesh. It is then mixed with silica, vanadium pentoxide, or ammonium bivanadate according to a specific formula. The calcined semifinished product is processed by wet grinding, washing, dehydration, drying, sieving, and color matching.

14.4.2 Bismuth vanadate

Bismuth vanadate is used as a temperature indicator pigment. Bi_2O_3, V_2O_5, $Mg(OH)_2$, and CaO can be mixed to produce $BiVO_4$—0.1MgO and $BiVO_4$—0.1CaO following similar processes as those described earlier. These materials perform better than $BiVO_4$ alone and have good reversibility. At 25°C, the pigment is yellow or light yellow, at 140°C, it is orange-red, while at 350°C, it is orange-red to pink; however, there is no clear transition temperature. $BiVO_4$—0.1MgO and $BiVO_4$—0.1CaO start to melt at 844°C and 853°C, respectively.

Traditional yellow pigments, such as lead chromate and cadmium yellow, contain chromium, lead, cadmium, and other toxic heavy metals. Hence, increasingly strict environmental protection regulations are being developed, which have banned their use in many countries. Yellow bismuth vanadate pigments are environment-friendly and can be used as a substitute for cadmium yellow and chromium yellow. Bismuth vanadate pigments have good performance and are used for plastic, rubber, ceramics, paint, and printing inks. In particular, they are used for yellow paint for cars and high-performance architectural coatings.

14.4.3 Compound pigments

Vanadium pentoxide (0.02%) is added to glass to prevent transmission of high-energy ultraviolet rays that can damage eyes and fade fabric colors. Vanadium is also used in tinted yellow-green glass, where green glass is made by adding a mixture of vanadium oxide and cesium oxide to reduce the intensity of UV radiation. Oxides and vanadates are used in the production of printing ink, which react to produce a black resin coating. Addition of ammonium vanadate can produce fast drying ink. A small amount of vanadium pentoxide is also used in the textile printing industry. Vanadium helps oxidize aniline and produce a strong black dye that does not fade.

14.4.4 Vanadium carbide powder

The purity requirement for VC is >99.0%. The coating thickness should have a strength of 2090 HV, melting point of 2810°C, with high hardness and wear resistance. VC is an important component of hard tool steels that have carbide or other metal−ceramic hard coatings.

14.5 Functional materials

14.5.1 Superconducting and photosensitive materials

V−Ga alloy (V_3Ga) can be used to make superconducting electromagnets that generate strong magnetic fields with a strength up to 175,000 Gs. In addition, $YVO_4:Eu^{3+}$ is used as an optical material in fluorescent lamps. Yttrium europium vanadate phosphate ($Y(VO_4)PO_4$:Eu^{3+}) and yttrium europium vanadate (YVO_4:Eu^{3+}) are efficient red-light-emitting materials that are used as red phosphors in color televisions and are still widely used in fluorescent lamps and high-pressure mercury vapor discharge lamps to adjust the chromaticity of the light-emitting materials.

The addition of Eu^{3+} ions to the vanadate forms $LnVO_4$:Eu^{3+} (Ln = La, Y) that can emit different wavelengths of light by adjusting the composition and concentration of the cation. These materials are used in fluorescent lamps and various special light sources, including cathode ray tubes and TV screens, ionizing radiation detection crystals, X-ray fluorescent screens and intensifying screens, as well as a variety of electroluminescent flat, numbers, symbols, and image displays. In addition, these vanadates are used to produce luminescent rubber, plastic, resin, paint, and ink products, which can store energy and have a long afterglow.

14.5.2 Film materials

As VO_2 has a reversible first-order phase change around 68°C, where the structural differences before and after the phase change lead to its reversible transformation from transmission to reflection of infrared light, respectively; this can be applied in the field of intelligent temperature control films. With high reproducibility of the method of vacuum on the surface of the coating VO_X, the substrate is a sol–gel solution prepared from alkoxy vanadium that is deposited as a thin film using spin coating or impregnation. The film can be uniformly coated over a large area and on curved surfaces, up to a thickness of 1 μm. To reduce transmitted infrared light in space shuttle applications, transparent and opaque vanadium oxide films are used, which prevent infrared light and electromagnetic radiation from being transmitted during heating. The optical transmittance of the film changes with changes in the ambient temperature. Functional materials containing VO_2 are often produced using sol–gel coating technology, where the optical, electrical, and other special properties after the phase transition are applied to control devices such as solar infrared radiation thermometers, thermal resistors, heat switches, variable mirrors, infrared pulse laser protective films, optical disc materials, holographic storage materials, point to change color display materials, filters, color change materials, nonlinear or linear resistance materials, temperature sensors, tunable microwave switches, infrared optical modulation materials, high-sensitivity strain sensors, transparent conductive materials, and antistatic coatings.

The sol–gel method requires the use of precursor components with high chemical activity. These raw materials are homogeneously mixed in the liquid phase and then hydrolyzed and condensed to form a stable transparent sol system in solution. The sol is slowly polymerized between aged colloidal particles to form a gel with a three-dimensional network structure. Materials with molecular and even nanometer substructure can be achieved after drying and sintering. The chemical process occurring during the sol−gel method begins with the raw material being dispersed in the solvent, followed by the active monomer being formed via a hydrolysis reaction. Then, the active monomer is polymerized to form a sol, and a gel with a certain spatial structure is generated. Nanoparticles and the required materials are prepared via drying and heat treatment. The basic processes are as follows:

1. Hydrolysis reaction:

$$M(OR)n + xH_2O \rightarrow M(OH)x(OR)n - x + xROH \tag{14.21}$$

2. Polymerization reaction:

$$-M-OH + HO-M- \rightarrow M-O-M- + H_2O$$

$$-M-OR + HO-M- \rightarrow -M-O-M- + ROH \tag{14.22}$$

The sol−gel method is classified into three types depending on the reaction mechanism:

1. Traditional colloidal method. By controlling the precipitation of metal ions in solution, the formed particles do not agglomerate into large particles, and a stable and uniform sol is obtained by precipitation, while the gel is obtained by evaporation.
2. Inorganic polymer method. Metal ions are evenly dispersed in the gel through a sol process where the polymer is soluble in water or an organic phase. Common polymers are polyvinyl alcohol and stearic acid.
3. Complex method. Metal ions form complexes in the presence of complexing agents that participate in the sol−gel process to form complex gels.

The sol−gel method has many unique advantages compared with other methods: (1) since the raw materials used in the sol−gel method are first dispersed in the solvent to form a low-viscosity solution, molecular-level uniformity is obtained in a very short time. When the gel is formed, the reactants are generally homogeneously mixed at the molecular level; (2) after the solution reaction step, it is easy to achieve uniform doping at the molecular level by adding trace elements; (3) the chemical reaction is more easily performed than solid-phase reactions. In addition, it requires a lower synthesis temperature, as the diffusion of components in the sol−gel system is thought to be in the range of nanometers, while diffusion in the solid-phase reaction is in the range of microns; (4) various new materials can be prepared under appropriate conditions.

The metal compounds are solidified in the solution as sol and gel structures, while the final nanoparticles are generated by low-temperature heat treatment. It is featured with reaction species, uniform particle products, and process control and is suitable for preparation of oxides and II−VI compounds. The sol−gel method is important for the synthesis of inorganic materials at low temperature or mild conditions. It has been widely used in the preparation of glass, ceramic, film, fiber, and composite materials. Sol−gel and dip-coating of V–W materials were used to coat glass substrates with vanadium pentoxide films. The phase composition of the thin films was analyzed using XRD and XPS.

When the temperature changes, there is a phase change from vanadium dioxide to vanadium trioxide, where atomic rearrangement occurs. The phase transition is accompanied by considerable changes in the magnetic, electrical, and optical properties of the material. Hence, vanadium oxide and vanadium trioxide can be used to manufacture various materials for electronic and optical devices, such as current-limiting elements, thermal-sensitive devices, and intelligent window coatings.

14.5.3 Hydrogen storage materials

The main functions of hydrogen storage alloys include: (1) energy conversion; (2) hydrogen separation, purification, and recovery; (3) hydrogen isotope separation; (4) catalysis; (5) alloy sensors; (6) alloy batteries; and (7) hydrogen storage and transport. The main hydrogen-absorbing alloys and their hydride properties are shown in Table 14.2.

Vanadium is the only metal that can absorb and release hydrogen at room temperature. Many countries have developed various V-based solid-solution BBC hydrogen storage alloys. The use of expensive vanadium metal as the alloying element makes the production cost too high, which has limited its application.

V-based hydrogen storage alloys with a solid-solution microstructure have been developed, which showed hydrogen absorption greater than 3.3%, hydrogen discharge greater than 2.3%, and a hydrogen discharge temperature below 100°C. Vanadium has a body-centered

Table 14.2: Main hydrogen-absorbing alloys and their hydride properties.

Type	Alloy	Hydride	Hydrogen absorption capacity (mass%)	Hydrogen release pressure (temperature) (MPa)	Heat of hydride formation (kJ/mol H_2)
AB5	$LaNi_5$	$LaNi_5H_{6.0}$	1.4	0.4 (50)	− 30.1
AB2	$LaNi_{4.5}Al_{0.4}$	$LaNi_{4.5}Al_{0.4}H_{5.5}$	1.3	0.2 (80)	− 38.1
AB	$MmNi_5$	$MmNi_5H_{6.3}$	1.4	3.4 (50)	− 26.4
A2B	$MmNi_{4.5}Mn_{0.5}$	$MmNi_{4.5}Mn_{0.5}H_{6.6}$	1.5	0.4 (50)	− 17.6
	$MnNi_{4.5}Al_{0.5}$	$MmNi_{4.5}Al_{0.5}H_{4.9}$	1.2	0.5 (50)	− 29.7
	$CaNi_5$	$CaNi_5H_4$	1.2	0.04 (30)	− 33.5
	$Ti_{1.2}Mn_{1.8}$	$Ti_{1.2}Ni_{1.8}H_{2.47}$	1.8	0.7 (20)	− 28.5
	$TiCr_{1.8}$	$TiCr_{1.8}H_{3.6}$	2.4	0.2–5 (−78)	—
	$ZrMn_2$	$ZrMn_2H_{3.46}$	1.7	0.1 (210)	− 38.9
	ZrV_2	$ZrV_2H_{4.8}$	2.0	10^{-9} (50)	− 200.8
	$FeTi$	$TiFeH_{1.95}$	1.8	1.0 (50)	− 23.0
	$TiFe_{0.8}Mn_{0.2}$	$TiFe_{0.8}Mn_{0.2}H_{1.95}$	1.9	0.9 (80)	− 31.8
	Mg_2Ni	$Mg_2NiH_{4.0}$	3.6	0.1 (253)	64.4

cubic crystal lattice structure, while VH has a body-centered tetragonal crystal lattice, and VH_2 has a face-centered cubic lattice. After hydrogen absorption, lattice changes result in a (alpha + beta) two-phase area. Low hydrogen contents result in the formation of the body-centered tetragonal lattice. At 313K, there is a balance between VH and VH_2, and the equilibrium dissociation pressure is 303.9 kPa. At room temperature, hydrogen atoms in vanadium jump between two adjacent lattices at a frequency of 2×10^{12}/s in the temperature range of 435K−620K. Hydrogen has a high diffusion coefficient in vanadium, especially at 435K−620K.

V−Ti, V−Ti−Cr, and other V-based solid-solution alloys (e.g., V−Ti and V−Ti−Cr) react with hydrogen to form VH and VH_2. These materials are used for hydrogen purification, where hydrogen and impurities can be extracted and separated from isotopes as a result of the isotope effect and characteristics of hydrogen absorption. During hydrogen absorption and release, large amounts of heat are released and absorbed, respectively. This property has been used to design new heat exchangers.

The amount of hydrogen in the V-based solid solution of the hydrogen storage alloy depends on the V content of the alloy, as it is the vanadium metal that reacts with hydrogen to form VH_x ($x = 0-2$). The reaction between vanadium and hydrogen is generally a four-step process: hydrogen gas molecules are converted to hydrogen atoms that then diffuse in the alloy, produce hydride, and then hydrogen is diffused to the hydride layer. The alloy can absorb and release hydrogen reversibly at appropriate temperature and pressure, storing 1000 times its own volume and theoretically absorbing 3.8% hydrogen. The fastest hydrogen diffusion occurs in the hydride. In the developed hydrogen storage alloys, the V−H alloy has a large coexistence area. At temperatures above 420K, hydrogen atoms are distributed irregularly in the metal lattice spacings, and the dissolved hydrogen content is low. At temperatures below 420K, hydrogen atoms are distributed regularly and densely in the metal lattice spacings, and the dissolved hydrogen content is high. When the H:V atomic ratio in the alloy is <0.4, V−H phases coexist as (alpha + beta). Due to the irregular alpha state, the hydrogen storage capacity is small. However, for H:V > 0.4 and reaches 1.5−2.0, the alpha phase disappears, and the hydrogen storage quantity increases to the theoretical value.

Hydrogen storage alloys already in use include the rare-earth AB5 type with hydrogen storage <1.3% and the Zr−Ti AB2 type with hydrogen storage <2.0%. Other types of hydrogen storage alloys are difficult to use because of their high cost or harsh conditions required for hydrogen uptake and release.

14.5.4 Battery materials

As secondary battery cathode materials, V-based solid-solution alloys generally do not show good electrode performance. However, adding suitable amounts of catalytic compounds and optimizing the heat-treatment process to produce a three-dimensional network structure in the alloy can increase the electrode activity, charge and discharge ability, and cycle stability of the alloy. Adding elements such as Al, Si, Mn, Fe, and Co can improve the stability to a certain extent and reduce the initial discharge. Hydrogen storage alloys can replace Ni–Cd cathodes in Cd batteries. For example, Ni/MH Ovonic battery companies (United States) developed Zr–V–Ti–Cr–Ni-based hydrogen storage alloys that can be used as 30–250 Ah large-capacity battery cathodes. The energy density reached 80 Wh/kg, the cycle life exceeded 1000 times in the case of 100% deep discharge.

A mesoporous microsphere structure was formed using electrostatic matching of metal ions on cell walls, inducing gene and enzyme catalysis. Using yeast cells as a template for self-assembly, deposition and mineralization of lithium vanadium phosphate nanoparticles were controlled at the molecular level. Using the function of gene recognition regulation and sequencing of yeast cells, lithium vanadium phosphate nanoparticles were assembled into mesoporous microspheres. High-performance mesoporous microspheres containing lithium vanadium phosphate/carbon powder were synthesized by in situ addition of a conductive agent via yeast cell carbonization and crystallization heat treatment to achieve precise regulation of the structure and electrochemical properties. This composite is a fast ion conductor with high energy density, good safety, and long cycle life. It is environment-friendly and could be a preferred anode material for Li-ion batteries and promote the research, production, and application of various electronic products and electric vehicles while alleviating the dependence of petroleum-driven automobiles and environmental pollution. Bionic synthesis of lithium vanadium phosphate/carbon composite materials catalyzed by microorganisms can achieve precise control and assembly of nanostructures at the molecular level. This process is inexpensive, achieves a good structure and compositional repeatability, has high yield, mild reaction conditions, and is a simple process. In addition, it produces no pollution, has low energy consumption, and could easily be upscaled to an industrial scale to synthesize large quantities of material.

CHAPTER 15

Vanadium catalysts

Chapter outline
15.1 Catalyst types 416
    15.1.1 Homogeneous catalysts 416
    15.1.2 Heterogeneous catalysts 416
    15.1.3 Biocatalysts 417
15.2 Manufacturing methods 417
    15.2.1 Mechanical mixing 417
    15.2.2 Precipitation 417
    15.2.3 Impregnation 418
    15.2.4 Spray evaporation 418
    15.2.5 Thermal melting 418
    15.2.6 Leaching 419
    15.2.7 Ion exchange 419
    15.2.8 Other methods 419
15.3 Sulfuric acid production and flue gas desulfurization catalysts 420
    15.3.1 Synthesis methods 420
    15.3.2 Manufacturing of typical vanadium catalysts 422
    15.3.3 Sulfuric acid production and flue gas desulfurization 424
    15.3.4 Requirements for sulfuric acid production catalysts 426
    15.3.5 Factors affecting the vanadium catalyst 427
15.4 Flue gas denitrification catalysts 429
    15.4.1 Denitrification reactions 430
    15.4.2 Types and compositions of denitrification catalysts 431
    15.4.3 Typical denitrification catalysts 433
    15.4.4 Catalyst maintenance 435
    15.4.5 Catalyst regeneration 439
15.5 Vanadium catalysts for organic synthesis 439
    15.5.1 Important catalysts 439
    15.5.2 Main vanadium components 440
    15.5.3 Organic synthesis applications 440

A catalyst is a substance that changes the rate of a reaction without changing the total standard Gibbs free energy. Catalysts are single compounds, complex compounds, or mixtures. Catalysts have very strong selectivity, and different catalysts are used for different reactions. Vanadium pentoxide (V_2O_5) is an important catalyst in the chemical industry. Vanadium catalysts are a series of catalysts containing vanadium compounds as active components. The active components of vanadium catalysts commonly used in industry include V-containing oxides, chlorides, and complexes and heteropoly salts. The most common active component is V_2O_5 with one or more additives. Catalysts with V_2O_5 as the main component are effective for almost all oxidation reactions. Vanadium catalysts play an important role in the modern chemical industry and are used as special catalysts for sulfuric acid production, rubber synthesis, petroleum cracking, and synthesis of some high-molecular compounds [5,14–16,35,72].

15.1 Catalyst types

There are many types of catalysts that can be classified into liquid or solid catalysts. The reaction systems are classified as homogeneous or polyphase catalysts depending on their physical state. Homogeneous catalysts include acid, alkali, soluble transition metal compounds, and peroxide catalysts. Heterogeneous catalysts include solid acid catalysts, organic alkali catalysts, metal catalysts, metal oxide catalysts, complex catalysts, rare earth catalysts, molecular sieve catalysts, biological catalysts, and nanoscale catalysts. According to the reaction type, the catalysts are divided into polymerization, condensation, esterification, acetal, hydrogenation, dehydrogenation, oxidation, reduction, alkylation, isomerization, and other catalysts. According to their effect, they can be classified as either the main catalyst or promoter.

15.1.1 Homogeneous catalysts

If the catalyst and reactant are in the same phase without a phase boundary, the reaction is called homogeneous catalysis, where the catalyst is referred to as a homogeneous catalyst. Homogeneous catalysts include liquid acids, alkali catalysts, solid acids, and alkaline catalysts, and soluble transition metal compounds (salts and complexes). Homogeneous catalysts operate independently via molecules or ions with uniform active centers that have high activity and selectivity.

15.1.2 Heterogeneous catalysts

Heterogeneous catalysts are in a different phase than that of the reactants. For example, in the production of margarine, unsaturated vegetable oils and hydrogen can be converted into saturated fats in the presence of solid nickel catalyst. Solid nickel is a heterogeneous catalyst that catalyzes the liquid (vegetable oil) and gaseous (hydrogen) reactants. A simple heterogeneous catalytic reaction consists of a reactant (substrate), where plastid is adsorbed on the surface of the catalyst. The bonds within the reactant are very fragile, which leads to new bonds. The bond between the product and the catalyst is weak, and the product is formed. The sites of the various possible adsorption reactions are known for many surface reactions.

15.1.3 Biocatalysts

Enzymes are biocatalysts and are catalytic organisms (mostly proteins) produced by plants, animals, and microbes. In addition, some RNA also acts as a biocatalyst. Enzyme catalysis is also selective. For example, enzyme catalysis aids starch hydrolysis into dextrin and maltose, while protease catalysis facilitates protein hydrolysis into peptides. Living organisms use enzymes to speed up chemical reactions in the body. Without enzymes, many of the chemical reactions in an organism would be slow and difficult to sustain. Many enzymes operate best around 37°C (body temperature). If the temperature is above 50°C–60°C, the enzyme will be damaged and lose their functionality. Therefore enzymes are used to break down stains on clothes as a biological detergent, which is most effective when used at low temperatures. Enzymes are of great importance in physiology, medicine, agriculture, and industry.

15.2 Manufacturing methods

Each method for preparing catalysts is actually a combination of a series of process steps. For convenience, the name of the characteristic operating step is used as the name of the manufacturing method. Traditional methods include mechanical mixing, precipitation, impregnation, solution evaporation and drying, hot melting, dissolution (leaching), and ion-exchange methods. Newly developed methods include chemical bond and fibrosis methods.

15.2.1 Mechanical mixing

More than two substances are combined in the mixing equipment in this simple method. For example, when manufacturing absorption- and conversion-type desulfurization catalysts, the active component (e.g., manganese dioxide, zinc oxide, or zinc carbonate) is mixed with a small amount of binder, such as magnesium oxide or calcium oxide powder. The mixture is continuously added to a rotary table with an adjustable speed and inclination, while a measured amount of water is sprayed at the same time. The powder is mixed and bonded by rolling, forming pellets of uniform diameter, which are then dried and roasted. Fe–Cr–K–O catalysts for dehydrogenation of ethylbenzene during styrene making are prepared by forming and roasting iron oxide, potassium chromate, and other solid powders. When using this method, attention should be paid to the particle size and physical properties of the powder.

15.2.2 Precipitation

This method is used to manufacture catalysts that require high dispersion and contain one or more metal oxides. During the manufacture of multicomponent catalysts, suitable precipitation conditions are very important to ensure a high-quality product with uniform composition. In this method, usually, one or more metal salt solutions are added as a precipitating agent (such as sodium carbonate or calcium hydroxide), and the end product is produced by precipitation, washing, filtering, drying, molding, and roasting (or activation). Attachment precipitation is the method where an insoluble substance (such as diatomite) is placed in the sedimentation bucket, and metal oxides or carbonates are deposited on the insoluble substance. Sedimentation requires efficient filtration and washing equipment to save water and avoid material leakage.

15.2.3 Impregnation

During impregnation, a high-porosity carrier, such as diatomite, alumina, or activated carbon (AC), is immersed in a solution containing one or more metal ions, which is maintained at a certain temperature to allow the solution to be absorbed by the carrier. The carrier is drained, dried, and calcined, resulting in a layer of solid metal oxides or salts forming on the inner surface of the pores. This method can produce catalytically active components that are highly dispersed and evenly distributed on the surface of the carrier. Catalysts containing precious metals (such as platinum, gold, osmium, and iridium) are usually prepared using this method, where the metal content is usually below 1%. This method is also used to prepare expensive nickel and cobalt catalysts. Most of the carriers are preformed, so the final shape of the catalyst follows that of the carrier. Another method is to load spherical carrier particles into a variable speed drum and then spray a solution or slurry containing the active component onto the carrier or coat the surface of the carrier.

15.2.4 Spray evaporation

The catalysts used in fluidized beds with particle diameters ranging from tens to hundreds of microns are prepared using spray evaporation. For example, catalysts for oxidation and ammoniation are used during preparation of dimethyl acrylic from metaxylene in a fluidized bed. First, the desired concentrations of the metavanadate and chromium salt aqueous solution are mixed and then combined with a freshly made quantity of silica gel. This mixture is then pumped into a spray dryer. After spray atomization, moisture is evaporated under the action of a hot air flow, forming microsphere catalyst particles that are continuously extracted from the bottom of the spray dryer.

15.2.5 Thermal melting

Melting methods are used to prepare some special catalysts, where high temperature is used to melt and mix the components to obtain an evenly distributed mixture. With the necessary follow-up processing, excellent performance can be obtained from the catalyst. Such catalysts have high strength, activity, and thermal stability, along with a long service life. They are mainly iron catalysts for ammonia synthesis. The selected magnetite and related materials are fused, cooled, crushed, and screened at high temperature and then reduced in a reactor.

15.2.6 Leaching

Porous catalysts can be prepared by removing some of the material from a multicomponent system with an appropriate solution (or water). For example, during manufacture of nickel catalyst skeleton, quantitative nickel and aluminum are fused in an electric furnace, and the molten material forms an alloy after cooling. The alloy is crushed into small particles and then leached in aqueous sodium hydroxide solution to dissolve most of the aluminum, where sodium aluminate is produced and a highly active porous nickel skeleton is formed.

15.2.7 Ion exchange

Metallic cations (such as Na) from certain crystalline materials (such as synthetic zeolites) can be exchanged with other cations. When placed into a solution containing ions of other metals (such as rare earth elements and certain precious metals), other metal ions are exchanged with Na at controlled concentrations, temperatures, and pH. Because the ion-exchange reaction takes place on the surface of the exchange agent, precious metals such as Pt and Pd can be dispersed on the limited exchange group in an atomic state, so as to make full use of it. This method is often used to prepare cracking catalysts, such as rare earth and molecular sieve based.

15.2.8 Other methods

The chemical bond method is now widely used to manufacture polymerization catalysts, where the aim is to solidify homogeneous catalysts. A carrier capable of bonding with a transition metal complex has certain functional groups (or chemically bonded functional groups) on its surface, such as −X, −CH$_2$X, and −OH groups. This kind of carrier reacts with phosphine, arsine, or amine, where unpaired electrons on the surface of the phosphor, arsenic, or nitrogen atoms coordinate with metal ions and transition metal complex centers to form a chemically bonded solid catalyst. Such carriers are used for liquid-phase polymerization of propylene and Ziegler−Natta catalysts.

The fibrosis method is used to manufacture carrier catalysts containing precious metals. Borosilicate pulled into glass-fiber filament is etched by a concentrated HCl solution, forming a porous glass-fiber carrier. The fibers are then soaked with chloroplatinate solution to deposit Pt components. Depending on the application, the fibrous catalyst is pressed into various shapes with the required degree of compactness, such as catalysts for oxidation of automobile exhaust, which are pressed into short round tubes. Carbon fiber can also be used for applications that are not oxidation processes. Manufacturing of fiber catalysts is a complicated process, and the cost is high.

15.3 Sulfuric acid production and flue gas desulfurization catalysts

Vanadium catalysis was discovered in 1880, where research experiments began in 1901, and the use in factories was popular by the 1930s. Vanadium catalysts were first used by the Germany Baden Aniline Soda Corporation in 1913. Since the 1930s, they have completely replaced platinum catalysts for sulfuric acid production.

15.3.1 Synthesis methods

Some chemical industries use the vanadium pentoxide produced by specialized factories to process their own vanadium catalysts, such as sodium metavanadate, ammonium metavanadate, ammonium—sodium metavanadate, and vanadium trichloride. The preparation technology is an important aspect of vanadium catalyst research as it directly affects the structure and properties of the catalyst. There are still few research reports regarding the basic theory and technology related to catalyst production and innovation (from China or around the world). Vanadium catalysts are generally prepared using mixed-grinding, impregnation, and sol—gel methods. In practice, these methods are often used together.

15.3.1.1 Mixed grinding

The mixed-grinding method is used to prepare catalysts by mechanical mixing of two or more substances. It has the advantage of being a simple method that is easy to operate and automate. However, the disadvantage is that the heterogeneous mixture and operation process greatly influence the catalyst performance. Multicomponent catalysts undergo this process before they are formed. Currently, vanadium catalysts used in the oxidation SO_2 and production of sulfuric acid in China are prepared by mixed grinding using a wheel-grinding machine. This preparation process is shown in Fig. 15.1.

Natural diatomite is pretreated to remove impurities and improve the physical structure to obtain refined natural diatomite. Then, V_2O_5 is dissolved in KOH solution to remove impurities. H_2SO_4 is used to neutralize material and obtain a $V_2O_5-K_2SO_4$ slurry. Finally, the slurry, refined diatomite, and an appropriate amount of sulfur are added to the wheel-grinding machine. The mixture is ground well to obtain a plastic material that is then fed into a screw extruder for forming. The catalyst products are obtained after posttreatment.

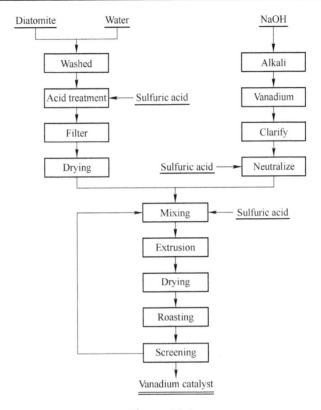

Figure 15.1
Process diagram of producing vanadium catalysts by mixed grinding.

15.3.1.2 Impregnation

This method of catalyst manufacturing includes impregnation as the key step. The impregnation process for producing vanadium catalysts is shown in Fig. 15.2. The basic process involves dipping a carrier material in a liquid- or sol-containing active vanadium substance. After the impregnated material reaches equilibrium, it is removed from the remaining liquid and then dried, roasted, and activated. The vanadium catalyst is obtained after other postprocessing procedures. Impregnation is a simple and economical method that takes advantage of the existing shape and size of the carrier, without requiring additional

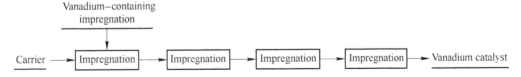

Figure 15.2
Impregnation method used to produce vanadium catalysts.

steps to form a catalyst structure. The specific surface area, pore radius, and mechanical strength of the catalyst can be tailored to the application. The active component has high utilization rate and low dosage. Disadvantages include a nonuniform distribution of active catalyst over the cross-section of the carrier, while other components can be absorbed by the carrier during impregnation along with the active components.

15.3.1.3 Sol–gel method

The sol–gel method is a newly developed process for catalyst preparation, which was first used to prepare oxide films by hydrolysis and gelation of metal alcoholate, and was later extended to catalysts. First, precursors are dissolved in water or organic solvents to form homogeneous solutions. Hydrolysis or alcoholysis reactions occur between the solute and solvent, where the reaction products accumulate to form nanoscale sol particles that condense into a transparent gel. After drying and heat treatment of the gel, the required catalyst is formed. The advantages of this method include a high-purity product with a small particle size, easy control of the reaction process, easy operation, simple equipment, and a low roasting temperature. In the preparation of vanadium catalysts, V-containing compounds or complexes are usually pretreated to obtain the precursors. In addition, a posttreatment process is required to obtain the vanadium catalyst.

15.3.2 Manufacturing of typical vanadium catalysts

Catalysts containing vanadium oxide can be used alone, loaded on a carrier, or used to form a composite system with oxides of Mo, Ti, Cr, P, K, or S. Such catalysts are used in the production of sulfuric acid, naphthalene or *o*-xylene oxidation to anhydride, and benzene or *n*-butene oxidation to maleic anhydride, anthraquinone. Common carriers include silica, titanium dioxide, alumina, silicon carbide, diatomite, and pumice.

In V-based catalysts, V_2O_5 is used as an active component, with alkali metal sulfate (K_2SO_4 or Na_2SO_4) as cocatalyst, and diatomite as a carrier in multicomponent catalysts in the V–K(Na)–Si system. Vanadium catalysts can be prepared using impregnation or wet-mixing methods. The mixing process is performed using a grinding wheel at an applied temperature of 400°C–600°C, where the V_2O_5–K_2SO_4 components on the carrier are in a molten state, catalytic reaction essence is in the molten liquid layer. After preliminary mixing the V_2O_5–K_2SO_4 component and carrier are mixed during a roasting process (500°C–550°C) for further homogenization. The use of a mixed-grinding process can meet the quality requirements of the catalyst. This preparation process is shown in Fig. 15.1.

Natural diatomite is purified by removing impurities such as Al_2O_3 and Fe_2O_3 by washing, acid treatment, filtration, rinsing, and drying, which also improve its physical structure. V_2O_5 is purified by ablation and then precipitated to remove iron impurities:

$$2KVO_3 + H_2SO_4 = V_2O_5 + K_2SO_4 + H_2O \qquad (15.1)$$

$$2K_3VO_4 + 3H_2SO_4 = V_2O_5 + 3K_2SO_4 + 3H_2O \qquad (15.2)$$

$$3KOH + H_2SO_4 = K_2SO_4 + H_2O \qquad (15.3)$$

V_2O_5–K_2SO_4 farrago and purified diatomite are accurately measured according to the desired composition and then loaded into a wheel-grinding mill with an appropriate amount of water. The material is ground to prepare a plastic material, which is then formed using screw extrusion to prepare 5-mm-diameter pellets, which are dried in a chain belt (chain plate) dryer. The dried material is fed into a storage hopper and then roasted in a drum roaster and passed through a vibrating screen (alloy kiln). The calcination temperature is 500°C–550°C, and the roasting time is ∼90 min. Finally, the products are obtained after cooling and screening and then packaged in airtight containers.

The main factors influencing the catalyst quality include the milling time and raw materials mixed in the wheel mill. The rolling performs two roles: mixing and evacuation. After crushing the bulk materials (diatomite and vanadium pentoxide) are loose and well dispersed, and the components are mixed uniformly. This helps achieve a more uniform multicomponent system. During a certain period of mixed grinding, the activity and strength increase with increasing grinding time. After this optimal time is exceeded, the strength continues to increase, but the activity decreases. Other important factors affecting the quality of vanadium catalysts are the roasting time and temperature. The roasting temperature should be above 500°C, while the roasting time is generally 90 min. Roasting can increase the mechanical strength of the catalyst, remove impurities (e.g., sulfur and organics) from the pore-forming agent, and ensure an excellent pore structure. At this temperature, V_2O_5 and K_2SO_4 form a eutectic and are redistributed on the carrier, where the catalyst has presaturated SO_2 stabilization activity. For catalysts with $K_2O/V_2O_5 = 2$–3, calcination at 500°C–550°C ensures that V_2O_5–K_2SO_4 components melt.

Vanadium catalysts for oxidation of sulfur dioxide to prepare sulfuric acid have a low melting temperature of 430°C. A high temperature is beneficial for their mutual diffusion and can overcome limitations related to a low degree of evacuation. If the temperature is too low, or the time is too short, organic impurities are not completely removed, and the sulfur is not sufficient, resulting in low activity. Long processing times at high temperature may cause sulfur and organic impurities to oxidize rapidly and release large amounts of combustion heat, resulting in serious sintering of the catalyst.

Diatomite is often used as a carrier for vanadium catalysts. Its chemical composition, mechanical strength, specific surface area, pore size distribution, and entrance to the catalyst activity, the application of life, and the loading quantity are the main factors affecting the quality of vanadium catalysts. The diatomite carrier of the vanadium catalyst should have sufficient purity and a suitable physical structure. Generally, it is required that w(SiO_2) > 85%, w(Fe_2O_3) < 1.0%, and w(Al_2O_3) = 2.0%–4.0%, the pore volume is >1.2 mL/g, and the specific surface area is >30 m^2/g.

15.3.3 Sulfuric acid production and flue gas desulfurization

Sulfuric acid is an important inorganic raw material. Its production is classified depending on the raw material into pyrite acid production, sulfur acid production, and smelting flue gas acid production. The conditions of the various processes are quite different, but the core step is the transformation of SO_2. Vanadium is a solid catalyst used in sulfuric acid industries globally. The catalyst is based on an active V_2O_5 component, supported by alkali metal sulfates (such as K_2SO_4) and silicides (usually diatomite). Such V−K−Si catalysts are classified into moderate-temperature (S101, S101Q, S101-2H), low-temperature (S107, S108, S107Q), wide-temperature (S109-1, H109-2), and arsenic-resistant (S106) types. The catalyst particles can have a bar, ball, ring, or daisy shape. The technology for producing vanadium catalysts is advanced, and the performance of the product is excellent; for example, the LP series made by Monsanto (the United States) and the VK series made by Topsoil (Denmark).

The function of the vanadium catalyst is to accelerate the reaction rate of sulfur dioxide oxidation to sulfur trioxide. Vanadium catalyst products are usually cylindrical particles with a diameter of 5 mm, where vanadium oxide and potassium oxide are attached to a porous diatomite carrier. During fabrication the active components of the vanadium catalyst were fused and distributed on the inner surface of the carrier micropores to form a liquid film of a certain thickness. The molten salt mixtures are composed of vanadium oxides dissolved in various pyrosulfates of alkali metals. The normal service life of a vanadium catalyst is at least 5–10 y. The loading fraction of catalyst on the production device is generally 180–300 L/t of sulfuric acid per day, which depends on the gas composition, design conditions, and exhaust gas requirements. The preparation is usually made by mixing V_2O_5 with the diatomite carrier, adding sodium salt or potassium salt to prepare a slurry, followed by forming, drying, and calcining. Some physical and chemical properties of the vanadium catalyst affect the catalytic performance, including the particle size, bulk density, porosity, mechanical strength, industrial combustion temperature, and contents of V_2O_5, K_2SO_4, and Na_2SO_4.

In the sulfuric acid production process the furnace gas is treated by a wet purification system and cooled to around 40°C. To ensure that the transform reaction occurs the gas must be heated using a heat exchanger, where the hot gas is indirectly heated to the reaction temperature. In the converter, reaction heat is released by oxidation of sulfur dioxide, resulting in an increase in the temperature of the catalyst layer of sulfur dioxide, which decreases the equilibrium conversion efficiency. Temperatures over 650°C can damage the catalyst. Such converters are divided into 3−5 layers, where interlayers are exchanged by indirect or direct cooling to ensure that they maintain an appropriate reaction temperature, and hence, high conversion efficiency in the catalyst layer and a high response speed.

The sulfur trioxide produced in the transformation process is absorbed in a packed absorption tower after cooling. The absorption reaction combines sulfur trioxide and water:

$$SO_3 + H_2O \rightarrow H_2SO_4 \quad \Delta H = -132.5 \text{ kJ} \tag{15.4}$$

However, it cannot be absorbed using water; otherwise, a large amount of acid vapor will form. In industry, 98.3% sulfuric acid is used as the absorbent, instead of water. As sulfur trioxide sulfuric acid in the liquid surface has the lowest total vapor pressure, the absorption efficiency is the highest. Absorption of sulfur trioxide in the absorption tower results in an increase in the sulfuric acid concentration. Hence, water must be added to the absorption tower circulation tank containing 98.3% sulfuric acid, where the acid flows between the drying tower and absorption tower to ensure a constant acid concentration. The finished acid product is extracted from the tower circulation system.

Modern sulfuric acid production involves two conversion processes. First, the gas flows through two or three layers of catalyst into an intermediate absorption tower, which absorbs the generated sulfur trioxide. After heating the residual gas is returned through the catalyst layer for a second transformation and then enters the final absorption tower for another absorption cycle. The intermediate absorption and removal of the reaction products result in higher conversion efficiency of the second transformation. Hence, the total conversion efficiency can reach more than 99.5%. Some older factories still use the traditional transformation process, where the gas is passed through all catalyst layers only once. The highest conversion efficiency of this process is only ∼98%.

The most important application of V_2O_5 is as a catalyst for the oxidation of sulfur dioxide into sulfur trioxide when the contact process is used for sulfuric acid production. Sulfuric acid can be obtained by further reaction of sulfur trioxide with water:

$$2SO_2 + O_2 \rightarrow 2SO_3 \tag{15.5}$$

Reaction (15.5) is a reversible reaction, which generally occurs at 400°C−620°C. At this temperature, V_2O_5 is not catalytically active. Above this temperature, V_2O_5 may decompose. The catalytic reactions are as follows:

$$SO_2 + V_2O_5(s) \rightarrow SO_3(g) + 2VO_2(s) \tag{15.6}$$

$$2VO_2(s) + \frac{1}{2}O_2(g) \rightarrow V_2O_5 \tag{15.7}$$

By optimizing the distribution of the catalyst in the converter and the transformation temperature, the conditions are made favorable for the catalytic transformation reaction, resulting in a total conversion rate of at least 99.73%. Adopting the "3 + 1" two transfer and absorption process, a high conversion efficiency of SO_2 and absorption efficiency of SO_3 are achieved.

15.3.4 Requirements for sulfuric acid production catalysts

Vanadium catalyst can only increase the reaction rate but cannot change the chemical equilibrium of the reaction. The composition and chemical properties of the vanadium catalyst remain unchanged after it participates in the reaction. The catalysts used in the sulfuric acid industry should have the following properties: (1) high activity, large range of temperature, and low temperature of activity; (2) good selectivity and ability to control side reactions; (3) good mechanical properties (not easily crushed); (4) good resistance to heat and poisoning, and a long service life; (5) large specific surface area and large porosity, and low resistance to gas diffusion; and (6) abundant sources of raw materials and low manufacturing costs.

The current global trend in vanadium catalysts for sulfuric acid production is the development of low-temperature, long-life, and high-vanadium catalysts that satisfy the requirements related to changes in raw materials and SO_2 emission control standards. Currently, research in this field is focusing on the development of various types of highly active catalysts that operate at low temperature and pressure, with low density and high porosity.

15.3.4.1 Vanadium catalyst morphology

Vanadium catalysts can have granular, lamellar, cylindrical, spherical, circular column, or tooth ring morphology. Spherical catalysts have the advantages of low resistance and low vanadium consumption for the same volume of catalyst; however, the manufacturing process is complicated. The most commonly used vanadium catalyst in the sulfuric acid industry is cylindrical.

15.3.4.2 Chemical composition

Although there are many types of vanadium catalysts, the chemical compositions are generally based on vanadium pentoxide, silicon dioxide, and alkali metal potassium salt. Only the gas content and preparation methods vary. Vanadium catalysts are mainly composed of 7%–12% V_2O_5. If the V_2O_5 content is too low, the activity of the vanadium catalyst will be poor, and its service life will be short. However, if the V_2O_5 content is too high, the cost is increased, while the activity of the catalyst may not be significantly increased, which is not economically viable. Vanadium pentoxide alone has very low activity and is not an effective catalyst. However, when a certain amount of alkali metal salt is added, its activity can be increased a 100 times.

Silica can be used as the catalyst carrier, although diatomite is commonly used in industry. The main function of the carrier is to have the catalyst loaded on its surface, resulting in a composite catalyst with a good structure, and high internal surface area for providing a high contact area between the active component of the catalyst and the reaction gas; this ensures that a good conversion efficiency is achieved. Diatomite is used as a carrier mainly because of its loose texture and large area per unit volume. It does not change with temperature, nor interact with the mixed gas. In addition, it is abundant, inexpensive and has other advantages.

Without the presence of potassium or sodium compounds, the addition of silica to V_2O_5 reduces its activity. The addition of potassium or sodium compounds, in the presence of silica, results in intense activation of the vanadium catalyst. For example, the $V_2O_5-8SiO_2-0.1K_2SO_4$ catalyst has an activity 250 times greater than that without K_2SO_4. The composition of a typical vanadium catalyst is shown in Table 15.1.

Table 15.1: Composition of a typical vanadium catalyst.

Compounds	Unit	Content
Vanadium pentoxide (V_2O_5)	%	8.09–9.55
Silicon dioxide (SiO_2)	%	54.04–65.37
potassium oxide (K_2O)	%	6.42–9.88
Sulfur trioxide (SO_3)	%	9.97–19.00
Sodium oxide (Na_2O)	%	3.22–5.12
Aluminum oxide (Al_2O_3)	%	2.33–2.70
Phosphorus pentoxide (P_2O_5)	%	1.98–2.33

15.3.5 Factors affecting the vanadium catalyst

15.3.5.1 Temperature

Temperature has a great influence on the vanadium catalyst. In the temperature range of 400°C–750°C, the chemical composition and activity of the catalyst change dramatically with changing temperature. At temperatures above 470°C, the vanadium catalyst has high stable activity and maintains a certain chemical composition. When the temperature decreases below 470°C, the catalytic activity begins to drop significantly, the color of the catalyst changes from yellow to green as vanadium pentoxide is converted into inactive vanadyl sulfate. At very high temperatures (above 600°C) the activity of the catalyst decreases significantly as the catalyst structure is damaged, and the porosity and surface area decrease. The maximum and minimum temperatures depend on the specific composition of the vanadium catalyst.

15.3.5.2 Initiation temperature

The light-off temperature of a catalyst refers to the lowest temperature necessary to initiate a catalytic reaction, where the catalyst can be rapidly heated by the reaction heat. In other words, it is the lower limit of the temperature range of the catalyst activity. The light-off temperature is the lowest suitable inlet temperature for sulfur dioxide gas entering the catalyst layer. In practice, the inlet temperature is usually slightly higher than the combustion temperature. The light-off temperature is an indicator of the activity of the catalyst. A high light-off temperature indicates that the required reaction speed can only be achieved at higher temperature; hence, the catalyst has low activity. A low light-off temperature indicates that the catalyst has sufficient reaction speed at a lower temperature, that is, higher catalytic activity.

It is preferable that the catalyst has a low light-off temperature, because it ensures that (1) the preheating temperature of the gas entering the catalyst layer is low, minimizing the heat transfer area and heating time; (2) the catalyst has good activity at low temperature. Hence, the end stage of the reaction can be performed at low temperature, which increases the equilibrium conversion efficiency of the later stage reaction and increases the total (final) conversion efficiency; and (3) the catalyst has high activity, which can increase its utilization efficiency.

The light-off temperature is not only related to the composition of catalyst but also dependent on the concentration of oxygen in the gas and its velocity in the air tower. The light-off temperature of mesotherm catalysts is commonly 400°C–420°C, where addition of alkali metal salt catalysts can lower the light-off temperature. The Chinese S_{107} product has the low-temperature catalyst of toxin immunity, with a light-off temperature of 380°C–400°C. The light-off temperature of the M_{615} catalyst (United States) is about 370°C, while that for the British-made ICI_{33-2} and ICI_{33-4} catalysts is 370°C–390°C.

15.3.5.3 Heat-resistance temperature of the catalyst

The upper temperature limit for catalytic activity is called the heat-resistance temperature of the catalyst. If the catalyst exceeds this temperature or is used at this temperature for a long time, the catalyst will become burnt out or lose its activity due to rapid aging. Therefore the heat-resistance temperature of the catalyst is also an indicator of the catalyst performance. The decrease in catalyst activity at high temperature is irreversible. A high heat-resistance temperature of the catalyst indicates that the production process can use a wider range of temperatures, which is favorable for production. The heat resistance of the S101 catalyst is higher than that of S107. For reliable production, the transformation period at the highest temperature should be at a temperature slightly lower than the heat-resistance temperature of the catalyst. In general, temperatures over 600°C are not used for long periods of time.

At high temperatures the catalytic activity of vanadium catalysts decreases for several reasons. First, V_2O_5 and potassium sulfate in the catalyst react at high temperature to produce relatively stable, but inactive, vanadyl vanadate with a composition of $4V_2O_5-V_{2.4}-K_2$, $4V_{2.5}-V_{2.4}-2K_2$, or $5V_{2.5}-V_{2.4}-K_2$. Second, at temperatures above 600°C, potassium and silica in the catalyst react, which decreases the potassium content in the active substance, make V_2O_5 from vast of exhalation, resulting in a loss of catalytic activity. Studies have shown that increasing the potassium content can increase the heat-resistance temperature of catalysts. Currently, in the use of vanadium catalysts, the gas heat-resistance temperature is under 620°C for long-term operation to meet the requirements of industrial production. Third, below 600°C, a solid-phase reaction between V_2O_5 and the silica carrier slowly occurs, which converts some of the V_2O_5 into inactive silicates. In the case of China's most popular medium-temperature catalyst (S101), an operating temperature of 425°C–600°C is used in all stages, where the catalytic activity has reached that of most advanced catalysts in the world.

15.3.5.4 Surface area

The total surface area of the catalyst consists of two parts: the surface area and the internal surface area of the micropores inside the catalyst. The specific surface of a catalyst is the sum of the surface area and internal surface area of 1 g of catalyst; this value depends on the porosity and heap density. The catalyst porosity refers to the volume fraction of micropores in the catalyst. The specific surface area of the catalyst generally increases with increasing porosity and decreasing density of catalyst. Catalysis mainly occurs on the internal surfaces. Hence, larger specific surface areas result in higher catalytic activity. However, the activity is also related to the pore size, on the surface than that at the same time. Larger micropores result in faster catalytic reactions, while smaller micropores result in slower catalytic reactions.

15.4 Flue gas denitrification catalysts

Current industrial emissions often contain SO_x and NO_x, which are very harmful to human health and the ecological balance of nature. For every 1000 kW h of electricity produced, 2.1 kg of nitrogen oxides are emitted. Negative side effects of nitrogen oxides include damage to the ecological environment (e.g., acid rain, smog, and destruction of the ozone layer) and harm to human health. Hence, the study of flue gas desulfurization and nitrogen removal technology is a global focus. In SO_x treatment of flue gas the gas concentration depends on the actual conditions of the factory. There are several desulfurization methods, where SO_2 is oxidized into SO_3, and then sulfuric acid is produced by absorption by water or dilute sulfuric acid, which has been brought to the attention of the people, such as sulfuric acid production using vanadium catalysts. The flue gas desulfurization technology used in the Wood River power plant in St. Louis is a typical example. Selective catalytic reduction (SCR) denitrification involves the reduction of NO_x in flue gas to N_2 and H_2 using ammonia, C, or hydrocarbons as a reducing agent in the presence of a catalyst. Current coal-fired power plants commonly used NH_3 SCR technology. This is similar to the SO_x method; however, as the risks to human health from NO_x exceed those from SO_x, control of NO_x is more important. V_2O_5 is loaded on AC to produce V_2O_5/AC catalysts that can remove SO_2 and NO in the flue gas at the same time when used at 200°C. This mixed catalyst has a better effect than pure AC, where regeneration of the vanadium catalyst also enhances its desulfurization and denitrification capability.

15.4.1 Denitrification reactions

A diagram of a denitrification catalyst is shown in Fig. 15.3. In the denitrification SCR method, NO_x in flue gas is reduced to N_2 and H_2 using ammonia, C, or hydrocarbon reductants in the presence of a catalyst.

The denitrification reactions are as follows:

$$4NH_3 + 4NO + O_2 = 4N_2 + 6H_2O \tag{15.8}$$

$$2NH_3 + NO + NO_2 = 2N_2 + 3H_2O \tag{15.9}$$

$$8NH_3 + 6NO_2 = 7N_2 + 12H_2O \tag{15.10}$$

The oxidation reactions using ammonia gas are as follows:

$$4NH_3 + 3O_2 = N_2 + 6H_2O \tag{15.11}$$

$$2NH_3 + 2O_2 = N_2O + 3H_2O \tag{15.12}$$

$$4NH_3 + 5O_2 = 4NO + 6H_2O \tag{15.13}$$

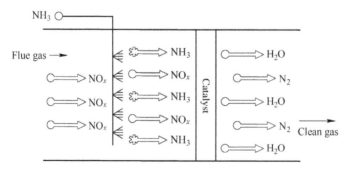

Figure 15.3
Diagram of a denitrification catalyst.

Flue gas desulfurization and denitrification catalysts are used in power plant SCR applications, in which the chemical reaction between the reductant and nitrogen oxides in the flue gas is selectively promoted at a certain temperature.

15.4.2 Types and compositions of denitrification catalysts

Current commercial SCR catalysts are mainly based on TiO_2, with V_2O_5 as the main active components, and WO_3 and MoO_3 as auxiliary components to prevent oxidation and poisoning. The catalyst types are classified into three types: plate, honeycomb, and corrugated plate structures. Plate catalysts use a stainless steel plate pressed into a metal mesh as the base material, onto which a mixture of TiO_2 and V_2O_5 is deposited. After pressing and roasting the catalyst plate is assembled into a catalyst module. Honeycomb catalysts are usually homogeneous catalysts. Mixtures of TiO_2, V_2O_5, and WO_3 are formed into catalyst elements with a section of 150 mm $\times$ 150 mm with different lengths using ceramic extrusion equipment, which are then assembled into a section of about 2 m and acute for standard 1 m modules.

Corrugated-board-type catalyst are reinforced with glass-fiber TiO_2 as a base material, which is dipped in WO_3, V_2O_5, and other active ingredients to coat the surface of the catalyst and improve its activity, to reduce the SO_2 oxidation efficiency. The catalyst is the core part of SCR technology, which determines the denitrification efficiency and economic feasibility of the system. The construction cost accounts for more than 20% of the flue gas denitrification project cost, while the operation cost accounts for more than 30%. In recent years, the United States, Japan, Germany, and other developed countries have devoted a great deal of man power, material resources, and money to the research and development of high-efficiency and low-cost flue gas denitrification catalysts. Great importance is placed on patented catalyst technology, technology transfer, and intellectual property protection during the process of product licensing.

The original catalysts were Pt, Rh, and metals such as Pt catalyst, combined with a monolithic ceramic, such as alumina carrier. Such systems have high activity and low reaction temperature, but their high cost limits their application in power plants. Since the late 1960s, three Japanese companies, Hitachi, Mitsubishi, and Takeda Chemical, have supported continuous research and development of TiO_2-based catalysts that have gradually replaced the Pt and Rh/Pt series catalysts. These catalysts are mainly composed of $V_2O_5(WO_3)$, Fe_2O_3, CuO, CrO_x, MnO_x, MgO, MoO_3, NiO, or a combination of these, usually with TiO_2, Al_2O_3, ZrO_2, SiO_2, or AC as the carrier. Liquid ammonia or urea is used as the reducing agent in the SCR system, where the reduction reaction occurs. Such materials have become mainstream catalyst products for SCR denitrification engineering applications. These catalysts are classified into three types: plate, honeycomb, and corrugated plate. Although all three catalyst types have proven successful in coal-burning SCR, the plate and honeycomb types are more effective than the corrugated plate type.

The catalyst is designed to achieve a certain reaction area and meet the requirements of the flue gas flow, temperature, pressure, component in the economizer outlet, and other basic SCR performance design requirements. In addition, a high denitrification efficiency and low ammonia escape rate need to be achieved. In an environment with variable ash content, antiblocking and antiwear performance are critical to ensure long-term safety and stable operation of the SCR equipment.

Considering the susceptibility of the catalyst to blocking by ash, for a certain reactor section and the same catalyst pitch, the flow area of the plate catalyst is the largest (generally >85%), followed by the honeycomb and corrugated-board-type catalysts (commonly around 80%). Under the same design conditions the honeycomb catalyst with a large pitch is selected appropriately, where its antiblocking effect is similar to that of the plate catalyst. In terms of the structure of the three catalysts, the wall surface of the plate type has the lowest wall angle, largest circulation area, and it is the most difficult to stop dust. Cellular catalyst circulation area commonly is small, where the catalyst wall angle is 90 degrees. Under bad flue gas conditions, it is easy to produce ash by pass and block the catalyst. The circulation area of corrugated plate catalyst is general, but its wall angle is low and its quantity is relatively large. The most commonly used catalysts are from the $V_2O_5-WO_3(MoO_3)/TiO_2$ series, where TiO_2 is the main carrier, and V_2O_5 is the main active component.

Catalysts are the core part of SCR denitrification technology, where their catalytic activity directly affects the overall denitrification efficiency. Most SCR catalysts currently used in China use imported TiO_2 as the catalyst carrier that accounts for about 80% of the total weight of the catalyst. The high cost and low utilization efficiency of active components are existing challenges. Monolithic catalysts include the active component and carrier as the substrate, which are mixed with adhesive, pore-forming agent, and lubricant and then formed by extrusion molding, drying, and calcining. Currently, in catalyst molding processes, there are still different degrees of catalytic activity to reduce.

Using pure Chinese TiO_2 as the carrier, V_2O_5 as the active component, and WO_3 as the auxiliary material, a series of vanadium catalysts can be prepared by impregnation. In addition, polyacrylamide (PAM), poly(vinyl alcohol) (PVA), and kaolin clay (kaolin) are added as forming agents to prepare monolithic catalysts of $V_2O_5/WO_3/TiO_2$ (VWTi). When the VWTi catalyst contains 0.4–1 wt.% V_2O_5, the WO_3 content is 8 wt.%. In the temperature range of 300°C–400°C and O_2 concentration >1%, NH_3/NO is 0.8–1, and the space velocity is 10000 h^{-1}. The denitrification efficiency can exceed 85%, and for different initial NO concentrations, which has good ability to adapt. The addition of MoO_3 does not contribute significantly to increasing the de-NO_x activity of the VW8Ti catalyst, but it improves the thermal stability and sulfur resistance of the catalyst. The mechanical strength of the catalyst can be increased if the specific surface area of the catalyst does not decrease. Inorganic-forming agents can significantly increase the mechanical strength of the catalyst but can affect the crystal structure of the catalyst carrier. PAM and kaolin greatly affect the activity of the VWTi catalyst, where the catalytic activity decreases gradually with their increasing content; when this content reaches 4 wt.%, the maximum denitrification efficiency of the catalyst is ~75%. The PVA content affects the VWTi catalyst activity. At 400°C the catalytic activity can exceed 85%, with a compressive strength above 0.7 MPa. The PVA-molding agent is the best as it can help widen the window of catalytic activity and improve the physical and chemical properties of the catalysts.

15.4.3 Typical denitrification catalysts

Vanadium oxide denitrification catalysts include vanadium/tungsten—titanium catalysts, TiO_2—Al_2O_3 composite carriers, supported catalysts, and multicomponent composite active oxide catalysts. The Brunauer, Emmett, and Teller (BET)-specific surface area and pore characterization of typical catalysts are shown in Table 15.2.

Table 15.2: BET-specific surface area and pore structure of typical catalysts.

Sample	Surface area (m^2/g)	Porosity (cm^3/g)	Average pore size (nm)
TiO_2	11.47	0.038	13.3
W_8Ti	12.17	0.036	12.13
$V_{0.4}W_8Ti$	11.72	0.037	12.33
$V_{1.2}W_8Ti$	10.07	0.029	11.77
$V_{1.6}W_8Ti$	9.23	0.040	17.37

The process for preparing V/W—Ti catalysts is shown in Fig. 15.4. The WO content is 3—8 wt.%, which is tested in NO = NH_3 = 500 ppm, O_2 = 2%, air speed 10,000 h^{-1}, with N_2 as the equilibrium gas. No V_2O_5, the highest activity of ~40%, occurs at 400°C for a vanadium content in the catalyst of 0.4%. When the catalyst is used at 350°C—400°C, the highest denitrification efficiency is close to 99%. For vanadium contents of 0.4% and 0.8%, the activity window of the catalyst is 250°C—450°C. When the vanadium content is above 0.8%, the denitrification efficiency significantly decreases.

Representative data from an activity test using V_2O_5—WO_3/TiO_2 catalyst is shown in Table 15.3. Fig. 15.5 shows a flowchart for V/Fe—Ti catalyst preparation. The addition of iron oxide increases the activity of the catalyst to some extent, although the V_2O_5 content in the catalyst was too high.

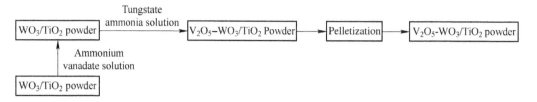

Figure 15.4
Preparation process for vanadium/tungsten–titanium catalyst.

Table 15.3: Results of V_2O_5–WO_3/TiO_2 catalyst activity test.

V_2O_5/ content (%)	Active starting temperature (°C)	Temperature where activity drops (°C)	Maximum active temperature point (°C)	Highest denitrification activity (%)
0	350	425	400	49.7
0.4	250	450	400	98.3
0.8	250	425	350	96.1
1.2	250	425	350	91.5
1.6	250	375	300	86.4
2.0	250	350	300	82.2
2.4	250	375	300	86.8

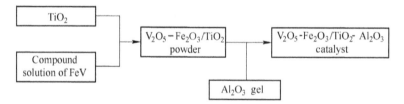

Figure 15.5
Preparation process of V_2O_5–WO_3/TiO_2–Al_2O_3 catalyst.

Aluminum oxide was chosen to modify and prepare composite carriers for titanium dioxide, where V_2O_5–WO_3/TiO_2–Al_2O_3 catalysts supported by a composite carrier were prepared, as shown in Fig. 15.5.

The process for preparing V_2O_5–MoO_3–WO_3/TiO_2 catalyst is shown in Fig. 15.6. Changing the Mo content significantly increased the low-temperature activity of the catalyst during denitrification. Typical compositions of multicomponent oxide catalysts are shown in Table 15.4. The effects of multicomponent oxides on the denitrification efficiency of catalysts are complex. The denitrification efficiency is also affected by the inert equilibrium material.

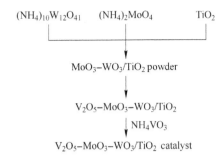

Figure 15.6
Preparation process of $V_2O_5-MoO_3-WO_3/TiO_2$ catalyst.

Table 15.4: Typical compositions of multicomponent oxide catalysts.

No.	Catalyst	Composition (wt.%)	Equilibrium composition (%)	Maximum temperature (°C)	NO removal efficiency (%)
1	M10701	Fe_2O_3, 27.53; MnO_2, 6.83; Cr_2O_3, 1.67; V_2O_5, 1.29; TiO_2, 7.98	Al_2O_3, 27.35; SiO_2, 27.35	300	96.09
2	M10702	Fe_2O_3, 27.53; MnO_2, 6.83; Cr_2O_3, 1.67; V_2O_5, 1.29; TiO_2, 7.98	Al_2O_3, 27.35; SiO_2	350	89.56
3	M10703	Fe_2O_3, 27.53; MnO_2, 6.83; Cr_2O_3, 1.67; V_2O_5, 1.29; TiO_2, 7.98	Al_2O_3, 43.75; SiO_2, 10.94	350	88.97
4	M10704	Fe_2O_3, 27.53; MnO_2, 6.83; Cr_2O_3, 1.67; V_2O_5, 1.29; TiO_2, 7.98	Al_2O_3, 36.46; SiO_2, 18.23	300	95.26
5	M10705	Fe_2O_3, 27.53; MnO_2, 6.83; Cr_2O_3, 1.67; V_2O_5, 1.29; TiO_2, 7.98	Al_2O_3, 43.75; SiO_2, 10.94	300	93.11
6	M10801	Fe_2O_3, 27.53; MnO_2, 6.83; Cr_2O_3, 1.67; V_2O_5, 1.29; TiO_2, 7.98; K_2O, 0.14	Al_2O_3, 27.28; SiO_2, 27.28	300	91.20
7	M10802	Fe_2O_3, 27.53; MnO_2, 6.83; Cr_2O_3, 1.67; V_2O_5, 1.29; TiO_2, 7.98; Na_2O, 5.90	Al_2O_3, 24.40; SiO_2, 24.40	300	81.77

15.4.4 Catalyst maintenance

As catalyst deactivation can occur as a result of many factors, studying the deactivation mechanism of denitrification catalysts is of great importance for extending the lifetime of the catalyst and reducing the operating costs of SCR flue gas denitrification systems.

15.4.4.1 Catalyst poisoning

The main poisoning phenomena and their causes are discussed here.

15.4.4.1.1 Sintering of the catalyst

Sintering is one of the important causes of catalyst deactivation as it is irreversible. In general, the flue gas temperature is above 400°C, which can result in sintering. According to the design of the conventional catalyst, when the flue gas temperature is below 420°C, the catalyst sintering rate is acceptably low. When the reactor inlet flue gas temperature is above 450°C for an extended period, the lifetime of the catalyst will be quickly and significantly reduced. Currently, SCR catalysts are generally V_2O_5–WO_3–TiO_2 catalysts, where V_2O_5 is the active component, WO_3 is the stable component, and TiO_2 is the carrier material. In SCR flue gas denitrification systems, anatase TiO_2 is used, which is transformed into rutile after sintering. As a result, the crystal size doubles, resulting in a sharp decrease in the number of micropores and active catalysis sites.

Increasing the WO_3 content of the catalyst can improve its thermal stability and limit its sinterability. At present, SCR flue gas denitrification systems in China have no bypass. Once the flue gas temperature entering the SCR system exceeds the maximum that the catalyst can withstand, the flue gas can only flow through the catalyst. In the case of failure conditions, such as abnormal blowing of furnace ash, which substantially increases the smoke temperature at the entrance of the SCR system, the boiler load should be reduced to protect the catalyst to avoid sintering losses.

15.4.4.1.2 Arsenic poisoning

Arsenic is an element found in most types of coal. Arsenic poisoning of the SCR catalyst is caused by the accumulation of gaseous arsenic compounds that block the active pathways within the catalyst. The main form of gaseous arsenic in the flue gas is As_2O_3 that forms some As_3O_5 or As_4O_6 in the operating temperature range of the SCR catalyst. As_2O_3 mainly deposits and blocks the medium-sized pores (diameter of $0.1-1.0$ μm) of the catalyst. The arsenic concentration of the gas phase depends on the furnace type and chemical composition of the coal. The concentration of gaseous arsenic in flue gas produced by liquid slag discharge furnaces is much higher than that of solid slag discharge furnaces. However, despite the furnace type, arsenic poisoning of the catalyst occurs. At each stage of combustion, various physical and chemical methods are used to reduce the arsenic content in the flue gas and effectively reduce arsenic poisoning of the catalyst. For example, physical and chemical methods are used to reduce the arsenic content in the raw coal before combustion. During combustion, the formation of gaseous arsenic is inhibited by spraying calcium into the furnace. In addition, the physical and chemical properties of the catalysts can be modified to avoid arsenic poisoning and improve the chemical properties of catalysts. First, the acid sites on the catalyst surface can be changed to make the catalysts inactive to arsenic and not adsorb arsenic oxides. Second, mixed oxides of V and Mo are used to obtain stable catalysts after high-temperature calcination so that the sites of arsenic adsorption do not affect the catalytically active sites. TiO_2–V_2O_5–MoO_3

catalysts were prepared with $V_9Mo_6O_{40}$ as the precursor, which has strong resistance to arsenic poisoning. Compared with catalysts with the same V and Mo loads, the arsenide absorption capacity increased significantly, and the surface tension was changed by the distribution of active components and the formation of new materials during the preparation process. This enhanced the performance of the catalyst against arsenic poisoning.

15.4.4.1.3 Calcium poisoning

The influence of alkaline earth metals on SCR catalysts is mainly manifested in the oxide on the surface of the catalyst, the reaction of sedimentary and further congestion of the pore structure. $CaSO_4$ is the main alkaline earth metal compound deposited on the surface of the catalysts, followed by $Ca_3Mg(SiO_4)_2$, $CaCO_3$, along with $CaSO_4$ and $CaCO_3$ produced by reaction between CaO and SO_3 or CO_2, respectively. Like K, Ca can also influence NH_3 adsorption on acid sites and $V^{5+}=O$.

CaO in flue gas can solidify As_2O_3 in the gas phase, but excessively high CaO concentrations can enhance blockage of the catalyst by $CaSO_4$. Under certain arsenic concentrations, with increasing CaO content in the coal, the catalyst lifetime first increases and then decreases. When the CaO content is low, the catalyst lifetime is mainly affected by arsenic poisoning, while when the CaO content is higher, the catalyst lifetime is mainly affected by $CaSO_4$ blockage.

15.4.4.1.4 Alkali poisoning

Alkali metal elements are the largest class of toxic elements for SCR catalysts, where their toxicity decreases in the order of $Cs_2O > Rb_2O > K_2O > Na_2O > Li_2O$. In addition to alkali metal oxides, alkali metal sulfates and chlorides can also result in deactivation of the catalyst. In coal-fired power plants the effect of K is the most significant because of its high content. The chemical mechanism for alkali metal poisoning is a reaction between K and VOH acid sites on the catalyst surface to form V—OK, which reduces the ability of the catalyst to absorb NH_3. In this way the amount of adsorbed NH_3 involved in the NO reduction reaction decreases, and its activity in the SCR reaction also decreases. The concentrations of the various types of potash on the surface of the catalyst are different. The deactivation rate from K poisoning of the catalyst is much higher than that of the specific surface area reduction.

15.4.4.1.5 SO_3 poisoning

SO_2 in the flue gas can also poison the catalyst. Under the effect of the vanadium catalyst, SO_2 is oxidized to SO_3, which reacts with NH_3 and water vapor in the flue gas, generating various ammonium salts. This both wastes the NH_3 and leads to the active sites of the catalyst becoming covered, resulting in catalyst deactivation. In addition, SO_2 reacts with the active metal components in the catalyst to produce metal sulfates that deactivate the catalyst. In the case of high-CaO coal, calcium sulfate is the main source of catalyst deactivation in the presence of other toxic substances. In the reaction between CaO and SO_3, CaO is first deposited on the catalyst surface with a relatively slow deposition rate. The reaction between CaO deposited on the surface of catalyst and SO_3 in the flue gas is a solid–gas reaction. Because there are active substances on the catalyst surface to catalyze the oxidation of SO_3, its concentration is relatively high and the reaction rate is fast. The volume of $CaSO_4$ increases rapidly due to the fast reaction, blocking the catalyst surface and affecting the diffusion of reactants.

15.4.4.1.6 Phosphorus poisoning

Some compounds of phosphorus, including H_3PO_4, P_2O_5, and phosphate, also have passivating effects on SCR catalysts. When the denitrification activity was tested, the catalyst activity decreased with increasing P_2O_5 loading, but the effect was much smaller than that of alkali metals. The phosphorus poisoning mechanism of SCR catalysts involves V in V–OH and W in W–OH being replaced by P, which generates P–OH groups, which are not as acidic as V–OH and W–OH, but can provide weak Bronsted acids. Hence, for small loadings, phosphorus poisoning phenomenon of catalyst is not significant. P can also react with the V=O active site on the catalyst surface to form substances such as $VOPO_4$, thus reducing the number of active sites.

15.4.4.1.7 Water poisoning

Water occurs as vapor in the flue gas and can condense on the surface of the catalyst. This can intensify poisoning of the catalyst by K, Na, and other alkali soluble salts. In addition, when the temperature increases, water vapor condensed in the catalyst pores vaporizes and expands; this can damage the catalyst structure and lead to its collapse.

15.4.4.1.8 Pore clogging with ash

Of all the factors resulting in SCR catalyst poisoning, ash accumulation is the most complex and influential. Ash accumulation on the surface of the catalyst can be explained via the following mechanism. Fly ash particles containing elements such as K, Na, Ca, and Mg, as well as oxides, are deposited on the catalyst surface as the flue gas enters the SCR reactor. The fly ash particles react with CO_2 in the flue gas and some oxides to form carbonates. Meanwhile, the fly ash particles can be further salinized by sulfuric acid due to the oxidation of SO_2 on the catalyst surface to SO_3. Solid metal oxides are gradually melted with carbonate, sulfate, and catalyst surface materials, while some small particles infiltrate the catalyst interior. The active sites on the catalyst surface are gradually lost, and the inner pore structure is blocked, resulting in catalyst poisoning.

Most of the fly ash produced by coal combustion contains fine particles as the velocity of the flue gas through the catalytic reactor is low, generally around 6 m/s. The airflow is laminar, and the fine ash particles accumulate upstream of the SCR reactor and fall onto the surface of the catalyst. In addition to small ash particles, the flue gas can also contain some large fly ash particles, which are generally larger than the SCR catalyst channels and can directly block the catalyst. The composition of the generated flue gas can vary greatly depending on the type of coal-fired power plant boiler furnace and the coal quality. Fly ash from power plants burning lignite and bituminous coal usually contains a large amount of alkali and alkaline earth metal elements, mainly K, Na, Ca, and Mg. These minerals are mainly found in quartz, clay, carbonates, sulfates, sulfides, and phosphatite-containing materials. By analyzing the effect of the plant reactor type producing the fly ash on the flue gas composition and deposition of fly ash on catalyst surface, it was shown that the fly ash deposited on the catalyst surface mainly contained particles <5 μm in size. The flue gas contains more sulfate particles than fly ash particles. Elements such as As and Na are more likely to be enriched on small particles, which result in serious catalyst poisoning.

15.4.5 Catalyst regeneration

As the catalyst is the core component of SCR flue gas denitrification systems, its performance directly affects the overall denitrification performance. SCR catalyst and regeneration technology is now a very mature field, where rapid regeneration can be achieved without removing the catalyst. The denitrification catalyst can be regenerated by a process, including dust cleaning, demineralized water washing, active regeneration, demineralized water rinsing and drying, washing with a renewable liquid, including cleaning fluid and an active liquid, and finally cleaning with a fluid containing H_2SO_4 and emulsifying agent S-185. The active additive contains $VOSO_4$ and ammonium metatungstate. Ultrasonic cleaning methods are used. After regeneration the relative activity (K/K_0) of the catalyst is 0.95−0.99, where the conversion efficiency of SO_2/SO_3 is <1%.

15.5 Vanadium catalysts for organic synthesis

Vanadium compounds are some of the most important catalytic oxidation catalysts for industrial applications and are widely used in the production of sulfuric acid and organic chemical raw materials, such as phthalic anhydride, maleic anhydride. These catalysts are used in polymerization, alkylation, and oxidative dehydrogenation reactions.

15.5.1 Important catalysts

Vanadium catalysts are a series of catalysts containing vanadium compounds as active components. The active components of industrial vanadium catalysts include vanadium oxide, vanadium chloride, and complexes, as well as a variety of polyoxometalates. However, the most common active components contain one or more V_2O_5 additives. Catalysts with V_2O_5 as the main component are effective for almost all oxidation reactions. Some important V-based catalysts are shown in Table 15.5.

Table 15.5: Some important V-based catalysts.

Vanadium compound	Process	End use
Vanadium pentoxide (V_2O_5)	In the production of H_2SO_4, it is used as a catalyst to oxidize SO_2 to SO_3, and as a catalyst to oxidize cyclohexane to adipic acid	Used for producing sulfuric acid, phosphate fertilizer, and nylon
Ammonium metavanadate (NH_4VO_3)	In the production of H_2SO_4, as the catalyst used for oxidation of SO_2 to SO_3 and is used for oxidation of benzene to maleic anhydride	Used for the production of sulfuric acid, production of phosphate fertilizer, production of unsaturated polyester (polyester etc.), and production of polyvinyl chloride
Vanadium oxytrichloride ($VOCl_3$)	Used as a crossbonding between ethylene and propylene	Used in the production of ethylene, propylene, and rubber for the automotive industry
Vanadium tetrachloride (VCl_4)	Used as a crossbonding between ethylene and propylene	Used in the production of ethylene, propylene, and rubber for the automotive industry

The main vanadium component of vanadium catalysts include vanadium pentoxide (V_2O_5), vanadium trioxide (V_2O_3), ammonium metavanadate (NH_4VO_3), Vanadium oxytrichloride ($VOCl_3$), and vanadium tetrachloride (VCl_4). V_2O_5 is the standard product for vanadium extraction, while V_2O_3 can be obtained by reduction of V_2O_5 or ammonium polyvanadate using a reductant such as carbon, gas, or coke oven gas. $VOCl_3$ can be obtained by chlorination of low-valence vanadium oxide. VCl_4 can be produced by chlorination of vanadium pentoxide or ferrovanadium. Based on the main vanadium composition, auxiliary components are added, and the material is processed using special methods according to the requirements of the organic synthesis reaction and characteristics of the reaction device to obtain the desired catalyst.

15.5.2 Main vanadium components

The main components of vanadium in catalysts include vanadium pentoxide, vanadium trioxide, Vanadium oxytrichloride, and vanadium tetrachloride.

15.5.3 Organic synthesis applications

V-based catalysts are used as special catalysts for production of nylon, polyester, ethylene, propylene, rubber, and certain macromolecular compounds, and in the petroleum cracking industry.

15.5.3.1 Anhydride production

Phthalic anhydride is mainly used in the production of plasticizers, unsaturated polyester, alkyd resin, dyes, coatings, pesticides, pharmaceutical and instrument additives, and edible saccharin, which are very important chemical raw materials. Mercury salt and Cr_2O_3 were used as catalysts in early production processes. Currently, phthalic anhydride is mainly produced from benzene- or *o*-xylene by vanadium catalyst oxidation. The catalysts used in domestic production of phthalic anhydride by oxidation in fluidized bed include 0401, 0402, Dx1, and Dx11 types that are mainly composed of V_2O_5, K_2SO_4, and SiO_2, where the V_2O_5 is the active ingredient, K_2SO_4 mainly has the inhibitory effect of depth of oxidation and maintains the activity of catalyst, and SiO_2 is the carrier.

15.5.3.2 Maleic anhydride production

Maleic anhydride is an important organic chemical raw material that is widely used in coatings, medicine, and food additives. Maleic anhydride is commonly produced from benzene oxidation using catalysts with various carriers with different catalytic performance. Table 15.6 shows the typical vanadium catalyst used in the oxidation of benzene to produce maleic anhydride, which is rich in *n*-butane resources and inexpensive. Compared with benzene, *n*-butane is more difficult to oxidize, and the reaction conditions are harsh. Hence, developing a high-performance catalyst is important, which can be combined with an impregnation method of butane selective oxidation to produce maleic anhydride with a particular phase structure of vanadyl phosphate catalyst.

Lonza (Switzerland) successfully developed a new generation of catalyst (X4) for its proprietary fixed-bed anhydride production process. It is based on a traditional vapor-pressure osmometry catalyst that is modified using a proprietary dopant. Compared with the

Table 15.6: The typical vanadium catalyst used in the oxidation of benzene to produce maleic anhydride.

Catalyst	Reaction temperature (K)	Raw material concentration (%) (volume fraction)	Recovery of maleic anhydride (%) (the mass fraction)
V_2O_5/pumice stone	573–973	0.8	72
V_2O_5–MoO_3/α-Al_2O_3	748–813	0.662–82	62–80
V_2O_5–MoO_3/P_2O_5/TiO_2	653	0.86	84.2
V_2O_5–MoO_3/P_2O_5–Na_2O–Mo/α-Al_2O_3 M: Zn/Cu/Bi/Co	723	0.95	85
V_2O_5–MoO_3/P_2O_5–Na_2O–Mo/α-Al_2O_3 M: Co/Ni/Ca/Fe	678	1.25	101
V_2O_5–P_2O_5–WO_3–Na_2O–Mo/α-Al_2O_3 M: Mg/Ca/Zn/Ti/Mn/Ni	683	0.95	86.7
V_2O_5–WO_3–P_2O_5/α-Al_2O_3	653	0.658	98

previous catalyst, the selectivity of X4 is 4% higher, and the operating temperature was reduced to about 410°C, which greatly extends the service life of the catalyst.

15.5.3.3 Catalytic polymerization

Polymerization reactions are some of the important in the chemical industry and can be achieved by transforming simple organic compounds to polymers with various properties. Depending on the polymerization reaction, catalysts with different properties are required. Currently, vanadium catalysts are mainly used in the synthesis of ethylene propylene rubber (EPR) and the polymerization of partial olefin. The technology of the V−Al catalytic system used in the global EPR industry is quite mature and the product performance is excellent. Among the Ziegler−Natta catalysts, VCl_3−$AtEt_2$ and $VOCl_3$−$AtEt_2$ are the most attractive as they can catalyze polymerization to produce random alternating copolymers, while also catalyzing ethylene−propylene copolymerization to obtain EPR. $V(acac)_3$−$Al(i-Bu)_2Cl$ is a highly dispersed polyphase colloidal catalyst that is highly active for catalytic polymerization of butadiene. Supported (2-acetyl-1-naphtholoxy) vanadium dichloride catalyst is prepared with $MgCl_2$ as the carrier, which has significantly better activity and stability compared with the original homogeneous catalyst for catalytic polymerization of ethylene.

Mesoporous alumina (m-Al_2O_3) has been loaded with vanadium catalyst (V/m-Al_2O_3). The mesoporous alumina has large specific surface area, narrow pore size distribution, and a two-dimensional hexagonal structure, which can ensure that the desired amount of active vanadium is loaded and is highly dispersed. Weak acid catalysis conditions are maintained, which increase the propane conversion and propylene selectivity.

15.5.3.4 Organic synthesis

Vanadium-based catalysts loaded on carriers are currently used in the production of propane by oxidative dehydrogenation of propane and are studied the most widely as they have high catalytic activity. Common carriers include MgO, TiO_2, ZrO_2, Al_2O_3, and SiO_2. It is generally thought that the isolated V−O tetrahedron structure is beneficial for oxidization and dehydrogenation of propane to produce propylene. The resulting structure of the vanadium is related to the loading capacity and physicochemical properties of the carrier. Vanadium loaded on SiO_2 can easily form V_2O_5 crystal phases, while on MgO, it can easily form metal vanadates. In addition, vanadium can be well dispersed on Al_2O_3 carriers. The V/SBA-15 catalyst is used for the production of propane by oxidative dehydrogenation of propane, where the propane conversion and propylene selectivity are superior to V/SiO_2 catalysts. This is mainly due to the high specific surface area of SBA-15 that allows good dispersion of the active component, resulting in more isolated vanadium sites for increasing propylene selectivity. In contrast, it is generally thought that paraffin adsorption and activation on the catalyst is the limiting step in the oxidation and dehydrogenation of alkanes. However, the adsorption of alkanes requires that the catalyst has a certain Lewis acid center. Therefore mesoporous Al_2O_3 (m-Al_2O_3) material should be a good carrier for oxidation and dehydrogenation catalysts for alkanes.

Compared with commercial Al_2O_3, metal-doped ordered m-Al_2O_3 has a larger specific surface area, narrower pore size distribution, and larger pores, where the pore size can be adjusted over a certain region. Disordered m-Al_2O_3 loaded with V- or Mo-based catalysts showed higher activity and selectivity for oxidative dehydrogenation of ethane compared with use of γ-Al_2O_3 as the catalyst carrier. In addition, m-Al_2O_3 loaded with V- and Ni-based catalysts have high catalytic activity for propane oxidation and dehydrogenation to produce propylene, attributed to the large specific surface area and pore diameter of m-Al_2O_3.

Vanadium catalysts are also used in alkylation, oxidative dehydrogenation, and acetic acid production. Soluble vanadium derivatives are efficient reagents and catalysts for the oxidation of various organic compounds. When V_2O_5–Fe_2O_3 catalyst was used for alkylation of phenol and methanol to prepare xylenol, it showed high conversion efficiency and selectivity at low temperature and a long service life. Research into vanadium catalysts for oxidative dehydrogenation reactions showed that binary or ternary composite oxides of V_2O_5, Ag_2O, and NiO had high activity and selectivity for oxidation of toluene to produce benzaldehyde. A vanadium catalyst for acetic acid production was reported in a US patent. The catalyst VO_x/CuSBA-15 can selectively catalyze O_2 molecules and oxidize benzene to produce phenol. Compared with direct dehydrogenation, the oxidation and dehydrogenation of alkanes can prevent accumulation of carbon due to the existence of O_2 or CO_2, where the catalyst stability can be increased.

Chapter 16

Vanadium batteries

Chapter Outline
- 16.1 Technical background of vanadium cell development 446
- 16.2 Vanadium battery systems 449
  - 16.2.1 Electric reactor technology 450
  - 16.2.2 Electrolyte technology 451
  - 16.2.3 Control technology 451
  - 16.2.4 System integration technology 451
  - 16.2.5 Technical features 452
  - 16.2.6 Operating principle 453
- 16.3 Applications of vanadium batteries 456
  - 16.3.1 Features of vanadium batteries 457
  - 16.3.2 Potential vanadium cell market 458
- 16.4 Key materials for vanadium batteries 462
  - 16.4.1 Electrolyte materials 462
  - 16.4.2 Membrane materials 466
  - 16.4.3 Electrode material 466
- 16.5 Vanadium battery assembly 468
  - 16.5.1 Structure of a single vanadium cell 468
  - 16.5.2 Static cells 468
  - 16.5.3 Flow-type cells 469

Vanadium belongs to the VB group elements and has a valence electron structure of 3 d^3s^2. It can form ions with four different valence states (V^{2+}, V^{3+}, V^{4+}, and V^{5+}) that have active chemical properties. Valence pairs can be formed in acidic medium as V^{5+}/V^{4+} and V^{3+}/V^{2+}, where the potential difference between the pairs is 1.255 V. The electrolyte of REDOX flow batteries requires such metal ions with different valence states. The stored electrolyte circulates during charging and discharging. Vanadium batteries are known as vanadium redox batteries (VRBs), which are a type of redox battery with circulating liquid and active substances. Different solutions of vanadium ions have been used as the active materials for the positive and negative electrodes. The solution is pumped from an external storage tank to the cell stack to complete the electrochemical reaction and then returned to the storage tank. The liquid with active substances is continuously circulated. The active material of vanadium liquid flow batteries is stored in liquid form in the external storage tank. The flow of active material minimizes concentration polarization. The battery capacity depends on the amount of external active material and can be adjusted. The standard potential difference between positive and negative electrodes of vanadium batteries is 1.26 V, and the solution concentration of the active substances at both the positive and negative electrodes is 1 mol/L. As the solution concentration increases, the potential difference increases correspondingly; this is of great practical value [14,25].

16.1 Technical background of vanadium cell development

The power supply of the current electrical grid is in a process of dynamic balance. Power transmission and transformation of grid power require the addition of stable load balancing system between the supplier and consumer. General economic activities have periodic variations in electricity load that can impact the power grid, resulting in large fluctuations in the electricity supply. To maintain a balanced base load and adapt to supply fluctuations in the grid, intermediate energy storage systems are required that can cut the peaks and stagger the valleys in the power supply. Furthermore, to adapt to relatively rich electrical energy conversion in remote areas and provide a storage solution for power produced by distributed generation (especially wind power, solar energy, and small hydropower), self-sustaining power supply systems are required, which can also provide power for islands, remote mountainous areas, and machine stations. Here, we will focus on the transformation of new energy sources, such as wind, solar, and tidal power.

Secondary batteries can be developed for intermediate energy storage and conversion systems. Redox flow batteries were studied for the first time at NASA's Lewis Research Center in 1973. In 1975 L.H. Thaller identified the redox couples Fe^{2+}/Fe^{3+} and Cr^{2+}/Cr^{3+}, which can be used as positive and negative active substances. In the batteries, Fe^{2+}/Fe^{3+} electric pairs are used as the positive electrode of the liquid flow battery, and Cr^{2+}/Cr^{3+} electric pairs are used as the negative electrode. A diaphragm divides the battery into two sections. Fe/Cr liquid flow batteries have poor reversibility of the Cr half battery, and the hydrogen evolution reaction occurs at the negative electrode, resulting in a voltage drop loss. At the same time the separation provided by the ion-exchange membrane is not ideal, and cross contamination of different metal ions between the two half-cell electrolytes is a problem. Although some improvements have been made, the inherent defects of Fe/Cr REDOX flow batteries have not yet been fully resolved. To overcome the cross contamination of different metal ions, a single metal ion solution has been used for battery electrolytes, mainly Cr, Ce, and V systems.

At the University of New South Wales, Australia in 1984, M. Skyllas-Kazacos proposed the concept of a VRB, where V^{5+}/V^{4+} and V^{3+}/V^{2+} were applied to REDOX flow batteries. It was found that V^{5+} is stable in a sulfuric acid medium. A patent for an all-vanadium battery system was obtained in 1986. Through systematic research, the vanadium battery diaphragm, conductive polymers, and graphite felt material were studied to finalize the design, and several related patents were published. In 1993 vanadium batteries were used with solar energy systems, where 300 sets of 4 kW vanadium batteries were installed. In 1994 vanadium batteries were used in golf carts. In addition, 4-kW vanadium batteries are used in submarines as backup power sources. After the development and transfer of technology, these batteries were further studied in Australia, Japan, and Canada. At present, VRB Power Systems of Canada and Sumitomo Electric of Japan have entered the stage of practical application of all-vanadium liquid flow battery technology.

The VRB-ESS energy storage system was developed by VRB Power Systems based on the VRB technology proposed by researchers from the University of New South Wales. In this system, chemical energy can be converted into electrical energy and vice versa. The chemical energy is stored by the different valence states of the vanadium ions in the vanadium ion sulfuric acid electrolyte solution. The electrolyte flow is controlled by pumping it from two independent plastic storage tanks into the two half battery units, where a proton-exchange membrane (PEM) is used as the battery diaphragm. The electrolyte solution flows parallel to the electrode surface and electrochemical reactions occur. The current is collected and conducted by double-electrode plates. The reaction is reversible, consisting of charging and discharging cycles. The VRB-ESS system consists of two electrolyte storage tanks with vanadium ions in different oxidation states, specifically V(IV)/V(V) at the positive and V(II)/V(III) at the negative oxidation−reduction electrodes. The electrolyte is circulated by the pump between the storage tank and the reactor. The reactor consists of multiple battery packs, each of which has two half batteries separated by a PEM. In a semibattery the electrochemical reaction is carried out on a carbon plate electrode, generating a current to charge and discharge the battery.

Since 1985, Sumitomo Electric Co., Ltd., Japan has cooperated with Kansai Electric Power Plant, Japan to develop vanadium batteries as a dedicated fixed peak storage system for a power station. Mitsubishi Chemical obtained permission from the University of New South Wales in 1993 and in 1994 developed a photoelectric transformation system with vanadium battery energy storage. They built a 50 kW × 50 h vanadium battery system, containing single piles of 2 kW in groups of five to form 10-kW piles. The current density was 100 mA/cm^2, with an output power density of 1.2 kW/cm^2. In 1997 Sumitomo Electric built a 450-kW vanadium battery with a cycle of 170. In 1997 Sumitomo Electric built a 200 kW × 4 h vanadium battery system for peak regulation in the Kashima power plant with a cycle of 650 times. In 1999 Sumitomo Electric built a 450-kW vanadium battery for peak regulation of a power station, which operated for 5 years with 1819 cycles. In 2001 250- and 520-kW vanadium batteries used for studying energy storage systems test were used commercially in Japan; after 8 years of use the 25 kW laboratory vanadium battery pile reached 16,000 cycles. The service life of the battery diaphragm is limited to a certain extent. Other components, including the electrolyte, can be recycled.

With increasing maturity of the technology, vanadium batteries are constantly adapting to different energy storage requirements. In March 2001 the Institute of Applied Energy installed a stable vanadium battery system for storing wind turbine output of AC 170 kW × 6 h. In April 2001 Obayashi Corporation installed an AC 1.5 MW × 1 h vanadium peak-regulation battery. The Obayashi Corporation installed a DC 30 kW × 8 h vanadium battery for a photovoltaic system. In November 2001 Kwansei Gakuin University installed an AC 500 kW × 10 h peak-regulation vanadium battery. In February 2000 the office was equipped with a 100 kW × 8 h balanced-load vanadium cell. In April 2001 the semiconductor factory was equipped with a 1500 kW × 1 h uninterrupted power supply (UPS) balanced-load vanadium cell. In April 2001 the wind power station was equipped with a 170 kW × 6 h vanadium battery to stabilize wind turbine output. The golf club of Obayashi Corporation was equipped with a 30 kW × 8 h vanadium battery in April 2001 to balance their photovoltaic system.

The Chinese Academy of Engineering Physics took the lead in developing vanadium batteries in China in 1995. They successfully developed 20-, 100-, and 500-W vanadium battery model systems and constantly provide breakthroughs in the key vanadium battery technology. They successfully developed a tetravalent vanadium solution, conductive plastic molding, mass production methods, medium-sized battery assemblies, and debugging methods. In 1998 a 500-W vanadium battery prototype was used to drive an electric car, and now an 800 W model has been developed. The main parameters are as follows: monomer number: 10; electrode area: 784 cm^2; thickness of a single cell: 13 mm; electrolyte concentration: 1.5 M $VOSO_4$ + 2 M H_2SO_4; electrolyte volume: 10 L; theoretical capacity: 200 A h; maximum charging current: 80 A (current density: 102 mA/cm^2); charging voltage (50% charging state); 40 A charging voltage of 15.0 V and 80 A charging voltage of 16.5 V; charging capacity: 40 A h; maximum charge current: 80 A (current density: 102 mA/cm^2); charge voltage (50% discharge state): 40 A charge voltage of 11.5 V and 80 A charge voltage of 10 V; discharge capacity: 30 A h; charge and discharge utilization efficiency: $\geq 80\%$; and maximum power of electric reactor: over 800 W.

In 2011 the Chinese Academy of Engineering Physics and Panzhihua YinJiang Jinyong Industry and Trade Co., Ltd. successfully demonstrated vanadium batteries that were integrated with a solar energy system to form a power supply for a 1000 m^2 office building. In 2012 a demonstration project of vanadium batteries was built in Panzhihua for Ganbatang Woodland Protection, with power-driven pumping up to 320 m elevation. In 2013 a vanadium battery demonstration project was built in Panzhihua for Yisarah Travel Protection, with power-driven pumping up to 820 m elevation.

The Dalian Institute of Chemical Physics of the Chinese Academy of Sciences studied ferrochrome liquid flow storage batteries in the late 1990s. In 2000 they began research and development of vanadium flow batteries for energy storage. They have made significant progress in the preparation of electrodes with a double-plate design, distribution of the electrolyte solution, and the design of the battery pack public pipeline design, assembly and battery systems, and integration technology. They successfully developed all-vanadium flow battery storage systems with a size of 10 and 100 kW. The developed all-vanadium oxidation−reduction flow battery modules had a rated output power of 10.1 kW and maximum discharge power of 28.8 kW; the system ran stably, with an energy efficiency of 80.4%. Since July 6, 2007, the research and development of the vanadium flow battery demonstration system showed automatic trouble-free running 105 days in a row (over 2500 h). The demonstration system consisted of three parts: a kilowatt battery module, system control module, and LED screen. With this system configuration the energy storage battery can be used to store the power at night. During the daytime the LED screen can be charged. The energy efficiency of the battery was 87%, and no attenuation was observed during operation. The Dalian Institute of Chemical Physics continues researching the attenuation mechanism of the battery energy storage capacity. To improve the stability of the battery capacity and optimize the design of high-power battery modules, they are developing the third generation of 5-kW battery modules that show a conversion efficiency of over 80% during charge and discharge cycles.

Several Chinese universities, including Tsinghua University, Northeastern University, Beijing University of Science and Technology, Chongqing University, and Central South University, have also studied vanadium batteries. Tsinghua University studied and assembled a 10-kW vanadium battery−electric reactor and found that under a current density of 50 mA/cm^2, the energy efficiency of the battery reached 82.35%. Pan Steel cooperated with Central South University to assemble a 5-kW vanadium battery (first generation) and build a vanadium battery power demonstration system (second generation). Beijing Puneng Century Technology Co., Ltd. was China's first company devoted to vanadium battery technologies; they perform a lot of battery research and have many development personnel. In vanadium battery−electric reactor integration technology, the progress in key material research and development and electrolyte preparation has been made. Considering electric reactor integration, this company has designed and developed a series of 50 W to 5 kW vanadium battery prototypes that have good sealing and simple processing and assembly technology.

16.2 Vanadium battery systems

The VRB is a new type of clean energy storage device that has been applied and tested in the United States, Japan, Australia, and other countries. Compared with lead−acid batteries and nickel hydride batteries that are currently on the market, VRBs have obvious technical advantages, including high power, long lifetime, frequent support large current charge and discharge, and green pollution-free technology. VRBs are mainly used for renewable energy storage integrated with the grid, urban power grid energy storage applications, and remote power supplies, such as for islands and UPS systems. The main technical systems include electric reactor technology, electrolyte technology, and sealing technology.

16.2.1 Electric reactor technology

The reactor is a system that includes the membrane, electrode materials, bipolar plate, flow field design, and sealing components.

16.2.1.1 Membrane

The membrane is considered the core of the vanadium cell, as it generally determines the lifetime and efficiency of the battery. The membrane used in vanadium batteries does not limit the use of a certain kind of film. The key is to use a membrane that is corrosion resistant (providing a long lifetime), has sufficient ion-exchange ability (providing high battery efficiency), and is reliable. The membrane is used after pretreatment. For example, a DuPont Nafion117 ion-exchange membrane was treated with 2% hydrogen peroxide solution and then heated to 80°C for 1 h to remove organic impurities, then cleaned with hot distilled water four to five times, treated with 0.5 M sulfuric acid solution, and heated to 80°C for 1 h to remove metal impurities, cleaned again with hot distilled water 4−5 times, and finally, the diaphragm seal is cleaned, and the membrane is stored in distilled water.

16.2.1.2 Electrode materials

At present, electrode materials for vanadium batteries mainly include graphite felt and carbon felt. The graphite felt is prepared at high temperature and have a high degree of graphitization. Carbon felt is produced at lower temperature and has a lower degree of graphitization. These felts have different conductivities and costs. The specific electrode materials chosen depend on the design of the vanadium battery reactor. Good electrode materials can increase the current density of the vanadium cell and provide some protection against corrosion of the bipolar plate.

16.2.1.3 Bipolar plate

Several requirements are considered when selecting bipolar plate materials, including the corrosion resistance, area, toughness, strength, conductivity, and price. In vanadium batteries, commonly used bipolar plate materials include graphite plates (including hard graphite and soft graphite) and conductive polymers. Although many groups have studied metal composite bipolar plates, only graphite plates and conductive polymers are currently available. The choice of electrode material should consider the specific design of the vanadium battery stack, where testing of products from different manufacturers considering the various requirements for the bipolar plate is required; this can be time-consuming. In particular, the bipolar plate of current vanadium battery stacks is a large proportion of the total cost of vanadium batteries and is the key to industrialization of vanadium batteries.

16.2.1.4 Flow field design

The design of the flow field has a great influence on the performance of vanadium cells and may also affect the lifetime of the electric reactor.

16.2.1.5 Sealing technology

It is important to integrate dozens of single-chip batteries with an area of thousands of square centimeters or even thousands of square centimeters without any leakage and to ensure effective sealing in any scenario for 10 years.

16.2.2 Electrolyte technology

In REDOX flow batteries, energy is stored by chemical changes in the electrolyte. This fluid contains soluble substances, which can store energy by electrochemical oxidation and reduction reactions. The electrolyte determines the storage capacity of the vanadium cell and is an important component of the overall cost. The electrolyte system depends on the composition, where the goal is to improve power density and thermal stability. Second, the production of effective electrolyte at a low cost is important. The composition of the electrolyte affects the membrane lifetime, electrode lifetime, and battery efficiency. The key technology of electrolyte production lies in the source of the raw materials, which determines the production cost of the electrolyte, purification target, purification process, and related environmental issues. The cost of the electrolyte will play an important role in the market competitiveness of vanadium batteries.

16.2.3 Control technology

The control system for vanadium batteries is very important for their long-term stable operation, where the temperature, flow, flow distribution, charge and discharge voltages, and current are all controlled. Compared with fuel cells, the control system of vanadium batteries is relatively simple.

16.2.4 System integration technology

First, system integration technology needs to consider selection of the vanadium battery system and integration with the energy supply, for example, the selection of pumps, pipelines, valves, and controllers to ensure long-term stable operation of the system. Other components include the charger, current and voltage controllers of the high-power system, along with integrated controllers for wind power or solar power generation systems. Since an advantage of using vanadium batteries is their high power, many system integration technologies belong to the group of engineering technologies. To facilitate transportation, electric reactors are usually assembled using large containers as the shell, where the corresponding design considers the weight, volume, channels, pipelines, wire bundles, and various interfaces. Field installation engineering is required for installing the large electrolyte storage tank, connecting the pipelines between the electric reactor and application end, and installing lightning protection, rain protection, waterproofing, and remote monitoring systems.

16.2.5 Technical features

The technical characteristics of vanadium cell are well known. (1) The energy is stored in the electrolyte, and the battery capacity can be increased by increasing the volume of the electrolyte storage tank or the concentration of the electrolyte. That is, vanadium batteries with the same power output can be adjusted as desired according to the demand, which is necessary for large-capacity energy storage applications. (2) The output power is determined by the area of the cell stack; by increasing or decreasing the size of the single cell and series and parallel connections between cells, power batteries have been adjusted to meet different requirements. Recent commercial demonstration of a vanadium battery in the United States achieved a power of 6000 kW. (3) Charging and discharging does not involve solid-phase reactions. The theoretical service life of the electrolyte is unlimited, and hence, it can be used for a long time. During charging of lead−acid batteries, the lead ions in solution are converted into solid lead oxide deposits on the electrode surface. During discharge, solid lead oxide is dissolved in the liquid phase. The charging and discharging processes are accompanied by liquid/solid-phase transformations of the plate material. To ensure the stability of solid lead oxide electrode crystals, the degree of charge/discharge of the battery must be strictly controlled. The change in the electrode structure leads to gradual deterioration of electrochemical performance, which results in a limited number of charge/discharge cycles and the battery life. (4) The reaction speed is fast, and the battery can be started in an instant. It only takes 0.02 s for the charge and discharge states to switch during operation, where the response speed is 1 ms. (5) In theory the ratio of charging time to discharging time is 1:1, although during practical application it is 1.5−1.7:1. This supports frequent high-current charge and discharge cycles, and the depth of the charge and discharge has little influence on battery life. During charging and discharging of the battery, the active materials of the positive and negative electrodes of the battery are in the liquid phase. In nickel−metal hydride, lithium-ion, and other similar batteries, there is no danger of the growth of dendritic crystals on the electrode penetrating the diaphragm and causing short circuit of the battery. (6) The battery reactor can be separated from the electrolyte, and the energy stored in the electrolyte can be preserved for a long time, without self-discharge losses. (7) The energy circulation efficiency is high, and the energy conversion efficiency of charge and discharge can be above 75%. This is much higher than the value of 45% for lead−acid batteries. The electrolyte is not consumed in the process of charge and discharge, and repeated cycling does not affect the battery capacity. (8) The energy reserves can be accurately measured. (9) The use of an electrolyte with the same metal ion at the positive and negative electrodes avoids the problem of cross contamination of the electrolyte and increases the efficiency and lifetime of the battery. (10) The electrolyte flow increases the consistency and reliability of each single cell in the battery pack. (11) The lifetime of the system can be increased by increasing the electrolyte or replacing the electrolyte. By changing the electrolyte, instant recharging can be realized, similar to refueling a car. (12) These batteries have a simple structure, are easily replaced and maintained, and have a low operating cost and small maintenance workload. (13) The system is fully automatic and has closed operation, no noise, no pollution, simple maintenance, and low operating cost. (14) The system can be charged and discharged simultaneously, and the charging and discharging mode can be adjusted according to different application requirements. One or more electrical inputs can be present simultaneously, or similarly, multiple voltages can be output simultaneously. If a series battery is used to discharge the battery, charging can be carried out at different voltages at other parts of the battery stack. (15) The system has a long service life, with a charging and discharging lifetime of more than 10,000 cycles, which is far longer than that of a fixed lead−acid battery. To date, the vanadium battery module with the longest running time was a commercial demonstration developed by VRB Power Systems in Canada, which has been running normally for more than 9 years, with a charging and discharging cycle life of more than 18,000 times. (16) The system has high safety. Vanadium batteries have no potential explosion or fire hazards, and even mixing the anode and cathode electrolytes is not dangerous, although the temperature of the electrolyte increases slightly. (17) In addition to the ionic membrane, the materials are cheap and readily available. No precious metal is required as the electrode catalyst, resulting in low costs. The cost of mass production is even lower than that of lead−acid batteries. (18) The electrolyte can be used for a long time and is considered environmentally friendly as it emits no pollution.

16.2.6 Operating principle

Vanadium is a VB group element with an electron structure of $3d^3s^2$. It can form vanadium ions with four different valence states, that is, V^{2+}, V^{3+}, V^{4+}, and V^{5+}, which have active chemical properties. Valence pairs can be formed in acidic medium with valence states of V^{5+}/V^{4+} and V^{3+}/V^{2+}, where the potential difference between the two electric pairs is 1.255 V. Fig. 16.1 shows the electrolyte color and mutual reactions between the solutions of vanadium ions with different valence. The different valence ions result in different colors, which rely on each other in the redox reaction to form different valence ion. In a dilute

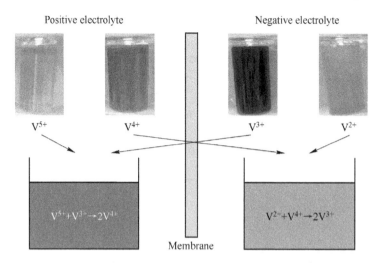

Figure 16.1
Color and reactions between vanadium ions of difference valence in an electrolyte.

sulfate solution, V^{5+} ions form a bright yellow solution, while V^{4+} ions form a bright blue solution, V^{3+} ions form a light green solution, and V^{2+} ions give a bright purple solution.

The electrolyte of redox flow batteries needs to contain metal ions with different valence. The stored electrolyte circulates during charging and discharging. VRBs are a type of redox battery with circulating liquid and active substances. Adopt different vanadium ion solution as the battery is the cathode active material. The solution is pumped from an external storage tank into the cell stack to complete the electrochemical reaction and then returns to the storage tank. The active material in the liquid of a vanadium liquid flow battery is continuously circulated and stored in the external storage tank. The flow of the active material minimizes concentration polarization. The battery capacity depends on the amount of external active material and can be adjusted. The standard potential difference between positive and negative electrodes of vanadium batteries is 1.26 V, and the solution concentration of active substances in both the positive and negative electrodes is 1 mol/L. As the solution concentration increases, the potential difference increases correspondingly.

In an acid medium the potential of vanadium is as follows:

$$0.999 \quad 0.314 \; - \; 0.255 \; - \; 1.17$$

$$VO_2^+ \leftarrow \rightarrow VO^{2+} \leftarrow \rightarrow V^{3+} \leftarrow \rightarrow V^{2+} \leftarrow \rightarrow V$$

Vanadium batteries, commonly known as VRBs, are a type of redox flow battery with a flow of active substances in a liquid. VRBs are a type of flow battery and a chemical energy storage technology. Compared to traditional lead−acid batteries and Ni−Cd batteries, they have a unique design, and their performance is applicable to various industrial applications, such as alternatives to oil machines and standby power supplies. Using VRB technology, the Vanadium Energy Storage System was designed and manufactured. The design and operating characteristics based on VRB were optimized, and the system integrated much intelligent control and automation components to manage the operation of the device. Vanadium batteries convert the energy stored in the electrolyte into electricity by exchanging electrons between two different types of vanadium ions separated by a membrane. The electrolyte is a mixture of sulfuric acid and vanadium and is as acidic as a traditional lead—acid battery. Since this electrochemical reaction is reversible, VRB batteries can be charged or discharged. As the concentration of the two vanadium ions changes during charging and discharging, conversions between electrical and chemical energy occur.

The VRB battery consists of two electrolyte tanks and stacks of cells. An electrolyte tank is used to hold two different electrolytes. Each cell consists of two "half-cells" separated by a diaphragm, where an electrode is used to collect the current. The two different half-cells contain electrolytes with vanadium in different ionic states. Each electrolyte tank is equipped with a pump to deliver the electrolyte to each half-cell via a closed pipe. When the charged electrolyte flows through cell stacks, electrons flow to an external circuit, which is the discharge process. When electrons flow from the outside to the inside of the battery, the reverse process occurs, and the electrolyte in the cell is charged and pumped back into the cell. In VRBs the electrolyte flows between multiple battery cells. The total voltage of 1.2 V is the sum of the voltage of each cell in series. The current density is determined by the surface area of the current collector in the cell, although the supply of current depends on the flow of electrolyte between cells, not the battery layer itself. Fig. 16.2 shows a diagram of the structure of a vanadium redox flow cell.

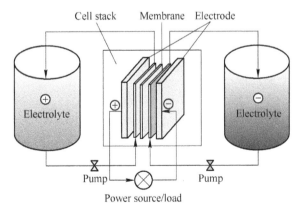

Figure 16.2
Diagram of the structure of a vanadium redox flow cell.

The positive electrolyte consists of solutions of V(V) and V(IV) ions, while the cathode electrolyte consists of solutions of V(III) and V(II) ions. After charging the battery the anode material contains the V(V) ion solution, while the cathode contains the V(II) ion solution. After discharging the battery the positive and negative electrolytes contain V(IV) and V(III) ions, respectively, in solution, due to H^+ conduction. V(V) and V(IV) ions in acidic solution are in the form of VO^{2+} and VO_2^+ ions, respectively. In the vanadium battery the negative reaction can be formulated as follows:

$$\text{Cathode reaction:} VO_2^+ + 2H^+ + e \leftarrow \rightarrow VO^{2+} + H_2O \tag{16.1}$$

$$\text{Anodic reaction:} V^{3+} + e \leftarrow \rightarrow V^{2+} \tag{16.2}$$

One of the most important features of VRB battery technology is that the peak power depends on the total surface area of the battery layer, and the battery power depends on the amount of electrolyte. In conventional lead–acid and nickel-based batteries, the electrodes and electrolytes are placed together, and the power and energy are strongly dependent on the polar plate area and electrolyte capacity. However, this is not the case for VRB batteries, where their electrodes and electrolytes do not have to be placed together. Hence, the storage of energy is not dependent on the outer shell of the battery. In terms of the supply of electricity, different levels of energy can be obtained by providing sufficient electrolyte to the different cell units or cell groups in the battery layer. Charging and discharging of the battery layer does not necessarily require the same voltage. For example, a VRB battery can discharge at various voltages at the battery layer, while charging can be achieved at different voltages at other parts of the battery layer.

All-vanadium flow batteries are a new type of energy storage device with high efficient conversion. The different valences of vanadium ions in solution act as the positive and negative active materials and are stored in separate electrolyte storage tanks. The electrolyte is pumped into the battery stack by an external pump and circulates in a closed loop between the different storage tanks and semibatteries. An ion-exchange membrane is used as the diaphragm of the battery pack. The electrolyte solution flows through the electrode surfaces in parallel and produces the electrochemical reaction. The current is collected by double-electrode plates, and the chemical energy stored in the solution is converted into electrical energy. This reversible process allows vanadium batteries to be successfully charged and discharged repeatedly. Fig. 16.3 shows a schematic diagram of the vanadium cell reaction principle.

16.3 Applications of vanadium batteries

VRB battery systems are economical storage options and can provide large-scale power according to the demand. The main mode is fixed. VRB systems have long lifetimes, low

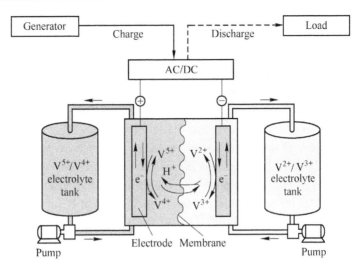

Figure 16.3
Schematic diagram of the vanadium cell reaction principle.

cost, low maintenance requirements, and high efficiency, which enable upscaling of power and energy storage capacity. VRB energy storage systems realize optimal matching of supply and demand by storing electrical energy, which is especially effective for renewable energy suppliers, power grid enterprises, and end users.

All-vanadium flow battery storage system can be applied to each link of the value chain in the power supply and can convert intermittent renewable energy sources, such as wind and solar power, into a stable power output. Such batteries are also optimal solutions for power supplies in remote areas. The fixed investment is deferred in the power grid, and peaks and valleys in the supply can be smoothed. VRB systems can also be used as a backup power source for substations and communication base stations. The VRB system is environmentally friendly and has the lowest impact on the environment of all energy storage technologies. In addition, it does not use lead or cadmium as the main reactant.

16.3.1 Features of vanadium batteries

Vanadium batteries are used as energy storage systems and have the following characteristics: (1) The power output of the battery depends on the size of the stack, where the energy storage capacity depends on the concentration of the electrolyte reserves. Hence, the design is very flexible. To achieve a certain output power, the energy storage capacity can be increased by increasing the electrolyte storage tanks or electrolyte concentration. (2) The active substances in the vanadium battery are in the liquid phase, and the electrolyte ions are only vanadium ions. Hence, there is no common phase change of other batteries during charging and discharging, resulting in a long battery life. (3) Good charging and discharging performance, where deep discharge can occur without damaging the battery. (4) Low self-discharge; when the system is in the closed mode, there is no self-discharge of the electrolyte in the storage tank. (5) Vanadium batteries have great flexibility in site selection, and the system can be fully automatic and closed. In addition, it does not produce pollution, is easy to maintain, and has low operating costs. (6) The battery system has no potential explosion or fire hazards and is highly safe. (7) Most of the battery components are carbon-based materials and engineering polymers, which are abundant materials and easy recovery. No precious metals are needed for the electrode catalyst. (8) The batteries have high energy efficiency, up to 75%−80%, and very high cost performance; (9) Fast start-up speeds are common; if the reactor is full of electrolyte, it can start within 2 min, and only 0.02 s is needed to switch between charging and discharging during operation.

The disadvantages of current all-vanadium liquid flow batteries are as follows. (1) A low energy density. Currently, the energy density of advanced products is only about 40 Wh/kg. Lead−acid batteries have a value of ∼35 Wh/kg; (2) a large area is required because of the low energy density; and (3) the current international range of working temperatures is 5°C−45°C, need to adjust the too high or too low.

16.3.2 Potential vanadium cell market

Vanadium batteries with high power, large capacity, high efficiency, low cost, long service life, low environmental impact and a series of unique advantages are suitable for large-scale energy storage of wind and photovoltaic power. In addition, they have widespread potential applications in power grid peak shaving, distributed power stations, military energy storage, transportation, municipal systems, communication stations, UPS, and other areas.

As the overall vanadium flow battery system can provide continuous stable, safe, and reliable power output when coupled with wind, solar, and other renewable energy power generation systems, the discontinuous and unstable characteristics of these supplies can be solved. When used in power systems, VRBs can adjust the load balance for end users and ensure stable operation of an intelligent power grid. They are used in charging stations of electric cars, which can avoid the impact on the power grid caused by the large current charging of the electric car. In addition, they can be used in industries with high energy consumption to reduce production costs, in telecommunication base stations, and standby power stations in important departments of the state.

16.3.2.1 Wind power generation

Wind power generation is inherently random and intermittent, and its large-scale development is expected to have a significant influence on the safe operation of the power grid load and overall system. Hence, advanced energy storage technologies are required.

Research has shown that if the proportion of installed wind power in the total installed capacity is less than 10%, the security of the power grid can be guaranteed by relying on traditional power grid technology and increasing hydropower and gas units. However, if this proportion exceeds 20%, the peak capacity and safe operation of the power grid will face great challenges. To reduce the impact on the grid, each turbine now needs to be equipped with a backup battery of 4% of its power. In addition, a battery equivalent to about 1% of the power of the turbine is required to protect the air collector in the case of an emergency.

The demand for power grid stability with integrated wind power has become the bottleneck for wind power development. With the rapid development of wind power, the trade-off between increasing wind power output and power grid stability is becoming increasingly prominent. If more than 90% of the power output of wind power needs to be smoothed, energy storage batteries with ~20% of the rated power need to be installed for a wind power plant. To smooth peaks and valleys in the supply from wind power plants, they need to be equipped with a dynamic storage battery equivalent to 40%–50% of the output power. If the turbine is off the grid, a larger proportion of dynamic energy storage is required. Turbines currently use lead–acid batteries, with small capacity, short lifetime, poor stability, laborious maintenance, and high level of pollution. Hence, vanadium batteries can provide significant advantages and could completely replace the existing lead–acid batteries as the main body of dynamic wind energy storage systems.

China's wind power resources have been estimated to be about 1 billion kW to 10-m high turbines. The capacity of onshore wind power resources is 235 million kW, while there are 750 million kW of coastal wind power resources. It expands to a height of 50 m, which is 2–2.5 billion kW. According to the national medium- and long-term energy plans, the wind power installation target is 20 million kW by 2020. By the end of 2008, China had installed 12.15 million kW of wind power, with 6.25 million kW of new capacity that year. It is estimated that the national wind power installation will exceed 100 million kW by 2020, accounting for about 10% of the national electricity generation.

The rapid development of the wind power industry poses severe challenges to the operation and control of the power grid, especially as the majority of China's wind power plants belong to "large-scale centralized development and long-distance transportation." Large-capacity energy storage products are the key factor for solving the trade-off between power grid stability and wind power capacity. Even considering the minimum requirements of wind power regulation, the demand for energy storage batteries will reach 5 million kW in 2020, corresponding to 5% of the wind power storage capacity. If more than 90% of the wind power output needs to be smoothed, the need for storage batteries is expected to more than triple.

16.3.2.2 Photovoltaic power generation

In 2008 the total amount of solar energy installed worldwide reached 15 million kW. That year, the new installed capacity reached over 5.5 million kW, more than 80% of which was in Europe. In 2008 China's solar cell production reached about 2.6 million kW, accounting for 32.9% of the global output. The installed capacity of Chinese photovoltaic cells is 40,000 kW, which is planned to increase to 20 million kW in 2020.

As photovoltaic power generation depends on the sun, and most current large photovoltaic power plants are connected to the grid, there are strict requirements for peak load regulation of the grid. To date, electricity from coal accounts for a large proportion of the generated power in China. Nuclear power and thermal power units cannot be used for peak load regulation. Although hydropower and gas generation provide good peak load regulation of the power grid, the proportion of these sources is low. Increasing the proportion of photovoltaic power generation will result in grid control challenges.

The function of energy storage batteries in photovoltaic power generation systems is to store the electricity generated when the solar cell arrays are illuminated and supply the power to the load at any given time. The basic requirements of photovoltaic power generation for energy storage batteries are (1) low self-discharge rate, (2) long service life, (3) strong deep discharge capacity, (4) high charging efficiency, (5) low or no maintenance requirements, (6) wide operating temperature range, and (7) low cost. Currently, photovoltaic power matching uses lead–acid batteries for energy storage due to the power and capacity they provide. However, their lifetime cannot meet the demands of forming a complete set of photovoltaic power generation. Vanadium batteries will serve as the photovoltaic energy storage battery choice for the future.

16.3.2.3 Power grid peak regulation

The main peak-regulation technology is pumped-storage power stations. As it is necessary to build two reservoirs in the pumped-storage power station, it is not easy to build in flat areas due to the limitations in the geographical conditions. To cope with urban peak loads, significant new investments are being made in the power system every year for the construction of power grid and power reserve capacity. However, the utilization efficiency is very low. In Shanghai, for example, over the period of 2004–06, the city had 183.25 h of peak load, which required a yearly investment of over 20 billion yuan for grid regulation. This form of transmission and distribution has an annual average utilization efficiency of less than 2%.

Vanadium batteries are used to replace pumped-storage power stations. High-capacity energy storage batteries can manage urban peak loads, free of geographical restrictions, require less land area, and have lower maintenance costs. Batteries can also improve the efficiency of energy utilization and save a huge amount of investment for the country.

The effects of land saving, energy saving, and emission reduction are incomparable to those of other peak adjustment measures.

16.3.2.4 Communication base stations

Communication base stations and machine rooms need batteries as backup power supplies, where the time usually cannot be less than 10 h. For the communication operators, safety, stability, reliability, and service life are the most important factors. In this field, vanadium batteries have unmatched advantages over lead–acid batteries.

Vanadium and lead–acid battery technologies are comparable to the obvious advantages in network communication applications: their long life, simple maintenance, high energy storage stability, precision of control, and self-discharge can be advantageous for adjusting the energy storage capacity, with a low overall cost. Vanadium batteries have obvious advantages of low energy storage costs for communication applications. Diesel generators are commonly used in base station power systems in communication networks to provide long periods of power during power outages. Diesel engines account for a large part of the investment in backup power systems and require continuous mechanical maintenance to ensure their reliability. In practical applications the utilization efficiency of diesel engines is very low, so its cost per unit time is relatively high. The lead–acid batteries frequently used in such systems also need constant maintenance due to their tendency to self-discharge.

Vanadium batteries can completely replace the combination of lead–acid batteries and diesel engines in the power system, providing a highly reliable energy storage solution for DC power supply systems. Vanadium batteries are also compatible with the wide geographical distribution and large number of solar cells used in network communication systems. They can replace the lead–acid batteries commonly used in the current solar power systems, while reducing maintenance requirements and costs and increasing productivity.

16.3.2.5 Uninterrupted power supply

The continuous rapid development of China's economy has led to decentralization of user demand for UPS systems, which has led to more industries and enterprises producing such systems. Compared with lead–acid batteries, vanadium batteries have definite advantages considering the power, safety, stability, and service life. Compared with traditional lead–acid or Ni–Cd batteries, vanadium batteries are considered superior as the single energy storage system can provide different voltages at the same time depending on the application requirements.

16.3.2.6 Distributed power stations

One of the flaws of large power grids is that it is difficult to guarantee the quality, efficiency, safety, and reliability of the power supply. Hence, important industries and enterprises often require a double power supply to provide backup power and security. Distributed power stations can reduce or avoid power outages due to power grid failures or accidents. Hospitals, command and control centers, data processing and communication centers, commercial buildings, entertainment centers, key government departments, pharmaceutical and chemical material plants, and the precision manufacturing industry are the focus of distributed power station development. Vanadium batteries are expected to play an important role in the development of distributed power stations.

16.3.2.7 Urban transportation

Currently, automobile exhaust pollution is a leading source of pollution in big cities. There has been consensus to develop energy-saving and environment-friendly electric vehicles to replace traditional fuel vehicles. The large capacity, large current charge, and simultaneous charge and discharge feature of vanadium batteries mean that they are ideal for charging electric vehicles, bicycles, ships, and other equipment.

16.4 Key materials for vanadium batteries

The key materials for vanadium cells include the vanadium electrolyte, membrane, and electrodes. Strict technical control and testing of these components are required during their preparation.

16.4.1 Electrolyte materials

Initially, the vanadium battery electrolyte was made by dissolving $VOSO_4$ directly in H_2SO_4. However, due to the high price of $VOSO_4$, the field began investigating other vanadium compounds, such as V_2O_5 and NH_4VO_3. Currently, there are two methods to prepare the electrolyte: mixed heating and electrolysis. The mixed heating method is suitable for the preparation of 1 mol/L electrolyte, while electrolysis is used for 3–5 mol/L electrolytes.

16.4.1.1 Preparation of vanadium electrolyte

$VOSO_4$ can be dissolved to directly produce the vanadium electrolyte, which can also be obtained by dissolution of high-purity V_2O_5 and V_2O_3 in sulfuric acid. After fine grinding of a V_2O_5 and V_2O_3 mixture with a certain mole ratio, it is added to sulfuric acid to prepare a 3–8 mol/L solution, which is heated and then boiled for several hours. The vanadium compounds are dissolved as much as possible, and then the liquid is cooled and filtered to obtain the vanadium electrolyte.

The specific chemical reactions are as follows:

$$V_2O_3 + 3H_2SO_4 \rightarrow V_2(SO_4)_3 + 3H_2O \tag{16.3}$$

$$V_2O_5 + H_2SO_4 \rightarrow (VO_2)_2SO_4 + H_2O \tag{16.4}$$

$$V_2(SO_4)_3 + (VO_2)_2SO_4 \rightarrow 4VOSO_4 \tag{16.5}$$

Reaction (16.3) shows the dissolution reaction of V_2O_3 in sulfuric acid. Reaction (16.4) shows the V_2O_5 being dissolved in sulfuric acid, while reaction (16.5) shows V(V) and V(III) generated by the V(IV) redox reaction. By controlling the V_2O_5 and V_2O_3 molar ratio, a mixed electrolyte with both V(III) and V(IV) can be obtained. The V_2O_5 and V_2O_3 molar ratio is controlled to 1:1 to obtain the V(IV) electrolyte or 1:3 to obtain a mixture with equal parts of V(IV) and V(III) ions. Vanadium sulfate is a blue crystalline powder, which is easily soluble in water, slightly soluble in ethanol and ether, insoluble in benzene and xylene, and has good fluidity.

16.4.1.2 Testing the vanadium electrolyte composition

The vanadium ion concentrations are characterized using potentiometric titration methods. The V(II), V(III), and V(IV) ions are tested using standard liquid titration for potassium dichromate. In this method a sample of the solution is placed in a beaker and diluted with 10 mol/L phosphoric acid solution to 30 mL. The solution is stirred and then titrated with standard potassium dichromate solution, where the following chemical reactions occur:

$$Cr_2O_7^{2-} + 6V^{2+} + 14H^+ = 2Cr^{3+} + 6V^{3+} + 7H_2O \tag{16.6}$$

$$Cr_2O_7^{2-} + 6V^{3+} + 2H^+ = 2Cr^{3+} + 6VO^{2+} + H_2O \tag{16.7}$$

$$Cr_2O_7^{2-} + 6VO^{2+} + 2H^+ = 2Cr^{3+} + 6VO_2^+ + H_2O \tag{16.8}$$

In the initial stage of reacting V(II), V(III), and V(IV) ions by titration with the standard solution, the solution potential changes slowly, until some valence ions disappear from solution, and an abrupt change in the solution potential is observed. In the case of V(II) and V(III) ions, this abrupt potential change occurs at -150 to 150 mV, while it is 500–900 mV for V(III) and V(IV), and 1000–1100 mV for V(II) and V(IV). These potentials are calculated as follows: for V(II), V(III), and V(IV), where the concentrations are in mol/L:

$$C_{V(II)} = 6V_1C_1/V \tag{16.9}$$

$$C_{V(III)} = 6V_2C_1/V \tag{16.10}$$

$$C_{V(IV)} = 6V_3C_1/V \tag{16.11}$$

where C_1 is the standard solution concentration of potassium dichromate, and V_1, V_2, and V_3 are the titration volumes of V(II), V(III), and V(IV), respectively, required to fully consume the potassium dichromate standard solution. V is the volume of the sample used for the titration experiment.

V(V) is determined using a standard solution of ammonium ferrous sulfate. The test solution is placed in a 50 mL beaker with 10 mL of 10 mol/L phosphoric acid solution, followed by addition of 10 mL of distilled water. It is then stirred well and titrated with a standard ammonium ferrous sulfate solution. The following chemical reaction occurs:

$$VO_2{++} + Fe^{2+} + 2H^+ = VO^{2+} + Fe^{3+} + H_2O \qquad (16.12)$$

The potential slowly decreases during titration of the solution until the V(V) ions are transformed into V(IV) ions and an abrupt change in the solution potential is observed in the range of 750–450 mV. The concentration of V(V) in solution (mol/L) is calculated using:

$$C_{V(V)} = V_4 C_2 / V \qquad (16.13)$$

where C_2 is the Fe(II) concentration of the standard solution, V_4 is the volume of standard solution consumed during titration, and V is the volume of vanadium solution being tested.

The total vanadium concentration is quantified by placing a certain volume of testing solution in a beaker with 10 mL of 10 mol/L phosphate solution, which is diluted with distilled water and stirred until the potential of the solution is stable. Then, 2.5% potassium permanganate solution is added dropwise until the solution obtains a reddish color, followed by the addition of two excess drops and stirring for 10 min. The V(II), V(III), and V(IV) ions in the electrolyte are oxidized and converted into V(V). Then, a few drops of 20% urea solution is added, followed by slow addition of 1% $NaNO_2$ solution until a faded red color is achieved, followed by one additional drop and mixing for 5 min to eliminate the influence of excessive potassium permanganate. Then, the total V (mol/L) concentration is calculated by titrating the standard solution of ammonium ferrous sulfate to the point of abrupt change in the solution potential:

$$C_{V(total)} = V_5 C_2 / V$$

where C_2 and V_5 are the concentration and consumed volume at the end of titration, respectively, of the standard solution of ammonium ferrous sulfate, and V is the volume of the solution to be measured.

The vanadium electrolyte is a sulfuric acid solution containing vanadium ions with different valence. The sulfuric acid in solution is both a conductive medium and also the supporting electrolyte. The acidity has a great influence on the solubility of the vanadium ions; hence, the sulfuric acid root is commonly used to determine the acidity with excess barium chloride, where barium sulfate precipitation occurs. After drying fully the amount of sulfuric acid in the solution is determined, and the acidity of the solution is calculated by deducting the sulfate combined with vanadium.

16.4.1.3 Additives

Sorbitol, methane, sulfonic acid, taurine, and citric acid are used as additives to stabilize the solution, where the additive content is generally 2%. For a total vanadium concentration of 2.08 M, V(II) and V(III) ions account for 50% of the total. Testing once every 10 days showed that the total vanadium ion concentration decreased over time, although the degree of reduction varied. The control electrolyte sample with no additive began to fall faster and then decreased more slowly. This is mainly because the electrolyte is initially supersaturated, which is caused by unstable. For vanadium electrolyte containing sorbitol, methane, sulfonic acid, taurine, and citric acid, the total vanadium electrolyte ion concentration decreased slower than that of the blank sample, where the additives played a role in stabilizing the solutions. The total ion concentration of the vanadium electrolyte containing methane and sulfonic acid showed the slowest decrease, while that of the solution containing citric acid showed the fastest decrease.

Citric acid promoted crystallization of V(III), so they cannot be used as mixed electrolytes. Sorbitol, methane, sulfonic acid, and taurine are relatively stable. Of these, methane and sulfonic acids are the most stable and increase the activity of the solution, making them suitable mixed electrolyte additives. Sorbitol addition has little effect on the conductivity of the electrolyte.

16.4.1.4 Testing the performance of electrochemical solutions

Cyclic voltammetry applies a linear scanning voltage to the electrode. Starting at a voltage φi, the voltage is scanned at a constant rate until it reaches φr, at which point the reverse scan is started, which finishes at φi. During the voltage scan the response current is recorded. Assuming that the electrode reaction in solution is a simple first-order reaction, $O + 1 \leftarrow \rightarrow R$, forward scanning is carried out starting at voltage φi. Initially, only non-Faradaic current flows due to the presence of the double-layer structure. With gradual increase in the potential, the active material O at the electrode surface begins to react, resulting in a Faradaic current. With further increase in the potential, the electrode reaction accelerates, and the current increases rapidly. Meanwhile, the reactant around the electrode is consumed gradually, which reduces the reaction rate and gradually decreases the peak current. When scanning to voltage φr, the direction of the potential scan is reversed, and scanning is performed in the opposite direction. Material R near the electrode is increasingly oxidized and the electrode reaction is accelerated, resulting in a gradual increase in current. As the reaction continues, there is a continuous decrease in R, resulting in a lower reaction rate and a decrease in current flow, generating a peak current. Subsequently, the concentration of R significantly decreases, resulting in a drop in the current. Cyclic voltammetry curves have two important parameters, the peak current ratio *Ipc/Ipa* and the peak electrical potential difference $\varphi pc - \varphi pa$. Experimental results have shown that the change is positively proportional to the concentration of the solution and is independent of the diffusion coefficient and scanning speed.

The AC impedance method is used to characterize the system near the equilibrium potential by using low-amplitude sine waves of the potential (or current) as the disturbance signal. The use of small-amplitude signals has two functions. The first is to ensure an approximately linear relationship between the applied signal and response signal. Hence, the obtained results are simple and convenient for mathematical treatment. Second, mechanical and electrical information can be obtained from impedance measurements.

16.4.2 Membrane materials

The membrane in the vanadium cell must inhibit the mixing of vanadium ions with different valence from the anode and cathode electrolytes while facilitating transfer of hydrogen ions through the membrane. This requires the selection of a cation ion−exchange membrane with good conductivity that allows hydrogen ions to pass easily. Cell membranes are generally cation-exchange membranes, similar to the expensive Nafion membrane (DuPont). Treatment of the cation-exchange membrane can increase the efficiency of the vanadium cell. Perfluorosulfonic acid ion-exchange membranes were first developed by DuPont under the trademark Nafion, which are considered the ion-exchange membranes with the best performance.

16.4.3 Electrode material

The electrode material is a key material as it is the site of the electrochemical reaction. A large specific surface area and good reactivity are required. All-vanadium flow batteries designed to achieve large energy storage capacity must use several single cells in series or parallel. In addition to the electrode, such basic request all electrodes made from bipolar electrodes. Due to the strong oxidizing nature of VO^{2+} and the strongly acidic sulfate, vanadium battery electrode materials must have strong oxidation and acid resistance, high conductivity, high mechanical strength, and good electrochemical activity. Vanadium battery electrode materials are mainly divided into three categories: (1) metal, such as Pb and Ti; (2) carbon, such as graphite, carbon cloth, and carbon felt; and (3) composite materials, such as conductive polymers and polymer composite materials.

16.4.3.1 Graphite felt treatment

In general, polypropylene (PAN)-based graphite felt has a loose porous structure, is abrasion resistant, and is suitable for vanadium battery electrode materials. Due to the manufacturing process of the graphite felt, the surface often contains carbonyl, carboxyl, lactones, quinones, phenols, and hydrocarbon groups that affect water absorption. Hence, the felt requires activation treatment to make the surface hydrophilic and electrochemically active.

Graphite felt was soaked in distilled water for 24 h under heat treatment and the weighed after drying to examine the water absorption. It was then soaked in electrolyte for 24 h and then weighed after drying to investigate the vanadium absorption rate. Graphite felt was also soaked in concentrated sulfuric acid for 24 h and then soaked in distilled water for 24 h. After the graphite felt was removed, it was placed in an oven and fully dried. After soaking in the electrolyte for 24 h and drying, the felt was weighed and the vanadium absorption rate was investigated.

16.4.3.2 Electrochemical performance tests

Cyclic voltammetry of graphite felt electrodes was performed using a three-electrode system. The graphite felt electrode was used as the working electrode, while a platinum electrode was used as an auxiliary electrode, and a saturated calomel electrode was used as a reference electrode. The scanning potential was −1.5 to 1.5 V, and the scanning speed was 1 mV/s. The graphite felt electrode was subjected to AC impedance tests. The graphite felt electrode was immersed in the test solution until the potential stabilized. The initial potential was 0.6 V, and the end potential was 1.1 V. Polarization testing of the graphite felt electrode was carried out. It was immersed in the test solution until the potential stabilized. The equilibrium potential was used as the initial potential. Testing was performed over a frequency range of 0.01−100,000 Hz, with an amplitude of 5 mV.

After heat treatment and acid treatment, the stabilized potential of the graphite felt electrode was smaller than that of the original felt. The reactivity and reversibility of the electrolyte of the positive and negative electrodes were good, the exchange current after heat treatment was large, and the polarization resistance after heat treatment was small (45.5% less than that of the original felt). The polarization resistance after acid treatment was the highest (an increase of 45.5% compared with that of the raw felt). The exchange current is proportional to the velocity constant and inversely proportional to the polarization resistance. Larger exchange currents indicate faster reactions, and hence, lower polarization resistance. The polarization resistance of the graphite felt electrode decreased and the reactivity increased after heat treatment.

The impedance spectra of different graphite felts showed two time constants. The heat-treated graphite felt showed a semicircle in the mid-to-high-frequency regions and a straight line in the mid-to-low-frequency regions. The acid-treated graphite felt showed a smaller compressed arc in the mid-to-high-frequency region, a larger semicircle in the mid-to-low-frequency region, and a straight line in the low-frequency region. The arc of the heat-treated sample was smaller than that of the acid-treated and raw felt.

16.4.3.3 Preparation of integrated composite electrodes

The conductive current collectors are made from graphite, high-density polyethylene, carbon fiber, and conductive carbon black powder as raw materials. They are mixed in the desired proportion at room temperature with anhydrous ethanol (as a dispersant), resin powder, and conductive filler under magnetic stirring. The mixture is loaded into a stainless-steel evaporating dish and then placed in an oven for drying at 100°C for ~20 min and then pressed in a mold. The finished product is placed in a drying oven at 150°C for ~15 min and then removed for air cooling.

The integrated composite electrode consists of two parts, the active electrode material and the conductive collector plate, which are bonded using a conductive adhesive layer. The binder, water, and ethanol are mixed into a paste and applied evenly on the conductive current collector. Then, the processed graphite felt is pressed on top and bonded under an applied pressure at 160°C for 10 min, followed by removal from the oven and cooling to produce the integrated composite electrode. Fig. 16.4 shows a schematic diagram of the composite electrode.

16.5 Vanadium battery assembly

Vanadium batteries typically consist of one or more individual batteries that form a battery pack. The main components of the single battery need to be professionally tested.

16.5.1 Structure of a single vanadium cell

The internal structure of a single vanadium cell is shown in Fig. 16.5. Each battery includes an end plate, collector plate, composite electrode, sealing washer, liquid flow frame with inlet and outlet, and diaphragm. The battery pack can be formed by connecting single batteries in series or parallel. A high output voltage is obtained by series connection, while a high output current can be obtained by parallel connection.

16.5.2 Static cells

In this type of battery the vanadium electrolyte is static. During battery operation the positive and negative halves of the battery are connected using an inert gas that reduces electrolyte polarization. As the negative half of the battery has no oxygen, oxidation of the

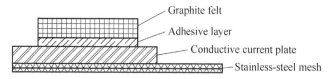

Figure 16.4
Schematic diagram of the composite electrode.

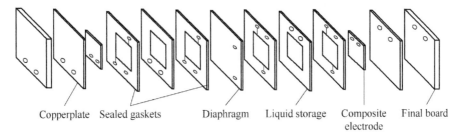

Figure 16.5
Internal structure of a single vanadium cell.

negative electrolyte in the vanadium ion solution is prevented. Since the electrolyte is static, concentration polarization is likely to occur, and the battery volume is limited, so the capacity of the battery cannot be changed.

16.5.3 Flow-type cells

A schematic diagram of a flow-type vanadium redox vanadium cell is shown in Fig. 16.6. The biggest difference between a flow-type and static-type vanadium cells is that the electrolyte is flowing during charging and discharging in the former, which reduces concentration polarization. The capacity of the battery can be adjusted using two electrolyte storage tanks. The electrolyte is driven by two peristaltic pumps, where pump power consumption is 2%−3% of the total battery energy.

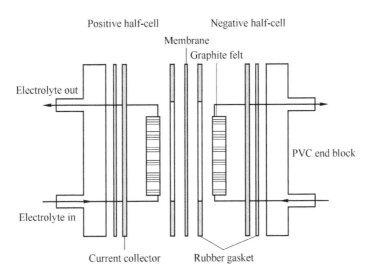

Figure 16.6
Schematic diagram of flow-type vanadium redox cell.

CHAPTER 17

Vanadium carbonitride and vanadium−aluminum alloy

Chapter Outline
17.1 Vanadium carbide 472
    17.1.1 Production process 472
    17.1.2 Reactions 475
    17.1.3 Production system 479
17.2 Vanadium nitride 479
    17.2.1 Production process 480
    17.2.2 Reaction principles 481
    17.2.3 Production system 481
17.3 Nitrided ferrovanadium 482
    17.3.1 Reactions 482
    17.3.2 Preparation methods 482
    17.3.3 Compositional control 483
17.4 Aluminum−vanadium alloy 483
    17.4.1 Reactions 483
    17.4.2 Production methods 484

Vanadium oxide reduction follows a step-by-step transformation process. The stability of vanadium oxide follows the order: $VO > V_2O_3 > VO_2 > V_2O_5$. The critical reduction temperature is calculated from the thermodynamics for VO (1866.9K), V_2O_3 (1464.9K), VO_2 (600.5K), and V_2O_5 (229.9K). Vanadium pentoxide is used as a raw material for producing vanadium carbonitride, where the reduction, carbonization, and nitriding processes are interwoven. The general test production process is designed considering the temperature, atmosphere, carbon content, desired products, and reactor. Sometimes, part of the reduction function is separated, where V_2O_3 and V_2O_4 are produced. Vanadium carbonitride is prepared in an integrated reactor, or sometimes in a separate reactor. Usually, V_2O_5 is first reduced to VO_2, then VO_2 is reduced to V_2O_3, and finally V_2O_3 is reduced to VO. Nonmetallic reductants for vanadium oxides include carbon, hydrogen, gas, silicon, and natural gas. The carbon thermal reduction processes are classified into carbon thermal reduction of vanadium oxide, carbon thermal reduction of vanadium chloride, and vacuum carbon thermal reduction. In general, solid carbon reduction is used, with graphite or carbon powder as the reductant. In addition, gas reduction can be used with gas or natural gas as the reductant. Coke oven gas mainly contains H_2, CH_4, and CO and contains more complex minor components than natural gas, which depend on the coal used for coking, the coking process, and many other factors.

Vanadium nitride is produced from V-containing materials, such as vanadium oxides (V_2O_5 and V_2O_3), ammonia vanadate (NH_4VO_3), and ammonia polyvanadate. Carbon, hydrogen, ammonia, and CO are used as the reducing agents, and the reduction is performed at high temperature under vacuum. Nitrogen or ammonia is added for nitrogenization. The vanadium nitride preparation method is classified according to the preparation system and material, the use of high-temperature vacuum, or high-temperature antivacuum conditions. These categories are further classified depending on the use of different raw materials for the preparation of vanadium nitride using solid reactants, Here is the raw material classification, which roughly includes C and V_2O_5, C and V_2O_3, C and NH_4VO_3.

During the preparation of vanadium–aluminum alloy, V_2O_5 is used as a raw material, with aluminum as the reducing agent. The process methods include aluminothermic reduction, aluminothermic reduction and vacuum induction melting, and auto-combustion methods, which produce products with high purity, grade, and density [7,9,14,25,80,82–85,87].

17.1 Vanadium carbide

Vanadium and carbon combine to form two types of carbides, VC and V_2C. VC [w(C) = 19.08%] contains 43–49 at.% C. Vanadium carbide (VC) has a face-centered cubic structure similar to that of NaCl. V_2C has w(C) = 10.54% and contains 29.1–33.3 at.% C, which forms a dark crystal with a hexagonal-close-packed-like lattice. VC has a density of 5.649 g/cm^3 and melting point of 2830°C–2648°C. V_2C has a density of 5.665 g/cm^3, melting point of 2200°C, and is slightly harder than quartz. $V_{84}C_{13}$ products have a density of 4.0 t/m^3, heap weight of 1.92 t/m^3, and lumpiness of 38 × 31 × 19 mm.

17.1.1 Production process

Different VC production methods are used in China and around the world, which are generally classified depending on the temperature, atmosphere, carbon addition, required products, and reactor. Sometimes, part of the reduction process is separated, where V_2O_3 and V_2O_4 are produced. VC can be prepared in an integrated reactor, or in a separate reactor.

17.1.1.1 Main methods

The main methods are described here: (1) vanadium trioxide, Fe powder, and iron scale are used as raw materials, with carbon powder as the reducing agent. VC is formed at high temperature under vacuum, which is cooled by argon gas or in the vacuum furnace. (2) V_2O_5 is used as the raw material and VC_xO_y is produced in a rotary kiln, followed by treatment at high temperature under vacuum to produce VC, which is cooled by inert gas. (3) Vanadium pentoxide or vanadium trioxide is used as the raw material with carbon black as the reducing agent. VC is produced in an Ar atmosphere at high temperature in a small rotary kiln or crucible. (4) VC is prepared by reducing V_2O_5 at high temperature with carbon (charcoal, coal coke, or electrode) as the reducing agent. (5) The raw material is vanadium pentoxide, which is used to produce the carbide via propane reduction with nitrogen plasma. (6) Beijing University of Science and Technology uses V_2O_5 and activated carbon as raw materials to produce VC in a high-temperature vacuum molybdenum wire furnace. The former Jinzhou Ferroalloy factory used a vacuum method to manufacture VC in a vertical furnace using ammonium polyvanadate and carbon powder as raw materials.

VC is a high-carbon additive containing vanadium and iron. It is used in the production of steel and V-containing alloys and has a composition of 83%–86% V, 10.5%–13% C, and 2%–3% Fe. During production, natural gas is supplied to a rotary kiln, where V_2O_5 is reduced to V_2O_4 at 600°C, which is then reduced in a rotary kiln at 1000°C using gas to produce vanadium carbon oxides. Vanadium carbohydrate is added to coke or graphite in an appropriate amount, where the calculated carbon content is over 1% compared with the calculated carbon oxygen reaction value. The mixture of vanadium carbon oxide and graphite is pressed into blocks. In the vacuum furnace, it is heated to 1000°C and converted into VC. Typical components of VC produced by United Carbide (United States) are 84.5% V, 12.25% C, 0.05% Si, 0.005% Al, 0.004% S, 0.004% P, and 2.5% Fe. The prepared alloy is usually a flat cake shape with a density of 4.58 g/cm^3. The VC is packaged into bags containing an equivalent of 25 lb of vanadium and 3.7 lb of carbon which can be added directly to furnaces or steel ladles without analysis. Compared with ferrovanadium, VC is cheaper, dissolves quickly after adding the liquid steel, and gives a higher vanadium yield.

17.1.1.2 Carbon addition calculation

Pure VC is prepared by carburizing vanadium hydride powder under vacuum. At 800°C, C:V = 0.72–0.87 (molar ratio), and the chemical reaction is:

$$V_2O_3 + 4C = V_2C + 3CO \tag{17.1}$$

The actual carbon content is 90%–110% of the theoretical amount.

17.1.1.3 Material and process parameter requirements

In Union Carbide, VC is produced using a vacuum furnace with V_2O_3, iron powder, and carbon powder as raw materials, which is heated to 1350°C, and kept at this temperature for 60 h under vacuum. South Africa's Vametco mineral company is a subsidiary of Strategic Minerals, United States (STRATCOR), which specializes in the production of VC. They use natural gas at 600°C to reduce V_2O_5 into V_2O_4 in a rotary kiln. Then, in another rotary kiln at 1000°C, V_2O_4 is reduced into VC_xO_y, which is then mixed with carbon or graphite and pressed into a cake, and heated to 1000°C in a vacuum furnace to obtain VC. The components of such VC products are shown in Table 17.1.

US patent US3334992 describes the production of VC using a vacuum method. The raw materials should have a particle size <0.2 mm; generally, the particle sizes of vanadium trioxide is <0.2 mm, iron powder is <0.15 mm, iron scale is <0.074 mm, and carbon reductant powder is <0.074 mm. In addition, 1.5%–2% of a special adhesive and 15%–20% water is added, and the mixture is formed into briquettes of 51 × 51 × 38 mm, which are dried at 120°C to remove 95% of the water. The briquettes are then placed in a vacuum furnace, evacuated to 8–27 Pa, heated up to 1385°C, and treated for 18–60 h at a furnace pressure of 67–1600 Pa. When the pressure in the furnace is 8–24 Pa, the heating is stopped, and the furnace is cooled using argon gas or under vacuum. The VC product contains w(V) 85%, w(C) 12%, and w(N) 0.1%.

US patent US3342553 describes the production of VC using vacuum reduction, where flake V_2O_5 is used as a raw material, with a reduction temperature of 565°C–621°C and reduction time of 60–90 min. Then, VC_xO_y ($x = 0.4$–0.6 and $y = 0.4$–0.8) is reduced at 1040°C–1100°C for 100–180 min, which is then cooled in a nonoxidizing atmosphere. With no more than 10% of the theoretical carbon, carbon black, or graphite particles with a size <0.043 mm is >60%. A vacuum heating temperature of 1370°C–1483°C is used with continuous exhaust, where the vacuum is used to maintain the pressure. When the pressure reaches 7 Pa, heating is stopped and an inert gas is used for cooling. The VC products have a carbon content of 8%–15% and density of 4.0–4.5 g/cm³.

US patent describes the preparation of VC under a protective atmosphere using vanadium pentoxide or vanadium trioxide as raw materials, where their granularity is controlled to <0.2 mm. Carbon black is used as the reductant, with a granularity <0.074 mm. All raw materials are mixed and pressed into a cake and then placed in a crucible or small rotary kiln into which argon or other inert gases are injected. The furnace is heated to 1400°C–1800°C, and VC products are obtained with w(V) 77.31%–85.68%, w(C) 10.22%–21.66%, w(O) 1.58%–4.48%, and w(Fe) 1.7%–2.57%. The addition of various amounts of Fe powder (particle size <0.42 mm) can improve the strength and particle size of products to varying degrees.

Table 17.1: Components of vanadium carbide products.

Component	V	C	Al	Si	P	S	Mn
Mass fraction (w/%)	82–86	10.5–14.5	<0.1	<0.1	<0.05	<0.1	<0.05

US patent US3565610 describes a carbon-reduction method using carbon (charcoal, coal tar, or electrode) as a reductant, where V_2O_5 is reduced at high temperature to produce VC (V_2C) at a reduction temperature of 1200°C–1400°C. The VC product contains w(V), 70%–85%, 5%–20% w(C), and 2%–10% w(O).

Russia uses nitrogen plasma to produce VC by propane reduction, with a flow rate of 3.0 g/min of pure nitrogen carrier gas. The V_2O_5 powder is mixed with propane following the required proportions. The mixture travels along the axis of the reactor and is placed within the arc heater plasma collision zone at 2600K–3500K, where VC is generated. When the temperature is lower than 2600K, nitrogen is incorporated in the VC, generating VCN.

Beijing University of Science and Technology used a vacuum method for producing VC. Vanadium pentoxide and activated carbon are used as raw materials. In a high-temperature vacuum molybdenum wire furnace, VC (V_2C) is produced from a mixture of $r_{V_2O_5}:r_C = 1:5.5$ (mole ratio) at a reduction temperature of 1673K and pressure of 1.33 Pa. The V_2O_5 is evenly mixed with C, along with polyvinyl alcohol binder and 20% water, which is hydraulically pressed into pieces. The dry pieces are loaded in a graphite crucible and placed in the vacuum furnace. The obtained VC contains w(V) 86.41%, w(C) 7.10%, and w(O) 6.48%.

17.1.2 Reactions

Vanadium pentoxide is toxic, with a low melting point (650°C–690°C), and is volatile above 700°C. V_2O_5 is used as a raw material during carbonization at temperatures above 700°C. In general, due to volatilization losses, vacuum is required. Vanadium pentoxide is reduced to low-valence vanadium oxide for carbonization. V_2O_5 is reduced at low temperature to produce VO (melting point 1790°C), V_2O_3 (melting point 1970°C–2070°C), and VO_2 (melting point 1545°C–1967°C), which can effectively control the reduction and carbonization processes.

17.1.2.1 Direct reduction

Carbon and vanadium oxide are mixed with raw materials and direct reduction reaction is performed. The reactions involved in the carbon thermal direct reduction of vanadium oxide are as follows:

$$V_2O_5 + C = 2VO_2 + CO\uparrow \tag{17.2}$$

$$\Delta G_T^\theta(C) = 49,070 - 213.42\ T\,(J/mol)$$

$$2VO_2 + C = 2V_2O_3 + CO\uparrow \tag{17.3}$$

$$\Delta G_T^\theta(C) = 95,300 - 158.68T\,(J/mol)$$

$$V_2O_3 + C = 2VO + CO\uparrow \tag{17.4}$$

$$\Delta G_T^\theta(C) = 239,100 - 163.22T\,(\text{J/mol})$$

$$VO + C = 2V + CO\uparrow \tag{17.5}$$

$$\Delta G_T^\theta(C) = 310,300 - 166.21T\,(\text{J/mol})$$

$$V_2O_5 + 7C = 2VC + 5CO\uparrow \tag{17.6}$$

$$\Delta G_T^\theta(C) = 79,824 - 145.64T\,(\text{J/mol})$$

Important steps:

$$V_2O_5(s) + C(s) = V_2O_4(s) + CO(g)\uparrow \tag{17.7}$$

$$\Delta G_T^\theta(C) = 4940 - 41.55T + RT\cdot\ln(p_{CO}/p^0)\,(\text{J/mol})(345K - 943K)$$

$$VO(s) + 2C(s) = VC(s) + CO(g)\uparrow \tag{17.8}$$

$$\Delta G_T^\theta(C) = 186,697 - 151.42T + RT\cdot\ln(p_{CO}/p^0)\,(\text{J/mol})$$

17.1.2.2 Indirect reduction

Since the carbon thermal reduction reaction of vanadium oxide generates CO, indirect carbon thermal reduction and chemical reactions between vanadium oxide and CO occur:

$$V_2O_5(s) + CO(g) = V_2O_4(s) + CO_2(g)\uparrow \tag{17.9}$$

$$\Delta G_T^\theta(C) = 69,250 - 173.63T + RT\cdot\ln\left(p_{CO_2}^2/p^0\right)(\text{J/mol})$$

$$VO(s) + 3CO(g) = VC(s) + 2CO_2(g)\uparrow \tag{17.10}$$

$$\Delta G_T^\theta(C) = 139,380 - 188.40T + RT\cdot\ln\left(p_{CO_2}^2/p_{CO}^2\right)(\text{J/mol})$$

The initial temperature of vanadium oxide reduction is related to the p_{CO} and p_{CO_2}. According to the Burdur reaction,

$$C = CO + CO_2 \tag{17.11}$$

For temperatures above 710°C, CO_2 cannot exist as it is not stable. The reaction to the left can occur at high temperature, providing the excess carbon, p_{CO} is far greater than the p_{CO_2}, and the indirect reaction can proceed smoothly.

17.1.2.3 Vanadium carbide prepared by V_2O_3 reduction

V_2O_3 is used to produce VC. The overall chemical reaction is as follows:

$$V_2O_3 + 5C = 2VC + 3CO \quad (17.12)$$

$$\Delta G^\theta = 655,500 - 475.68T$$

$$\Delta G^\theta = 0 \; T = 1378K = 1105°C$$

$$\Delta G_T^\theta = 655,500 + (57.428 \; \lg p_{CO} - 475.68)T$$

The relationship between p_{CO} and the initial temperature of VC produced by V_2O_3 is shown in Table 17.2.

V_2O_3 is used to produce V_2C. The overall chemical reaction is as follows:

$$V_2O_3 + 4C = V_2C + 3CO \quad (17.13)$$

$$\Delta G^\theta = 713,300 - 491.49T$$

$$\Delta G^\theta = 0 \; T = 1451K = 1178°C$$

$$\Delta G_T^\theta = 713,300 + (57.428 \; \lg p_{CO} - 491.49)T$$

The relationship between p_{CO} and the initial temperature of V_2C produced by V_2O_3 is shown in Table 17.3.

17.1.2.4 Carbon thermal reduction of vanadium oxide

Vanadium oxide that has been subjected to a certain degree of carbon reduction can form stable VCs (VC or VC_2). The stability of the CO is higher than that of vanadium oxides,

Table 17.2: Relationship between p_{CO} and initial temperature of vanadium carbide produced by V_2O_3.

p_{CO} (Pa)	Initial temperature, T (K)
1.013×10^5	1378
1.013×10^4	1230
1.013×10^3	1110
1.013×10^2	1012
1.013×10^1	929
1.013×10^0	859

Table 17.3: Relationship between p_{CO} and initial temperature of V_2C produced by V_2O_3.

p_{CO} (Pa)	Initial temperature, T (K)
1.013×10^5	1451
1.013×10^4	1299
1.013×10^3	1176
1.013×10^2	1075
1.013×10^1	989
1.013×10^0	916

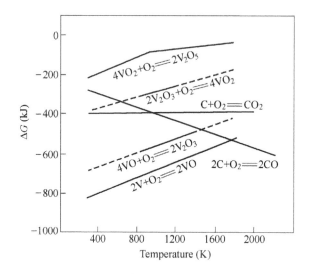

Figure 17.1
Formation free energy vanadium oxide and carbon oxide.

where the carbon thermal reduction of vanadium oxide follows the sequence $V_2O_5 \rightarrow V_2O_4 \rightarrow V_2O_3 \rightarrow VO \rightarrow V(O)s \rightarrow V$. The formation free energy of vanadium oxide and carbon oxide is as shown in Fig. 17.1.

The carbothermal reduction of VC follows the order: $VC \rightarrow VC2 \rightarrow V(C)s \rightarrow V$. The basic carbon thermal reduction reaction is as follows:

$$1/y \ V_xO_y + C = x/yV + CO \tag{17.14}$$

When the temperature is below 1000°C, the chemical reactions are as follows:

$$V_2O_5 + CO = 2VO_2 + CO_2 \tag{17.15}$$

$$2VO_2 + CO = V_2O_3 + CO_2 \tag{17.16}$$

When the temperature is above 1000°C, the chemical reactions are as follows:

$$V_2O_3 + 5C = 2VC + 3CO \quad (17.17)$$

$$2V_2O_3 + VC = 5VO + CO \quad (17.18)$$

$$VO + 3VC = 2V_2C + CO \quad (17.19)$$

$$VO + V_2C = 3V + CO \quad (17.20)$$

17.1.2.5 Vacuum carbon thermal reduction method

Vacuum carbon thermal reduction is one of the important methods of preparing VC. V_2O_5 is first reduced to V_2O_3 using hydrogen, and then mixed with carbon black. The carbon-reduction reaction of VO is as follows:

$$VO(s) + C(s) = V(s) + CO(g) \quad (17.21)$$

$$2V(s) + O_2(g) = 2VO(s)(1500-2000k) \quad (17.22)$$

$$\Delta G_{22}^{\theta} = -803,328 + 148.78T \, (J/mol)$$

$$2C(s) + O_2(g) = 2CO(g) \quad (17.23)$$

$$\Delta G_{23}^{\theta} = -225,754 - 173.028T \, (J/mol)$$

When V_2O_5 or V_2O_3 is reduced using a carbonaceous reductant under standard conditions, the highest reduction temperature $T_{initial} = 1794.77K$ (1521.77°C). To reduce $T_{initial}$ the pressure of the gas phase in the system can be reduced to reduce the partial pressure of CO (P_{CO}). Different P_{CO} values have different $T_{initial}$ values.

17.1.3 Production system

VC production generally uses small systems that mainly include a rotary kiln, small shaft furnace, TBY furnace, and/or vacuum furnace. The auxiliary system mainly includes power, measurement, mixing, control, hydraulic, block, bunker, and packing systems.

17.2 Vanadium nitride

Vanadium nitride is produced from V-containing materials, such as vanadium oxides (V_2O_5 and V_2O_3), ammonia vanadate (NH_4VO_3), and ammonia polyvanadate. Carbon, hydrogen, ammonia, and CO are used as the reducing agents, where the reduction is performed at high temperature under vacuum, and nitrogen or ammonia is added for nitrogenization. The vanadium nitride preparation method is classified according to the preparation system and material, and the use of high-temperature vacuum, or high-temperature antivacuum conditions. These categories are further classified depending on the use of different raw materials, such as C and V_2O_5, C and V_2O_3, and NH_4VO_3 for the preparation of vanadium nitride with a solid reactant.

Vanadium nitrogen alloy is a new type of alloy additive. Vanadium nitride has two crystal structures. The first is V_3N, with a hexagonal crystal structure, high hardness, microhardness of about 1900 HV, and unknown melting point. The second is VN, with a face-centered cubic crystal structure, microhardness of ~1520 HV, and melting point of 2360°C.

17.2.1 Production process

Vanadium nitride is produced via vanadium oxide reduction and carbonization. Union Carbide uses three methods to produce vanadium nitride. First, V_2O_3, iron powder, and carbon powder are used as raw material, which are heated to 1350°C for 60 h in a vacuum furnace to obtain VC. The carbide is then heated to 1100°C, where nitrogen gas is supplied for nitriding, and the furnace is cooled in a nitrogen atmosphere. The vanadium nitride contains 78.7% V, 10.5% C, and 7.3% N. In the second process a mixture of V_2O_5 and C is heated in a vacuum furnace to 1100°C–1500°C, evacuated, and then nitrogen gas is supplied for nitriding. This process is repeated several times to produce vanadium nitride with C and O contents below 2%. In the third process, vanadium compounds (V_2O_3 or V_2O_5 or NH_4VO_3) are reduced in NH_3 or N_2 and H_2 atmospheres. Some material is produced at high temperature with nitrogen vanadium, and oxygen, which is mixed with the carbide again and heated in an inert or nitrogen atmosphere in the vacuum furnace to obtain vanadium nitride with 7% C.

Vanadium nitride was prepared by Bhabha Atomic Research Institute, India at a high temperature of 1500°C. In the Netherlands Institute of Metallurgy, ammonium vanadate or V_2O_5 is used as the raw material, which is reduced in a fluidized bed or rotary tube at a temperature of 800°C–1200°C. The obtained vanadium nitride contains 74.2%–78.7% V, 4.2%–16.2% N, and 6.7%–18.0% C.

Traditional high-temperature processes for vacuum carbon thermal reduction to produce vanadium nitride have many limitations: (1) a reaction cycle of dozens of hours leads to low labor productivity; (2) large energy consumption due to the long reaction cycle; (3) the long high-temperature process results in large loss of equipment; and (4) the one-off investment in automation equipment is large, the production cost is high, and the product competitiveness is low. A new process for vanadium nitride production aims to omit vacuum and reduce the adaptability of the process.

17.2.2 Reaction principles

$$VO(s) + C(s) + \frac{1}{2}N_2(g) = VN(g) + CO(g) \quad (17.24)$$

$$\Delta G_T^\theta(C) = -64,830 - 7.36T + RT \cdot (\ln p_{CO}) - \frac{1}{2}\ln p_{N_2} \, (J/mol)$$

Vanadium nitride is a common nitride for nitriding steel. V_3N has a hexagonal crystal structure, high hardness, microhardness of about 1900 HV, and unknown melting point. VN has a face-centered cubic crystal structure, microhardness of $\sim$1520 HV, and melting point of 2360°C. Both nitrides have high wear resistance. Hence, V-containing steel after nitriding treatment has greatly improved wear resistance compared to steel.

When the pressed blocks of V–C–O compounds and graphite are combined with nitrogen in the furnace during sintering, vanadium nitride is produced with a composition of 78%–82% V, 11%–12% C, and $\geq$6.0% N. The carbides in the V–C system are V_2C and VC. When the carbon content is high, VC is the stable phase, along with a very small amount of vanadium metal. The reaction for producing VC and its standard free energy are as follows:

$$V(s) + C(s) = VC(s) \quad (17.25)$$

$$\Delta G_{13}^\theta = -102,090 + 9.581T \, (J/mol)$$

It can be seen that the carbonization of vanadium metal is an exothermic process.

In the V–N system, vanadium nitrides (VN_x) have a complex configuration, where $x = 0.5-1$. The specific x value is dependent on the nitrogen partial pressure P_{N2} and temperature T°C of the system. At a certain T°C, x increases with increasing P_{N2}. When P_{N2} is fixed and T°C increases, x decreases. It is generally thought that the stable phase in the V–N system is VN. The reaction for forming VN and its standard free energy ΔG_{14}^θ are:

$$V(s) + \frac{1}{2}N_2(g) = VN(s) \quad (17.26)$$

$$\Delta G_{14}^\theta = -214,639 + 82.425T \, (J/mol)$$

It can be seen that the nitriding process of vanadium is also an exothermic process.

17.2.3 Production system

VN production generally uses small systems that mainly include a rotary kiln, small shaft furnace, TBY furnace, and/or vacuum furnace. The auxiliary system mainly includes power, measurement, mixing, control, hydraulic, block, bunker, and packing systems.

17.3 Nitrided ferrovanadium

Nitrided ferrovanadium can be obtained by alloy nitriding methods, which include the addition of either solid or liquid nitrogen sources. The methods of solid-state nitrogen addition include drum, groove, decarburization, and reduction steps, while liquid nitrogen addition involves blowing–washing melt, metal thermal reduction, and blow-washing surface method, as well as the CBC method of nitrating (high-temperature synthesis by self-diffusion) during sintering. The CBC method of preparing nitrided ferrovanadium uses a closed container into which liquid nitrogen is bubbled at high pressure ($P'_{N2} = 10^5$ Pa). Heat emitted from the nitriding reaction helps the ferrovanadium powder form nitrides. The degree of nitriding decreases with increasing temperature. The CBC method does not require high temperature, although the nitrogen source needs to be controlled according to stoichiometric compositional requirements. The nitrided ferrovanadium produced from the high-temperature self-diffusion method has high nitrogen content and the alloy has high density. In this process the nitriding time is short, the power consumption is low, and it is possible to accurately control the nitrogen content at a high level.

17.3.1 Reactions

The formation of nitrides is based on the combustion of pure metal in a nitrogen atmosphere, where the equation is as follows:

$$XR + \frac{y}{2}\{N_2\} = R_xN_y \tag{17.27}$$

$$\frac{Lg\alpha R_xN_y}{\left(\alpha_R^x \cdot \alpha_N^{y/2}\right)} = \frac{-\Delta G_T^\theta}{(2.3RT)} \tag{17.28}$$

When $\alpha_{R_xN_y} = 1$, $p_{N2} = 10^5$ Pa, $p'_{N2} = 10^7$ Pa, the equilibrium constant for VN generation is calculated from the Gibbs free energy change ΔG_T^θ using lg $K = 9134/T - 4.38$. When $p_{N2} = 10^5$ Pa, the absorption temperature is 2085K and when $p'_{N2} = 10^7$ Pa, the absorption temperature is 2702K; $VN_{0.5}$ is generated when lg $K = 6780/T - 2.32$; $p_{N2} = 10^5$ Pa, the absorption temperature is 2992K; and $p'_{N2} = 10^7$ Pa, and the absorption temperature is 3725K.

17.3.2 Preparation methods

The equipment used in the production of nitrided ferrovanadium by CBC includes (1) a crusher for crushing the vanadium ferroalloy to a certain size, (2) air flow mill for further crushing of the alloy, (3) classifier and dust separator to obtain alloy powder with a uniform particle size; (4) storage hopper and discharge hopper, (5) a CBC reactor where the nitriding reaction is carried out to produce the nitrided alloy powder, and (6) compression system to pressurize the nitrogen.

Nitriding vanadium is an exothermic reaction. The initial layer of the material is stratified and nitrided by self-diffusion combustion. The high-temperature process helps part of the product melt and to accelerate the tight knot. Material with a uniform structure is obtained with no nitrogen gradient along the cross section. Nitrogen saturation occurs because of heat released by the nitriding reaction. The nitrating reaction of the initial product is a synchronous and instantaneous process. In a device with a capacity of 0.2 m^3 the nitrogen flow is 0.5 t/h, where the closed conditions avoid material loss and contamination.

17.3.3 Compositional control

Nitrided ferrovanadium products produced using the CBC method have a high density (6.2–7.0 g/cm^3); high nitrogen content of 10%–11%, and low porosity of 1%–3%.

17.4 Aluminum–vanadium alloy

Aluminum–vanadium alloy is silver-gray in color with a metallic luster and occurs in blocks. As the vanadium content in the alloy increases, its metallic luster, hardness, and oxygen content also increase. When the content of V is more than 85%, the product is not easy to break and a surface oxide film can easily be formed for long-term storage. The grain-size range of the VAl55–VAl65 products is 0.25–50.0 mm. The grain size range of VAl75–VAl85 is 1.00–100.0 mm.

17.4.1 Reactions

When Al is used as the reducing agent, the reduction reactions with vanadium oxide are as follows:

$$3V_2O_5 + 2Al = 3V_2O_4 + Al_2O_3 \quad (17.29)$$

$$3V_2O_4 + 2Al = 3V_2O_3 + Al_2O_3 \quad (17.30)$$

$$3V_2O_3 + 2Al = 6VO + Al_2O_3 \quad (17.31)$$

$$3VO + 2Al = 3V + Al_2O_3 \quad (17.32)$$

$$3V_2O_5 + 10Al = 6V + 5Al_2O_3 \quad (17.33)$$

Vanadium trioxide is also a stable vanadium oxide and can be used as a raw material for aluminum–vanadium alloy production. The reduction reaction of vanadium oxide is as follows:

$$3V_2O_3 + 2Al = 2V + Al_2O_3 \quad (17.34)$$

When potassium chlorate is used as catalytic thermostat, the reduction reaction of vanadium oxide is as follows:

$$KClO_3 + 2Al = KCl + Al_2O_3 \tag{17.35}$$

17.4.2 Production methods

Vanadium−aluminum alloy is a vanadium additive used in the production of titanium alloys, such as Ti−6Al−4V. It is divided into three grades according to the vanadium content: 50%, 65%, and 85%, where the remainder is aluminum. The gas content in V−Al alloys is low, and other impurities, such as Fe, Si, C, and B, should meet the requirements of titanium alloy. High-purity vanadium pentoxide is produced for use in the fabrication of aluminum alloy by purifying the industrial product to reduce the impurity content. The Al raw material is high-purity Al. The production site should be kept clean to avoid contamination.

First, 85% V−Al alloy is produced using the aluminum thermal method. Specifically, 85% V−Al alloy and aluminum are smelted to produce V−Al alloy with specified vanadium content in a 500-kg vacuum induction furnace. In addition to degassing the purpose of remelting is to homogenize the composition of the alloy, where the variation in vanadium content should be <0.5. During degassing the pressure inside the furnace is 0.266 Pa (2×10^{-3} Torr). After crushing and screening of the alloy ingot, undersize do not content, V−Al alloy particles with a grain size of 0.2−6 mm are removed by magnetic separation. The material is inspected by ultraviolet and fluorescence examination, alloy particles with nonmetallic inclusions are removed, and the product is packaged for sale.

17.4.2.1 Aluminothermic process

Vanadium pentoxide, Al, and slagging agent (CaF_2) are used as raw materials, which are mixed according to thermal mixing ingredients after the accurate match. These materials are placed in a reactor with a metal shell and refractory lining, where aluminum thermal reduction is carried out in air. Vanadium−aluminum alloy and reducing slag are produced after cooling. This method has a simple process, raw materials, and equipment; although the product uniformity and quality are low, high levels of some impurities can occur, and it produces a high level of environmental pollution.

17.4.2.2 Two-step method

The thermite method is combined with vacuum induction melting method in a two-step method. In 1957, Germany began studying V−Al alloy and established the two-step process in the 1960s. An Al−V intermediate alloy is produced by aluminum thermal method, which is then further purified. The V−Al alloy containing 85% V produced by the thermite method is treated in a secondary vacuum induction smelting step, where the Al content is adjusted and V−Al alloy is produced with the required aluminum alloy components. The advantages include uniform product quality and high purity, while disadvantages include an increase in the reprocessing procedure and a high cost.

17.4.2.3 Self-combustion method

Vanadium pentoxide and Al powder are used as raw materials for direct combustion under vacuum conditions. V−Al alloy with uniform composition, high purity, and high density can be prepared.

17.4.2.4 Patented technology

The German GFE Electric Metallurgical Co., Ltd. issued patent about the two-step method of V−Al alloy production in 1985, and in 1987 applied for a US patent "Intermediate alloy used for titanium alloy production and its preparation method." The intermediate alloy has a high Mo content, which can be melted and dispersed in the titanium alloy. The Mo-based alloy has a composition: 25%−36% Mo, 15%−18% V, 7% Ti, where the balance is Al. The Mo content is 1.4 times that of V. This alloy has a melting point of 1500°C.

In 1981, Reading Alloy companies (United States) applied for the US patent "Intermediate alloy of vanadium, ruthenium and aluminum and its preparation method." The alloy contains the refractory metal ruthenium and is produced using the thermite reaction between vanadium oxide and aluminum. Ruthenium is added during smelting to produce an ideal intermediate alloy of vanadium, ruthenium, and aluminum, with a composition of 59%−70% V, 29%−40% Al, and 1%−10% Ru.

In 1993, Teledyne industrial companies (United States) applied for the US patent "Intermediate alloy of vanadium−nickel−chromium alloy and its preparation process." Vanadium pentoxide and small amounts of nickel powder and chromium powder are used as raw materials in a thermite reaction. An intermediate alloy is obtained that contains 4%−17% V and 5%−12% Cr.

17.4.2.5 Major vanadium−aluminum alloy specifications

The Stratcor Vanadium Company (United States) is one of the most important manufacturers of V−Al alloy. Its production line is installed with a LumenX digital X-ray inspection system. They were awarded the ISO9002 quality certification and specialize in providing high-quality V−Al alloy for titanium alloy production. The technical specifications of their V−Al alloy products are (1) 1/4 × 70 mesh in size, packaged in a 450-kg barrel, 65% Al−V, 34%−39% Al, 60%−65% V, dark purple color, passed digital X-ray detection test; (2) 85% Al−V, size 8 × 50 mesh, packaged in a 450-kg barrel; (3) 65% Al−V, 13%−16% Al, 82%−85% V, dark purple color, passed digital X-ray detection system test.

Table 17.4: Comparison of vanadium–aluminum alloy produced by three methods.

Production method	Alloy particle size/mm	Alloy composition (%)					
		V	Fe	Si	O	C	Al
Self-combustion	0.83–3	55.47	0.18	0.14	0.06	0.03	rest
Two-step	0.25–6	50.50	0.27	0.17	0.06	0.01	rest
Thermite	0.83–3	56.74	0.24	0.25	0.13	0.07	rest

Chinese V–Al intermediate alloy brands include AlV50, AlV70, and AlV80. The particle size of lumpy AlV50 is 1–50 mm, while that of AlV70 and AlV80 is generally no more than 100 mm, while the weight of the 1×1 mm particles is no more than 3% of the total weight.

17.4.2.6 Comparison of alloy quality from different processes

A comparison of vanadium–aluminum alloy produced by the three different methods is shown in Table 17.4.

CHAPTER 18

Development of the vanadium industry

Chapter Outline
18.1 Development of vanadium industry technology 489
    18.1.1 Vanadium industry technology 489
    18.1.2 Vanadium industry value chain 493
18.2 South African vanadium industry 494
    18.2.1 Main processes and equipment of the Highveld company 496
    18.2.2 Vametco Minerals Corp 501
18.3 Russian vanadium industry 501
    18.3.1 Nizhny Tagil iron and steel company 503
    18.3.2 Catchikara Vanadium Mining Co., Ltd 504
    18.3.3 Chousoff steel works 504
    18.3.4 Tola Vanadium Ferrous Metallurgical Corporation 505
18.4 New Zealand vanadium industry 505
    18.4.1 Steel production technology and equipment 506
    18.4.2 Vanadium extraction equipment 508
18.5 China's vanadium industry 508
    18.5.1 Panzhihua vanadium industry 510
    18.5.2 Chengde vanadium industry 514
    18.5.3 Jinzhou ferroalloy factory 515
    18.5.4 Vanadium extraction from stone coal 518

The global vanadium industry has developed greatly in four stages: recognition of the industrial value of vanadium, development and optimization of industrial processes, streamlining of the vanadium industrial chain, and expansion of the market and applications. Competition and cooperation has occurred in the industry, which has embraced the world resources, technology and equipment, product, environment, and market. The development of the vanadium industry began when the value of vanadium for various applications was recognized. In the initial stages, little attention was paid to vanadium. The value of vanadium was demonstrated for various applications, including catalysis and steel alloying. In addition, increasing awareness of the economic and social benefits of vanadium resulted in growth and development of the vanadium industry. In the case of shortages of the resource, new technology and equipment for exploration and enrichment of vanadium resources were developed. Innovation of vanadium extraction technology has accelerated the development of regional competition and resources cooperation. With the discovery of large vanadium deposits and the development of large-scale iron and steel technologies, the range of applications and market for vanadium expanded, and directly promoted production capacity, at the same time, will develop the field of application. Vanadium is a consumable and nonrenewable alloy material. When used in other applications, vanadium gradually accumulates in the industrial cycle, forming a secondary resource. When the accumulation of vanadium reaches a certain level in the cycle, the vanadium industry is affected, which results in raw material structure changes. In addition, the urgency for extracting vanadium from mineral ores for applications is reduced, where the driving force of industrial development shifts to the products, environment, and market.

Following the regulation of modern products, public safety must be ensured by strictly controlling the use of toxic materials during production and use. The application properties and modality of vanadium are in variable, natures and influences are also changing dynamically. Therefore the application and storage of vanadium must be designed to ensure stability of the material. Vanadium products that cannot be cured and whose uses change should be recalled for recycling. During production and use, internal and external circulation should be added to balance energy and materials. All intermediate products and tailings related to vanadium extraction must be stored and used carefully. Vanadium and its associated toxic elements in wastewater from vanadium extraction must be discharged following appropriate standards.

The world production of vanadium is highly monopolized by a group of five companies [South African Highveld, Swiss Glencore International AG (XSTRATA), Russian Tula-Chousoff, Panzhihua Steel—Chengde Steel in China, and US Strategic Minerals], which have a combined production capacity of over 80% of the global value. South African Highveld and US Strategic Minerals was Russia's Evraz holding company (Evraz) holdings, which is actually a competition of three. Currently, the main vanadium resources include V-bearing titanoferrous magnetite (vanadium slag), oil ash, desulfurization waste sulfuric acid catalyst, spent catalyst and other secondary resources, stone coal, and other raw materials, for which diversified. The proportion of secondary vanadium resources is increasing as resource recycling is accelerating. Developed countries deliberately protect their own resources as strategic materials and buy vanadium products from underdeveloped countries. The potential for industrial development of vanadium resources development is the largest in Russia. China has large vanadium resources in the Panzhihua-Xichang region, while new resources were also discovered in Chengde. Development of stone coal vanadium resources and use of secondary vanadium resources broke the geographical boundary that the vanadium industry developed. There has been a growth in vanadium resources under internal control, along with export-oriented development, which has strengthened the driving force for industrial upgrade [1−4,14−16,20,21,25,39,67,80,84,85,87].

18.1 Development of vanadium industry technology

In 1801 Spanish mineralogist A.M. del Rio discovered vanadium during his study of lead ore. He named it the red element, and it was mistaken for a compound of chromium. In 1830 the Swedish chemist N.G. Sefström separated an element in the process of smelting iron, and named it vanadis. The German chemist F. Wöhler proved that the new element found by Sefström was the same as that discovered by del Rio. In 1867 Henry Enfield Roscoe first obtained pure vanadium in the form of hydrochloric acid vanadium (III) by hydrogen reduction.

18.1.1 Vanadium industry technology

The technological process of extracting vanadium was designed for standard vanadium pentoxide products. The process begins with processing raw materials containing vanadium and ends with producing the standard vanadium pentoxide products. The technical parameters and processing capacity of equipment are based on the requirements of the standard vanadium pentoxide that is used to produce downstream products. Vanadium extraction processes are designed based on the raw materials and are classified into main extraction and subextraction processes. The aim of the main process is to extract vanadium, in which all V-containing materials are enriched. Subextraction processes involve extracting vanadium via the enrichment of other by-products. Vanadium extraction processes using chemical additives are generally divided into alkali extraction and acid treatment extraction processes. Alkali methods include the sodium and calcium salt methods, where the material can also be calcined without an additive and then roasted to extract vanadium. The conversion of vanadium is classified into alkaline leaching, acid leaching, and hot-water leaching processes. For treatment of low-grade vanadium slag, it is necessary to develop a simple and feasible process, which can be adjusted to suit regional conditions and be integrated into other industries. Then, vanadium extraction is performed after enrichment and transformation.

In 1882 basic slag was treated with 0.1% HCl at low temperature and successfully prepared vanadium products. In 1894 a method for reduction of vanadium oxides was invented by Jose Maweng. Later, in 1897, Gordon Schmitt invented the alumina-thermal reduction process for vanadium oxide. In 1906 in Peru, patronite was discovered, and the United States mined and extracted this mineral for 50 years, which accounted for a quarter of the global production at the time. In the United States, this process was mainly used to extract vanadium from carnotite. In 1911 Ford first successfully used vanadium in a steel alloying in the United States. In 1912 Bleecker published a patent regarding the use of sodium salt roasting combined with a water leaching process to recover vanadium.

Vanadium extraction by sodium salt roasting was a landmark process and is still in use. Generally, vanadium extraction from vanadium slag involves five technological steps, including roasting, leaching, purification, vanadate precipitation, and calcination, where vanadium pentoxide products are finally obtained. The key technology is roasting conversion. Roasting vanadium slag (including raw material for other vanadium) is an oxidation process, where trivalent vanadium V(III) in the ore (or slag) is oxidized to tetravalent vanadium V(IV) and pentavalent vanadium V(V). To break down the mineral structure of the vanadium slag (or ore), additives need to be used during vanadium extraction to facilitate oxidation of V(III) and its conversion into soluble vanadium salt. Two kinds of additives are commonly used: sodium salt and calcium salt additives. For vanadium slag with high Ca contents, phosphate can be used for CaO reduction during sodium salt roasting. High-Ti vanadium slag can be treated by sulfide roasting to extract vanadium. Molten iron containing high phosphorus and sulfur contents is used to produce soda-treated vanadium slag. High-temperature roasting with multicomponent additive systems can increase the efficiency of vanadium transformation. The most mature roasting process is still sodium salt roasting, although it results in serious environmental pollution. The application of calcium salt roasting in Russia has been successful and has solved the problem of ammonia and nitrogen in the wastewater, which hindered development of the vanadium industry. Trial production of calcium salt roasting processes for vanadium extraction is under way in China. Some V-containing materials are transformed by salt-free roasting. The vanadate structure is conducive to the dissolution of vanadium, especially using the acid-dissolution process.

Before 1936 the Soviet Union obtained vanadium slag from vanadium containing iron ore and then extracted vanadium products from vanadium slag. In the 1970s they successfully applied the Ca-salt treatment process to extract vanadium. In 1955 Finland began to extract vanadium from magnetite. In 1957 South Africa began to produce vanadium slag and vanadium oxide from vanadium containing titano ferrous magnetite. In the 1960s and 1980s the United States, Japan, Canada, and Peru extracted and recovered vanadium from oil ash and waste catalyst materials, and the monopoly of regional vanadium resources was gradually broken.

There are three main decomposition methods for vanadium ores. (1) In the acid treatment method, $(VO_2)_2SO_4$ or VO_2Cl are obtained after treatment with sulfuric acid or hydrochloric acid. (2) In the alkali treatment method, sodium hydroxide or sodium bicarbonate are used during high-temperature roasting, which react with V_2O_5 in the ore to produce $NaVO_3$ or Na_3VO_4. (3) During the sodium chloride roasting method, sodium chloride is used to obtain $NaVO_3$ by roasting the ore with salt. Vanadium-extraction processes generally focus on producing standard vanadium pentoxide products.

The extraction of vanadium pentoxide from vanadium slag is a system-selection process. To a certain extent, it is the typical mainstream process, which represents an era, a new industry, and a product of the development of the advanced level and its characteristics. It is necessary to reflect the core ideas of vanadium extraction by systematic theoretical analysis, along with the consideration of production experience. This will enhance the valuable industrial products and resources. Depending on the technical means and parameter selection, the core idea runs through all-round runs through the whole process. Comprehensive consideration of the actual supply of raw materials should be first addressed, followed by preparation of a complete set of equipment satisfying the requirements of advanced technology, economic viability, and controllability. Finally, the process must satisfy the market demand for high-end products and achieve environmental and safety values of the industry at the same time.

Depending on the vanadium content of mineral resources from different regions, different vanadium extraction technologies are required to process the raw material, which can have multiple complex and changeable compositions and be classified. Vanadium extraction technology includes three phases: extracting vanadium from low-grade raw materials, extracting vanadium from high-vanadium raw material, and a combination of extracting vanadium and precious metal recycling, which results in balanced enrichment, transformation, and recycling. All vanadium extraction technologies have characteristics of chemical metallurgy and metallurgical chemistry. The vanadium extraction process requires that suitable parameters are chosen to convert the vanadium components into compounds that are soluble in salts, alkali, acid, or water. Considering the solubility of sodium vanadate, vanadium extraction uses roasting and calcination steps to convert vanadium oxide from a low-valence to high-valence state. Sodium salt and vanadium oxide are mixed and reacted at high temperature, where conversion and structural transformation occur, and soluble vanadate as a stable intermediate compound is formed. Vanadium is separated from other mineral components, where soluble vanadium compounds enter the liquid phase, and the insoluble material remains in the slag. This process is suitable for vanadium as the main target of vanadium raw materials.

The vanadium content of minerals and comprehensive materials is generally low, while some vanadium raw materials are also rich in precious metals or other valuable materials. Hence, vanadium extraction needs to be comprehensively planned to include recovery of both vanadium and other valuable metals. Vanadium generally occurs in mixed compositions, where some vanadium is in the form of the primary mineral, while the remainder occurs as secondary vanadium-containing raw materials. Hence, it is difficult to balance different process functions of recycling and extraction. Acidolysis and acid leaching processes can produce a unified liquid phase, which can be separated by precipitation of different metal salts for inorganic and organic extraction processes, depending on the characteristics of the liquid. All of these processes can be successfully used for extraction and enrichment of vanadium products and recovery of valuable metallic elements. Vanadium extraction is a chemical metallurgy process that can achieve high recovery in a short processing time, although it requires a high grade of raw materials. It played an important role in the vanadium industry before the widespread development of South Africa's vanadium industry in the 1960s.

Vanadium oxide has good catalytic performance and vanadium-based catalysts show unique behaviors. The production of vanadium pentoxide greatly increased the catalytic production of sulfuric acid and facilitated scale-up of this industry by replacing use of noble metal catalysts. Vanadium-based catalysts proved effective in the petrochemical industry, accelerating chemical reaction processes and resulting in an increase in the reliability and stability of organic synthesis reactions. In 1880 vanadium catalysts were identified, and experimental research in this field began in 1901. In 1913 vanadium catalysts were introduced for the first time in processes in the Baden Aniline Soda Company, Germany, and by 1930, they were commonly used in the factory. The popularity of vanadium catalysts increased throughout the 1930s as a replacement for platinum catalysts used in sulfuric acid production. Vanadium catalysts mainly consist of vanadium compounds, such as oxides, chlorides, and complexes, as well as a variety of forms, such as salts of heteropoly acids. The most common active component contains one or more of the following V_2O_5 additives. V_2O_5 is the main active component of most catalysts for oxidation reactions. Vanadium compounds are important industrial catalysis for oxidation and are widely used in the sulfuric acid industry; for the synthesis of organic chemicals such as phthalic anhydride and maleic anhydride; and for catalytic polymerization, alkylation reactions, and oxidative dehydrogenation. Vanadium catalysts have been used to perform double catalysis for desulfurization and denitrification, as well as for environmental purification for removing inorganic and organic chemicals. Recently, vanadium batteries are promoting the development of clean energy applications. Vanadium has been an "industrial MSG," boosting the industrial economy.

The emergence of vanadium pentoxide as an alternative to precious metal catalysts changed the history of sulfuric acid production, where sulfuric acid production increased by 100 times. Vanadium pentoxide subsequently performed well in the petrochemical industry, accelerating the chemical reaction process and increasing the reliability and stability of the organic synthesis reactions. In 1889 Professor Araud at the University of Sheffield began studying the special role of vanadium in steel. In the early 20th century, Henry Ford discovered a new way of applying vanadium bearing steel and good performance of key components of Ford motors was obtained by vanadium alloying. Vanadium pentoxide is used to produce ferrovanadium, vanadium nitride alloy, and vanadium metal, which are used in the steel industry to achieve steel products with complex functions. The addition of vanadium can refine the grains in the steel matrix, and comprehensively improve the performance of the steel product, including enhancing the strength, toughness, ductility, and heat resistance. Track steel, bridge steel, alloy steel, and construction steel have achieved the strength requirements while reducing the weight of high-rise buildings and structures for carrying heavy vehicles and high-speed locomotives. The steel is strengthened by dissolved nitrogen and fixed nitrogen in vanadium−nitrogen alloys. The introduction of nitrogen saves vanadium in steel-making. Vanadium aluminum alloys are used in the manufacture of titanium alloy Ti6Al4V. The micro-vanadium treatment can increase the strength of aluminum alloy and the microstructure of copper alloy, resulting in an increase in the strength of cast aluminum, copper, and titanium products. British scientist Roscoe obtained vanadium metal for the first time via hydrogen reduction of vanadium chloride. Applications for vanadium metal have expanded to include radiation shielding and superconductivity materials as its special properties are recognized.

Vanadium-based solid melt alloys can reversibly absorb and release hydrogen under appropriate temperature and pressure, with a hydrogen storage capacity of 1000 times its own volume. The theoretical hydrogen absorption capacity is 3.8%. Hydrogen diffuses quickly in hydrides, and hydrogen storage alloys have been developed to form new energy materials for hydrogen storage. A vanadium redox battery (VRB) is a mobile battery that converts energy stored in an electrolyte into electricity by exchanging electrons between two different types of vanadium ions separated by a membrane. The electrolyte is a mixture of sulfuric acid and vanadium. Since this electrochemical reaction is reversible, VRBs can be both charged and discharged. As the concentration of the two vanadium ions changes during charging and discharging, conversion between electrical and chemical energy occurs. The VRB consists of two electrolyte tanks (for the two different electrolytes) and stacks of cells.

18.1.2 Vanadium industry value chain

Vanadium is the common link in the vanadium industry value chain. Through continuous discovery and exploration of the value of vanadium for various applications, the technologies used for vanadium exploration and extraction have been diversified and developed. Various vanadium resources are currently being developed and used. The advance of chemical technologies laid the foundation for the early development of the vanadium industry. The value of vanadium for catalysis and steel alloy applications has been demonstrated, which resulted in cross-industry international trade during technological development of the vanadium industry. This led to fine division of labor and an increase in the economic and social impact of the entire industry. Businesses related to vanadium production and application were developed, along with multinational companies. In addition, peripheral support for the vanadium industry appeared in the fields of education, scientific research, consulting, sales, logistics, and government regulations.

There are three typical production structures. First, some vanadium factories produce vanadium slag and other resources, often sell these materials as commodities, and are classified as forward integration companies. Second, in addition to selling vanadium products, some manufacturers also perform processing and production of special alloys, catalysts, special materials, and other downstream products (e.g., the strategic mineral companies in the United States and some vanadium factories in Japan) and are classified as backward integration companies. Third, a few manufacturers rely on purchasing raw materials to produce vanadium products for downstream manufacturers and belong to typical vanadium processing production enterprises.

The vanadium extraction industry must also be integrated with upstream and downstream industries, while competing with peers and substitutes, as shown in Fig. 18.1. Currently, vanadium products are mainly traded in three ways: (1) companies directly selling products to fixed users, (2) searching for users through intermediaries or metal trading markets (the largest proportion), and (3) companies with backward integration consume some vanadium products to produce downstream products. The dealer is considered a special kind of user. When selecting a product from a specific manufacturer to distribute, the price, transportation, service quality, relationship with the manufacturer, and end-user preferences are considered. Data from countries and vanadium producers indicate that direct sales and those through a middleman account for the largest proportion of transactions.

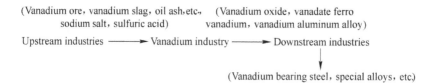

Figure 18.1
Schematic diagram of upstream and downstream companies in the vanadium industry.

18.2 South African vanadium industry

The Bushveld composite ore in South Africa is the world's largest known vanadium ore reserve and occurs on upper spiral ore layer and neck of V-bearing titanoferrous magnetite deposits. This huge lamellar immersion is located in the Tlatuwa province of South Africa, where the entire ore body contains significant vanadium. The main V-bearing titanoferrous magnetite is the upper dribbling layer, where the high V_2O_5 content is in the range of 1.6% ± 0.2% h. The lamellar ore body extends hundreds of kilometers outward in an oval shape and has outcrops in places such as Roshenica. The main ore bed with high titanium magnetite content is composed of closely packed almost equivalent particles, while secondary silicates exist in the aperture. The titanium in the ore mainly exists as a solid solution in a rich titanium magnetite phase (titanium spinel Fe_2TiO_4). A small amount of titanium occurs as ilmenite, which appears as elongated intergranular bodies with individual grains or as segregated chips of minerals arranged parallel to the magnetite octahedra. Vanadium in the ore mainly occurs in the internal solid solution of the titanium magnetite spinel, where the V^{3+} replaces Fe^{3+}. Vanadium is uniformly distributed in magnetite particles coated with ilmenite chips, not as an independent mineral. When exposed to weathering, magnetite is oxidized to produce V-containing magnetic hematite $(TiFe)_2O_3$, while the small amount of hematite does not change the ore structure.

In 1949 W. Bleloch successfully smelted V-bearing titanoferrous magnetite in a submerged arc furnace, where carbon was added as reductant, and Fe and V were preferentially reduced. In this process, fully carburized cast iron with low-Ti content is obtained, where TiO_2 is enriched in the slag. Vanadium-rich slag can be recovered from the low-Ti pig iron using a side-blown air converter.

In May 1960 the Anglo-American company of South Africa formed the Highveld joint development company. After 3 years of exploration, they confirmed the existence of 200 million tons of ore (average composition of Fe 56%, TiO_2 13%, and V_2O_5 1.5%−1.9%). The process used at the Mapochs mine in South African includes prescreening the ore into 32−6 and −6 mm ore products. Two different separation processes are used. The lump ore (32−6 mm) is reduced in a rotary kiln and smelted in a submerged arc furnace, producing V-rich iron. The vanadium slag is then recovered using oxygen blowing and a vibrating ladle. After cooling the vanadium slag is crushed and the Fe is removed by magnetic separation to increase the grade of the vanadium slag. The composition of the vanadium slag is V_2O_5 25%, SiO_2 16%, Cr_2O_3 5%, MnO 4%, Al_2O_3 4%, CaO 3%, MgO 3%, with a rest of iron oxides and iron. The powder ore (−6 mm) is ground by wet ball milling to a size of −200 mesh (60%), which is followed by magnetic separation to produce V-containing iron concentrate, which is mixed with sodium salt and roasted in a rotary kiln (or multihearth roasting furnace). The calcined product is leached to produce flake V_2O_5 using a rotary drying system. This production process was designed based on that for processing pig iron and vanadium slag, which was successfully demonstrated in 1949. An intermediate test was carried out in 1961. After the ore is prereduced, vanadium slag and molten iron are obtained by deep reduction in an electric furnace. This method was first used for production in 1968. They successfully established a rotary kiln pretreatment system, a submerged arc furnace for smelting molten iron, a vibration tank and oxygen blowing system for recovering vanadium slag, mineral processing systems, and combined dressing and smelting technology. The roasting and leaching of −6 mm powder ore is achieved using sequential grinding, magnetic separation, and enrichment steps. The raw materials are prepared in a rotary kiln, followed by sodium salt oxidization roasting, polyvanadate leaching, and drying to extract the vanadium.

The Highveld Steel and Vanadium Company developed the open-air Mapochs mine in 1967 and began production. This mine is situated in the town of Roshenica in the eastern province of Tlatuwa. The main magnetite ore is tilted about 13 degree toward the west, in line with mine terrain structure. The gravel of the magnetite particles and boulders were produced in the eastern part of the outcrops. Weathered pitching occurs in the middle of the main outcrop and rocks, which is about 0.75–1.0 m, with an average composition of 53%–57% TFe, 1.4%–1.9% V_2O_5, 12%–15% TiO_2, 1.0%–1.8% SiO_2, 2.5%–3.5% Al_2O_3, and 0.15%–0.6% Cr_2O_3.

Off-balance-sheet seam is uncovered about 3 m. The pitching ore is broken mechanically by blasting or hydraulic rock crushers. The exposed solids and purer seams are processed by drilling and blasting and then moved to a storage site by bulldozer. Then, the ore is transported to a concentrator using a loader, and ore with a particle size of 4.5–25 mm and <4.5 mm are obtained by magnetic separation and enrichment. The massive ore is sent to a steel plant, and the powdery mineral is sent to a wet-process vanadium extraction plant.

Highveld, South Africa performed a test from April 1963 to May 1964 using a 15 t/day semiindustrial test facility; after 10 months of research, iron, steel, and vanadium products were produced simultaneously. In January 1965 design and construction were started. The integrated steel plant was built in 1968 which began operation in April 1969, with annual capacity of 300,000 t steel (rail, joist steel, steel columns, angles steel, and flat steel), rolling and continuous casting billet of 120,000 t, and production of 26,000 t standard vanadium slag (containing 25% V_2O_5). The company's Vantra vanadium plant is the world's largest vanadium producer. Before 1972, it produced ammonium metavanadate. The second phase of the project began in 1983 and went into operation in the summer of 1985. Currently, the production capacity of steel billet is 1 million t/a, and the production capacity of vanadium is 180,000 t of standard vanadium slag and 22,000 t of V_2O_5 (equivalent).

18.2.1 Main processes and equipment of the Highveld company

The ironmaking process and equipment used in the Highveld plant are shown in Fig. 18.2.

18.2.1.1 Direct reduction and electric melting plant

Highveld, South Africa is equipped with 13 sets of rotary kilns for reduction with a diameter of 4 m and length of 61 m, and rotational speed of 0.40–1.25 rpm. Seven submerged arc furnaces were used for smelting, two with a power of 45 MVA, and the other two of 33 MVA, with a diameter of 14 m, smelting period of 3.5–4 h, and furnace capacity of 70 t of iron/furnace. Another 63-MVA furnace has a diameter of 15.6 m, a smelting period of 3.5–4 h, and capacity of 80 t of iron/furnace.

Development of the vanadium industry 497

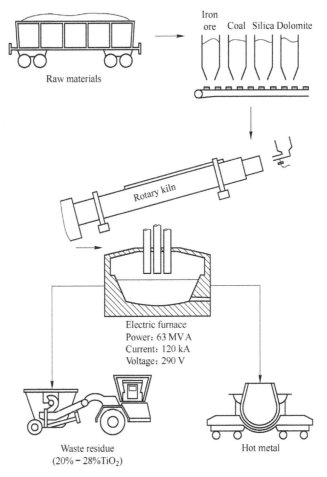

Figure 18.2
Ironmaking process and equipment used in the Highveld plant.

18.2.1.2 Vanadium extraction and steelmaking plant

For shaking ladle vanadium recovery, Highveld is equipped with 4 shaking ladle platforms and 16 shaking ladles, which are 5.5 m high, with 4.3 m inner diameter of the shell, and a standard capacity of 75 t of molten iron. The mineral composition (%) is as follows: 53–57 TFe, 1.4–1.9 V_2O_5, 12–15 TiO_2, 1.0–1.8 SiO_2, 2.5–3.5 Al_2O_3, and 0.15–0.6 Cr_2O_3. The composition of molten iron (%) is as follows: 3.95 C, 1.22 V, 0.24 Si, 0.22 Ti, 0.22 Mn, 0.08 P, 0.037 S, 0.29 Cr, 0.04 Cu, and 0.11 Ni. Fig. 18.3 shows the process flow of vanadium blowing and steelmaking. The titanium slag composition (%) is as follows: 32 TiO_2, 22 SiO_2, 17 CaO, 15 MgO, 14 Al_2O_3, 0.9 V_2O_5, and 0.17 S. Three sets of oxygen converter equipment have a capacity of 75 t and size of Φ 4.8 m $\times$ 7.1 m. Five sets of continuous casting machines have a flat slab size of 180 mm $\times$ 230 mm. The main technical indicators were oxidation of vanadium was 93.4%, recovery of vanadium was 91.6%, recovery of semisteel was 93%, total blowing time was 52 min, vibrating time was 59 min, total cycle time was 90 min/furnace, temperature of molten iron before oxygen blowing was 1180°C, metal melting temperature was 1270°C, oxygen nozzle diameter was 2 in., height of oxygen blowing tube from static pool surface was 76.2 cm, normal oxygen flow rate was 28.3 N m^3/min, oxygen flow rate was 42.5 N m^3/min, and oxygen blowing tube pressure (under normal flow) was 160 kPa. The semisteel composition (%) is as follows: 3.17 C, 0.07 V, 0.01 Si, 0.01 Ti, 0.01 Mn, 0.09 P, 0.040 S, 0.04 Cr, 0.04 Cu, and 0.11 Ni. The nonmagnetic vanadium slag composition (%) is 27.8 V_2O_5, 22.4 FeO, 0.5 CaO, 0.3 MgO, 17.3 SiO_2, 3.5 Al_2O_3, 2.5 C, and 13.0 Fe. The magnetic slag composition (%) is 1.3 V_2O_5, 96.5 Fe, and 89.6 free Fe. The V_2O_5 content in all slag was 26.1%.

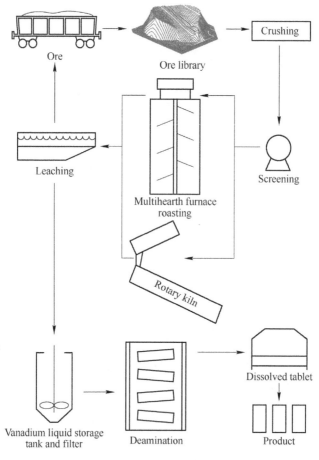

Figure 18.3
Highveld vanadium production process.

18.2.1.3 Division of Highveld steel and vanadium company—Vantra factory

The first vanadium recovery production plant was completed and put into operation in 1957 in Witbank, South Africa. It was constructed by a subsidiary of a mineral investment construction engineering company in South Africa, which was controlled by Colorado mineral engineering company of the Rockefeller. Ammonium vanadate and vanadium pentoxide were produced using roasting and leaching technology. In 1959 British and American companies took over management of the factory in South Africa. This acquisition was completed after 1 year, and the factory was renamed Vantra. Before 1972, raw materials were provided by surrounding mines producing Bushveld composite ore, where Kennedy's Warley mine was the main supplier. After 1972, Vantra was mainly supplied by fine-grained ores from Mapochs mines. In the Vantra factory, ore is ground by wet grinding. Some gangue rich in SiO_2 and Al_2O_3 is dissociated from the ore and separated from the pulp using magnetic separation. After slurry dehydration the cake composition (%) is 56.4 TFe, 1.65 V_2O_5, 14.1 TiO_2, 1.2 SiO_2, 3.1 Al_2O_3, and 0.4 Cr_2O_3. Sodium bicarbonate, sodium sulfate, or a mixture of these is homogeneously combined with finely ground ore in accurately measured quantities. This mixture is then roasted using a multihearth furnace and rotary kiln.

Before 1974, sodium chloride was used as an additive for multihearth furnace roasting. However, the roasting effect was not superior to that of sodium carbonate or sodium sulfate. Use of sodium sulfate can achieve V-rich solutions, while using sodium carbonate can give solutions with low vanadium concentration. Mainly water-soluble sodium aluminate, sodium chromate, and sodium silicate are formed, which affect the vanadium concentration in solution. To prevent the ringing of the rotary kiln and agglomeration of the burden in the multihearth furnace, the burden structure and kiln parameters must be considered. When impure minerals are used with sodium carbonate additive, the roasting temperature is low. When high-quality minerals are used with sodium sulfate additive, the roasting temperature is higher. In addition, higher additive dosages can decrease the roasting temperature but must avoid burning mineral block and ensure good contact between the oxygen and mineral grains.

The hot clinker from the roasting furnace is transported by a chain conveyor to a cooling box in the upper part of the leaching pool where sodium vanadate is dissolved into water. When the V_2O_5 content in solution reaches 50–60 g/L, the V-rich solution is pumped to the liquid storage tank. After several washing cycles the leached clinker is discharged from the leaching pool and stored in the tailings pond.

Before 1972, ammonium metavanadate was produced. In 1973 they began the production of ammonium polyorthovanadate. In 1974 the production of ammonium metavanadate was stopped. The roasting equipment includes four sets of multihearth furnaces with 10 layer $\times$ Φ 6.1, three sets of rotary kiln Φ 1.52 $\times$ (outer) $\times$ 18.3 m, and one set of rotary kiln

Φ 2.6 (outer) × 36.5 m. Heat is provided by burning coal powder. Sodium chloride was used as an additive before 1974, which released HCl and converted using ammonia to ammonium chloride. In the Highveld Vantra plant, vanadium pentoxide is produced from vanadium slag or ore, where 60% of the particles are below 200 mesh. After magnetic separation, vanadium slag or ore are mixed with salt and roasted in rotary kiln or multihearth furnace. Coal is used as fuel. Clinker is delivered by a chain conveyor to the quenching tank and fed into the leaching pool, which is filled with clinker and water. When the V_2O_5 concentration of the leaching solution reaches 50–60 g/L V_2O_5, the leaching solution is pumped into a mother liquor storage tank. After continuous washing and leaching in the pool the residue is disposed as tailings. Fig. 18.4 shows the process flow of vanadium extraction in Highveld.

The ammonium metavanadate is precipitated with excessive ammonium chloride, heated with steam, and stirred by air. The overflow is passed to a second reactor and then finally into the thickener. The ammonium metavanadate is clarified and raked to the center of the discharging mouth. The waste liquid is reduced to a V_2O_5 content of 1 g/L and pumped through a box-type filter. This material is washed to produce ammonium hexavanadate precipitate. The overflow or vanadium-depleted liquid is pumped into a scintillating evaporator that is heated by coal. The two-stage evaporator recycles the waste liquid and the resulting concentrated ammonium chloride solution is returned to the precipitation plant for reuse. The concentrated wastewater solution is returned to the roasting leaching system, and the crystallized sodium sulfate obtained by centrifugation is returned to the roasting step. Ammonium metavanadate is sent to the deamination reactor using a tubular screw device for external heating. Finally, vanadium pentoxide powder is obtained. Vanadium pentoxide powder is melted at 850°C in a furnace and then released from the furnace mouth to a cooling wheel, where flake vanadium pentoxide powder is produced.

Production of V_2O_3 began in July 1993, where ammonium polyorthovanadate precipitation was used with a batch quantity of 20 t of solution. In this process, sulfuric acid is used to adjust the solution pH to 5.5. Then, ammonium sulfate is added and the pH is adjusted to around pH to 2, followed by steam heating. When the V_2O_5 content of the mother liquor decreased to 0.5 g/L, precipitation is stopped, and the slurry is filtered and washed to produce ammonium polyorthovanadate. This is the same process used for ammonium metavanadate precipitation. V_2O_3 is obtained by reducing APV with natural gas, where the minimum V_2O_3 content of the products was 66%. The wastewater from vanadium extraction is treated with a new two-stage vacuum evaporator. The recovered ammonium sulfate is directly returned to the precipitation step, and sodium sulfate is returned to the roasting cycle after scintillation and drying.

The standard vanadium slag production is 190 kt/a, or a V_2O_5 equivalent of 22 kt/a, including direct vanadium extraction capacity from the ore of ~6 kt/a. When the market demand for vanadium is low, the vanadium slag is sold to Vametco Minerals Corp. and Xstrata Alloys vanadium factories, or the vanadium production is reduced to regulate market supply and demand.

Highveld is also involved in VRB research. They purchased the intellectual property for vanadium batteries from Australia and have developed two prototypes (1 and 2 MW systems) that are used in the United States. The solubility of the $VOSO_4$ in the liquid is 1.6 mol. Now 2.5 mol has been reached.

18.2.2 Vametco Minerals Corp

Vametco uses mineral ore, vanadium slag, or mixtures of these as raw material for vanadium extraction. Usually when the market is good, only minerals are available. When the market is bad, Highveld supplies them with vanadium slag, thus reducing the market supply of vanadium. The production process is shown in Fig. 18.5, since most processes are similar. To treat the wastewater from vanadium extraction, Vametco uses evaporation to recycle sulfates. Vametco Minerals Corp. is now a holding company of Stratcor. The company was founded in 1965 by the USAR mineral company, and Union Carbide acquired from Flderale Volksbelgging, at the last in 1986, Stratcor acquired from Union Carbide as its holding company. The mine was built in 1967 and is located about 12 km southwest of Britz in Bophuthatswana, which has two regional locations.

The V-bearing titanoferrous magnetite used in Vametco is from the upper area of igneous rock in the Bushveld deposit, which is mined by open-cut methods. The deposit has a length of 3.5 km, with an ore deposit inclination angle of ~ 20 degree and depth of ~ 10 m. Vametco built concentrator of Bon Accord in the early 1970s and began to expand the extraction facility in 1976, which was finally completed in 1981. The concentrator and V_2O_5 extraction factory were acquired in 1986 by Stratcor and were improved significantly after the acquisition. Vametco bought an electric furnace and has been developing V_2O_3 deep processing technology to produce ferrovanadium and vanadium nitride since 1993. Then, the production of traditional V_2O_5 products was stopped and they currently only produce ferrovanadium and vanadium nitride with a production scale of 3500 t/a of vanadium. In 1996 the company was awarded the ISO9002 certification, and in 2003 passed ISO2000 certification. Most of their products are sold to the United States, Europe, and Japan.

18.3 Russian vanadium industry

Most titanoferrous magnetite from Russia contains vanadium. Russia's vanadium production began in 1936, when the Chousoff steel works founded a chemical plant for processing vanadium slag from open-hearth furnaces and began production. The factory was the first enterprise in the former Soviet Union involved in industrialized production of

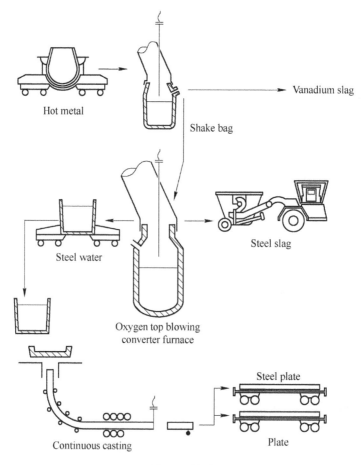

Figure 18.4
Highveld vanadium blowing and steel making process.

ferrovanadium, and the only supplier of vanadium products. In 1937 the output of ferrovanadium reached 500 t at the Chousoff steel works. However, the output was limited by the technical conditions at the time, where the capacity of the blast furnace and converter were limited. Due to the use of low-vanadium containing titanoferrous magnetite from first Uralsk and Kucinski over a long period of time, ferrovanadium production in the Chousoff steel works was low. No vanadium ore has been found in other parts of the former Soviet Union, and the rich V-bearing titanoferrous magnetite in the Urals region is underexploited. Mass production of vanadium products has important significance for the iron, steel, and chemical industries.

In the mid-1960s, experts from Russian Urals Metallurgy Research Institute collaborated with researchers from the Nizhny Tagil iron and steel company to develop a new process for vanadium extraction during steelmaking with a 120 t converter. This process has two steps. First, the molten iron produced from smelting Catchikara V-bearing titanoferrous magnetite was loaded in one converter, where its chemical composition (%) was 0.4–0.5 Si, 0.25–0.35 Mn, 0.40–0.48 V, 0.05–0.11 P, and 0.03–0.05 S. Then, vanadium extraction was carried out until the content of C in the semisteel was 3.2%–3.8%, and the content of V was reduced to 0.02%–0.04%. After extracting 90%–95% of the vanadium from the molten iron into vanadium slag, the semisteel was placed in another converter for refining to produce finished steel. Nizhny Tagil iron and steel company can produce V-containing pig iron, V-containing steel, and vanadium slag in large quantities using Catchikara V-bearing titanoferrous magnetite. Depending on the V content in the molten iron, the composition fluctuates over the ranges of 14%–20% V_2O_5, 15%–20% SiO_2, and 45%–55% iron oxide. The former Soviet Union became the main vanadium-producing country in the world after the new No. 2 ferrovanadium plant of Chousoff steel plant was put into operation in 1964.

There are nine enterprises in Russia currently involved in vanadium production, including four big companies. Catchikara Vanadium Mine Company produces iron concentrate from V-bearing titanoferrous magnetite and pellet. Nizhny Tagil iron and steel company produces V-containing pig iron and vanadium slag. Chousoff steel works produces V-containing pig iron, vanadium slag, vanadium pentoxide, and ferrovanadium, where the plant has been restructured into a company. Tola Vanadium Ferrous Metallurgical Corporation produces vanadium oxide and ferrovanadium.

18.3.1 Nizhny Tagil iron and steel company

The predecessor of Nizhny Tagil iron and steel Co., LTD. was Nizhny Tagil iron and steel company, which was founded in 1940. The company is located in the city of Nizhny Tagil, Sverdlovsk in the Russian Urals region, which is an important industrial city. Nizhny Tagil iron and steel company is Russia's fifth largest steel company and Russia's main producer of steel containing V and Ti, and alloys for rail, wheel, and tire applications. Their factory is equipped with six blast furnaces. Each of the No. 1 and No. 2 blast furnaces has capacity of 1242 m^3 and was put into production in 1940. The No. 1 and No. 2 blast furnace are now ready to retire after the completion of the renovation of the No. 5 blast furnace (1719 m^3) and No. 6 blast furnace (2700 m^3). The No. 6 blast furnace was built and put into operation in 1963, but due to its unstable performance during smelting of V-bearing titanoferrous magnetite, its use was discontinued in 1996. The overhaul of the No. 5 and No. 6 blast furnaces is expected to guarantee an output of 5.45 million t/a of V-containing pig iron, while reducing the coke consumption to 60 kg/t of iron.

The Nizhny Tagil iron and steel company was the first large steel company in the former Soviet Union to use an oxygen converter to make steel. In 1963 the first 130-t oxygen converter was built in the former Soviet Union. Currently, the company operates four 160-t oxygen converters, two ladle refining furnaces, one RH vacuum furnace, and two continuous casting machines. The converter plant uses two-step vanadium extraction in a steelmaking process to smelt V-containing molten iron, which produces vanadium slag containing ~18% commercial products, along with high-quality steel containing native vanadium. The production of vanadium from vanadium slag produced by Nizhny Tagil iron and steel company accounts for 80% of Russia's vanadium production.

18.3.2 Catchikara Vanadium Mining Co., LTD

The Catchikara iron mine is located in the state of Sverdelov in the Ural region of Russia, about 100 km from Nizhny Tagil. Catchikara iron mine has large ore reserves, where the sulfur and phosphorus content of the ore is low. The deposit is suitable for open-pit mining. This company has two deposits of V-bearing titanoferrous magnetite, with an Fe content of only 15.8%; other deposits have iron contents in the ore of 30%–50%. The Catchikara mine began production in 1963. According to the technical specifications of TY14-9-93-90, iron concentrate containing 60.3% Fe was produced. According to the technical specifications of TY14-00186933-003-95, the Fe content is 61.0%, and pellets without flux are produced. According to the technical specifications of TY 14-00186933-005-95, high-basicity sinter with an Fe content of 53.0% is produced. The particle size of the tailings produced by the company after mining is used for gravel slag production of 5–10 and 10–40 mm.

18.3.3 Chousoff steel works

Chousoff steel works is old iron and steel enterprise in the Russian Ural region and was founded in 1879. It has two 114 m^3 blast furnaces and 12, 15, and 18 t basic open-hearth furnaces. During the First World War, Chousoff steel works had two sets of 122 m^3 blast furnace and four sets of 30 t open-hearth furnaces, which represented the scale and level of advanced equipment of the steel works. In 1936 Chousoff steel works began the former Soviet Union's first industrialized production of vanadium products. In 1961 they built the No. 2 ferroalloy plant, which used the most advanced technology of the time to produce the first batch of ferrovanadium. Vanadium products from this plant entered the international market in the 1960s, which had the only full process (sintering, blast furnace, converter, hydrometallurgy and electricity, and metallurgy) successfully used to produce high-quality vanadium products from Catchikara V-bearing titanoferrous magnetite. Currently, the main equipment for ironmaking and steelmaking in Chousoff steel works includes a 16 m^2 sintering machine, 225 m^3 and 1033 m^3 blast furnaces, 250 t mixed iron furnace, three sets of 20 t Bessemer (for vanadium extraction), and two sets of 250 t open-hearth furnaces. The rolling equipment mainly consists of 250, 370, 550, and 800 rolling machines. In the steelmaking factory of Chousoff steel works the converter has a processing capacity of 2100 t/d of V-containing hot metal, open-hearth furnaces with steelmaking capacity of 1200–1300 t/d, a steel rolling plant with production capacity of 350,000 t/a, and ferrovanadium plant with production capacity of 7000 t/a.

18.3.4 Tola Vanadium Ferrous Metallurgical Corporation

The Tola Vanadium Ferrous Metallurgical Corporation was established in the former Soviet Union and specializes in the production of ferrovanadium, vanadium pentoxide, silicon–calcium–vanadium intermediate alloy, ferromanganese silicon nickel alloy, chromium dicarbide, calcium carbide, and other products. The plant was built in the early 1970s and started operation in 1974. It was the largest factory in the former Soviet Union for vanadium production. Before 1976 the Soviet government awarded the Lenin Prize to the research group from this plant for their new vanadium products. Because the blast furnace at Tola cannot smelt V-bearing titanoferrous magnetite, V-containing raw materials for vanadium production are purchased from Nizhny Tagil iron and steel company, which is located thousands of kilometers away.

The new Tola iron and steel plant was built by the former Soviet government in February 1931. The No. 1 blast furnace began producing iron in June 1935, and the No. 2 blast furnace was completed and put into operation in August 1938. In April 1953 the steelmaking plant of the new Tola iron and steel plant was completed and put into production. In December of the same year the world's first vertical continuous casting machine was put into production. In 1955 the steelmaking plant was built and put into operation, which was the first oxygen converter in the former Soviet Union. In 1960 the sintering plant was built and the No. 3 blast furnace began operation in 1962. Tola Vanadium Ferrous Metallurgical Corporation is one of the biggest enterprises in Russia for manufacturing vanadium products, with an annual output of 16,000 t (vanadium pentoxide equivalent). They are one of the major international suppliers of vanadium products. In 1996 the calcium salt vanadium extraction process was implemented. Tola is equipped with two sets of 6 t furnaces with a power of 4000 kW for ferrovanadium production; one is used to produce 50 VFe with the silicon thermal method, while the other uses the thermite method to produce 80 VFe.

18.4 New Zealand vanadium industry

The Tasman coast of New Zealand is rich in iron–vanadium–titanium resources, mainly from volcanic magma deposits, with open-type placer deposits containing Fe, V, and Ti.

The titanium resources of New Zealand steel companies are mainly in the Waikato mining area. Titanium tailings are made up of natural sand mounds for more than 10 km. There are reserves of hundreds of millions of tons of TiO_2 in the tail ore 4.5%−6%, which mostly occurs in granular ilmenite. The existing reserves are estimated to be tens of millions of tons.

18.4.1 Steel production technology and equipment

The process is divided into two parts: ironmaking and steelmaking Fig. 18.5.

18.4.1.1 Ironmaking

There are four multihearth furnaces, four rotary kilns, and two rectangular electric smelting furnaces. The multihearth furnace has 12 layers. After mixing iron ore, coal, and lime in the raw material area, the mixture is moved by a belt conveyor and fed into the multihearth

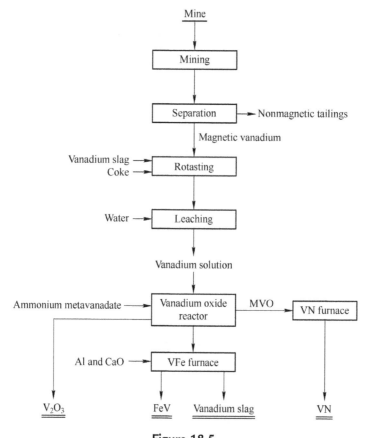

Figure 18.5
Vanadium production process used by Vametco Minerals Co.

furnace. The top temperature of the multihearth furnace is 500°C, middle layer temperature of 5 and 6 is 900°C, and discharge temperature is 600°C. The furnace charge in the multihearth furnace is not reduced or roasted; the furnace charge is simply preheated to remove moisture and volatile compounds.

The rotary kiln is 65 m × 4.6 m Φ. The furnace charge discharging from the multihearth furnace is conveyed using a hopper hoist and fed into the rotary kiln for reduction and calcining with limestone. The metallization rate of the rotary kilns is 78%, and its waste gas is used to generate electricity. The rectangular melting furnace has an external dimension of 20 m × 7.6 m × 7.5 m and includes six self-baked electrodes. The maximum power of each electrode is 42 MW, and the working voltage is 60 V. Twelve feeding ports are distributed on either side of the electrode. There are two iron outlets and two slag outlets. Three transformers each supply two electrodes. The charge in the rectangular smelting furnace is lifted using 8-t material tanks. The processing capacity is 65 t/h, and the amount of iron produced is 42 t/h. Every 4 h, one batch of Fe is produced. Heavy iron tanks are transferred to the steel mills using forklifts. The gas produced by the rotary kiln and rectangular smelting furnace is used for boiler power generation.

18.4.1.2 Steelmaking

The steelmaking equipment includes two sets of vanadium recovery hot metal ladle units (VRU) (one set for production and one spare set), one crawler loader, 60 t K-OBM top-bottom blowing converter, two refining stations, and one machine, and one flow slab continuous casting machine. The factory originally had an electric furnace and a small billet continuous casting machine, but these have been removed.

During vanadium extraction from molten iron the V-containing molten iron supplied by rectangular smelting furnace is loaded in the ladle. Vanadium is extracted as vanadium slag by blowing oxygen into the ladle. The iron scale used as the coolant, and N_2 is used for stirring. When needed, ferrosilicon is added to remove oxygen during vanadium slag blowing. After blowing a hydraulic slag grilled machine is used to move the slag in the ladle on the original car.

During converter steelmaking a 60 tK-OBM top-bottom blowing converter is used. N_2, O_2, Ar, and CaO powder are blown in the bottom, and slag splashing is used for protection furnace, which is set up with temperature measurement, sampling, automatic control, primary wet dust removal, secondary BDC (bag dust-cleaning), and roof dust-removal systems. During steelmaking, iron is mixed first, and then scrap steel is added. The powder materials added to the converter include lime, limestone, high-magnesium lime, silica sand, and fluorite, where coke is used for heating. The total amount of slag produced during steelmaking is 100–120 kg/t steel. The quantity of steel in the converter is ∼73 t and the steel tapping temperature is ∼1680°C. All produced steel is low-carbon steel. Furnace age is 1300–1500, laying lining in different places. The whole body is replaced once every 6–7 weeks. The process flow of vanadium extraction from molten iron is shown in Fig. 18.6.

18.4.2 Vanadium extraction equipment

Molten iron ladle systems for vanadium extraction include a molten iron tank car with electric hydraulic tilting system, molten iron ladles, top-blowing oxygen lance and its lifting equipment, top-blowing nitrogen lance and its lifting equipment, automatic temperature measurement system, sampling lance and lifting equipment, grilled slag machine with hydraulics, and dust-removal system. The flue gas from vanadium extraction from molten iron is introduced into the secondary steelmaking converter dust-removal system.

The weight of a new molten iron ladle is 40 t, while the used ladle weighs 52–55 t. The ladle is lined with high-Al brick and has an inner diameter of 2490 mm, and height of 3550 mm. The molten iron tapping temperature from the rectangular furnace is 1500°C–1520°C. The molten Fe is sent to the vanadium extraction station in the steel plant at a temperature of 1380°C–1420°C. The clearance of the molten iron ladle is 500–800 mm. The molten iron composition is shown in Table 18.1.

18.5 China's vanadium industry

The production of vanadium in China was started by the Jinzhou Iron Institute in the period of Japanese control, where vanadium concentrate was used as the raw material for vanadium extraction. After the founding of the People's Republic of China, the administration of the iron and steel industry within the heavy industry ministry issued

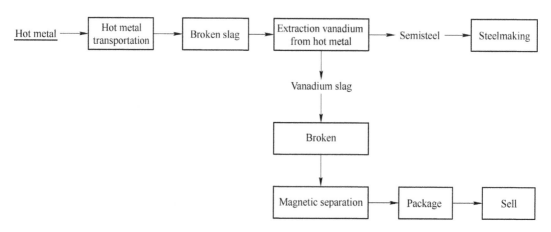

Figure 18.6
Present process flow of the New Zealand Steel Company.

Table 18.1: Molten iron components and temperature.

Amount of molten iron (t)	C (%)	Si (%)	Mn (%)	P (%)	S (%)	Ti (%)	V (%)	Cr (%)	Temperature (°C)
74	3.3	0.20	0.40	0.06	0.032	0.25	0.49	0.045	1400
62–86	3.0–3.8	0.06–0.40			0.024–0.048	0.08–0.40	0.45–0.53		1380–1420

scientific research tasks in 1954 including "extracting vanadium and smelting ferrovanadium from titanoferrous magnetite containing vanadium of Chengde great temple by pyro-metallurgy." Projects completed in 1955 provided the basis for the construction of China's ferrovanadium (FeV) industry. In 1958 the Jinzhou ferroalloy plant resumed production of FeV from concentrates of Chengde titanoferrous magnetite containing vanadium. In 1959 vanadium slag was first used to produce ferrovanadium. In the 1960s the national key development focus was in Panzhihua, where Panzhihua Iron and Steel Company was founded in 1965 to smelt V-bearing titanoferrous magnetite. In 1978 the Panzhihua Iron and Steel Company introduced an original atomization method, which resulted in China becoming one of the key players in vanadium production in the world. Panzhihua has changed the distribution of technological resources of vanadium, as well as the economic geography of China and the world. The vanadium industry changed Panzhihua, which attracted global attention. Plants extracting vanadium from vanadium slag were built, including the Jinzhou ferroalloy plant, Nanjing ferroalloy plant, Panzhihua Steel and Emei Ferroalloy plant, which changed the structure of the construction steel industry. Vanadium has played the role of "industrial monosodium glutamate," boosting rapid development of the industrial economy. In the 1970s and 1980s, several small vanadium plants were built in southern China to process rich stone-like coal resources, which partially resolved the shortage of vanadium resources at the time and promoted the widespread development of the vanadium industry in China.

18.5.1 Panzhihua vanadium industry

The Panzhihua-Xichang V-bearing titanoferrous magnetite resource has been holistically developed in China, which has attracted worldwide attention over a long period. Since 1872, foreign geological exploration of the Panzhihua-Xichang region was undertaken by Leahy (Germany), Law (Hungary), Le Suntech (France), the harvest of Switzerland. Chinese geological surveys of the Panzhihua-Xichang area were performed by Wenjiang Ding, Xichou Tan, Chunli Li, Chengsan Li, Jiqing Huang, Longqing Chang, Zhixiang Liu, and Yuqi Cheng. In 1936 geologists Longqing Chang and Xuezhong Yin investigated the minerals of Panzhihua-Xichang, and discovered impregnated magnetite related to granite in the Daomakan area of Panzhihua. In August 1943 Wuhan University professors of geology Zheng Chen and Chengfeng Xue were asked by the central geological survey bureau chief Gengyang Li to conduct qualitative analysis of Ti and Fe in mining samples. Meanwhile, the analysis results of Shanbang Li and Xinling Qin are cited. The chemical analysis results of the mining samples from the survey showed 51% Fe, 16% TiO_2, and 9% Al_2O_3, which proved that the Panzhihua iron ore contains Ti and is titanium magnetite. Wengui Guo and Zhizheng Ye of the state resources committee visited Panzhihua-Xichang, by the way was dropping in survey of the Panzhihua area, proving that the ore deposit was formed from magma. They pointed out that the main minerals are magnetite, and ilmenite in small quantities. In 1944 Yuqi Cheng showed that titanium magnetite contains vanadium. Thus it was determined that Panzhihua iron ore is V-bearing titanoferrous magnetite.

Panzhihua iron mine completed geological exploration in the 1950s and supplementary exploration in the mid-1960s. Based on these surveys, many groups have performed studies of the regularity of ore formation, mineral prediction, associated elements, expansion prospects, and selection of reserve exploration bases. By 1980 the Panzhihua-Xichang area had 54 proven iron ore deposits, with total reserves of 8.1 billion tons, of which 23 were V-bearing titanoferrous magnetite, with total reserves of 7.76 billion tons. In 1985 proven reserves of V-bearing titanoferrous magnetite in the Panxi region had reached 10 billion tons, accounting for more than 80% of these types of iron ore reserves in China. The vanadium reserves account for 87% of the national total, while the titanium reserves account for 92% of the total. Panzhihua city has six major mining areas in Panxi, including Panzhihua, Baima, Hongge, Anning village, Zhonggan ditch, and Baicao, with total reserves of 7.53 billion t.

Panzhihua Iron and Steel Co., LTD. is a very large vanadium, titanium, iron, and steel enterprise, which was developed solely by China under extremely difficult conditions, relying on the abundant resources of V-bearing titanoferrous magnetite in the Panzhihua-Xichang area. The construction of Panzhihua Iron and Steel Co., LTD. started in 1965, where iron was first produced in 1970, and steel was produced in 1974. After more than 40 years of construction and development, Panzhihua Iron and Steel Co., LTD. has been developed into a transregional, cross-industry group, which produces vanadium, titanium, iron, and steel and has strong market competitiveness and high visibility. They have a total production capacity of 20,000 t of vanadium products, 300,000 t of ilmenite, 23,000 t of titanium white powder, 8.3 million t of iron, 9.4 million t of steel, and 8.9 million t of rolled steel. Panzhihua Iron and Steel Co., LTD. is China's largest and the world's second-largest vanadium production company; China's largest producer of titanium materials and main chain; China's largest railway steel producer and most complete varieties structure of seamless pipe production, which has the most advanced heavy rail production line, and a first-class seamless steel tube production line. Panzhihua Iron and Steel Co., LTD. has a line of titanium and vanadium products including vanadium oxide, ferrovanadium, vanadium nitride alloy, and titanium dioxide produced using the chlorination process and sulfuric acid method. Their series of iron and steel products include 310 E steel, heavy rail, auto truck beam, cold rolled steel sheet, IF steel, seamless steel pipe, and military steel products.

The development of Panzhihua Iron and Steel Co., LTD. has experienced three important stages. The period from 1965 to 1980 was an important historical period for the company, during which they carried out the arduous first phase of construction and entrepreneurship and developed the local industry. The construction of Panzhihua Iron and Steel Co., LTD. began in 1965. Iron was first produced in 1970, steel was produced in 1971, and rolled steel was produced in 1974. In 1980 the technical and economic indexes of the main products had reached or exceeded the expected levels, with a total production capacity of 1.5 million t steel. The second phase of construction occurred from 1981 to 2000, which involved expansion to new level, adjustment of the product structure, and transformation from a "billet company" to "steel company," was an important historical period of strategic transformation. In this phase the new No. 4 blast furnace, slab continuous casting plate, and other equipment were commissioned. The overall equipment has reached level by the end of the 1980s and reached an advanced international level by the early 1990s. The production of iron, steel, and billet increased by 1 million t to an output of 4 million t/a.

Panzhihua Iron and Steel Co., LTD. is one of the largest vanadium producers in China. According to the output of V_2O_5, the V-containing raw materials produced by this company once accounted for ~74% of the national total and 18% of the world total. The rise of the vanadium industry in China was mainly due to the development and use of Panzhihua V-bearing titanoferrous magnetite. With the development of atomization technology for vanadium extraction by Panzhihua Iron and Steel Co., LTD. in 1972, vanadium was produced in China from scratch. In 1980 China changes from an importer of vanadium to an exporter of vanadium. Panzhihua V-bearing titanoferrous magnetite was dressed to produce V-containing iron concentrate; after sintering, it is smelted in a blast furnace, and molten iron containing vanadium is obtained. Vanadium is recovered from hot metal by atomization or converter extraction to produce vanadium slag. Panzhihua began extracting vanadium from molten iron in 1972 using the atomization technology that they invented. This technology was used in their plant and developed with independent intellectual property rights. Before 1995 the output of vanadium slag was 75,000 t/a, which was used to produce 1.5 million t/a of V-containing molten iron. With the expansion of steel production a converter vanadium production process was developed to improve handling of V-containing molten iron and the vanadium recovery yield. The molten iron processing capacity has recently reached 5 million t/a, and the vanadium slag production is ~230,000 t/a. The oxidation rate of vanadium was increased from <85% to ~90% using the converter, while the residual vanadium in the semisteel was decreased to <0.04%, and the technical indexes were greatly improved.

The V_2O_5 plant of Panzhihua Iron and Steel Co., LTD. was completed and put into operation in March 1990. The current production capacity is about 38,000 t/a (including the production capacity of Xichang branch of Panzhihua Iron and Steel Co., LTD.), where the vanadium yield of this process is 85%. To reduce production costs and increase production efficiency, V_2O_3 production technology was developed in 1998 and a V_2O_3 plant with the relevant equipment and technology was built. This plant had an annual capacity of 3350 t. The V_2O_3 output was 2180 t in 1999, and in 2000, the production reached its designed capacity. In 1991 Panzhihua Iron and Steel Co., Ltd. developed their own FeV80 production technology and equipment and began commercial production with an annual production capacity of 1300 t and vanadium yield over 95%. In 1993 the company introduced the FeV80 smelting thermite method and equipment from Luxembourg. In Beihai, Guangxi a second FeV80 production line was built with a production capacity of 2000 t to meet the different needs of domestic steel mills, Panzhihua Iron and Steel Co., LTD. has also produced FeV50 since 1998.

In 1991 Panzhihua Iron and Steel Co., LTD. developed vanadium carbide in the laboratory using ammonium vanadate, ferrovanadium, and nitrided ferrovanadium as raw materials. In 1997 they cooperated with Northeastern University to develop a process for producing vanadium nitride using vanadium trioxide as the raw material. Under normal the reaction $V_2O_3 + 4C = V_2C + 3CO$ occurred smoothly, where the carbonization reaction time of 40–60 h was reduced to 5 h, where there is potential to further shorten the carbonization time. The products have reached the international technical standards for similar products, and key breakthroughs have been made in the production process. As industrial batch applications achieved good results, a subsequent industrial production line of 4000 t/a was built.

Panzhihua Iron and Steel Co., LTD. took advantage of Panzhihua's vanadium resources and the developed technology. After 10 years of continuous efforts to develop dozens of varieties of V-containing steels, low-vanadium alloyed steel and vanadium micro-alloyed steel production accounted for more than 50% of the total steel production. Panzhihua Iron and Steel Co., LTD. is equipped with five blast furnaces and six converters for steelmaking and vanadium extraction. The V-containing molten iron production is 6 million t/a and the standard vanadium slag production is 250,000 t/a. Five production lines for V_2O_5, V_2O_3, ferrovanadium, and vanadium nitride alloy were installed, where the main production equipment includes four ball mills, four multihearth furnaces with 10 layers, three melting furnaces, five reduction roasting kilns, an electric furnace for ferrovanadium smelting, six pushed slab kilns, and two wastewater treatment facilities. The plant has a vanadium oxide (V_2O_5, V_2O_3) production capacity of 20,000 t/a, ferrovanadium (FeV80/50) capacity of 16,000 t/a, and vanadium nitride alloy capacity of 4000 t/a. The process flow for Panzhihua V-bearing titanoferrous magnetite is shown in Fig. 18.7.

Panzhihua Iron and Steel Co., LTD. constructed Panxi resources comprehensive utilization project in 2009. The main projects include a 10,000 t/a vanadium extraction line using calcium salt roasting, ferrovanadium production line, two 360 m^2 sintering machines, three 1750 m^3 blast furnaces, one 200 t converter for vanadium extracting, and two 200 t steelmaking converters, with a production capacity of vanadium slag of 227,000 t/a, and 18,800 t/a ferrovanadium.

Vanadium oxide production uses "clean production process of calcifying roasting—sulfuric acid leaching." The main processing steps include vanadium slag is ground in a ball mill and sieved to remove Fe by magnetic separation; calcium carbonate is added and mixed for calcification, followed by roasted in a rotary kiln; the clinker is acid leached; the V-

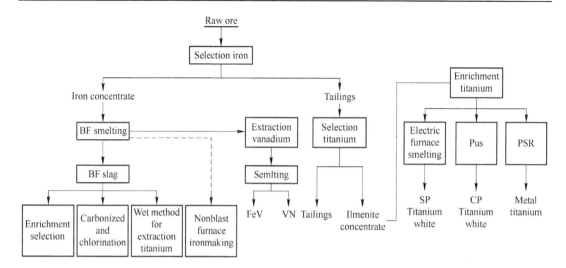

Figure 18.7
Process flow for Panzhihua V-bearing titanoferrous magnetite.

containing solution is continuously precipitated, filtered by a plate and frame filter press, and then air dried; finally, gas reduction is performed. This technology solved the challenge of treating wastewater from leaching and solid waste (sodium sulfate), where low-cost wastewater treatment can be used and manganese resources can be recycled. The production of ferrovanadium uses electric aluminum thermal reduction of vanadium trioxide. The intermediate products are 12,625 t/a V_2O_3 and 4000 t/a V_2O_5, and the product is 18,800,000 t/a FeV50 (or FeV80).

18.5.2 Chengde vanadium industry

Chengde Iron and Steel Group Co., LTD. was founded in 1954 as one of the 156 projects in China funded by the former Soviet Union. This leading company was the birthplace of China's vanadium and titanium industry and is located in Luanhe river town, Shuangluan district, Chengde city, Hebei province, China. The company developed and improved vanadium extraction and smelting technology using Chengde V-bearing titanoferrous magnetite over the last 50 years. They gradually developed a complete production line for vanadium and titanium products, where smelting and rolling of V-Ti containing low alloy steel is their main business, along with mixed smelting, refining together.

The Damiao deposit of V-bearing titanoferrous magnetite is located in the Xuanhua−Chengde−Beipiao deep-fault zone at the eastern end of the Inner Mongolia earth axis, where the basic-ultrabasic rock intruded into the pre-Sinian stratum. Residual slurry separated from late ore-bearing slurry was formed by penetrating structural fractures. titanium magnetite ore bodies were produced in lenticular, vein, or cystic forms from anorthosite or anorthosite contact parts of the fracture zone, and surrounding rock boundaries. The ore body in gabbro is usually disseminated or veined and has a gradual relationship with the surrounding rocks. The ore body is generally 10–360 m long, with a depth of 10–300 m. The ore has two types: dense block and disseminated. The main ore minerals include magnetite, ilmenite, hematite, and rutile. Magnetite and ilmenite are separated as solid solutions. Vanadium occurs as an isomorphism in V-bearing titanoferrous magnetite. The average grade of V_2O_5 ore is 0.16%–0.39%. In iron concentrate the V_2O_5 content is 0.77%.

In 1960 Chengde Iron and Steel Group successfully developed the air side blown converter vanadium extraction process. Research regarding the technology for smelting V-bearing titanoferrous magnetite and high titanium in a blast furnace was completed in 1965, while development of a new industrial-scale vanadium extraction technology by water leaching was completed in 1967. In 1972 a new technology for smelting vanadium slag by direct alloying in an electric furnace was developed. Following the British standard BS4449, Chengde Iron and Steel Group produced V-containing high-strength reinforced bar in 1980. In 1984 a new technology for direct alloying converter vanadium slag was developed, which has since achieved success. In 2009 Chengde Iron and Steel Group achieved a steel capacity of 8 million t, a vanadium slag capacity of 360,000 t, and a vanadium product capacity of 30,000 t. Large-scale equipment has been developed and the main equipment has been greatly modernized. Their main V-containing products include twisted HRB500, HRB400, HRB335 steel, low-V alloyed round steel, strip steel, high-speed wire rod, vanadium pentoxide (tablet, powder), ferrovanadium, and high-grade titanium concentrate.

This company has six blast furnaces with a total volume of 4179 m^3, with individual volumes of 2500, 1260, 450, 380, 315, and 274 m^3. In the 30-t converter system, there are five converters, including one 80 t vanadium extraction converter and four 30 t steelmaking converters. In the 120 t converter system, there are three 120 t converters, including one vanadium extraction converter and two steelmaking converters. In the 150 t converter system, there are three 150 t converter, including one vanadium extraction converter and two steelmaking converters. There is a 30,000 t/a vanadium pentoxide production line, a VN production line, and a ferrovanadium production line. The production process used by Chengde Iron and Steel Group is shown in Fig. 18.8, while the production process for vanadium products is shown in Fig. 18.9. The production capacity of vanadium slag is 360,000 t/a and vanadium pentoxide is 30,000 t/a.

18.5.3 Jinzhou ferroalloy factory

The Jinzhou ferroalloy factory was formerly known as the Jinzhou smelting institute of Japan (Manchuria) Special Iron Ore Corporation, which was founded in 1940. In 1953 the Chinese government and the Soviet Union signed a construction assistance project

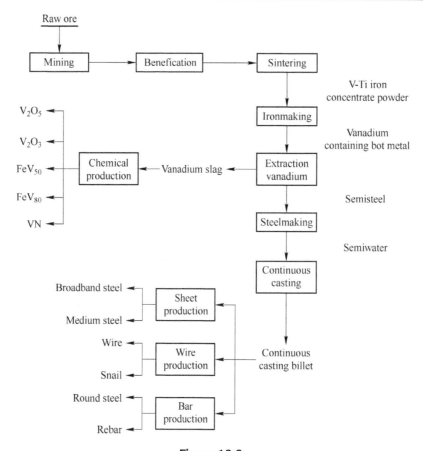

Figure 18.8
Production process of Chengde iron and steel group company.

agreement, including the former Jinzhou Smelting Institute. After the founding of the People's Republic of China, the administration of the iron and steel industry within the heavy industry ministry issued scientific research tasks in 1954 including "extracting vanadium and smelting ferrovanadium from titanoferrous magnetite containing vanadium of Chengde great temple by pyro-metallurgy." Projects completed in 1955 provided the basis for the construction of China's ferrovanadium (FeV) industry. In 1957 the Ministry of the Metallurgical Industry of China issued a document regarding the Jinzhou Ferroalloy Factory. In 1958 Jinzhou ferroalloy plant resumed the production of ferrovanadium from concentrate of Chengde V-bearing titanoferrous magnetite, which began to produce ferrovanadium in 1959. In 1959 vanadium slag was used to produce ferrovanadium. In 1959 the trial production of ferrosilicon, ferronitride, ferroboron, electrolytic chromium powder, and metal vanadium was successfully demonstrated and production was begun. From 1961 to 1969, trial production of metallic chromium, vanadium aluminum alloy, high titanium

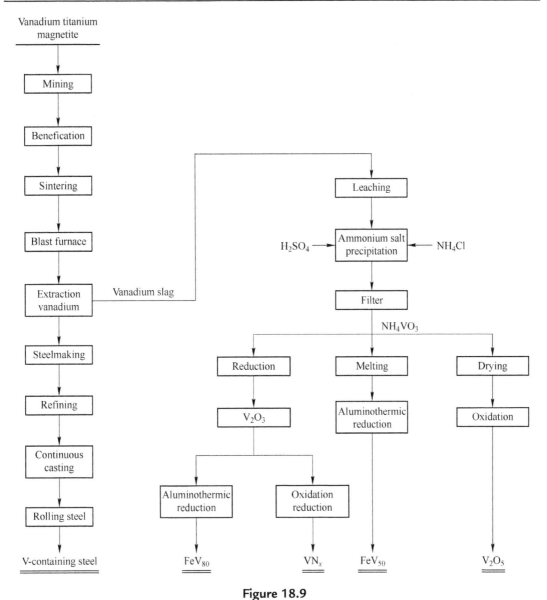

Figure 18.9
Production process of vanadium products in Chengde iron and steel group company.

slag, rutile, zirconium carbide, and sponge zirconia was successfully demonstrated and upscaled to production lines. From 1970 to 1977, trial production of metallic titanium, manganese silicon alloy, ferromanganese, sponge titanium, electric vacuum zirconium powder, and molybdenum oxide block was successfully performed. In 1980 ferrovanadium won the national quality silver award. From 1981 to 1982 the development of hot cell chromium powder and sodium vanadate was successful.

Jinzhou ferroalloy factory was one of the first key national ferroalloy producers identified by the state, including Nanjing and Emei ferroalloy factories, which are considered pioneers and practitioners of China's industrial vanadium production. Bold and effective attempts have been successfully made to use different raw materials, processes, and equipment. Various vanadium technologies have been developed and promoted in China. Jinzhou ferroalloy factory has produced metal alloys and metal oxides of eight elements, including vanadium, titanium, chromium, manganese, molybdenum, magnesium, zirconium, and hafnium; they produce a total of 66 species with more than 90 brands, and a total production capacity of 70,000 t/a. The ferrovanadium, vanadium pentoxide, vanadium aluminum alloy, ferrotitanium, metal chromium, molybdenum oxide, zirconium powder, sponge zirconium, and sponge hafnium products are important for local industries. Quality brand products account for 88% of the total ferroalloy produced.

18.5.4 Vanadium extraction from stone coal

Stone coal is a type of highly metamorphic sapropelic or algal coal, which generally has high ash and sulfur contents, and low heat and hardness. In addition to organic carbon, it also contains silica, calcium oxide, and small amounts of iron oxide, aluminum oxide, and magnesium oxide. It looks like stone and is not easily distinguished from limestone or carbon shale. It is a combustible and kata-metamorphism organic mineral with high ash content (generally greater than 60%). High-quality stone coal with high-carbon content is black, with a semibright luster, few impurities, and relative density of 1.7–2.2. Stone coal with a low-carbon content is grayish and a little dull and contains significant amounts of pyrite, quartz vein, and phosphorus. The calcium nodules have a relative density of 2.2–2.8. The calorific value of stone coal is not high, between 3.5 and 10.5 MJ/kg, and is hence a type of low-calorific fuel.

V-bearing stone coal can be used for the extraction of vanadium pentoxide. The V_2O_5 content in the stone coal is low, usually about 1.0%. In V-bearing stone coal, most vanadium occurs as $V(III)$, with some $V(IV)$, and rarely as $V(V)$. As the ionic radii of V(III) and Fe(V) are equal (74 pm), and that of Fe(III) is similar (64 pm), $V(III)$ does not occur as a single-metal mineral but occurs in the silica tetrahedral structure of ferroaluminum minerals such as roscoelite and kaolin by isomorphism.

Stone coal resources are widely distributed in China in Sichuan, Chongqing, Shanxi, Gansu, Hubei, Hunan, Jiangxi, Zhejiang, Guangxi, Guangdong, and Anhui provinces and have become an important raw material. In the vanadium extraction process, there are scattered, messy, small and micro phenomenon, products are vanadium powder, which lack integrity process, main control technology is generally poor, ability and market mix, production capacity is erratic. In heyday, stone coal vanadium plant reached forty at home.

18.5.4.1 Vanadium extraction by pyrolysis

Vanadium stone coal occurs with a silicon oxygen tetrahedron structure of clay minerals with trivalent by isomorphism, which is strongly combined and insoluble in acid and alkali solutions. At high temperature, additives are introduced to transform trivalent vanadium into soluble pentavalent vanadium; meanwhile, carbon in the stone coal is removed. Hence, roasting and transformation of V-bearing stone coal is an indispensable process.

The main steps in V_2O_5 extraction from shale by pyrogenic processes are transformation and roasting of the mineral. Roasting processes include blank roasting and doping roasting. Blank roasting does not use any additives, and high concentrations of acid are required for leaching and decomposition. Additives (such as Na, Ca, Fe, Ba and other salts, and sulfuric acid) are used in doping roasting to produce sodium vanadate, calcium vanadate, and other vanadates that are soluble in water or acid.

In traditional pyrogenic processes for extracting V_2O_5 from shale, NaCl and Na_2CO_3 are used as roasting additives. This results in large amounts of Cl_2, HCl, and other poisonous and harmful gas (e.g., SO_2) being produced, which leads to serious smoke pollution. The salinity of the wastewater is high, and the vanadium recovery is generally only about 50%. In addition, severe resource wastage occurs and the production environment is poor. Many sodium ions are present in the leaching residue, which cannot be scale multipurpose utilization. Calcium roasting does not produce Cl_2, HCl, or poisonous and harmful gases, such as SO_2. However, this roasting process is greatly influenced by the mineral type and its properties, and the roasting temperature, time, atmosphere, and dosage of calcium salt. If there is insufficient process control, calcium roasting can easy produce insoluble silicates, where some vanadium particles are coated with silica, or some of the vanadium reacts with other elements, such as Fe and Ca, to produce insoluble calcium vanadate and iron metavanadate compounds. The calcification processing residue can be large-scale multipurpose utilization.

Blank roasting is mainly used for stone coal decarburization and low-valence vanadium oxidation and has certain limitations related to the mineral structure. However, the roasting equipment is simply traditional shaft kilns and fluidized beds. The production scale is limited, and the vanadium crystal structure is not completely converted during the roasting process, which limits the recovery of vanadium. This process has poor adaptability to stone coal resource utilization. Sulfation roasting can enhance the mineral decomposition process. The sulfation roasting temperature is 200°C–250°C, roasting time is 0.5–1.5 h, and leaching is achieved using an aqueous solution with calcined clinker with pH 1.0–1.5, which results in a high use of sulfuric acid. The boiling point of sulfuric acid is 338°C and its concentration is 98.3%. The gas from roasting is mainly water vapor, so cleaning the gas is easy. Sulfation roasting of stone coal at low temperature is simple without oxidation.

The leaching step during pyrogenic extraction of vanadium pentoxide from shale is classified into three types: water leaching, alkali leaching, and acid leaching. Water leaching is only applicable for sodium vanadate, which is soluble in water, and this process has been widely used with sodium salt roasting. Alkali leaching is suitable for the calcium salt roasting process. It is highly selective and can be used for circulating treatments. It is suitable for the treatment of stone coal with more alkaline gangue. Acid leaching processes are classified into concentrated-acid leaching and weak-acid leaching methods. Concentrated-acid leaching uses large amounts of acid but can leach more impurities; the acidity of the reacted acid is high, and its recovery is low. The distinctive features of weak-acid leaching include a long reaction time, moderate leaching of impurities, low acidity, and low recovery of the reacted acid. The leaching method can also be divided into powder leaching and pellet leaching. Powder leaching is faster than pellet leaching, although similar leaching efficiencies are obtained.

The purification and enrichment of the leaching solution use adsorption and extraction of a resin, where resin adsorption is only suitable for neutral leaching solutions. Extraction is divided into tetravalent and pentavalent vanadium extraction.

High-carbon stone coal needs to be decarbonized, and the vanadium is enriched as ash, which is used as a raw material for extracting V_2O_5. Some of the enriched ash has high contents of V_2O_5, which is suitable for extracting V_2O_5 in combination with vanadium slag or recycled V-containing materials.

18.5.4.2 Wet vanadium extraction process

The wet vanadium extraction process does not require roasting. Acid leaching via a wet process is mainly used for weathered stone coal. To obtain a higher V_2O_5 leaching efficiency, a large amount of H_2SO_4 is consumed. In production the H_2SO_4 dosage is commonly 25%–40% of the ore quantity. The V_2O_5 leaching efficiency is generally 65%–65%, more than 80% of the efficiency is very few, while the V_2O_5 recovery is generally not more than 70%. Purification of the acid leach liquor to removing impurities is difficult. Fe(III) reduction and pH adjustment consumes a large amount of additives, especially ammonia, leading to the production of wastewater containing ammonia and nitrogen that is difficult to treat.

Important schedule

See Tables A.1 and A.2 for physical properties of elements.

Table A.1: Physical properties of elements.

Elemental symbol	Element name	Melting point (°C)	Boiling point (°C)	Heat capacity [J/(kg K)]	Density (20°C) (g/cm³)
Ag	Silver	960.15	2117	234	10.5
Al	Aluminum	660.2	2447	900	2.6984
Ar	Argon	−189.38	−185.87	519	1.7824×10^{-3}
As	Arsenic	817 (12.97 MPa)	613	326 (sublimation)	2.026 (yellow) 4.7 (black)
Au	Gold	1063	2707	130	19.3
B	Boron	2074	3675	1030	2.46
Ba	Barium	850	1537	192	3.59
C	Carbon	4000 (6.83 MPa)	3850 (sublimation)	711 519	2.267 (graphite) 3.515 (diamond)
Ca	Calcium	861	1478	653	1.55
Ce	Cerium	795	3470	184	6.771
Cl	Chlorine	−101.0	−34.05	477	2.98×10^{-3} (gas)
Co	Cobalt	1495	3550	435	8.9
Cr	Chromium	1990	2640	448	7.2
Cu	Copper	1683	2582	385	8.92
F	Fluorine	−219.62	−188.14	824	1.58×10^{-3}
Fe	Iron	1530	3000	448	7.86
H	Hydrogen	−259.2	−252.77	1.43×10^4	0.8987×10^{-3}
Hg	Mercury	−38.87	365.58	138	13.5939
K	Potassium	63.5	758	753	0.87
Mg	Magnesium	650	1117	1.03×10^3	1.74
Mn	Manganese	1244	2120	477	7.30
Mo	Molybdenum	2625	4800	251	10.2
N	Nitrogen	−209.97	−195.798	1.04×10^3	1.165×10^{-3}
Na	Sodium	97.8	883	1.23×10^3	0.97
Ni	Nickel	1455	2840	439	8.90
O	Oxygen	−218.787	−182.98	916	1.331×10^{-3}

(Continued)

Table A.1: (Continued)

Elemental symbol	Element name	Melting point (°C)	Boiling point (°C)	Heat capacity [J/(kg K)]	Density (20°C) (g/cm³)
P	Phosphorus	44.2 59.7 610	280.3 431 (sublimation) 453 (sublimation)		1.828 (white) 2.34 (red) 2.699 (black)
Pb	Lead	327.4	1751		11.34
Pt	Platinum	1774	−3800	130	21.45
Re	Rhenium	3180	5885	138	21.04
Rh	Rhodium	1966	3700	243	12.41
S	Sulfur	112.3 114.6 106.8	444.60	732	2.68 (α) 1.96 (β) 1.92 (γ)
Sb	Antimony	630.5	1640	20	6.684
Si	Silicon	1415	2680	711	2.33
Sn	Tin	231.39	2687	218	7.28 (white)
Ti	Titanium	1672	3260	523	4.507 (α) 4.32 (β)
V	Vanadium	1919	3400	481	6.1
W	Tungsten	3415	5000	134	19.35
Zn	Zinc	419.47	907	285	7.14
Zr	Zirconium	1855	4375	276	6.52 (mixed)

Table A.2: Physical properties of elements.

Elemental symbol	Element name	Thermal conductivity [W/(m K)]	Electrical resistivity (Ω m)	Heat of fusion (kJ/mol)	Heat of vaporization (kJ/mol)
Ag	Silver	4182	1.6×10^{-8}	11.95	254.2
Al	Aluminum	211.015	2.6×10^{-8}	10.76	284.3
Ar	Argon	0.016412		1.18	6.523
As	Arsenic	817 (12.97 MPa)	3.5×10^{-7}		
Au	Gold	293.076	2.4×10^{-8}	12.7	310.7
B	Boron		1.8×10^{-4}		
Ba	Barium		6.0×10^{-7}	7.66	149.32
C	Carbon	23.865	1.375×10^{-5}	104.7	326.6 (sublimate)
Ca	Calcium	125.604	4.5×10^{-8}	9.2	161.2
Ce	Cerium		7.16×10^{-7}		
Cl	Chlorine		>10 (liquid state)	6.410	20.42
Co	Cobalt	69.082	0.8×10^{-7}	15.5	398.4
Cr	Chromium	66.989	1.4×10^{-7}	14.7	305.5
Cu	Copper	414.075	1.6×10^{-8}	13.0	304.8
F	Fluorine			1.56	6.37
Fe	Iron	75.362		16.2	354.3
H	Hydrogen			0.117	0.904

(Continued)

Table A.2: (Continued)

Elemental symbol	Element name	Thermal conductivity [W/(m K)]	Electrical resistivity (Ω m)	Heat of fusion (kJ/mol)	Heat of vaporization (kJ/mol)
Hg	Mercury	10.476	9.7×10^{-7} (liquid) 2.1×10^{-7} (solid)	2.33	58.552
K	Potassium	97.134	6.6×10^{-8}	2.334	79.05
Mg	Magnesium	157.424	4.4×10^{-8}	9.2	13.9
Mn	Manganese			14.7	224.8
Mo	Molybdenum	146.358	0.5×10^{-7}		
N	Nitrogen			0.720	5.581
Na	Sodium	132.722	4.4×10^{-8}	2.64	98.0
Ni	Nickel	58.615	6.8×10^{-8}	17.6	378.8
O	Oxygen			0.444	6.824
Sb	Antimony	22.525	3.9×10^{-7}	20.1	195.38
Si	Silicon	83.736		46.5	297.3
Sn	Tin	64.058	1.15×10^{-7}	7.08	230.3
Ti	Titanium		0.3×10^{-7}		
V	Vanadium		5.9×10^{-7}		
W	Tungsten	167.472	5.48×10^{-8}		
Zn	Zinc	110.950	5.9×10^{-8}	6.678	114.8
Zr	Zirconium		4.0×10^{-7}		

See Table A.3 for physical properties of common oxide.

Table A.3: Common oxide physical properties.

Oxide	Mass fraction of oxygen (%)	Density (g/cm³)	Melting temperature (°C)	Evaporation temperature (°C)
Fe_2O_3	30.057	5.1–5.4	1565	
Fe_3O_4	27.640	5.1–5.2	1597	
FeO	22.269 (partly stability) 23.239–23.28 (stable)	5.163 (oxygen content 23.91%)	1371–1385	
SiO_2	53.257	2.65 (quartz)	1713 (silica 1750)	2590
SiO	36.292	2.13–2.15	1350–1900 (sublimation)	1990
MnO_2	36.807	5.03	535 (predecomposition)	
Mn_2O_8	30.403	4.30–4.80	940 (predecomposition)	
Mn_3O_4	27.970	4.30–4.90	1567	
MnO	22.554	5.45	1750–1788	
Cr_2O_3	31.580	5.21	2275	
TiO_2	40	4.26 (rutile) 3.84 (anatase)	1825	3000
P_2O_5	49	2.39	569 (pressurized)	350 (sublimation)
TiO	56.358	4.93	1750	

(Continued)

Table A.3: (Continued)

Oxide	Mass fraction of oxygen (%)	Density (g/cm³)	Melting temperature (°C)	Evaporation temperature (°C)
V_2O_5	25.038	3.36	663–675	1750 (decomposition)
VO_2	43.983	4.30	1545	
V_2O_3	38.581	4.84	1967	
VO	32.024	5.50	1970	
NiO	25.901	6.80	1970	
CuO	21.418	6.40	1148 (decomposition), 1062.6	
Cu_2O	20.114	6.10	1235	
ZnO	19.660	5.5–5.6	2000 (5.629 MPa)	1950 (sublimation)
PbO	7.168	9.12 ± 0.05 (22°C) 7.794 (880 °C)	888	1470
CaO	28.530	3.4	2585	2850
MgO	39.696	3.2–3.7	2799	3638
BaO	10.436	5.0–5.7	1923	−2000
Al_2O_3	47.075	3.5–4.1	2042	2980
K_2O	16.986			766
Na_2O	25.814			890

See Table A.4 for relationship between free energy of common chemical reactions and temperature.

Table A.4: Relationship between free energy of common chemical reactions and temperature.

	$\Delta G^0 = A + BT$ (J/mol)			
Reactions	A (J/mol)	B [J/(mol K)]	Error (kJ)	Temperature range
$Al_{(s)} = Al_{(l)}$	10,795	−11.55	0.2	660 (melting point)
$Al_{(l)} = Al_{(g)}$	304,640	−109.50	2	660–2520 (boiling point)
$Al_{(s)} + 1.5O_2 = Al_2O_3$ (s)	−1,675,100	313.20		22–660 (melting point)
$Al_{(l)} + 1.5O_2 = Al_2O_3$ (s)	−1,682,900	323.24		660–2024 (melting point)
$Al_{(l)} + 1.5O_2 = Al_2O_3$ (l)	−1,574,100	275.01		2042–2494 (boiling point)
$Al_{(g)} + 1.5O_2 = Al_2O_{3(l)}$	−2,106,400	468.62		2494–3200
$Al_{(l)} + 0.5O_2 = Al_2O_{(g)}$	−170,700	−49.37	20	660–2000
$Al_{(l)} + O_2 = Al_2O_{2(g)}$	−470,700	28.87	20	660–2000
$3Al_{(l)} + 3 C = Al_4C_{3(S)}$	−265,000	95.06	8	660–2200 (melting point)
$Al_{(l)} + 0.5N_{2(g)} = AlN_{(S)}$	−327,100	115.52	4	660–2000
$Al_2O_{3(s)} + SiO_{2(s)} = Al_2O_3 \cdot SiO_{2(s)}$	−8800	3.80	2	25–1700
$2Al_2O_3 + 2SiO_2 = 3Al_2O_3 \cdot 2SiO_{2(s)}$	−8600	−17.41	4	25–1750 (melting point)
$Al_2O_3 + TiO_2 = Al_2O_3 \cdot TiO_{2(s)}$	−25,300	3.93		25–1860 (melting point)

(Continued)

Table A.4: (Continued)

	$\Delta G^0 = A + BT$ (J/mol)			
Reactions	A (J/mol)	B [J/(mol K)]	Error (kJ)	Temperature range
$C_{(s)} = C_{(g)}$	713,500	−155.48	4	1750–3800
$C_{(s)} + 0.5O_2 = CO_{(g)}$	−114,400	85.77	0.4	500–2000
$C_{(s)} + 2H_{2(g)} = CH_{4(g)}$	−91,044	110.67	0.4	500–2000
$Ca_{(s)} = Ca_{(l)}$	8540	−7.70	0.4	839 (melting point)
$Ca_{(l)} = Ca_{(g)}$	157,800	−87.11	0.4	839–1491 (melting point)
$Ca_{(l)} + F_{2(g)} = CaF_{2(s)}$	−1,219,600	162.3	8	839–1484
$CaF_{2(g)} = CaF_{2(l)}$	2970	−17.57	0.4	1418 (melting point)
$CaF_{2(l)} = CaF_{2(g)}$	308,700	−110.0	4	2533 (boiling point)
$Ca_{(l)} + 2C_{(s)} = CaC_{2(s)}$	−60,250	−26.28	12	839–1484
$Ca_{(l)} + 0.5S_{2(g)} = CaS (s)$	−548,100	103.85	4	839–1484
$3CaO + Al_2O_3 = 3CaO \cdot Al_2O_{3(s)}$	−12,600	−24.69	4	500–1535
$CaO + Al_2O_3 = CaO \cdot Al_2O_{3(s)}$	−18,000	−18.83	2	500–1605
$CaO + 2Al_2O_3 = CaO \cdot 2Al_2O_{3(s)}$	−16,700	−25.52	3.2	500–1750
$CaO + 6Al_2O_3 = CaO \cdot 6Al_2O_{3(s)}$	−16,380	−37.58	1.7	1100–1600
$CaO + CO_{2(g)} = CaCO_{3(s)}$	−161,300	137.23	1.2	700–1200
$CaO + Fe_2O_3 = CaO \cdot Fe_2O_{3(s)}$	−29,700	−4.81	4	700–1216 (melting point)
$2CaO + Fe_2O_3 = 2CaO \cdot Fe_2O_{3(s)}$	−53,100	−2.51	4	700–1450 (melting point)
$3CaO + SiO_2 = 3CaO \cdot SiO_{2(s)}$	−118,800	−6.7	12	25–1500
$3CaO + 2SiO_2 = 3CaO \cdot 2SiO_{2(s)}$	−236,800	9.6	12	25–1500
$2CaO + SiO_2 = 2CaO \cdot SiO_{2(s)}$	−118,800	−11.3	12	25–2130 (melting point)
$CaO + SiO_2 = CaO \cdot SiO_{2(s)}$	−92,500	2.5	12	25–1540 (melting point)
$3CaO + 2TiO_2 = 3CaO \cdot 2TiO_{2(s)}$	−207,100	−−11.51	10	25–1400
$4CaO + 3TiO_2 = 4CaO \cdot 3TiO_{2(s)}$	−292,900	−17.57	8	25–1400
$CaO + TiO_2 = CaO \cdot TiO_{2(s)}$	−79,900	−3.35	3.2	25–1400
$CaO + MgO = CaO \cdot MgO$	−7200	0.0	1.2	25–1027
$3CaO + V_2O_5 = 3CaO \cdot V_2O_{5(s)}$	−332,200	0.0	5	25–670
$2CaO + V_2O_5 = 2CaO \cdot V_2O_{5(s)}$	−264,800	0.0	5	25–670
$CaO + V_2O_5 = CaO \cdot V_2O_{5(s)}$	−146,000	0.0	5	25–670
$Cr_{(s)} + 1.5O_2 = CrO_{3(s)}$	−580,500	259.2		25–187 (melting point)
$Cr_{(s)} + 1.5O_2 = CrO_{3(l)}$	−546,600	185.8		187–727
$Cr_{(s)} + O_2 = CrO_{2(l)}$	−587,900	170.3		25–1387
$2Cr_{(s)} + 1.5O_2 = Cr_2O_{3(s)}$	−1,110,140	247.32	0.8	900–1650
$2Cr_{(s)} + 1.5O_2 = Cr_2O_{3(s)}$	−1,092,440	237.94		1500–1650
$3Cr_{(s)} + 2O_2 = Cr_3O_{4(s)}$	−1,355,200	264.64	0.8	1650–1655 (melting point)
$Cr_{(s)} + 0.5O_2 = CrO_{(l)}$	−334,220	63.81	0.8	1665–1750
$Fe_{(s)} = Fe_{(l)}$	13,800	−7.61	0.8	1536 (melting point)
$Fe_{(l)} = Fe_{(g)}$	363,600	−116.23	1.2	1536–2862 (boiling point)
$Fe_{(s)} + 0.5O_2 = FeO_{(s)}$	−264,000	64.59	0.8	25–1377
$Fe_{(l)} + 0.5O_2 = FeO_{(l)}$	−256,060	53.68	2	1377–2000
$3Fe_{(s)} + 2O_2 = Fe_3O_{4(s)}$	−1,103,120	307.38	2	25–1597 (melting point)
$2Fe + 1.5O_2 = Fe_2O_{3(s)}$	−815,023	251.02	2	25–1462
$Fe_{(s)} + 0.5O_2 + V_2O_{3(s)} = FeO \cdot V_2O_{3(s)}$	−288,700	62.34	1.2	750–1536
$Fe_{(l)} + 0.5O_2 + V_2O_{3(s)} = FeO \cdot V_2O_{3(s)}$	−301,250	70.0	1.2	1536–1700
$3Fe_{(\alpha)} + C_{(s)} = FeC_{3(S)}$	29,040	−28.03	0.4	25–727

(Continued)

Table A.4: (Continued)

	$\Delta G^0 = A + BT$ (J/mol)			
Reactions	A (J/mol)	B [J/(mol K)]	Error (kJ)	Temperature range
$3Fe_{(\gamma)} + C_{(s)} = FeC_{3(S)}$	11,234	−11.0	0.4	727–1137
$Fe_{(\gamma)} + 0.5S_{2(g)} = FeS$ (s)	−336,900	224.51	4	630–760
$2FeO + SiO_2 = 2FeO \cdot SiO_{2(s)}$	−36,200	−61.67	4	25–1220 (melting point)
$2FeO.SiO_{2(s)} = 2FeO \cdot SiO_{2(l)}$	92,050	−61.67	4	1220 (melting point)
$2FeO + TiO_2 = 2FeO \cdot TiO_{(s)}$	−33,900	5.86	8	25–1100
$FeO + TiO_2 = FeO \cdot TiO_{(s)}$	−33,500	12.13	4	25–1300
$Fe_{(l)} + 0.5O_2 + V_2O_{3(s)} = FeO \cdot V_2O_{3(s)}$	−288,700	62.34	1.2	750–1536
$Fe_{(l)} + 0.5O_2 + V_2O_{3(s)} = FeO \cdot V_2O_{3(s)}$	−301,250	70.0	1.2	100 (boiling point)
$H_2O_{(l)} = H_2O_{(g)}$	41,086	−110.12	0.12	25–2000
$H_2 + 0.5O_2 = H_2O_{(g)}$	−247,500	55.86	1.2	25–2000
$H_2 + 0.5S_{2(g)} = H_2S_{(g)}$	−91,630	50.58	1.2	649 (melting point)
$Mg_{(s)} = Mg_{(l)}$	8950	−9.71	0.4	649–1090 (boiling point)
$Mg_{(l)} = Mg_{(g)}$	129,600	95.14		25–649 (melting point)
$Mg_{(s)} + 0.5O_2 = MgO_{(s)}$	−601,230	107.59		649–1090 (boiling point)
$Mg_{(l)} + 0.5O_2 = MgO_{(s)}$	−609,570	116.52		1090–1727
$Mg_{(g)} + 0.5O_2 = MgO_{(s)}$	−732,700	205.99		25–1400
$MgO_{(s)} + Al_2O_{3(s)} = MgO \cdot Al_2O_{3(s)}$	−35,600	−2.09	3.3	700–1400
$MgO_{(s)} + Fe_2O_{3(s)} = MgO \cdot Fe_2O_{3(s)}$	−19,250	−2.01	3.3	25–1500
$MgO_{(s)} + Cr_2O_{3(s)} = MgO \cdot Cr_2O_{3(s)}$	−42,900	7.11	5	25–1898 (melting point)
$MgO_{(s)} + SiO_{2(s)} = MgO \cdot SiO_{2(s)}$	−67,200	4.31	6	25–1577 (melting point)
$2MgO_{(s)} + SiO_{2(s)} = 2MgO \cdot SiO_{2(l)}$	−41,100	6.10	6	25–1500
$MgO_{(s)} + TiO_{2(s)} = MgO \cdot TiO_{2(s)}$	−25,500	1.26	2	25–1500
$MgO_{(s)} + TiO_{2(s)} = MgO \cdot 2TiO_{2(s)}$	−26,400	3.14	3	25–1500
$MgO_{(s)} + 2TiO_{2(s)} = MgO \cdot 2TiO_{2(s)}$	−27,600	0.63	3.3	25–670
$2MgO_{(s)} + V_2O_{5(s)} = 2MgO \cdot V_2O_{5(s)}$	−721,740	0	6	25–1200
$MgO_{(s)} + V_2O_{5(s)} = 2MgO \cdot V_2O_{5(s)}$	−53,350	8.4	3	1244 (melting point)
$Mn_{(s)} = Mn_{(l)}$	12,130	−7.95		1244–2062 (boiling point)
$Mn_{(l)} = Mn_{(s)}$	235,800	−101.17	4	25–1277
$Mn_{(s)} + 0.5O_2 = MnO_{(s)}$	−385,360	73.75		25–1277
$3Mn_{(s)} + 2O_2 = Mn_3O_{4(s)}$	−1,381,640	334.67		25–1277
$2Mn_{(s)} + 1.5O_2 = Mn_2O_{3(s)}$	−956,400	251.71		25–727
$Mn_{(s)} + O_2 = MnO_{2(s)}$	−519,700	180.83		527–1277
$MnO_{(s)} + Al_2O_{3(s)} = MnO \cdot Al_2O_{3(s)}$	−48,100	7.3	6	25–1291 (melting point)
$MnO_{(s)} + SiO_{23(s)} = MnO \cdot SiO_{2(s)}$	−28,000	2.76	12	25–1345 (melting point)
$2MnO_{(s)} + SiO_{2(s)} = 2MnO \cdot SiO_{2(s)}$	−53,600	24.73	12	25–1360
$MnO_{(s)} + TiO_{2(s)} = MnO \cdot TiO_{2(s)}$	−24,700	1.25	20	25–1450
$2MnO_{(s)} + TiO_{2(s)} = 2MnO \cdot TiO_{2(s)}$	−37,700	1.7	20	98–675 (melting point)
$MnO_{(s)} + V_2O_{5(s)} = MnO \cdot V_2O_{5(s)}$	−65,900		6	98–801 (melting point)
$2Na_{(l)} + 0.5O_2 = Na_2O_{(s)}$	−514,600	218.8	12	98–883 (melting point)
$Na_{(l)} + 0.5Cl_2 = NaCl_{(s)}$	−411,600	93.00	0.4	850–2200
$2Na_{(l)} + C_{(s)} + 0.5O_2 = Na_2CO_{3(s)}$	−1,227,500	273.54		250–884 (melting point)
$2Na_{(l)} + C_{(s)} + 0.5O_2 = Na_2CO_{3(l)}$	−1,229,600	362.47		25–1089 (melting point)
$Na_2O_{(s)} + SO_{2(g)} + 0.5O_2 = Na_2SO_{4(s)}$	−651,400	237.3	12	25–974 (melting point)

(Continued)

Table A.4: (Continued)

	$\Delta G^0 = A + BT$ (J/mol)			
Reactions	A (J/mol)	B [J/(mol K)]	Error (kJ)	Temperature range
$Na_2O_{(s)} + SiO_{2(s)} = Na_2O \cdot SiO_{2(s)}$	−237,700	−3.85	12	25–1030 (melting point)
$Na_2O_{(s)} + 2SiO_{2(s)} = Na_2O \cdot 2SiO_{2(s)}$	−283,500	8.83	12	25–986 (melting point)
$Na_2O_{(s)} + TiO_{2(s)} = Na_2O \cdot TiO_{2(s)}$	−209,200	−1.26	20	25–1128 (melting point)
$Na_2O_{(s)} + 2TiO_{2(s)} = Na_2O \cdot 2TiO_{2(s)}$	−230,100	−1.7	20	25–527
$Na_2O_{(s)} + 3TiO_{2(s)} = Na_2O \cdot 3TiO_{2(s)}$	−234,300	−11.7	20	25–627
$Na_2O_{(s)} + V_2O_{5(s)} = Na_2O \cdot V_2O_{5(s)}$	−325,500	−15.06	16	25–527
$2Na_2O_{(s)} + 2V_2O_{5(s)} = 2Na_2O \cdot 2V_2O_{5(s)}$	−536,000	−29.3	20	25–627
$3Na_2O_{(s)} + 2V_2O_{5(s)} = 2Na_2O \cdot 2V_2O_{5(s)}$	−721,740	−0	20	25–670
$Na_2O_{(s)} + Fe_2O_{3(s)} = Na_2O \cdot Fe_2O_{3(s)}$	−87,900	−14.6		25–1132
$P_{(s,白)} = P_{(l)}$	657	−2.05	0	44 (melting point)
$P_{(s,\square)} = 0.25P_{(g)}$	32,130	−45.65	1.2	25–431
$2P_{2(g)} = P_{4(g)}$	217,150	−139.0	2	25–1700
$0.5P_{2(g)} + 0.5O_2 = PO_{(s)}$	−77,800	−11.59		25–1700
$0.5P_{2(g)} + O_2 = PO_{2(s)}$	−385,800	60.25		25–1700
$2P_{2(g)} + 5O_2 = P_4O_{10(s)}$	−3,156,000	1010.9		358–1700
$S_{(s)} = S_{(l)}$	1715	4.44	0	115 (melting point)
$S_{(l)} = 0.5S_{2(g)}$	58,600	68.28	2	115–445 (boiling point)
$S_{2(g)} = 2S_{(g)}$	469,300	−161.29	2	25–1700
$S_{4(g)} = 2S_{2(g)}$	62,800	−115.5	20	25–1700
$S_{6(g)} = 3S_{2(g)}$	276,100	305.0	20	25–1700
$S_{8(g)} = 4S_{2(g)}$	397,500	−448.1	20	25–1700
$0.5S_{2(g)} + 0.5O_2 = SO_{(g)}$	−57,780	−4.98	1.2	445–2000
$0.5S_{2(g)} + O_2 = SO_{2(g)}$	−361,660	72.68	0.4	445–2000
$0.5S_{2(g)} + 1.5O_2 = SO_{3(g)}$	−457,900	163.34	1.2	445–2000
$Si_{(s)} = Si_{(l)}$	50,540	−30.0	1.6	1412 (melting point)
$Si_{(l)} = Si_{(g)}$	395,400	−111.38	4	1412–3280 (boiling point)
$Si_{(s)} + 0.5O_2 = SiO_{(g)}$	−104,200	−82.51		25–1412
$Si_{(l)} + O_2 = SiO_{2(S)}$	−907,100	175.73		25–1412 (melting point)
$Si_{(s)} + O_2 = SiO_{2(\alpha,\beta)}$	−904,760	173.38		25–1412 (melting point)
$Si_{(l)} + O_2 = SiO_{2(\alpha,\beta)}$	−946,350	197.64		1412–1723 (melting point)
$Si_{(L)} + O_2 = SiO_{2(L)}$	−921,740	185.91		1723–3241 (boiling point)
$Ti_{(s)} = Ti_{(l)}$	15,480	−7.95		1670
$Ti_{(l)} = Ti_{(g)}$	426,800	−120.0		1670–3290 (boiling point)
$Ti_{(s)} + 0.5O_2 = TiO_{(\alpha,\beta)}$	−514,600	74.1	20	25–1670
$Ti_{(l)} + O_2 = TiO_{2(s)}$	−941,000	177.57	2	25–1670 (melting point)
$2Ti_{(s)} + 1.5O_2 = Ti_2O_{3(s)}$	−1,502,100	258.1	10	25–1670
$3Ti_{(s)} + 2.5O_2 = Ti_3O_{5(s)}$	−2,435,100	420.5	20	25–1670
$V_{(s)} = V_{(l)}$	22,840	−10.42		1920 (melting point)
$V_{(l)} = V_{(g)}$	463,300	−125.77	12	1920–3420 (boiling point)
$V_{(s)} + 0.5O_2 = VO_{(S)}$	−424,700	80.04	8	25–1800
$2V_{(s)} + 1.5O_2 = V_2O_{3(S)}$	−1,202,900	237.53	8	20–2070
$V_{(s)} + O_2 = VO_{2(S)}$	−706,300	155.31	12	25–1360 (melting point)
$V_2O_{5(s)} = V_2O_{5(l)}$	64,430	−68.32	3.3	670 (melting point)

See Table A.5 for Standard free energy of dissolution of certain elements in molten iron ($\Delta G^0 = A + BT$).

Table A.5: Standard free energy of dissolution of certain elements in molten iron ($\Delta G^0 = A + BT$).

Reactions	γ^θ_i	$\Delta G^0 = A + BT$ (J/mol)
$Al_{(l)} = [Al]$	0.029	$-631,80 - 27.91T$
$C_{(s)} = [C]$	0.57	$22,590 - 42.26T$
$Cr_{(l)} = [Cr]$	1.0	$-37.70T$
$Cr_{(s)} = [Cr]$	1.14	$19,250 - 46.86T$
$1/2H_{2(g)} = [H]$		$36,480 + 30.46T$
$1/2H_{2(g)} = [H]$		$36,480 - 46.11T$
$Mg_{(g)} = [Mg]$	91	$117,400 - 31.4T$
$Mn_{(l)} = [Mn]$	1.3	$4080 - 38.16T$
$Mo_{(l)} = [Mo]$	1	$-42.80T$
$Mo_{(s)} = [Mo]$	1.68	$27,510 - 52.38T$
$Ni_{(l)} = [Ni]$	0.66	$-23,000 - 31.05T$
$1/2N_{2(s)} = [N]$		$3600 + 23.89T$
$1/2O_{2(g)} = [O]$		$-117,150 - 2.98T$
$1/2P_{2(g)} = [P]$		$-122,200 - 19.25T$
$1/2S_{2(g)} = [S]$		$-135,060 + 23.43T$
$Si_{(l)} = [Si]$	0.0013	$-131,500 - 17.61T$
$Ti_{(l)} = [Ti]$	0.074	$-40,580 - 37.03T$
$Ti_{(s)} = [Ti]$	0.077	$-25,100 - 44.98T$
$V_{(l)} = [V]$	0.08	$-42,260 - 35.98T$
$V_{(s)} = [V]$	0.1	$-20,710 - 45.6T$
$W_{(l)} = [W]$	1	$-48.1T$
$W_{(s)} = [W]$	1.2	$31,380 - 63.64$

The standard state is 1% solution (mass fraction).

References

[1] J. Chen, J.Q. He, G.L. Li, et al., Vanadium and Vanadium Metallurgy, Leading Group Office of Panzhihua Resources Comprehensive Utilization, 1983.
[2] China Encyclopedia General Editorial Committee (Ed.), China Encyclopedia (Mining and Metallurgy), China Encyclopedia Publishing House, 1984.
[3] S.M. Liao, T.L. Bo, Vanadium Metallurgy Abroad, Metallurgical Industry Press, Beijing, 1985 (in Chinese).
[4] H. Ryakishev, et al., Vanadium and Its Application in Ferrous Metallurgy (K.Z. Cui, Trans.), Chongqing Branch of Science and Technology Literature Press, 1987.
[5] Z.L. Wang, Zinc Smelting Electric Furnace, Metallurgical Industry Press, 2011.
[6] Q.L. Zhang (Ed.), English-Chinese Dictionary of Steel Smelting, Beijing Press, 1993.
[7] Z.M. Li (Ed.), Nonferrous Metal Metallurgy, Chemical Industry Press, 2010.
[8] L.S. Wu, G. Bai, Z. Yuan (Eds.), Minerals and Rocks, Chemical Industry Press, 2005.
[9] L.C. Zhao, Q.X. Zhang (Eds.), Practical Manual of Ferroalloy Production, Metallurgical Industry Press, 2010.
[10] J.Y. Yang, D.X. Liu (Eds.), Extraction, Metallurgical Industry Press, 1988.
[11] J.Y. Sun, H.Y. Du (Eds.), Manufacturing and Application of Inorganic Materials, Chemical Industry Press, 2001.
[12] P.X. Wang (Ed.), Powder Metallurgy, Chemical Industry Press, 1997.
[13] Design Manual Editorial Board of Nonferrous Metallurgical Furnace (Ed.), Design Manual of Nonferrous Metallurgical Furnace, Metallurgical Industry Press, 1999.
[14] S.Z. Yang (Ed.), Vanadium Metallurgy, Metallurgical Industry Press, 2010.
[15] S.L. Yang, Non-Blast Furnace Smelting Technology of Titanoferrous Magnetite Containing Vanadium, Metallurgical Industry Press, 2012.
[16] C.F. Yang, et al., Metallurgical Principle and Application of Vanadium Bearing Steel, Metallurgical Industry Press, 2012.
[17] D.X. Huang, et al., Vanadium Extraction and Steelmaking, Metallurgical Industry Press, 1999.
[18] J.Y. Chen (Ed.), Handbook of Hydrometallurgy—Hydrometallurgy of Vanadium and Chromium, Metallurgical Industry Press, 2005.
[19] J.S. Zhu (Ed.), Mineral Processing Experimental Research and Industrialization, Metallurgical Industry Press, Beijing, 2004.
[20] Panzhihua Science and Technology Committee, Scientific Research History of Panzhihua Titanoferrous Magnetite Containing Vanadium, Panzhihua Science and Technology Committee, 1999.
[21] X. Zhu, China Mining, Science and Technology Press, Beijing, 1999.
[22] H. Cheng, China Natural Resources Handbook, Science Press, Beijing, 1990.
[23] H.G. Du, et al., Principle of Smelting Titanoferrous Magnetite Containing Vanadium in Blast Furnace, Science Press, Beijing, 1990.
[24] Y.W. Mao, et al., Slag Atlas, Metallurgical Industry Press, Beijing, 1996.
[25] S.L. Yang, et al., Vanadium and Titanium Containing Materials, Metallurgical Industry Press, 2007.
[26] Y.D. Jin, A.S. Feng, Guidelines for the Utilization of Metallic Minerals, Science and Technology Press, Beijing, 2007.
[27] J.A., Dean, Lange's Handbook of Chemistry (J.F. Shang, et al., Trans.), Science Press, Beijing, 1991.

[28] X.Z. Jing, et al., Application Manual of Metal Materials, Shaanxi Science and Technology Press, Xi'an, 1989.

[29] R.J. Ma, S.W. Xiao, Separation of Metals by Ion Exchange, Metallurgical Industry Press, Beijing, 2003.

[30] H.S. Wang, Technical analysis on vanadium extraction by sodium additive roasting from stone coal, Min. Metall. Eng. 14 (1994).

[31] T. Jiang, X.H. Yang, Development status of titanoferrous magnetite containing vanadium abroad, in: Panzhihua Vanadium−Titanium, 6 2012, The First.

[32] L.Y. Zheng, et al., Study on the kinetics of vanadium extracting by alkali leaching from stone coal, Min. Metall. Eng. (2011).

[33] S.S. Liu, Review on vanadium extraction from stone coal, Eng. Des. Res. 23 (1995).

[34] W.X. Chen, Research progress on resource utilization of waste residue in vanadium extraction from carbonaceous shale containing vanadium, Hydrometallurgy 30 (2011).

[35] H. Zhang, Synthesis of n-butyl acrylate catalyzed by sulfonated silica gel, Ningxia Eng. Technol. 110 (2011); (in Chinese).

[36] S.B. Yang, et al., Thermodynamic Study of Vanadium Distribution Between Molten Iron and Converter Slag, Iron and Steel 53 (2006).

[37] X.B. Wang, et al., The construction principle and application of the relation diagram between free energy change of oxidation reaction and temperature, J. Fuyang Normal Univ. (Nat. Sci. Ed.) 21 (2004).

[38] W. Liu, A brief analysis on the production process and development prospect of vanadium extraction from vanadium containing steel slag by wet process, China Nonferrous Metall 38 (2009).

[39] Panzhihua Iron and Steel Group Co., Ltd., Proceedings of International Academic Exchange Conference on Comprehensive Utilization of Vanadium and Titanium Resources, 2005, p. 4.

[40] Changsha Mine Design and Research Institute of Ferrous Metallurgy, Ministry of Metallurgy Industry, Research Report on Titanium Supply and Demand in the World, 1983, p. 8.

[41] Z.Y. Huang, Research and application of operation method of vanadium pentoxide melting furnace, Ferroalloy (4)(1988).

[42] X.J. Wang, W.H. Liu, D.L. Pu, Analysis of vanadium loss in the melting process of ammonium polyvanadate and countermeasures, Sichuan Nonferrous Met. (4)(2006).

[43] R.L. Wang, Study on ammonium polyvanadate decomposition and vanadium pentoxide melting process, Ferroalloy 5 (1986) 21−25; (in Chinese).

[44] Z.Y. Zeng, Preparation of powdery vanadium pentoxide by static thermal decomposition of ammonium polyvanadate, Inorg. Salt Ind. 37 (8) (2005) 21.

[45] J. Li, X.L. Zhang, Review of vanadium resources and vanadium extraction process, Hubei Agric. Mechanization 1 (2009).

[46] Y.Y. Gu, S.X. Zhang, S.A. Zhong, et al., Study on the nonpolluting roasting process for extracting vanadium from silicate vanadium ore, Rare Met. 1 (2007).

[47] P. Zhang, F.H. Jiang, Q.R. He, The feasibility of calcified roasting of low-grade vanadium ore to extract vanadium, Iron Steel Vanadium Titanium 2 (1993).

[48] S.F. Ma, G.X. Zhang, Study on the process of vanadium extraction from calcium additive calcined vanadium clay, Rare Met. 6 (2007).

[49] J.M. Fu, J.L. Du, F.Z. Sun, H.M. Shang, Experimental study on extraction of vanadium by combined top-bottom blowing, Iron Steel Vanadium Titanium 15 (1994).

[50] S.F. Pan, Current status of vanadium extraction from oxygen top-blown converter at home and abroad and discussion on vanadium extraction from converter at Panzhihua Iron & Steel Co., Ltd, Iron Steel Vanadium Titanium 1 (1995).

[51] J.C. Wang, H.S. Chen, G.S. Li, Q. Xie, X. Deng, Research on V_2O_5 production process from vanadium slag in converter of Panzhihua Iron and Steel Co., Ltd, Iron Steel Vanadium Titanium 4 (1998).

[52] D.X. Cheng, Production of high vanadium containing pig iron from vanadium extracting tailings, Ferroalloy 2 (2000).

[53] M.A. Abdel-Latif, Recovery of vanadium and nickel from petroleum fly ash, Miner. Eng. 15 (2002).

[54] R. Ressel, M. Hchenhofer, et al., Processing of vanadiferous residues to ferrovanadium, in: EDP Congress, 2005.
[55] P.K. Tripathy, R.H. Rakhasia, R.C. Hubli, A.K. Suri, Electrorefining of carbothernic and carbonitrothermic vanadium: a comparative study, Mater. Res. Bullethu 38 (2003).
[56] G.Q. Li, P.K. Zhang, C.Y. Zhu, L. Zhang, Z. Sui, Recovery of vanadium containing pig iron from vanadium containing steel slag by carbon thermal reduction, Sichuan Metall. 5 (2005).
[57] M.C. Amiri, Recovery of vanadium as sodium vanadate from converter slag generated at Isfahan steel plant, Iran. Trans. Inst. Min. Metall. (Sect. C: Mineral Process. Extr. Metall.) 14 (1999).
[58] L.J. Gu, Progress and present situation of vanadium extraction by pyrolysis in China, Vanadium Titanium 21 (1992).
[59] Department of Integrated Utilization of Maanshan Iron and Steel Co. Ltd, Vanadium enrichment from high ratio steel slag, Metall. Environ. Prot. 21 (1999).
[60] H.S. Chen, Study on the extraction process of V_2O_5 by roasting vanadium slag with lime, Iron Steel Vanadium Titanium (1992) 6.
[61] K. Borowiec, Recover of Vanadium From Slag by Suiphiding, 30.
[62] L.F. Huo, Extraction of vanadium pentoxide from stone coal, Inorg. Salt Ind. 26 (6) (1994).
[63] L. Fu, P. Su, Study on the extraction of vanadium from stone coal with compound roasting additive, J. Guangxi Inst. Nationalities 12 (2) (2006).
[64] G.W. Li, W.S Ge, Effect analysis of vanadium extraction with 339 oxygen gun and 435 oxygen gun, Sichuan Metall. 20 (5) (1998); (in Chinese).
[65] G.H. Chen, Theoretical study on the chemical formation of vanadium slag, Vanadium Titanium 22 (4) (1993).
[66] R.G. Chen, et al., Vanadium extraction by atomization at Panzhihua Iron and Steel Co., Ltd, Iron Steel Vanadium Titanium 10 (2) (1989).
[67] G.H. Li, Investigation report on vanadium extraction from converter in former USSR, Titanium Vanadium 22 (5) (1993).
[68] B. Rollman, Translated K.H. Zhou Kehua Vanadium from South Africa, Titanium Vanadium 15 (6) (1986).
[69] S.B. Yang, W.S. Ge, et al., Analysis of vanadium flow in the process of vanadium extraction from converter at Panzhihua Iron and Steel Co., Ltd, Vanadium Titanium 19 (2) (1998).
[70] Z.S. Li, C.S. Xu, Oxidation kinetics of vanadium when oxygen top blows low vanadium iron solution, Titanium Vanadium 21 (4) (1992).
[71] L.J. Chen, Complete decomposition and comprehensive utilization of vanadium containing steel slag, Sci. Res. Ma-Steel 2 (2) (1990).
[72] G.Z. Liu, Z.T. Sui, Study on the extraction of vanadium and molybdenum from HDS waste catalyst, Miner. Compr. Util. 23 (2) (2002).
[73] W.C. Dai, Q.J. Zhu, Q.B. Chen, et al., The study of the new technology of comprehensive utilization of vanadium extraction from coal stone, Non-Ferrous Metals (Concentration) 52 (3) (2000).
[74] X.Y. Zou, Y.Z. Ouyang, Q.J. Peng, et al., Study on the production of vanadium pentoxide by salt-free roasting and acid leaching of stone coal containing vanadium, Chem. World 42 (3) (2001).
[75] Z.H. Lu, B.N. Zhou, Z.X. Yu, et al., Study on extraction of vanadium from oxidized and roasted stone coal with dilute alkali solution, Rare Met. 18 (5) (1994).
[76] Z.H. Zhang, Y.H. Wang, A new process for vanadium extraction by calcification roasting of siliceous vanadium ore, Chem. World 41 (6) (2000).
[77] Y.S. Wang, G.L. Li, Q.Y. Tong, Recovery of vanadium and molybdenum from carbonaceous shale by solvent extraction, Rare Met. 19 (4) (1995); (in Chinese).
[78] P.N. Zheng, Study on the application of ion exchange in the extraction of vanadium from stone coal, Eng. Des. Res. 20 (6) (1992).
[79] P. Zhang, F.H. Jiang, Extraction of vanadium pentoxide from stone coal with caustic mud as roasting additive, Rare Met. 24 (2) (2000).
[80] J.C. Zhou, China vanadium industry development, in: International Conference on Vanadium, Guilin, 1999.

[81] H.Q. Zhao, Industrial test of metallized pellets produced by rotary hearth furnace, Yunnan Metall. 2 (2010).
[82] E.X. Wu, L.W. Yan, M.Z. Hu, et al., Thermodynamic analysis of direct carbonization in preparation of vanadium carbide, Powder Metall. Mater. Sci. Eng. (2004) 9.
[83] X.F. Xu, et al., Preparation of vanadium nitride from vanadium pentoxide, Iron Steel Vanadium Titanium (3)(2003).
[84] H.S. Chen, Vanadium carbide and vanadium nitride, Iron Steel Vanadium Titanium 21 (2000).
[85] H.Y. Du, Vanadium industry in Russia and its development prospect, Iron Steel Vanadium Titanium 22 (1) (2001).
[86] C.X. Cai, et al., Ammonium polyvanadate precipitation, Ferroalloy 17 (1980).
[87] B. Rollman, Vanadium, S. Afr. J. Afr. Inst. Mineral. Metall. 30 (1985).

Index

Note: Page numbers followed by "*f*" and "*t*" refer to figures and tables, respectively.

A

Abinabithi ore, 234–235
AC. *See* Activated carbon (AC)
AC impedance method, 466
Acid leaching, 5–6, 491–492.
    See also Direct acid leaching; Leaching
  equipment and process parameters, 184
  extracting vanadium from V-containing steel slag by, 182–184
  extraction
    process, 182
    reactions, 183–184
  process of extracting vanadium pentoxide, 217
Acid neutralization, 17
Acidolysis, 5–6, 491–492
Activated carbon (AC), 418
Active furnace cylinder work, 107–108
Additives, 465
Adenylate cyclase, 28–29
Adhesion, 99
Adirondack deposits, 80
Afrikaner deposit of Calley-cora, 78
Air-side converter blowing, 138–140
Alkali
  leaching, 223
  metals, 26
  poisoning, 437
Alkaline leaching, extracting vanadium pentoxide by, 217–218
Alkaline rock, 41–42

Alloy, 1, 243–244
  composition, 252, 252*t*
  high carbon and phosphorus contents in, 254
  steel, 269, 271
Alloying elements, 270–275
  and alloy steel, 271
  influence
    of nitrogen on steel, 274–275
    on grain size and hardenability, 273
    on phase transition during heating and cooling, 272
    on phase transition point, 271–272
  in structural steel, 273
Alum anhydride. *See* Vanadium pentoxide (V_2O_5)
Alum pentoxide. *See* Vanadium pentoxide (V_2O_5)
Alumina, 50, 418
  alumina-thermal reduction process, 1
  extracting vanadium pentoxide from alumina-rich slag, 234
Aluminite, 50–51
Aluminothermic reduction method, 338
Aluminum (Al), 201, 270
  bauxite, 50
  reductant, 248
  soil rocks, 36
  thermal method, 255–262, 258*f*
    factors affecting process, 261–262
    furnace characteristics, 259–260
    process details, 256

    process operation and technical indexes, 260–261
    raw materials and ingredient requirements, 257–259
    reduction process, 255–256
    reduction reactions, 257
  thermal process, 243–244
  thermal reduction, 258*t*
    reaction, 246–247
    smelting ferrovanadium by, 260
Aluminum–vanadium alloy, 483–486, 486*t*
  production methods, 484–486
    aluminothermic process, 484
    self-combustion method, 485
    two-step method, 484–485
    patented technology, 485
  reactions, 483–484
  specifications, 485–486
Ammonia polyvanadate, 472
Ammonia vanadate (NH_4VO_3), 472
Ammonium hexavanadate, 17
Ammonium metavanadate (NH_4VO_3), 24, 373, 399–400, 440, 499–500
  appearance, 399–400
  applications, 400
  Chinese national standard, 373*t*
  material safety data for, 374*t*
  preparation method, 400
Ammonium polyvanadate ((NH_4)$_2V_6O_{16}$), 155, 373–375, 400
  Chinese national standard for, 375*t*

Ammonium polyvanadate
    (($(NH_4)_2V_6O_{16}$) (Continued)
  identification information for, 375t
  physical properties, 374t
Ammonium salts
  vanadate precipitation by,
    166–168
  of vanadium, 25–26
Ammonium sulfate, 17, 221
Ammonium vanadate
    (($(NH_4)_3VO_4$), 25–26,
    373–375
  ammonium metavanadate, 373
  ammonium polyvanadate,
    373–375
Ammonium vanadate peroxide
    (($(NH_4)_6H_4V_2O_{10}$), 27
Amphibole, 76–77
AMR, 239
Anhydride production, 441
Anhydrous vanadic acid.
    See Vanadium pentoxide
    (V_2O_5)
Animals, biological effects of
    vanadium in, 30–31
Annealed steel, 269
Anorthosite, 61
APV. See Ammonium
    polyvanadate
    (($(NH_4)_2V_6O_{16}$)
Arsenic poisoning, 436–437
Arsenite, 40
Asia-Pacific, V-bearing
    titanoferrous magnetite in,
    45–46
Asphalt, 33–34
Atomization furnace
  blowing, 138–140
  for extracting vanadium,
    146–147
  method, 142
Austenite precipitation, 276–277
Austenitic steel, 269
Australia, V-bearing titanoferrous
    magnetite in, 46

B

Bag dust-cleaning (BDC),
    507–508
Baima deposit, 69

Bainite
  microalloyed nonquenched and
    tempered steel, 289
  precipitation, 279
Bainitic steel, 269
Ball mills, 158–159, 197
Banks Island deposit in British
    Columbia, 80–81
Barambi mines, 46
Barium–vanadium mica, 40
Barnesite, 49
Barra mine, 46
Basic rock, 51–52
Basicity of vanadium slag,
    132–133
Basite, 51–52
Batch grinding, 65, 66f
Battery materials, 412–413
Bauxite, extracting vanadium
    pentoxide from, 236–237
Bayer aluminum extraction
    process, 236–237
BDC. See Bag dust-cleaning
    (BDC)
Biocatalysts, 417
Biolite, 201
Bipolar plate, 450–451
Bismuth vanadate, 408
Bismuth vanadate ($BiVO_4$),
    402–403
Bismuth vanadate yellow.
    See Bismuth vanadate
    ($BiVO_4$)
Bismuth yellow. See Bismuth
    vanadate ($BiVO_4$)
Bismuth yellow 184. See Bismuth
    vanadate ($BiVO_4$)
Bituminous sand, 46–47
Black Mountain deposit, 74–75,
    75t, 76t
Black shale, 52, 201
Blank roasting, 206, 519
Blast furnace, 116–117. See also
    Electric furnaces; Shaft
    furnace
  chemical reactions during
    smelting, 99–102
  gas, 190
  pellet production, 96–97
  reduction

  of titanium, 103–105
  of vanadium, 102–103
  sintering
    in China, 94–96
    in practice, 93–96
    process, 90–93
    system, 116–117
  slag, 98–99
  smelting, 90–108
    characteristics, 98–99,
      105–108, 106t
  system installation, 117
Blowing
  method, 129
  temperature, 126–128
  time, 129
  vanadium slag, 137–138
Blue-black vanadium tetroxide
    (V_2O_4), 13
Boiler ash, 240–241
Brown lead ore. See Vanadinite
Bushveld composite ore in South
    Africa, 494–495

C

Caid deposit of Calley-cora, 78
Caius deposit in Keliri-Cora,
    41–42
Calcination process, 5
Calcite ($CaCO_3$), 64, 403–404
Calcium carbonate, 202
Calcium ferrite, 90–91
Calcium metavanadate (CaV_2O_6),
    169, 202
Calcium orthovanadate ($Ca_3V_2O_8$),
    202
Calcium phosphate ($CaPO_4$),
    235–236
Calcium poisoning, 437
Calcium pyrovanadate ($Ca_2V_2O_7$),
    175–177, 202
Calcium salt
  treatment process, 202
  vanadate precipitation by, 169
Calcium salt roasting, 150–151,
    214–218, 215f. See also
    Sodium salt roasting
  consumption of materials, 218
  direct recovery of vanadium,
    217–218

extracting vanadium pentoxide from vanadium slag by, 175–181
extraction
  characteristics, 178–179
  process, 175
  reactions, 175–178
  leaching of Ca salt–roasted vanadium slag, 180
  process operation and technical indexes, 216–218
  process reactions, 214–216
  production process, 214
  roasting and leaching equipment, 181
  vanadate precipitation, 180
  of vanadium slag, 179–180
    influence of CaO/V_2O_3 ratio, 179–180
    influence of roasting temperature, 180
    influence of roasting time, 180
    roasting characteristics, 179
Calcium thermal reduction method, 337–338
  of vanadium chloride, 348–349
Calcium–vanadium garnet, 40
Canada
  deposits in, 80–81
  V-bearing titanoferrous magnetite in, 42–43
Carbide precipitation, 278
Carbon, 204, 245–246
  carbon-reduction method, 475
  reduction reactions, 262
    of VO, 341
  steel, 269
  thermal reduction method, 339–344
    preparation of vanadium by multistep carbon thermal reduction, 342–344
    reactions, 340–341
    vacuum carbon thermal reduction, 341–342
  of vanadium oxide, 477–479
Carbonaceous
  mudstone, 204
  reductants, 341–342

rocks, 36
shale, 201, 204–205
siliceous rock, 204–205
Carbonate-containing mudstone-type vanadium ore, 205
Carbonate-containing silica-type vanadium ore, 205
Carbon-bearing crude oil, 46–47
Carbonitride, 278
Carburization reactions, 101
Carnotite, 33–35, 48, 55
  extracting vanadium pentoxide from, 233, 234f
Cast iron, 268–269
  grades, mechanical properties, and uses of wear-resistant cast iron, 331t, 332t
  vanadium in, 319–332
Cast steel, 269
Catalyst(s), 415–417. See also Spent catalysts
  biocatalysts, 417
  from crude oil desulfurization, 239
  heterogeneous, 416
  homogeneous, 416
  poisoning, 435–439
  from sulfuric acid making, 239
Catalytic polymerization, 442
Catchikara deposit, 40–41, 60
Catchikara iron concentrate, 77–78, 78t
Catchikara Vanadium Mining Co., Ltd, 504
Cevennes deposit in Quebec, 81
Chain-and-grid conveyer rotary kilns, 232
Chalcedony, 204
Chemical bond method, 419
Chemical properties of vanadium, 12–13
Chengde deposit, 38
Chengde V-bearing titanoferrous magnetite, 73–74
Chengde vanadium industry, 514–515
China
  chemical composition
    of materials sintered in, 94t
    of sinter, 94t, 95t, 96t

  of sintered raw materials, 95t
  of sintering ore, 95t
  sintering in, 94–96
  vanadium deposits in, 35–36
    magmatic deposits, 36
    other deposits, 37
    outside of China, 40–46
    sedimentary deposits, 37
    volcanic deposits, 37
  vanadium industry, 508–520
    Chengde vanadium industry, 514–515
    Jinzhou ferroalloy factory, 515–518
    Panzhihua vanadium industry, 510–514
    vanadium extraction from stone coal, 518–520
  V-containing titanoferrous magnetite outside of, 77–82
Chinese national standard
  for ferrovanadium, 376–380
  for vanadium aluminum alloy, 392–394
  for vanadium nitride and carbide, 389
  for vanadium pentoxide, 372
Chinese regions mining V-bearing minerals, 38–40
  Chengde deposit, 38
  Panzhihua deposit, 38
  vanadium deposit in Yangjiabu, 40
  vanadium phosphate uranium deposit in Fangshankou, Gansu, 39–40
  vanadium-silver deposit in Baiguoyuan, Hubei province, 39
  Xialan deposit in Guangdong, 39
  Yaoshan deposit in Anhui, 39
Chloride roasting process, 4
Chlorite, 76
Chlorolization, 73
Chousoff steel works, 504–505
Chromite, 49
Chromium, 101–102, 270
Citric acid, 465

Index

Clay
  dioctahedron, 218–219
  minerals, 201
Coal, 33–34, 46–47, 190
  coal-based reduction pellets, 108–109
  gas, 190
Coal–water slurry (CWS), 242
Coarse vanadium, 353
Coates mine, 46
Coke oven gas, 190, 471–472
Cold brittleness, 270
Cold-drawn steel, 269
Combustion of solid carbon, 92
Communication base stations, 461
Composite polymetallic vanadium ore, 51
Compound pigments, 408
Compound salts of vanadium, 24–27
  ammonium salts of vanadium, 25–26
  pentavalent salt, 24
  sodium salts, 26
  tetravalent salts, 25
  vanadium bronze, 27
  vanadium oxygenates, 26
  vanadium peroxide, 27
  vanadium sulfates, 26–27
Continuous grinding, 65
Control technology of vanadium batteries, 451
Converter process, 142–143
Coolant, 129
Copan ore, 40–41
Copper, 270
Copper–nickel sulfide, 49
Corrugated-board-type catalyst, 431
Corundum slag, 257
Crude oil, 232
  catalyst from crude oil desulfurization, 239
Crushing equipment, 83
Cryptocrystalline quartz, 204
Crystallization of sodium metavanadate, 158
CWS. *See* Coal–water slurry (CWS)
Cyclic voltammetry, 465–466

D

Damiao deposit, 73, 74t
  in Chengde, 73–74
  of V-bearing titanoferrous magnetite, 38
Deanna comprehensive deposit in New York, 43–44
Decavanadates, 399
Decomposition
  of carbonate, 92
  of iron and manganese oxides, 92–93
Denitrification reactions, 430–431
Deoxidization of sodium metavanadate, 158
Dephosphorization, 157–158
Desiliconization, 157–158
Desulfurization, 97
Diatomite, 418, 423–424
Dilute sulfuric acid leaching, 223
Direct acid leaching, 218–222, 219f, 220f. *See also* Acid leaching
  process details, 218
  process operation and technical indexes, 221–222, 221t
  process principles, 218–221
  solutions, 238
Direct alloying of vanadium slag to produce ferrovanadium, 262–265
  factors affecting process, 265
  process operation and technical indexes, 263–265
  raw material and ingredient requirements, 263
  reactions, 262–263
  technology, 262
Direct recovery of vanadium, 217–218
Direct reduction, 117–121, 496
  direct-reduction–electric furnace–shaking ladle–vanadium slag method, 143–145
  electric furnaces, 120–121
  in practice, 110–116, 111t
  production line in New Zealand, 112
  by rotary hearth furnace in Panzhihua, China, 114–116
  rotary kiln reduction test lines in Panzhihua Steel, China, 113–114
  South African rotary kiln reduction production line, 112
  production line in New Zealand, 112
  rotary hearth furnace, 119
  by rotary hearth furnace in Panzhihua, China, 114–116, 114t, 115f
  chemical composition of reduced TiO_2 slag, 116t
  composition of molten iron produced by rotary hearth furnace, 115t
  rotary kiln, 117–118
  shaft furnace, 119–120
  tunnel kilns, 120
  of vanadium oxide, 475–476
Distributed power stations, 461–462
Dry ball milling, 84
Dunite, 61

E

Earth's crust, 10, 60, 201
Edulcoration reactions, 157–158
Electric furnaces, 120–121, 244. *See also* Blast furnace
  requirements, 255
Electric melting plant, 496
Electric reactor technology of vanadium batteries, 450–451
  bipolar plate, 450–451
  electrode materials, 450
  flow field design, 451
  membrane, 450
  sealing technology, 451
Electro-aluminum thermal reduction, smelting ferrovanadium by, 260–261
Electrode materials, 450

Index 537

electrochemical performance
    tests, 467
graphite felt treatment, 466–467
preparation of integrated
    composite electrodes, 468
of vanadium batteries, 466–468
Electrolyte
    materials of vanadium batteries,
        462–466
        additives, 465
        preparation, 462–463
        testing performance of
            electrochemical solutions,
            465–466
        testing vanadium electrolyte
            composition, 463–465
    technology of vanadium
        batteries, 451
Electro-silicon thermal process,
    244–255, 245f
    details, 244
    electric furnace requirements,
        255
    factors influencing production of
        ferrovanadium, 253–254
    operation and technical indexes,
        247–248
    problems and solutions
        adjusting silicon content of
            ferrovanadium, 254
        adjusting vanadium content of
            ferrovanadium, 254
        boiling and slag loss, 254
        high carbon and phosphorus
            contents in alloy, 254
        lining protection, 255
    raw materials and ingredient
        requirements, 247–250
    reactions, 244–247
    steps, 248–250, 252t
Enrichment of vanadium slag,
    124–135
    blowing
        method, 129
        temperature, 126–128
        time, 129
    coolant, 129
    main equipment, 145–147
    semisteel quality, 134–135
    slagging agents, 130

vanadium
    extraction methods, 141–145
    grade of molten iron, 126
    slag quality, 130–134
    variable-valence elements, 130
V-containing
    molten iron by sodium
        treatment and oxidation,
        140–141
    molten iron pretreatment for
        slag blowing, 137–140
    steel slag from molten iron by
        steelmaking, 136–137
Enzymes, 417
Epiphyllite, 77
EPR. See Ethylene propylene
    rubber (EPR)
Equipment, 82–88
    crushing, 83
    grinding, 84
    magnetic separation, 87–88
    screening and grading, 84–86
Ethylene propylene rubber (EPR),
    442
Eutectoid steel, 269

F

Feldspar, 201
Ferrite
    ferrite–pearlitic microalloyed
        nonquenched and tempered
        steel, 288
    precipitation, 277–279
Ferrodiorite, 49
Ferrosilicon, 247–248
    ferrosilicon–vanadium alloys,
        243–244, 266
Ferrovanadium (FeV), 1, 243–244,
    275–276, 375–388, 381t,
    508–510, 515–517.
    See also Nitrided
        ferrovanadium
    adjusting silicon content, 254
    adjusting vanadium content, 254
    aluminum thermal method,
        255–262
    chemical composition, 378t
    direct alloying of vanadium slag
        to produce ferrovanadium,
        262–265

electro-silicon thermal process,
    244–255
ferrosilicon–vanadium alloy,
    266
particle granularity, 378t
physical state, 376
properties, 375
smelting process, 249
standard, 376–388
FeV. See Ferrovanadium (FeV)
Fibrosis method, 419–420
Film materials, 409–411
Filter, 198–199
Fine grinding, 84
Finland, deposits in, 80
First Urals deposit, 41–42
Fixation, 97
Flow field design, 451
Flow-type cells, 469
Flue gas
    denitrification catalysts,
        429–439
        catalyst maintenance,
            435–439
        catalyst poisoning, 435–439
        catalyst regeneration, 439
        denitrification reactions,
            430–431
        types and compositions,
            431–433
    desulfurization catalysts,
        420–429
Fluidized bed, sodium salt roasting
    in, 231
Fluosilicic acid, 221
Fly ash, vanadium extraction from,
    239–242, 240f, 240t, 241t
Food sources of vanadium, 28
Forged steel, 269
Former Soviet Union, V-bearing
    titanoferrous magnetite in,
    40–42
Functional materials, 408–413
    battery materials, 412–413
    film materials, 409–411
    hydrogen storage materials,
        411–412
    superconducting and
        photosensitive materials,
        408–409

G

Gabbro, 49, 61
Gangue minerals, 38, 69
Gas
   fuel, 189–190
   gas–solid reduction process, 98
German industrial standard for vanadium aluminum alloy, 394
Gibbs free energy diagram of oxide, 99
Glauber's salt. See Sodium sulfate
Global vanadium deposits
   genetic types of, 35
      intrusive volcanic rock deposits, 35
      magmatic deposits, 35
      other deposits, 35
      sedimentary deposits, 35
Global vanadium reserves, 58
Grading
   equipment, 84–86
   method, 84–85
Granular ilmenite, 64
Graphite felt treatment, 466–467
Grinding equipment, 84
Groove feeder, 67–68
Guchevoe deposit, 40–41, 41

H

Hager mine in New Jersey, 78–79
Halogen vanadium oxide, 25
Harzburgite, 61
Heat
   capacity enthalpy change, 13, 14t
   of evaporation, 14, 15t, 16t, 18t
   heat-resistance temperature of catalyst, 428–429
   heat-resistant steel, 314–318
   heat-treated PC rod, 296–298
Hematite (Fe_2O_3), 38, 63
Heterogeneous catalysts, 416
High alumina slag, 234
High-alloy steel, 269
High-carbon steel, 269
High-carbon stone coal, 207, 212, 212t
   process for, 209
High-grade fine steel, 269
High-purity
   Al, 338, 353
   vanadium pentoxide Chinese standard, 371–372
High-quality steel, 269
High-quality stone-like coal, 52
High-strength V-containing steels, 2
High-Ti vanadium slag, 154–155
Highveld company
   direct reduction and electric melting plant, 496
   division of Highveld steel and vanadium company, 499–501
   main processes and equipment of, 496–501
   vanadium extraction and steelmaking plant, 497–498
Homogeneous catalysts, 416
Hongge deposit, 69
Hot-rolled bar, 291–296
Hot-rolled steel, 269
Hubei province, 39
Humans, biological effects of vanadium in, 30–31
Hydrated ions, 23
Hydrochloric acid, 1, 4
Hydrocyclones, 85
Hydrofluoric acid, 12–13
Hydrogen
   reduction, 344–345
      of vanadium chloride, 347–348
   storage materials, 411–412
Hydrolysis, vanadate precipitation by, 166
Hydromica, 201
Hydroxide, 201
Hypereutectoid steel, 269
Hypoeutectic steel, 269

I

Ilmenite, 38, 63–64
Impregnation, 418, 421–422, 421f
India, V-bearing titanoferrous magnetite in, 46
Indirect reduction of vanadium oxide, 476
Industrial
   ammonium chloride, 186–187
   ammonium sulfate, 185–186
   fuel, 189–193
   raw materials, 184–189
   soda, 185
   sodium chloride, 185
   sulfuric acid, 187
Initiation temperature, 428
Intermediate compounds, 22
Intrusive volcanic rock deposits, 35
Iodides, thermal dissociation of, 355–356
Iodine, 337
Ion(s)
   exchange, 211, 419
   of vanadium in solution
      hydrated ions, 23
      and oxygen ions, 23–24
      vanadic acid, 23
Iron, 92–93, 201, 267–269
   olivine decomposition by oxidation, 156
   vanadate precipitation by iron salt, 169
Ironmaking, 506–507. See also Steelmaking
   chemical reactions during direct reduction, 109–110
   by direct reduction, 108–116, 110f
   in practice, 110–116
   process characteristics, 108–109
Iron–vanadium spinel ($FeO \cdot V_2O_3$), 155

J

Jaw crusher, 83
Jinzhou ferroalloy factory, 2, 515–518
Jinzhou smelting institute of Japan Special Iron Ore Corporation. See Jinzhou ferroalloy factory

K

Kakoxene, extracting vanadium pentoxide from, 235–236
Kaolin clay, 432–433
Kaolinite clay rock, 64, 65t, 201
Kata-metamorphism organic mineral, 203
Kennedy Valley mine, 42
Kiln body, 117–118, 193
Kliuke Van mine, 44

L

Lac tio mine in Allard, Canada, 81
Lake Michekamo mine, 81
Lauronitrile vanadate, 23
Le Creuset steel plant, 4−5
Leaching, 419
  of Ca salt−roasted vanadium slag, 180
  liquor, 237−238
  residue, 228−229
  of roasted vanadium slag, 164−165
  solution, 224
Lewisite steel, 269
Light-off temperature of catalyst, 428
Limestone ($CaCO_3$), 48, 175, 188
Limonite ($Fe_2O_3 \cdot nH_2O$), 63, 204
Low-alloy steel, 269
Low-calorific value gas, 190
Low-carbon steel, 269
Lower Cambrian black rock series, 36

M

Macintyre ore dressing plant, 78−79
Magmatic deposits, 35−36
Magmatic segregation deposit, 50
Magmatism, 35
Magnesite
  magnesite−ultramafic ferric layered intrusions, 49
  rock, 41−42
Magnesium
  refractory, 146
  thermal reduction, 338−339
    of vanadium chloride, 349−350
Magnesium metavanadate, 210
Magnesium oxide, 210
Magnesium pyrovanadate, 210
Magnetic separation equipment, 87−88
Magnetic separators, 198
Magnetic tailings, 76, 77t
Magnetite, 33−34, 38, 63
Maleic anhydride production, 441−442
Manganese, 101−102, 170, 270
Manganese oxides, 92−93, 201
Martensitic microalloy nonquenched and tempered steel, 289
Martensitic steel, 269
Mechanical classifier, 82−83
Mechanical mixing, 417
Medium-alloy steel, 269
Medium-carbon steel, 269
Melting
  point, 337, 352−353
  process of sodium salt roasting, 171−172
Melting furnace, 199
Membrane, 450
  materials of vanadium batteries, 466
Mesoporous alumina (m-Al_2O_3), 442
Metabolic absorption, 28
Metallic/metal
  cations, 419
  elements, 22
  iron oxidation, 155
  minerals, 204
  oxides, 38, 210
  thermal reduction, 335−339, 336t
    aluminothermic reduction method, 338
    calcium thermal reduction method, 337−338
    magnesium thermal reduction, 338−339
    physical properties of metal oxides during, 337t
  vanadium, 1, 333−350
    carbon thermal reduction method, 339−344
    hydrogen reduction, 344−345
    thermal reduction of vanadium oxide, 334−339
    vanadium chloride reduction, 346−350
Metallogenic characteristics of vanadium ore, 34−46
  Chinese regions mining V-bearing minerals, 38−40
  genetic types of global vanadium deposits, 35
  vanadium deposits
    in China, 35−37
    outside of China, 40−46
Metallogenic rock, 61
Metamorphism, 35
Metatyuyamunite.
  See Tyuyamunite
Metavanadates, 19, 26, 399
Metaviev deposit, 40−41
Methane, 189−190, 465
Mica, 201
Microalloying, 285−287
Microvanadium alloying, 2
Mineral(s)
  characteristics of V-bearing titanoferrous magnetite, 62−64
    calcite, 64
    hematite, 63
    ilmenite, 64
    kaolinite clay rock, 64
    limonite, 63
    pyrrhotite, 64
    quartz, 64
    titanomagnetite, 63
    white titanium ore, 63
  composition of V-bearing titanoferrous magnetite, 62, 63t
  of vanadium, 46−51
    barnesite, 49
    carnotite, 48
    montroseite, 51
    patronite, 48−49
    roscoelite in sandstone, 47−48
    tyuyamunite, 49
    uvanite, 51
    vanadine, 50−51
    vanadinite, 47
    vanadium magnetite, 49−50
    vichlovite, 51
Mineralization of carbonate, 92
Mixed-grinding method, 420, 421f
MnS inclusions, 277
Molten iron tank method, 142
Molten salt electrolytic refining, 353−355
Molybdenum, 270

Molybdenum bismuth yellow. *See* Bismuth vanadate (BiVO$_4$)
Monolithic catalysts, 432
Montmorillonite, 201
Montroseite, 51
Mud, 204
Multicore oxygen anionic condensation, 23–24
Multihearth roasting furnace, 195
Mustavara ore, 44–45

N

NADPH oxidation reaction, 29–30
Nafion, 466
Nanjing ferroalloy plant, 2
New Zealand
    deposits in, 82
    direct reduction production line in, 112
    vanadium industry, 505–508
        steel production technology and equipment, 506–508
        vanadium extraction equipment, 508
    V-bearing titanoferrous magnetite in, 46
Nickel, 270
Niobium, 101–102
Nitric acid, 12–13
Nitrided ferrovanadium, 482–483
    compositional control, 483
    preparation methods, 482–483
    reactions, 482
Nitrides, 274
Nitrification of vanadium pentoxide, 157–158
Nitrogen, 274
    influence on steel, 274–275
    plasma, 475
Nitrogenization, 472
Nizhny Tagil iron and steel company, 503–504
Nonmetallic elements, 21–22
Nonquenched steel, 285–289. *See also* Quenched steel
    bainite microalloyed, 289
    ferrite–pearlitic microalloyed, 288
    grades, 290t
    martensitic microalloy, 289
    mechanical properties, 291t
    steel grades and chemical compositions, 292t, 293t
Norite, 49, 61
Normalized steel, 269
Northern Europe, V-bearing titanoferrous magnetite in, 44–45
Norway, deposits in, 82

O

Oil
    deposits, 55–56
    extracting vanadium pentoxide from oil refining slag, 236
    shale, 46–47
Olivine, 50, 77
Open-air Mabochis mine, 496
Orange vanadium pentoxide, 13, 15–19, 19t
Ordinary steel, 269
Ore
    from Chengde, China, 73–77
        mineral composition, 73–76
        mineral processing technology, 76–77
        structural characteristics, 73
    dressing, 82–83
    ore-bearing rock types, 61
    ore-smelting electric furnaces, 120–121
    from Panzhihua, China, 68–72, 70t, 72t
        mineral composition, 68–69
        mineral processing, 69–72
        structure of deposit, 68
    structure, 60–61
Organic vanadium deposits, 51
Organovanadium ore, 55
Orthopyroxene, 50
Orthovanadate, 26
Otanmaki dressing plant, 80
Otanmaki mine, 44–45, 45t
Oxidation, 97
    of iron and manganese oxides, 92–93
    of metallic iron, 155
    states, 12–13
    of vanadium slag, 131–132
Oxidizing atmosphere, 163
Oxygen
    ions, 23–24
    oxygen-containing compounds, 23
    reaction conditions for vanadate radicals, 24t

P

PAM. *See* Polyacrylamide (PAM)
PAN-based graphite felt. *See* Polypropylene-based graphite felt (PAN-based graphite felt)
Panzhihua deposit, 38
Panzhihua gabbro body, 68
Panzhihua Steel standard for vanadium nitrogen alloy, 390
Panzhihua vanadium industry, 510–514
Partial pressure of CO (P_{CO}), 341–342
Patronite, 48–49, 55, 489
Pearlite steel, 269
Pellet production, 96–97
    control, 97
    pellet characteristics, 96–97
PEM. *See* Proton-exchange membrane (PEM)
Pentavalent salt, 24
Pentavalent vanadium, 27
Perfluorosulfonic acid ion-exchange membranes, 466
Peridotite, 49, 61
Petroleum, 33–34
    bituminite deposits, 35
    vs. catalyst from petroleum cracking, 238
Phosphate rock, 36
Phosphorite, 35
Phosphorus, 101, 170, 270
    block rock ore, 46–47
    contents in alloy, 254
    poisoning, 438
Photosensitive materials, 408–409
Photovoltaic power generation, 460
Phthalic anhydride, 441
Phycomycetes, 203
Physical properties of vanadium, 12, 12t
    corrosion resistance of vanadium, 13t

mechanical properties of industrial metal vanadium, 12t
Physiological effects of vanadium, 27–30
  food sources, 28
  metabolic absorption, 28
  overdose, 29
  physiological functions, 28–29
  physiological processes, 29
  physiological requirements, 29
  vanadium deficiency, 28
  vanadium use by human organs, 29–30
Piedmont mine in North Carolina, 79–80
Pig iron, 268–269
Plagioclase, 73, 76–77
  plagioclase–gabbro complex, 49
  plagioclase–gabbro type, 36
Plagioclasite, 49
Plants, biological effect of vanadium on, 32
Platinum-group element deposits, 49
Poacher Island mine in British Columbia, 80–81
Poly(vinyl alcohol) (PVA), 432–433
Polyacrylamide (PAM), 432–433
Polymerization, 24
  reactions, 442
Polymetavanadate, 23–24
Polypropylene-based graphite felt (PAN-based graphite felt), 466–467
Polyvanadates, 19, 26
Pore clogging with ash, 438–439
Porous catalysts, 419
Potassium chlorate, 338, 484
Potassium metavanadate, 24
Potassium vanadate (KVO_3), 402
Power grid peak regulation, 460–461
Precipitation, 417–418
Production equipment, 116–121
  blast furnace, 116–117
  direct reduction system, 117–121
Protein kinase reactions, 28–29

Proton-exchange membrane (PEM), 447
Pudoone mine, 40–41
Purification of vanadium leaching solution, 165
PVA. *See* Poly(vinyl alcohol) (PVA)
Pyrite, 204
Pyrogenic process, 89, 205–207, 233
Pyrolysis, vanadium extraction by, 519–520
Pyrovanadate, 19, 23–24, 26
Pyroxenite, 49, 61
Pyrrhotite, 64

Q

Quality of vanadium slag, 133–134
Quartz (SiO_2), 64, 201
Quenched steel, 284. *See also* Nonquenched steel
  substitutes for, 284–285
  V-containing, 285

R

Rail steel, 298
Reductants, 333
Reduction, 92–93, 255–256, 257f
  reactions, 257
  of titanium, 103–105
  of vanadium, 102–103
Regional smelting, 356
Residual slurry, 38
Roasting vanadium slag, 150, 490
Rod, 296
  mills, 84
Roscoelite, 55
  in sandstone, 47–48
Rotary hearth furnace, 119, 119f
  in Panzhihua, China, 114–116, 114t
Rotary kilns, 117–118, 193–194
  reduction test lines in Panzhihua Steel, China, 113–114
  sodium salt roasting in, 231
Russia, deposits in, 77–78
Russian vanadium industry, 501–505

Catchikara Vanadium Mining Co., Ltd, 504
Chousoff steel works, 504–505
Nizhny Tagil iron and steel company, 503–504
Tola Vanadium Ferrous Metallurgical Corporation, 505
Rutile, 38

S

Salt. *See* Sodium chloride
Salt-free oxidizing roasting, 150–151
Salt-free roasting
  process, 223
  process operation and technical indexes, 116–121, 224t
  reactions, 223
Sandstone, 48
  roscoelite in, 47–48
SCR. *See* Selective catalytic reduction (SCR)
Scrap steel, 255, 257
Sealing technology, 451
Sedimentary deposits, 35, 37
Sedimentary vanadium ore, 36
Sedimentation, 35
Selective catalytic reduction (SCR), 429–430, 432
Semisteel, 137
  quality, 134–135
  steelmaking, 108–109
Serpentine, 77
Shaft furnace, 119–120. *See also* Blast furnace
  sodium salt roasting in, 230
Shale, 201
  by wet processes, 207
Shaped steel, 298–305
Shuangtashan ore dressing plant, 76
Sichuan–Yunnan metallogenic belt, 35–36
Sieving method, 84–85
Silica, 204, 427
Siliceous carbonaceous rock, 204–205
Siliceous carbonaceous vanadium ore, 205

Siliceous reductants, 248
Silicon, 270
  silicon–calcium alloy, 263
  thermal method, 243–244
  thermal reduction, 253–254
Silicon oxide, 246
Silver, 39
Single vanadium cell, structure of, 468
Sintering
  of catalyst, 436
  characteristics, 90–91
  chemical reactions during, 92–93
    combustion of solid carbon, 92
    decomposition, reduction, and oxidation of iron and manganese oxides, 92–93
    decomposition and mineralization of carbonate, 92
  operation and control, 93
  process, 90–93
Slag, 257
  loss, 254
  recycling, 163
  slag–metal–gas mass transfer process, 124, 125f
Slagging agents, 130
Smelting
  in blast furnace, 90
  chemical reactions during, 99–102, 100f
    chemical composition of vanadium pig iron, 102t
    composition of V-containing pig iron from Panzhihua and Chengde, 102t
  ferrovanadium by
    aluminum thermal reduction, 260
    electro-aluminum thermal reduction, 260–261
SO_3 poisoning, 437–438
Soda ratio, 162
Soda-treated vanadium slag (Na-treated vanadium slag), 181
  extracting vanadium pentoxide from, 181–182

Sodium aluminosilicate ($1/2Na_2O$ $1/2Al_2O_3$ $2SiO_2$), 151
Sodium chloride, 162, 499
Sodium hypochlorite ($NaClO_4$), 233
Sodium metavanadate ($NaVO_3$), 24, 26, 153, 201–202, 210, 401
  crystallization and deoxidization, 158
Sodium orthovanadate (Na_3VO_4), 24, 26, 201–202, 210, 401
Sodium phosphate, 189
Sodium polyvanadate ($NaVO_3$), 26
Sodium pyrovanadate ($Na_4V_2O_7$), 24, 26, 201–202, 210, 401
Sodium salt roasting, 151–174, 208–213, 209f. See also Calcium salt roasting
  calcining and sodium roasting with phosphate, 174
  in chain-and-grid conveyer rotary kilns, 232
  equipment, 160
  and process parameters, 172–174
  in fluidized bed, 231
  leaching of roasted vanadium slag, 164–165
  melting process, 171–172
  precipitation of vanadate, 166–171
  preparation of raw materials, 158–160
  process, 155
    details, 208
    for high-carbon stone coal, 209
    operation and technical indexes, 212–213
    reactions, 210–212
  purification of vanadium leaching solution, 165
  reactions, 155–158
    crystallization and deoxidization of sodium metavanadate, 158
    iron olivine decomposition by oxidation, 156

nitrification of vanadium pentoxide, 157–158
oxidation of metallic iron, 155
vanadium spinel decomposition by oxidation, 157
roasting of vanadium slag, 161–164
  oxidizing atmosphere, 163
  quality of vanadium slag, 161
  roasting additives, 162–163
  roasting in practice, 164
  roasting temperature, 161–162
  slag recycling, 163
in rotary kiln, 231
in shaft furnace, 230
Sodium salts, 26
Sodium sulfate, 163, 188
Sodium vanadate, 157, 206–207, 401
Sodium zoidization, 73
Soft-melting zone, 104
Sol–gel method, 422
Solid carbon combustion, 92
Solid fuel, 190–193
Solid nickel, 416
Sorbitol, 465
South Africa(n)
  deposits in, 80
  rotary kiln reduction production line, 112, 113f
  vanadium industry, 494–501
  vanadium titanomagnetite, 42
South America, V-bearing titanoferrous magnetite in, 46
South Shaanxi–West Hubei metallogenic belt, 35–36
Soybean production, 32
Spent catalysts
  catalyst from petroleum cracking, vs., 238
  from crude oil desulfurization, 239
  extraction process, 237–238
  from sulfuric acid making, 239
  vanadium extraction from, 237–239
Spiral classifiers, 85

Sponge iron, 108
Spray evaporation, 418
Sri Lanka, V-bearing titanoferrous magnetite in, 46
Stainless steel, 318–319
Static cells, 468–469
Steel, 267–268
  alloying elements, 270–275
  classification, 268–269
  industry, 268
  main elements, 270
  metallurgy process, 243–244
  vanadium in, 275–280
Steel Mountain mine, 42–43
  in Newfoundland, Canada, 80–81
Steelmaking, 507–508. *See also* Ironmaking
  furnaces, 255
  process, 15, 108
Stone coal, 201, 203
  extracting vanadium pentoxide from, 207–224
    calcium salt roasting, 214–218
    direct acid leaching, 218–222
    salt-free roasting, 223–224
    sodium salt roasting, 208–213
  vanadium extraction from, 518–520
    pyrolysis, 519–520
    wet vanadium extraction process, 520
Stone-like coal, 37, 52, 53t
  resources, 56
Stove, 196
Stratcor Vanadium Company, 485
Structural steel, alloying elements in, 273
Sulfide addition, extracting V_2O_5 from vanadium slag by, 184
Sulfoarsenide, 38
Sulfonic acid, 465
Sulfur, 188, 270
Sulfur trioxide, 425
Sulfuric acid, 4
  catalyst from sulfuric acid making, 239
  leaching, 233
  production catalysts, 420–429
  requirements for, 426–427
Superconducting materials, 408–409
Surface area of catalyst, 429
Synthesis methods, 420–422
  impregnation, 421–422, 421f
  mixed-grinding method, 420, 421f
  sol–gel method, 422
System integration technology of vanadium batteries, 451–452

T

Tank furnace method, 141–142
Tasman coast of New Zealand, 505–506
Taurine, 465
Temperature, 427
Tempered steel, 284–289
  bainite microalloyed, 289
  ferrite–pearlitic microalloyed, 288
  grades, 290t
  martensitic microalloy, 289
  mechanical properties, 291t
  steel grades and chemical compositions, 292t, 293t
  substitutes for, 284–285
  V-containing, 285
Tempering, precipitation during, 279
Ternary vanadium compounds, 22–23
Ternis deposit in Norway, 45
Tetravalent salts, 25, 25t
Tetravalent vanadium, 27–28
Thermal dissociation of iodides, 355–356
Thermal melting, 418–419
Thermal reduction of vanadium oxide, 334–339
  metal thermal reduction, 335–339, 336t
Thermite method, 484–485
Thermodynamic data for vanadium oxide metal reduction, 336, 336t
Thermodynamic properties of vanadium, 13–14
Thickener, 85, 197–198
  systems, 86
Thiovanadate, 26
Tianshan–Yinshan metallogenic belt, 35–36
Titanium, 270
  hematite, 91
  reduction, 103–105
Titanium bismuth yellow. *See* Bismuth vanadate ($BiVO_4$)
Titanoferrous magnetite, 41
  containing vanadium, 35
Titanomagnetite (($Fe,Ti)_3O_4$), 61, 63, 68–69
Tola Vanadium Ferrous Metallurgical Corporation, 505
Tool steels, 305–314
  chemical composition of low-alloy tool steels containing vanadium, 317t
  grades and chemical composition of alloy, 315t, 316t
  effect of vanadium on overall properties, 314t
  V-containing tool steels, 313t
Trisodium phosphate. *See* Sodium phosphate
Troctolite, 49
Tungsten, 270
Tunnel kilns, 120
Tyuyamunite, 49

U

Ultrabasic rock, 51–52
Ultramafic rock, 42
Uninterrupted power supply (UPS), 448, 461
United States (US)
  deposits in, 78–80
  V-bearing titanoferrous magnetite in, 43–44
UPS. *See* Uninterrupted power supply (UPS)
Ural open-pit mine, 40–41
Uranium, 233

Uranium (*Continued*)
  molybdenum vanadium ore, 235
  shale, 48
  uranium-bearing sandstone, 46–47
  uranium–calcium compound, 49
Urban transportation, 462
Uvanite, 51

V

V2N. *See* Vanadium nitride (VN)
V$_4$C$_3$. *See* Vanadium carbide (VC)
Vacuum carbon thermal reduction method, 341–342, 479
Vacuum method, 475
Vametco Minerals Corp, 501
Vametco mining company, South Africa, vanadium nitride from, 390
Vanadate precipitation
  of calcium salt roasting, 180
  of sodium salt roasting, 166–171, 167t
    by ammonium salt, 166–168
    by calcium salt, 169
    factors influencing, 170–171
    by hydrolysis, 166
    in industry, 169–170
    by iron salt, 169
Vanadates, 19–21, 153, 399–404
  ammonium metavanadate, 399–400
  ammonium polyvanadate, 400
  bismuth, 402–403
  physical and chemical properties of, 20t
  potassium, 402
  sodium, 401
  sodium metavanadate, 401
  yttrium, 403–404
Vanadic acid, 23
Vanadic acid anhydride. *See* Vanadium pentoxide (V$_2$O$_5$)
Vanadine, 50–51
Vanadinite, 47, 52–54
  extracting vanadium pentoxide from, 234–235
Vanadium, 1, 10–12, 101–102, 267–268, 270, 274

aluminum alloy, 391–394
  applications, 392
  Chinese national standard, 392–394
  German industrial standard, 394
  properties, 391
applications, 280–319
batteries, 445–446
  applications, 456–462
  assembly, 468–469
  features, 457–458
  key materials for, 462–468
  operating principle, 453–456
  potential vanadium cell market, 458–462
  systems, 449–456
  technical features, 452–453
  vanadium cell development, 446–449
behavior, 276–279
biological effects
  of vanadium in humans and animals, 30–31
  of vanadium on plants, 32
in cast iron, 319–332
chemical composition
  of 00Cr$_{22}$Ni$_{13}$Mn$_5$Mo$_2$N stainless steel, 330t
  of 500-MPa high-strength reinforcement, 295t
  of 60CrV$_7$ spring steel, 312t
  of 90Cr$_{18}$MoV steel, 330t
  of AISI high-vanadium high-speed steel, 323t
  of general high-speed steel, 322t
  of heat-resistant structural steel, 329t
  of heat-resistant V-containing steel rods, 324t
  of heat-resistant V-containing steel tubes for boilers, 326t
  of high-strength construction machinery steels, 304t
  of hot die steels, 318t, 319t
  of hot-rolled rail steel, 299t
  of low-vanadium die steels, 318t

  of powder metallurgy die steels, 323t
  of special wear-resistant die steel, 320t
  and specifications of V-containing wear-resistant steels, 306t
  of steel bars, 295t
  of V-containing Fe-based and Fe–Ni-based alloys, 330t
  of V-containing heat-resistant steel plates and steel strips, 325t
  of V3N nitrogenous super-high-speed steel, 323t
chemical properties, 12–13
Chinese wire rods
  basic mechanical properties, 298t
  chemical compositions of, 298t
chloride reduction, 346–350
  calcium thermal reduction, 348–349
  hydrogen reduction, 347–348
  magnesium thermal reduction, 349–350
composition of V-containing hard wire steel, 297t
compounds, 14–24, 395–396
  compounds of vanadium and metallic elements, 22
  intermediate compounds, 22
  ions of vanadium in solution, 23–24
  salts of vanadium, 24–27
  ternary vanadium compounds, 22–23
  V$_2$O$_5$, 15–19
  vanadates, 19–21
  of vanadium and nonmetallic elements, 21–22
diseases from vanadium pollution, 30
effects, 275t
  on mechanical properties of steel, 280
  on microstructure and heat treatment of steel, 279–280

Index 545

on physical, chemical, and technological properties of steel, 280
extraction
  equipment, 82–88
  from fly ash, 239–242
  mineral characteristics of V-bearing titanoferrous magnetite, 62–64
  from spent catalysts, 237–239
  typical mineral processing methods, 65–77
  V-bearing titanoferrous magnetite, 60–62
  V-containing titanoferrous magnetite outside of China, 77–82
grades
  and chemical compositions of Chinese high-speed tool steels, 321t
  of molten iron, 126
halides, 396–398
heat-and hydrogen-resistant steels, 302t
heat-resistant steel, 314–318
high-strength structural steels
  chemical compositions of, 310t, 311t
  mechanical properties, 310t
hydrate mica, 40
industry, 487–488
  China's vanadium industry, 508–520
  main processes and equipment of Highveld company, 496–501
  New Zealand vanadium industry, 505–508
  Russian vanadium industry, 501–505
  South African vanadium industry, 494–501
  technology, 489–494
  upstream and downstream companies in, 494f
  value chain, 493–494
  Vametco Minerals Corp, 501

low-cost V–N microalloyed high-strength reinforcement
  chemical composition of, 296t
  process and performance, 296t
metal, 391
  and alloy products, 333
  application, 357–358
  metallic vanadium, 333–350
microalloyed strip steel for automotive applications, 308t
microalloyed structural steel, 289–314
mineral resources
  distribution of vanadium resources, 55–58
  metallogenic characteristics of vanadium ore, 34–46
  minerals of vanadium, 46–51
  vanadium minerals for vanadium extraction, 51–55
nonquenched and tempered steel, 285–289
ore, 51
PD3 rail
  chemical composition, 299t
  mechanical properties, 299t
physical properties of vanadium, 12
physiological effects of vanadium, 27–30
pigments, 407–408
  bismuth vanadate, 408
  compound pigments, 408
  vanadium carbide powder, 408
  zirconium vanadium blue, 407
processing performance characteristics of V-containing steel, 280–281
properties of HSS carbides, 322t
quenched and tempered steel, 284
recovery, 234
  converter, 145–146
refining, 350–356
  methods, 351–353

molten salt electrolytic refining, 353–355
regional smelting, 356
thermal dissociation of iodides, 355–356
requirements for continuous casting of V-containing steel, 281
resources distribution
  global distribution of vanadium resources, 56–57
  global vanadium reserves, 58
  stone-like coal resources, 56
  type and distribution, 55–56
  V-bearing titanoferrous magnetite, 56
semihigh-speed steel compositions, 323t
shale extraction
  extracting vanadium pentoxide from stone coal, 207–224
  vanadium extraction processes from stone coal, 224
  V-bearing carbon shale, 202–207
shale structure, 202–203
slag, 6, 124
  applications, 360
  blowing, 138–140
  Chinese national standard for, 361t
  grades and chemical compositions, 363t
  properties, 360
  quality standards, 360–363
  technical standards, 363
slag quality, 130–134
  basicity of vanadium slag, 132–133
  oxidation of vanadium slag, 131–132
  quality of vanadium slag, 133–134
spinel decomposition by oxidation, 157
stainless steel, 318–319
in steel, 275–280
steel for nuclear applications, 303t

Index

Vanadium (*Continued*)
  stone coal, 205–206
  substitutes for quenched and tempered steel, 284–285
  thermodynamic properties of vanadium, 13–14
  thermomechanical control, 281–282
  thermostable tempering composition of V-containing steels, 328t
    for rotors, 330t
  thermostable tougheners for gas turbine blades, 327t
  toughening components of V-containing impact-resistant cutting tools, 317t
  vanadium-based catalysts, 2–3
  vanadium-based solid melt alloys, 493
  vanadium-bearing mica, 33–34
  vanadium-extraction technology, 227–228
  vanadium–iron–silicide, 244–245
  vanadium–nitrogen alloys, 275–276
  vanadium–nitrogen microalloyed steel
    chemical composition of precipitated, 294t
    chemical phase structure, 295t
  vanadium-silver deposit in Baiguoyuan, 39
  V-bearing quenched and tempered steel, 282–284
    composition, 282–283
    heat-treatment characteristics, 283–284
    performance requirements, 282
  V-containing
    alloy structural steels, 312t
    high-strength low-alloy steels, 300t
    high-strength low-alloy structural steels, 307t
    iron, 124
    quenched and tempered steel, 285
    shale, 205
    steels, 3–4
    weathering steels, 309t
Vanadium (acid) anhydride. *See* Vanadium pentoxide (V_2O_5)
Vanadium bronze, 27
Vanadium carbide (VC), 2, 13, 21–22, 276, 472–479
  carbon thermal reduction of vanadium oxide, 477–479
  Chinese national standard for, 389
  powder, 408
  preparing by V_2O_3 reduction, 477
  production process, 472–475
    carbon addition calculation, 473
    main methods, 472–473
    material and process parameter requirements, 473–475
  production system, 479
  reactions, 475–479
    direct reduction, 475–476
    indirect reduction, 476
  from United States, 390
  vacuum carbon thermal reduction method, 479
Vanadium carbonitride (VC_xN_y), 276
Vanadium catalysts, 415–416, 492
  chemical composition, 426–427, 427t
  factors affecting, 427–429
  flue gas denitrification catalysts, 429–439
  manufacturing methods, 417–420, 422–424
  morphology, 426
  for organic synthesis, 439–443
    applications, 440–443
    vanadium components, 440
  sulfuric acid production and flue gas desulfurization catalysts, 420–429
Vanadium deposits
  Catchikara deposit, 41
  in China, 35–37
  first Urals deposit, 41–42
  Guchevoe deposit, 41
  outside of China, 40–46
  South African vanadium titanomagnetite, 42
  V-bearing titanoferrous magnetite
    in Asia-Pacific, 45–46
    in Australia, 46
    in Canada, 42–43
    in former Soviet Union, 40–42
    in India, 46
    in New Zealand, 46
    in Northern Europe, 44–45
    in South America, 46
    in Sri Lanka, 46
    in United States, 43–44
  in Yangjiabu, 40
Vanadium dioxide (VO_2), 404–405
Vanadium extraction, 5–6, 89
  atomization furnace method, 142
  ball mills, 197
  converter process, 142–143
  direct-reduction–electric furnace–shaking ladle–vanadium slag method, 143–145
  equipment, 193–199
    rotary kilns, 193–194
  filter, 198–199
  from fly ash, 239–242
  magnetic separators, 198
  melting furnace, 199
  methods, 141–145
  molten iron tank method, 142
  multihearth roasting furnace, 195
  from spent catalysts, 237–239
  from stone coal, 224, 225t
  stove, 196
  tank furnace method, 141–142
  technology, 491
  thickener, 197–198
  vanadium minerals for, 51–55
    carnotite, 55
    organovanadium ore, 55
    patronite, 55
    roscoelite, 55

stone-like coal, 52
vanadinite, 52–54
V-bearing titanoferrous magnetite, 51–52
Vanadium magnetite, 49–50
Vanadium nitride (VN), 2, 13, 21–22, 289, 389–390, 472, 479–481
 Chinese national standard for, 389
 Panzhihua Steel standard for vanadium nitrogen alloy, 390
 production
  process, 480
  system, 481
 properties, 389
  of steel improved by vanadium nitride addition, 294t
 reaction principles, 481
 from Vametco mining company, South Africa, 390
Vanadium oxide(s), 1–3, 210, 210f, 218–219, 344, 404–407, 472, 492
 carbon thermal reduction of, 477–479
 reduction, 471–472
 thermal reduction of, 334–339
Vanadium oxide(V). See Vanadium pentoxide (V_2O_5)
Vanadium oxygenate, 26
 thiovanadate, 26
Vanadium oxytrichloride ($VOCl_3$), 338–339, 396–397, 440
Vanadium pentafluoride (VF_5), 398
Vanadium pentoxide (V_2O_5), 52, 150, 201–202, 227–228, 244, 265, 365–367, 406–407, 415–416, 440, 471–472, 492–493
 and chemical constituents, 369t
 chemical properties, 367
 Chinese national standard for, 367–371
  inspection rules, 370

packaging, labeling, storage, transport, and quality certificate, 370
 physical state, 369
 sampling, 370–371
 technical requirements, 367–368
 test method, 370
Chinese national standard for, 372
Chinese standard for chemical purity, 372t
extraction
 from alumina-rich slag, 234
 from bauxite, 236–237
 from carnotite, 233
 from kakoxene, 235–236
 from oil refining slag, 236
 from shale by pyrogenic processes, 92
 from shale by wet processes, 207
 from uranium molybdenum vanadium ore, 235
 from V-bearing limonite, 236
 from V-bearing titanoferrous magnetite, 228–232
 from V-containing oil residue, 232–233
 from vanadinite, 234–235
high-purity vanadium pentoxide Chinese standard, 371–372
material safety data for, 368t
nitrification of, 157–158
physical properties, 365–366
standard for, 372t
standard GB-3283-87 for vanadium pentoxide powder, 372t
Vanadium peroxide, 27
Vanadium phosphate uranium deposit in Fangshankou, Gansu, 39–40
Vanadium recovery hot metal ladle units (VRU), 507
Vanadium redox batteries (VRBs), 445–447, 454–455
VRB-ESS energy storage system, 447
Vanadium sulfate, 26

Vanadium tetrachloride (VCl_4), 13, 398, 440
Vanadium trioxide (V_2O_3), 13, 261–262, 265, 338, 363–365, 405–406, 440, 483–484
 first aid measures, 364–365
 material safety data for, 366t
 properties, 364
 risks, 364
 safety data, 365
Vanadium(III) chloride, 10
Vanadyl cation, 28
Vanadyl sulfate, 27
Vantra factory, 499–501
Variable-valence elements, 130
Vasospasm, 31
V-bearing carbon shale, 202–207
 classification of V-bearing carbonaceous shale, 203–205
 demand for V-containing shale to extract vanadium pentoxide, 205–207
 vanadium shale structure, 202–203
V-bearing limonite, extracting vanadium pentoxide from, 236
V-bearing stone coal, 202
V-bearing titanoferrous magnetite, 36, 38, 42, 51–52, 56, 60–62, 62t, 68, 89–90, 201–202, 228–232
 in Asia-Pacific, 45–46
 in Australia, 46
 blast furnace smelting, 90–108
 in Canada, 42–43
 extraction reactions, 229–230
 in former Soviet Union, 40–42
 in India, 46
 ironmaking by direct reduction, 108–116
 mineral
  characteristics of V-bearing titanoferrous magnetite, 62–64
  composition, 62
  structure, 61
 in New Zealand, 46

V-bearing titanoferrous magnetite (*Continued*)
  in Northern Europe, 44–45
  ore structure, 60–61
  ore-bearing rock types, 61
  process
    details, 228–229
    operation, equipment, and typical process indicators, 230–232
  production equipment, 116–121
  in South America, 46
  in Sri Lanka, 46
  in United States, 43–44
VC. *See* Vanadium carbide (VC)
V-containing molten iron
  pretreatment for slag blowing, 137–140
    blowing vanadium slag, 137–138
    main slag blowing equipment, 138–140
    by sodium treatment and oxidation, enriched vanadium slag from, 140–141
V-containing oil residue, 232–233
  extraction reactions, 233
  extraction technology, 232–233
V-containing shale
  to extract vanadium pentoxide, demand for, 205–207
  extraction of vanadium pentoxide from shale
    by pyrogenic processes, 205–207
    by wet processes, 207
  from Nevada, 213
V-containing steel slag
  from molten iron by steelmaking, 136–137
  reactions occurring during slag production, 136
  slag blowing equipment, 136–137

V-containing titanoferrous magnetite deposits
  in Canada, 80–81
  in Finland, 80
  in New Zealand, 82
  in Norway, 82
  in Russia, 77–78
  in South Africa, 80
  in United States, 78–80
  extracting V_2O_5 from vanadium slag by sulfide addition, 184
  extracting vanadium from V-containing steel slag by acid leaching, 182–184
  extracting vanadium pentoxide
    from Na-treated vanadium slag, 181–182
    from vanadium slag by calcium salt roasting, 175–181
  extraction of vanadium from, 150–151
  outside of China, 77–82
  raw materials, 184–193
    industrial fuel, 189–193
    industrial raw materials, 184–189
  sodium salt roasting, 151–174
  vanadium extraction equipment, 193–199
Venezuelan crude oil, 232
Vibrating feeder, 67–68
Vichlovite, 51
VN. *See* Vanadium nitride (VN)
Volcanic deposits, 37
Volkov ore, 40–41
VRBs. *See* Vanadium redox batteries (VRBs)
VRU. *See* Vanadium recovery hot metal ladle units (VRU)

W

Wasikell deposit, New Jersey, 43–44

Waste
  vanadium sulfide catalyst, 237
  vulcanized V-containing catalyst, 238
Water
  leaching process, 1
  poisoning, 438
  water-soluble calcium pyrovanadate, 216–217
  water-soluble sodium vanadate, 153
Wet ball milling, 84
Wet process, 89, 207
White titanium ore, 63
Wind power generation, 458–459
Wire, 296
Wyoming's Iron Mountain mine, 44

X

Xialan deposit in Guangdong, 39
Xuanhua–Chengde–Beipiao deep fault zone, 73

Y

Yangjiabu vanadium deposit, 40
Yaoshan deposit in Anhui, 39
184 yellow dye. *See* Bismuth vanadate ($BiVO_4$)
Yinshan–Tianshan mountains, 39–40
Yttrium europium vanadate, 408–409
Yttrium europium vanadate phosphate, 408–409
Yttrium vanadate (YVO_4), 403–404

Z

Zambia, 235
Zinnisk mine, 40–41
Zirconium vanadium blue, 407